Complete A Level Chem

Complete A Level Chemistry

Diana Kelly

**Head of Chemistry,
Stoke Damerel High School for Girls,
Plymouth**

Bell & Hyman

Published by
BELL & HYMAN LIMITED
Denmark House
37–39 Queen Elizabeth Street
London SE1 2QB

© D. Kelly 1984
First published by University Tutorial Press Limited 1984
Reprinted by Bell & Hyman Limited 1986

ISBN 0 7135 2735 8

Printed and bound in Great Britain by
Scotprint Limited, Musselburgh

Preface

Many Advanced level chemistry texts take the subject as far as is required in a first year university course. This text is intended to set out clearly and concisely the basic knowledge needed at Advanced level for those students unlikely to continue with chemistry as a major discipline, while at the same time providing at least a sound foundation for the student wishing to pursue the subject at university level.

The first part sets out the theoretical, physical and chemical principles. The text covers the chemistry of the more important elements, in terms of periodic and group trends; it also provides reference material for the student on the occurrence, properties and compounds of these elements. The carbon chemistry section concentrates on the behaviour of organic compounds, interpreted in terms of electron structure and reaction mechanism where appropriate. None of these sections is interrupted with details of experimental laboratory work, but important organic preparations, synthesis routes and industrial processes are described in the section on applied chemistry.

The text is based on the Advanced level requirements of the Universities of Cambridge, London and Oxford, and of the Joint Matriculation Board and the Associated Examining Board. The nomenclature is based on that proposed by the IUPAC and follows the recommendations of the ASE handbook *Chemical Nomenclature, Symbols and Terminology* (1979) and the Joint Statement to Schools by the GCE Boards (1973). A glossary covers the recommended terms for the traditional names of various compounds.

My sincere thanks are due to my husband, who not only checked and typed the script, but also gave me much valuable and constructive criticism. I have also to thank the staff of UTP for their help and encouragement.

D.K.

Acknowledgements

The author and publisher would like to thank the following Examination Boards for permission to reproduce questions from past examination papers:

The Associated Examining Board (AEB)

The University of Cambridge Local Examinations Syndicate (C)

The University of London University Entrance and School Examinations Council (L)

The Joint Matriculation Board (JMB)

The Oxford Delegacy of Local Examinations (O)

The answers to the numerical questions are, however, entirely the responsibility of the author and have not been approved by the Examination Boards.

Thanks are also due to the Nuffield Foundation for permission to use information from their Book of Data.

Contents

PART I THEORETICAL PRINCIPLES

1. Definitions and nomenclature	2
2. Structure of the atom	8
3. Nuclear chemistry	31
4. Bonding	43

PART II PHYSICAL PRINCIPLES

5. Kinetic theory and states of matter	66
6. Physical equilibrium	89
7. Solubility and colligative properties	108

PART III CHEMICAL PRINCIPLES

8. Thermochemistry	128
9. Rates of reaction	146
10. Chemical equilibrïum	163
11. Acids and bases	173
12. Electrochemical cells and redox reactions	187
13. Electrolysis	204

PART IV CHEMISTRY OF THE ELEMENTS

14. Periodicity	216
15. Hydrogen and water	224
16. The s-block elements	234
17. Boron and aluminium in Group III	247
18. Group IV	257
19. Nitrogen and phosphorus in Group V	272
20. Oxygen and sulphur in Group VI	288
21. The halogens	305
22. The d-block elements	319

PART V CARBON CHEMISTRY

23. Hydrocarbons	340
24. Isomerism	364
25. Alcohols, phenols and ethers	374
26. Aldehydes, ketones and carbohydrates	388
27. Carboxylic acids and derivatives	402
28. Alkyl and aryl nitrogen compounds	417
29. Alkyl and aryl halogen compounds	431

PART VI APPLIED CHEMISTRY

30. Laboratory preparations of organic compounds	448
31. Important reactions and synthetic pathways	457
32. Important industrial processes	468
Glossary of traditional names in common use	482
Answers to numerical questions	484
Index	486
Relative atomic masses of selected elements	501

Part I
Theoretical Principles

1 Definitions and nomenclature

1.1 The periodic classification of the elements

There are ninety-two naturally occurring elements, and the great variety of compounds they form make up the substance of the earth's crust. The elements are composed of particles called atoms; the atoms of each element are of a unique kind, different from the atoms of all other elements.

The elements can be collected into groups, having similar properties within each group, such as lithium, sodium and potassium, or chlorine, bromine and iodine. It was realised early in the nineteenth century that if the elements were arranged in order of atomic mass then their properties showed striking similarities at regular intervals. Various attempts were made to classify the elements according to the periodicity of their properties; the most successful was that of Mendeleyev, a Russian chemist, in 1869. The modern form of the Periodic Table (Fig. 1.1), first put forward by Thomson in 1885, is based on Mendeleyev's model. The elements are still arranged substantially in order of ascending atomic mass, but greater significance is now given to the atomic number of the element (page 12). Thus, for example, argon (atomic number 18) is placed before potassium (atomic number 19) although its atomic mass is slightly greater than that of potassium.

Elements of similar properties fall into vertical columns, called Groups:

Group I   The Alkali Metals. Lithium → Caesium
Group II   The Alkaline Earth Metals. Beryllium → Radium
Group III   Boron → Thallium
Group IV   Carbon → Lead
Group V   Nitrogen → Bismuth
Group VI   Oxygen → Polonium
Group VII   The Halogens. Fluorine → Astatine

The elements in the final group are known as the Group 0 elements; they are the noble gases which have exceptionally stable configurations of electrons and are consequently very unreactive.

Hydrogen, which has properties in common with both Group I and Group VII, is placed on its own.

The horizontal rows are called *periods*, and consist of three short periods followed by longer ones. The fourth period which is the first long period is lengthened by the occurrence of a set of elements between calcium and gallium. These have very similar properties and are known as the transition elements (page 319). Similar sets occur in the fifth and sixth periods, the latter being extended still further by the occurrence of the lanthanoids (58-71). The seventh period is extended by the actinoids (90–103), of which only the first three, up to uranium (92), occur naturally. The rest, up to lawrencium (103), have been produced in the laboratory.

Not only do the elements fall naturally into groups with similar properties, but it can be seen that the more metallic elements fall on the left hand side of the table while the non-metal elements fall mainly to the right of the

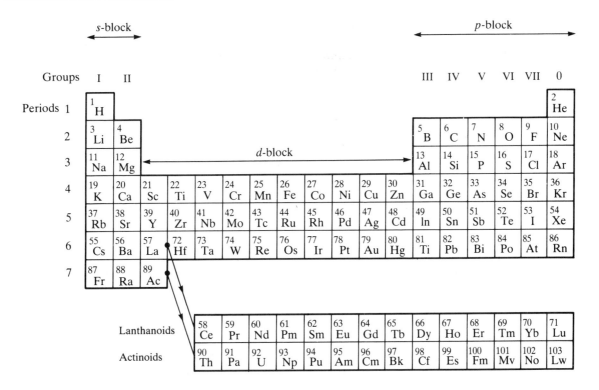

Fig. 1.1 Periodic classification of the elements

table. Metallic character, in fact, decreases across any period of the table and increases on descending a group. There are a number of elements towards the middle of the table which are neither strongly metallic nor strongly non-metallic; these are termed metalloids (page 257).

1.2 Atoms, ions, molecules and compounds

Atoms combine to make compounds either through the formation of ions or by combining covalently to form molecules.

Ions are charged particles formed from atoms by the loss or gain of one, two or three electrons (page 43). In general, metals and hydrogen form positive ions and non-metals form negative ions. Negative ions are also formed from small stable groups of non-metal elements, such as the sulphate ion SO_4^{2-} and the carbonate ion CO_3^{2-}. Compounds containing ions are mainly metal oxides and metal salts; the formulae indicate the ratios in which the atoms of the composing elements occur.

Molecules are formed by the covalent combination of non-metal atoms (page 45). They range from small discrete molecules such as carbon dioxide, CO_2, or oxygen, O_2, where the formula represents the whole molecule, to very large structures where the bonding is continuous (page 45). Here the formula indicates the ratios of the atoms present in the structure, as in silica, SiO_2, or polythene, $(-CH_2-)_n$.

1.3   Relative atomic, molecular and formula masses

The relative atomic masses of the elements were first defined by using the atom of hydrogen as the unit mass. The development of very accurate determinations of relative atomic mass by means of the mass spectrometer led to the discovery of isotopes (page 12) and to the recognition of the need for a more precise definition.

Today relative atomic mass is defined as the ratio of the average mass per atom of the naturally occurring mixture of isotopes of an element to 1/12 of the mass of an atom of the ^{12}C isotope of carbon.

Thus

Relative atomic mass of an element

$$= \frac{\text{average mass of one atom of the element}}{1/12 \text{ mass of one atom of } ^{12}C}$$

On this scale the relative atomic mass of hydrogen, $A_r(H)$, is 1.0079. Similarly $A_r(Cl) = 35.453$ and $A_r(Na) = 22.9898$. These very precise figures are usually rounded off to the nearest half-unit for illustrative discussion. Unless otherwise clearly stated, the symbols are taken to refer to the naturally occurring mixture of isotopes.

For compounds, the relative molecular mass or relative formula mass is also defined in terms of one-twelfth of the mass of an atom of the carbon-12 isotope of carbon.

$$\text{Relative molecular mass} = \frac{\text{mass of one molecule of compound}}{1/12 \text{ mass of one atom of } ^{12}C}$$

A similar expression is used to define the mass per unit formula of compounds that are not molecular, thus $M_r(K^+Cl^-) = 74.555$.

It follows that the relative molecular mass of a compound is the sum of the relative atomic masses of the elements of which it is composed. For example, the relative molecular mass of tetrachloromethane, $M_r(CCl_4)$ is $(12 + 35.5 \times 4)$, or 154.0.

1.4   The mole concept and the Avogadro constant

It is impossible to weigh out so small a quantity as one atom of an element or one molecule of a compound, but at the same time it is very important to work with a known number of atoms. Thus the concept of a mole quantity, which contains a precise number of particles (rather like a dozen or a gross) has been introduced. The *mole* is defined as *the amount of a substance that contains the same number of elementary entities as there are carbon atoms in twelve grams (0.012 kg) of carbon*-12. The entities may be atoms, molecules, ions, electrons or other particles or groups of particles. Since twelve grams is the relative atomic mass of carbon-12 expressed in grams, it follows from the definition that *the relative atomic mass in grams of any element contains one mole of atoms*. Similarly the relative molecular mass in grams of a compound contains a mole of molecules. The term molar mass is sometimes used to describe the mass of a mole of atoms, ions, molecules, or other particles. Hence:

> The mass of a mole of hydrogen atoms (H) is 1 g
> The mass of a mole of hydrogen molecules (H_2) is 2 g
> The mass of a mole of carbon atoms (C) is 12 g
> The mass of a mole of oxygen molecules (O_2) is 32 g
> The mass of a mole of carbon dioxide molecules (CO_2) is 44 g
> The mass of a mole of sodium ions (Na^+) is 23 g
> The mass of a mole of sodium chloride ion pairs (Na^+Cl^-) is 58.5 g
> The mass of a mole of sulphate ions (SO_4^{2-}) is 96 g

The actual number of particles in a mole of substance has been estimated using methods involving X-ray diffraction (page 86), electrolysis (page 208) and radioactive decay (page 32). The value found, to three significant figures, is 6.02×10^{23}; this is called the *Avogadro Constant*, and given the symbol *L*.

Since compounds are always formed from elements in mole ratios, it is usual to convert any given or measured mass of an element or compound to fractions of a mole by dividing the given mass by the mass of a mole of the atoms, molecules or other particles involved. Thus:

2.3 g of sodium contains $\dfrac{2.3}{23}$ or 0.1 mole of sodium atoms.

0.585 g of sodium chloride contains $\dfrac{0.585}{58.5}$ or 0.01 mole of sodium chloride ion pairs.

22 g of carbon dioxide contains $\dfrac{22}{44}$ or 0.5 mole of carbon dioxide molecules.

The mole concept is useful when calculating from equations. For example:

$$2Mg + O_2 \rightarrow 2MgO$$

can be read as two moles of magnesium atoms combining with one mole of oxygen molecules to make two moles of magnesium oxide. Hence if 80 grams or 2 moles of magnesium oxide is required, then 2 moles or 48 grams of magnesium must be used (O = 16, Mg = 24).

1.5   Concentration of solutions

Solutions of known concentration are frequently required in chemistry. The concentrations of these solutions are based on mole quantities of the solute. Strictly speaking the concentration should be expressed as moles per metre cubed ($mol\ m^{-3}$), but the cubic decimetre or litre is a much more convenient volume, so the concentration is usually expressed as $mol\ dm^{-3}$ ($mol\ l^{-1}$). A solution containing one mole of solute dissolved in a cubic decimetre of solution is commonly termed a 'one molar solution' or 1 M; similarly one of concentration $0.1\ mol\ dm^{-3}$ is designated tenth molar or 0.1 M (although this is not exactly in accordance with the IUPAC conventions that are in general use nowadays).

The advantage of using solutions of known concentration is that a known mass of solute can be measured out readily in solution. Thus:

1 litre of 1 M KCl has a concentration of 1 mol dm^{-3} and contains 56 g of potassium chloride.
1 litre of 0.1 M KCl has a concentration of 0.1 mol dm^{-3} and contains 5.6 g of potassium chloride.
500 cm^3 of 0.1 M KCl has a concentration of 0.1 mol dm^{-3} and contains 2.8 g potassium chloride.

Care has to be taken when considering the concentrations of ions in solution. For example, if one mole of calcium chloride Ca^{2+}(Cl$^-$)$_2$ is dissolved in one litre of solution, the concentration of calcium chloride is 1 mol dm^{-3}; the concentration of Ca^{2+} ions is also 1 mol dm^{-3}, but that of Cl$^-$ ions is 2 mol dm^{-3}, since two moles of chloride ions are released from each mole of calcium chloride:

$$Ca^{2+}(Cl^-)_2(aq) \rightarrow Ca^{2+}(aq) + 2Cl^-(aq)$$

1.6   Avogadro's law and gas volumes

The amounts of gases collected are most easily measured by determining their volumes. As early as 1811 Avogadro suggested that equal volumes of all gases, under the same conditions of temperature and pressure, contain equal numbers of molecules. It follows that one mole of any gas, since it contains a precise number of molecules, must occupy the same space as one mole of any other gas at the same temperature and pressure. This volume, *the gram-molecular or molar volume*, is 22.4 litres at 273 K and 1 atmosphere pressure, or 24.0 litres at 293 K and 1 atmosphere pressure. Since laboratory pressures and temperatures vary, volumes of gases collected have to be corrected to s.t.p. (273 K and 1 atmosphere pressure). This done, the volume can be equated to a mole quantity. For example, if 44.8 cm^3 of carbon dioxide are collected at s.t.p., then the amount of carbon dioxide is $\dfrac{44.8}{22\,400}$ or 0.002 mole.

In general:

$$\text{Mole quantity of gas collected} = \frac{\text{volume (cm}^3\text{) at s.t.p.}}{22\,400 \text{ cm}^3}$$

1.7   Nomenclature

As far as possible the names of compounds used in this book are those recommended in the ASE Report on Chemical Nomenclature (1979), which are based on the principles agreed by the IUPAC Commission on Symbols, Terminology and Units.

The oxidation states of metal ions and the central atoms of complex ions are denoted in the names by the use of roman numerals, except where the metal has only one oxidation state or where the oxidation state of the central atom of a complex ion is unambiguous. Thus: potassium chlorate

(V), copper(II) chromate(VI) and manganese(IV) oxide, *but* sodium chloride and calcium carbonate.

Oxidation states are more fully explained in Chapter 12. The naming of compounds containing liganded and complex ions is described in Chapter 4 and that of organic compounds in Chapter 23.

The traditional names of a number of compounds are still widely used; a selection of these, with their systematic names, is given in a glossary at the end of the book.

Questions

1. Find the relative molecular mass of sulphur dioxide (SO_2).
2. Find the mass of a mole of sodium sulphate (Na_2SO_4).
3. Calculate the mass of sodium sulphate that can be obtained by reacting 4.9 g of sulphuric acid (H_2SO_4) with sodium hydroxide (NaOH).
4. Find the number of moles of potassium bromide in 11.9 g of potassium bromide.
5. If 5.6 g of potassium hydroxide were dissolved in water to give a volume of 100 cm^3, what would be the concentration of the solution?
6. If you were given a 0.1 molar solution of barium hydroxide, what would its molarity be with respect to (*a*) barium ions (*b*) hydroxide ions?
7. How many moles of gas are present in (*a*) 89.6 cm^3 (*b*) 56 cm^3 (c) 112 cm^3 of gas collected at s.t.p?
8. What volume would be occupied by 6.4 g of sulphur dioxide (SO_2) at s.t.p?

2   The structure of the atom

2.1   Introduction

As early as 400 B.C. some Greek philosophers believed that matter was composed of minute particles or atoms. Development of the modern atomic theory started in 1803 with the great English chemist John Dalton, who considered that atoms were hard solid individual particles, and that the atoms of any one element were identical, but different from those of all other elements. Dalton did not suggest any structure for the atom, which he regarded as indivisible and indestructible, but since his time a detailed theory of the structure of the atom has developed.

The atom is now considered to consist of a small dense positively charged *nucleus*, in which the mass of the atom is concentrated, surrounded by a cloud-charge of negative electricity. The nucleus contains *protons*; which are particles of relative mass one, carrying one positive charge, and *neutrons*, which are particles also of relative mass one, but carrying no charge. Protons and neutrons are collectively called *nucleons*. The negative cloud-charge is made up of *electrons*, equal in number to the protons of the nucleus, and of almost negligible mass, but carrying one negative charge.

Table 2.1   The main particles of the atom

Particle	Relative mass	Charge
Proton	1	+1
Neutron	1	0
Electron	1/1840	−1

2.2   The electron

(a) Evidence for the electron as a particle

The concept of electrons as particles developed from the study of *cathode rays*. These are generated by applying a high potential difference between electrodes at the ends of an evacuated tube. A current flows through the tube in the form of invisible rays passing from the cathode towards the anode, causing the glass around the anode to glow. This glow is enhanced if the glass is coated with zinc sulphide.

An object placed in the path of the rays casts a sharp shadow, indicating that they travel in straight lines. They behave as though they were composed of negatively charged particles, being repelled from the cathode, attracted to the anode and deflected in the expected direction under the influence of a magnetic field.

In 1897 J. J. Thomson devised a method of measuring the ratio of charge to mass (e/m) of these particles (Fig. 2.2).

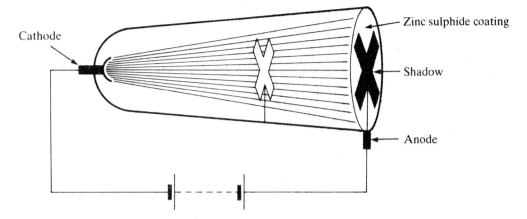

Fig. 2.1 Cathode ray tube

The rays are first deflected by the application of a magnetic field. A potential difference is then applied across the electrode plates of sufficient strength to bring the beam back to its original path. From the observed values of the magnetic and electrostatic fields, a value for the charge/mass ratio (e/m) for the negatively charged particles in the beam can be worked out. Thomson found that the values he obtained for e/m for these particles (now called electrons) were the same, irrespective of the material used for the cathode or the residual gas in the tube. From this he deduced that electrons are individual identical particles, present in the structure of all atoms.

Thomson later found values for both the charge and the mass of the electron separately. A very accurate method for doing this was devised by R. A. Millikan in 1909. He introduced a fine mist of spherical drops of oil of known density into the partially evacuated air space above the top electrode plate of the apparatus shown in Fig. 2.3.

Single drops descend through the hole in the top electrode plate under the influence of gravity, reaching a terminal velocity as a result of the drag effect of the remaining air molecules. The drops are illuminated, and their rate of fall measured with a microscope fitted with a scale. From the rate of fall the

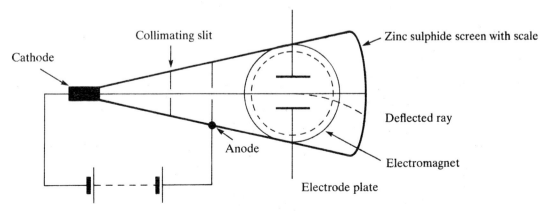

Fig. 2.2 Thomson's apparatus for measuring e/m for the electron

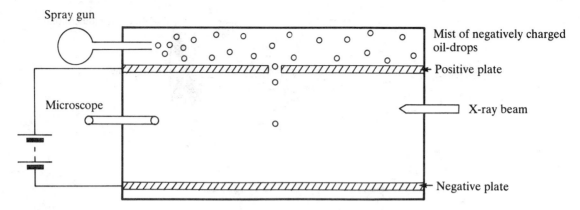

Fig. 2.3 Millikan's oil drop experiment

mass of a particular drop can be calculated. A potential difference is then applied to the electrodes, the top plate being made the positive. The drop, being negatively charged by friction on leaving the spray-gun, is attracted to the top plate and its fall is arrested. The potential difference is adjusted until the electrical field balances the gravitational pull and the drop remains stationary. From the values found the charge on a single drop can be calculated. Millikan varied the charge on the drops by irradiation with X-rays, and repeated his experiment using other liquids such as mercury. In all cases he found that the charges on the drops were integral multiples of 1.602×10^{-19} coulombs. This basic value is taken to be the charge on a single electron. From this figure, and the value of e/m, the mass of an electron is calculated to be 9.109×10^{-31} kg which is approximately 1/1840th of the mass of a hydrogen atom.

(b) The wave nature of the electron

It is a fundamental limit in nature that measurements of the velocity and determinations of the position of very small particles are affected by the method of measurement; it is not possible to determine precisely both the position and the velocity of an electron at a given instant. This is a statement of *Heisenberg's uncertainty principle*, fundamental to the study of wave mechanics which is beyond the scope of this book.

In 1924 De Broglie suggested that since a photon of light has both wave and particle properties it is reasonable to assume that an electron, being of similar size, may have the same duality of nature. De Broglie calculated the wavelength of electrons from the formula.

$$\lambda = \frac{h}{\text{momentum}} = \frac{h}{mv}$$

where m is the mass of the electron, v is its velocity, and h is a universal constant known as Planck's constant.

The concept of the wave nature of electrons was confirmed in 1928 by Davisson and Gerner who produced *electron diffraction patterns* similar to the diffraction patterns obtained by passing light waves through a diffraction grating. They directed a beam of electrons on to a nickel crystal; the atoms in the crystal lattice formed a diffraction grating of suitable dimensions to diffract waves of the wave-length of electrons and give diffraction patterns.

So, while it is sometimes convenient to regard electrons as negatively charged particles, much of the behaviour of electrons is better interpreted in terms of wave motion. For example, the electrons surrounding the nucleus of an atom are represented as existing in *orbitals* or electron clouds whose density in any region represents the probability of finding an electron in that location (page 26).

2.3   The nucleus

(a) Geiger and Marsden's experiment

The idea that the mass of an atom was concentrated into a small nucleus, rather than being distributed evenly throughout the atom, developed from an experiment carried out by Geiger and Marsden in 1909, under the direction of Professor Rutherford. They directed a beam of α-particles, which have a relative mass of four and carry two positive charges, at a thin metal foil, and observed the directions in which the α-particles emerged from it. Most of them passed through the foil with little or no deflection, indicating that they had not met anything solid enough to deflect or reflect them. However a few were deflected and a very small number actually reflected back from the foil; this indicated that they had collided or nearly collided with particles of comparable size, mass and charge as themselves. Rutherford deduced from the number and nature of the deflections and reflections that almost all the mass of each atom must be concentrated into a very small nucleus; he also calculated that these nuclei carried a positive charge, approximately equal to half the relative atomic mass of the element.

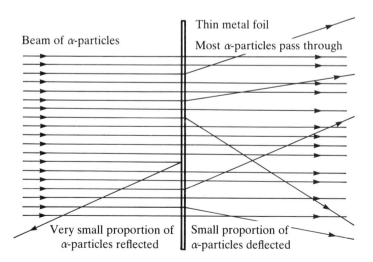

Fig. 2.4   Geiger and Marsden's experiment

(b) Moseley's experiment and atomic number

Further information about the nucleus of the atom was provided by the results of an experiment carried out by Moseley in 1913. This established that the atomic number of an element was a fundamental property of the element, and was numerically equal to the charge on the nucleus of the atom. Further work identified the atomic number of an element as the number of protons in the nucleus. Moseley's experiment involved bombarding metallic elements with an electron beam; he found that X-rays were produced of a frequency that depended only on the element used. Each element always produced X-rays of the same frequency whatever the conditions of the experiment, and the square root of this frequency varied directly as the atomic number of the element (Fig 2.5).

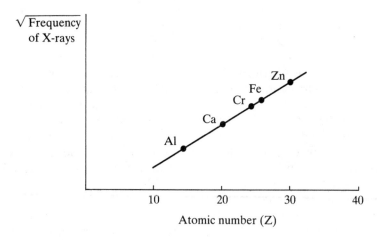

Fig. 2.5   Moseley's experiment

(c) Neutrons and isotopes

At about the same time as Moseley's work on atomic number J. J. Thomson discovered that some elements possessed atoms of slightly different masses. Since these atoms were chemically identical they all contained the same number of protons in the nucleus. The difference in mass indicated the presence of some other particle. Rutherford suggested that these were particles of relative mass 1 but carrying no charge, which he called neutrons. Neutrons could be present in different numbers in the nucleus without affecting the chemical nature of the atom. Later, very accurate determinations of relative atomic masses by Aston's mass spectrograph showed that many elements possessed atoms of different masses resulting from the presence of different numbers of neutrons in the nucleus. These atoms were called *isotopes*.

Hydrogen has two main isotopes—hydrogen and deuterium. Ordinary hydrogen, relative atomic mass 1, consists of a nucleus containing a single proton which is surrounded by one electron. The nucleus of deuterium contains one neutron as well as one proton, making its relative mass 2; its chemical properties are identical with those of hydrogen since these are determined by the presence of the one proton and the one surrounding electron. The structures of the two isotopes can be represented thus:

Hydrogen   Deuterium, or heavy hydrogen

The sum of the protons and neutrons in an isotope is known as its *mass number,* and the symbols of isotopes are conventionally written with the mass number and atomic number as prefixes. For example, the two isotopes of hydrogen are written $^{1}_{1}H$ and $^{2}_{1}H$, while the isotopes of chlorine, which have mass numbers of 35 and 37, are written $^{35}_{17}Cl$ and $^{37}_{17}Cl$, showing plainly that the nucleus of each isotope contains seventeen protons, but $^{35}_{17}Cl$ contains eighteen neutrons and $^{37}_{17}Cl$ twenty neutrons.

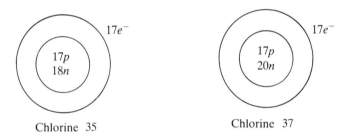

Chlorine 35   Chlorine 37

Chemically determined relative atomic masses give the average mass of the isotopes normally present. For hydrogen the average relative atomic mass is 1.008; measurements in the mass spectrograph show that 99.985% of natural hydrogen is $^{1}_{1}H$ and only 0.015% is $^{2}_{1}H$ or deuterium. The mass spectrum of chlorine shows that natural chlorine consists of 75% chlorine atoms of mass thirty-five, and 25% of chlorine atoms of mass thirty-seven; the average relative atomic mass calculated from these figures is $(35 \times 0.75) + (37 \times 0.25)$ or 35.5, which corresponds to the chemically determined value.

Now that it is clear that it is the atomic number (number of protons) that determines the order in which elements are placed in the periodic table, the placing of argon (relative atomic mass 39.95) before potassium (relative atomic mass 39.1) can be explained. Natural argon is composed of 99.5% of $^{40}_{18}Ar$ (18 protons and 22 neutrons) with only traces of $^{36}_{18}Ar$ and $^{38}_{18}Ar$, while natural potassium contains 93% of $^{39}_{19}K$ (19 protons and 20 neutrons) with 7% of $^{41}_{19}K$ and traces of $^{40}_{19}K$.

The term 'nuclide' is used to give greater precision to the concept of isotopes. It indicates a species of atoms of which each has an identical proton number (atomic number) and an identical nucleon number (mass number). Then isotopes or isotopic nuclides can be defined as nuclides having the same proton number but different nucleon number. A particular nuclide is fully specified by the notation:

$$\frac{\text{Nucleon number}}{\text{Proton number}} \text{Symbol} \frac{\text{Charge, oxidation state, excitation state}}{\text{Number of atoms in the entity}}$$

In practice, only the relevant information is written. Thus O_2, with no nucleon number indicated, denotes molecules of oxygen of the naturally occurring isotopic composition; $^{16}_{8}O_2$ denotes oxygen molecules containing atoms of proton number 8 and nucleon number 16. Similarly, ions of lead may be written Pb^{2+}, $^{207}Pb^{2+}$ or $^{207}Pb^{II}$, while helium may be denoted as natural helium He or as the excited nuclide $^{4}_{2}He^*$, where * indicates that the atoms are in an excited electron state.

(d) The mass spectrometer

The mass spectrometer, which was developed from Aston's mass spectrograph, is a powerful tool for detecting isotopes, determining the atomic mass of these isotopes, and determining the molecular mass and investigating the structure of organic molecules. Figure 2.6 illustrates the principles of the mass spectrometer.

The sample is vaporised at low pressure and passes as a stream of particles (atoms or molecules) into the main body of the apparatus, which is maintained under high vacuum. At B the particles pass through a beam of fast moving electrons which have sufficient energy to knock an electron out of particles they collide with, producing positive ions. These ions are accelerated to a uniform speed by an electric field at C, provided by two parallel charged plates with holes in them that allow a narrow beam of positively

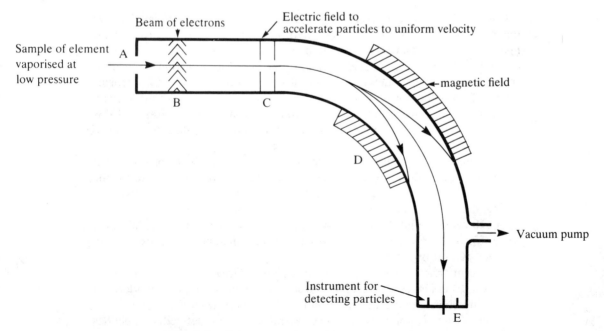

Fig. 2.6. Representation of mass spectrometer

charged particles, all of the same velocity, to pass into the magnetic field at D. Since the particles are travelling at the same speed they deflect to an extent that depends on their mass/charge ratio. If the magnetic and electric fields are kept constant, only particles of the same mass/charge ratio reach the detector E, which also measures the abundance of the particles reaching it. By adjusting the fields, particles of another mass/charge ratio can be directed on to the detector to be measured in their turn. A mass spectrum of the isotopes of an element and their abundance, or of the fragments of an organic molecule, can be constructed from the results. For example, the mass spectrum for the isotopes of lead (^{204}Pb, ^{206}Pb, ^{207}Pb and ^{208}Pb) is shown in Fig. 2.7.

From the Fig 2.7, the relative abundances and hence the percentages of the various isotopes of lead in a particular sample are found to be:

2% of isotopic mass 204
25% of isotopic mass 206
21% of isotopic mass 207
52% of isotopic mass 208

The relative atomic mass of lead, in this sample of the naturally-occurring metal, can be worked out by finding the weighted mean of these figures:

$$\frac{(204 \times 2) + (206 \times 25) + (207 \times 21) + (208 \times 52)}{100}$$

or 207.2

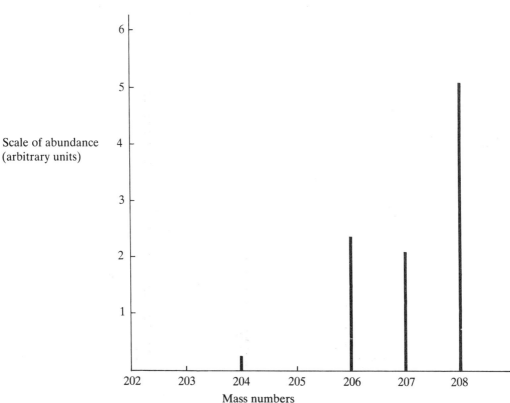

Fig. 2.7 Mass spectrum for isotopes of lead

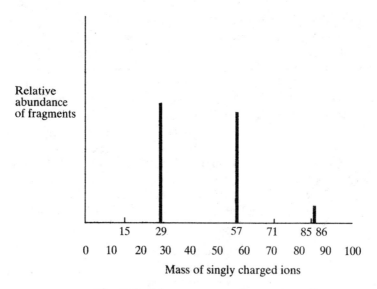

Fig. 2.8   Mass spectrum for pentane-3-one

The constitution of a molecule can be deduced from its molecular formula and the masses and relative abundance of the ionised fragments. The mass spectrograph analysis of pentane-3-one is shown in Fig. 2.8. Here the molecule is $C_5H_{10}O$ and all the ions are singly charged. Plainly the parent ion $C_5H_{10}O^+$, mass 86, appears in small quantity, while $C_5H_9O^+$ must be the

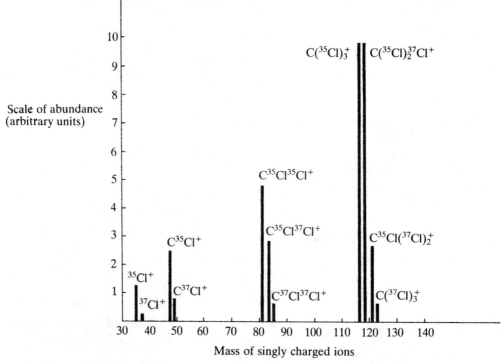

Fig. 2.9   Mass spectrum for tetrachloromethane

fragment at mass 85. At mass 29 the fragment must be $C_2H_5^+$, suggesting the presence of C_2H_5-groups in the molecule, and at mass 57 the fragment could be C_2H_5—CO^+. With these two fragments, occurring in equal quantities, forming the main products of the ionisation, it may be deduced that the original molecule breaks into $C_2H_5^+ + C_2H_5CO^+$ from $C_2H_5COC_2H_5$, which is indeed the accepted structure for pentane-3-one. Minor fragments of masses 15 and 71 could be CH_3^+ and CH_3CH_2—$\overset{\overset{\textstyle O}{\|}}{C}$—$CH_2^+$ respectively.

This very simplified example shows how mass spectrometer traces may be interpreted. As a somewhat more complex example, Fig. 2.9 shows the mass spectrum for tetrachloromethane, which is made up of the ions Cl^+, CCl^+, CCl_2^+ and CCl_3^+, each of which has more than one peak resulting from the presence of the isotopes of chlorine.

2.4 The arrangement of the electrons in the atom

(a) The emission spectra of the elements

When white light from the sun is analysed by passing it through a prism, it splits up into a number of colours. The separate colours correspond to electro-magnetic radiations of different wave lengths. Visible light and the wavelengths that compose it form only a small part of the total electro-magnetic spectrum which extends from radio waves of wavelength about 10^{11} nanometres through infra-red rays, visible light rays, ultraviolet rays and X-rays to γ rays of wavelength about 10^{-4} nanometres. The corresponding frequencies of the radiations range from 10^7 hertz to 10^{21} hertz (or vibrations per second, Hz).

When an element is 'excited'—supplied with energy by heating or electrical stimulation—it will emit electromagnetic radiation, which can be analysed spectroscopically, the spectrum exhibiting a number of lines corresponding to single frequencies.

A few metals in Group I and Group II emit electromagnetic frequencies in the visible range; for example, sodium compounds heated in a bunsen

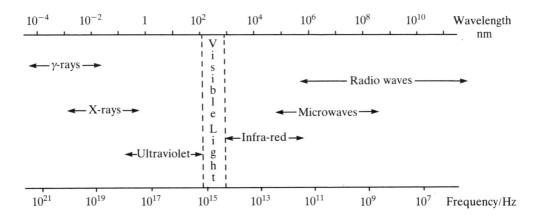

Fig. 2.10 The regions in the electromagnetic spectrum

flame give a golden light while potassium compounds give a violet light, which in the spectra appear as narrow lines of colour.

Most elements emit frequencies in parts of the spectrum other than the visible range, but in all cases the frequency corresponds to a definite energy change, which is thought to arise from the absorption and emission of precise amounts of energy by the electrons surrounding the nucleus of the atom. The study of the frequencies emitted and the energy changes they correspond to has led to an understanding of the arrangement of the electrons surrounding an atom.

The emission spectrum of hydrogen was first studied by Bohr in 1914; the situation in this case being simplified by the fact that the hydrogen atom has only one electron.

The spectrum of the hydrogen atom consists of five sets of discrete lines occurring in the ultra-violet, visible and infra-red regions of the electromagnetic spectrum. These sets are known as the Balmer, Paschen, Lyman, Brackett and Pfund series, after the men who detected them.

It was also found that the lines in the series were related to one another and could all be expressed in a single formula:

$$\frac{1}{\lambda} = R_H\left(\frac{1}{n^2} - \frac{1}{m^2}\right)$$

where λ is the corresponding wavelength
    R_H is a constant (the Rydberg constant)
    n and m are whole numbers related to the series as follows:

        Lyman series $n = 1$, $m = 2, 3, 4 \cdots$
        Balmer series $n = 2$, $m = 3, 4, 5 \cdots$
        Paschen series $n = 3$, $m = 4, 5, 6 \cdots$
        Brackett series $n = 4$, $m = 5, 6, 7 \cdots$
        Pfund series $n = 5$, $m = 6, 7, 8 \cdots$

To account for these series of discrete lines, Bohr suggested that the electron did not orbit the nucleus in a random fashion, but existed only in certain orbits (represented by the rings in Fig. 2.11) in which it did not radiate energy. These orbits he called stationary states and suggested that an electron in such an orbit possessed a particular amount of energy. Electromagnetic radiation by an 'excited' atom therefore resulted from the movement of an electron from an orbit of higher energy to any orbit of lower energy, the frequency of the radiation being determined by the difference in energy between the two levels.

Bohr's concept of a circular orbit for the electron has now been superseded by that of an *orbital*, meaning a charge-cloud within which the electron may be considered to be present, but his concept of definite energy levels has been retained. It can be seen from Fig. 2.12 that as the energy levels of the orbitals increase, so the differences between the values of the energies become less, with the result that they come closer and closer together. Finally a value, the convergence limit, is reached at which the electron can escape the influence of the nucleus and leave the atom altogether; this energy is the ionisation energy of the element.

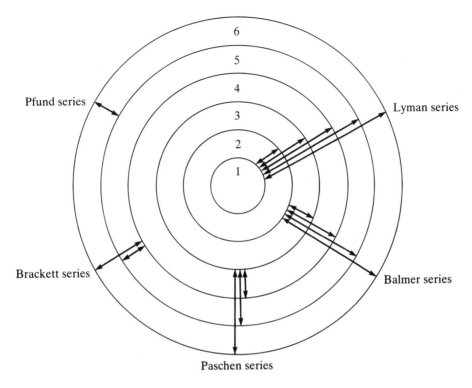

Fig. 2.11 Illustration of Bohr's circular orbits

Changes of orbital involve precise changes of energy, and these, resulting in exact frequencies of radiation, account for the line spectra that are obtained. For example, the lines in the Lyman series (Fig. 2.12) result from the falling back of 'excited' electrons from each of the higher levels to the ground state (the normal energy level for an unexcited electron), while those in the Balmer series result from a falling back to the second level giving rise to a spectrum in the visible region.

The frequency of the emitted radiation is related to the change of energy, or quantum of energy released, by the equation

$$E_2 - E_1 = h\nu$$

or

$$\Delta E = h\nu$$

where E_2 = energy of electron in higher orbital, kJ mol^{-1}
  E_1 = energy of electron in lower orbital, kJ mol^{-1}
  ΔE = change of energy or quantum of energy released, kJ mol^{-1}
  ν = frequency of the radiation,
  h = Planck's constant

Planck's constant has been found to have a value 3.99×10^{-13} kJ mol^{-1}.

The change in energy involved when an electron falls from one energy level to another can be calculated using this equation, since the frequency of the resulting electromagnetic radiation can be determined from the position of the line in the emission spectrum of the element.

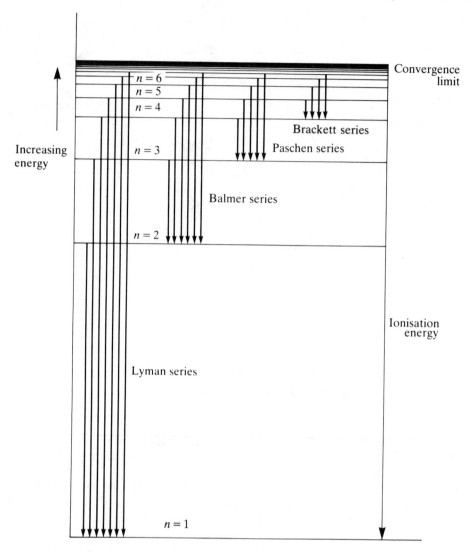

Fig. 2.12   Energy levels of the hydrogen atom (not to scale)

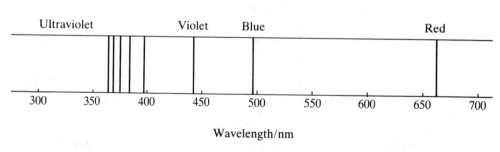

Fig. 2.13   Spectrum of hydrogen in the visible range

Examination and analysis of the emission spectra of other elements led to the extension of Bohr's ideas to the theory of the arrangement of electrons in orbitals in all elements.

(b) Energy levels and quantum numbers

The great number of lines observed in the electromagnetic spectra of the elements were accounted for by extending the number of possible orbitals in which an electron could exist in an atom. The orbitals are distinguished by *quantum numbers*. The principal quantum number (n) represents a group or shell of orbitals; the maximum number of possible electrons in the shell is $2n^2$. Thus if $n = 1$, the shell can contain up to two electrons; if $n = 2$ then the shell can accommodate up to eight electrons, and so on. The electrons in the shell occupy orbitals characterised by the subsidiary quantum number (l) whose value depends on (n), and known as the s, p, d and f orbitals.

Table 2.2

Value of n	Value of l	Subsidiary orbitals in the shell
1	0	s
2	0, 1	s, p
3	0, 1, 2	s, p, d
4	0, 1, 2, 3	s, p, d, f

The s, p, d and f orbitals are further characterised by magnetic and spin quantum numbers; these define the type of orbital and differentiate the pair of electrons in each subsidiary orbital. Thus in the appropriate shells there can be one s orbital, three p orbitals, five d orbitals and seven f orbitals. Each of these orbitals is filled with a maximum of two electrons. The final pattern is summarised for the lower quantum numbers:

Table 2.3

Principal quantum number (n)	1	2		3			4			
Subsidiary orbital	s	s	p	s	p	d	s	p	d	f
Number of electrons	2	2	6	2	6	10	2	6	10	14
	2	8		18			32			

The elements are arranged in the Periodic Table (Fig. 1.1) in order of increasing atomic number (number of protons in the nucleus), and consequently each successive element also has one more electron in its atom than its predecessor. In the natural, or ground state, of the element the electrons occupy the orbitals of the lowest energy levels available. The energy levels of the orbitals do not always increase in exact order. For example, d

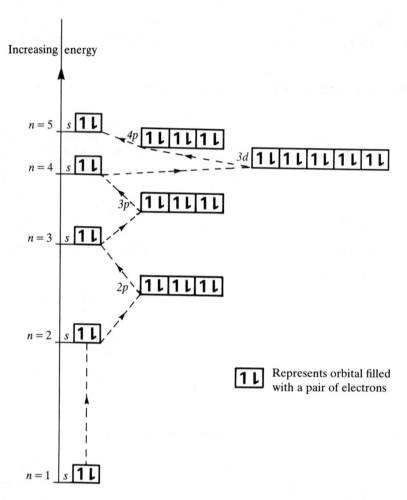

Fig. 2.14   Representation of energy levels (electronic structure of strontium). $\boxed{1\!\downarrow}$ represents orbital filled with pair of electrons

electrons in the quantum level ($n = 3$) have slightly higher energies than s electrons in the quantum level ($n = 4$). The energy levels of the electrons for the elements of low atomic number can be represented as in Fig. 2.14.

The electrons are represented in pairs since each orbital is fully occupied when filled by a pair of electrons. It is considered that, as the number of electrons increases with successive elements, the s orbitals in each shell are first filled by the available electrons; the next electrons occupy the p orbitals singly, and are only followed by a second or pairing electron when sufficient are available (Hund's rule). Thus carbon has six electrons; two fill the s orbital of the first quantum level; the next two fill the s orbital of the second quantum level, while the last two each occupy one of the available p orbitals. The next element is nitrogen with seven electrons; the first six

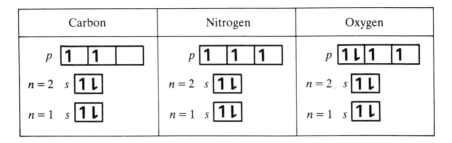

Fig. 2.15   Electron structures for carbon, nitrogen and oxygen

occupy the same orbitals as the six electrons of carbon and the seventh occupies the third available p orbital. The eighth electron of the next element, oxygen, pairs up in one of the p orbitals (Fig. 2.15).

In describing the position of the electrons in an atom, it is conventional to write the main quantum number followed by the letter denoting the orbital with the number of electrons as index;

Hydrogen	$1s^1$
Helium	$1s^2$
Carbon	$1s^2 2s^2 2p^2$
Nitrogen	$1s^2 2s^2 2p^3$
Oxygen	$1s^2 2s^2 2p^4$
Argon	$1s^2 2s^2 2p^6$
Manganese	$1s^2 2s^2 2p^6 3s^2 3p^6 3d^5 4s^2$
Iron	$1s^2 2s^2 2p^6 3s^2 3p^6 3d^6 4s^2$

The filling of the $3d$ orbitals characterises the transition elements of the first long period (page 319).

(c) Ionisation energies

(i) Successive ionisation energies
The ionisation energy of an element is defined as the energy required to remove a mole of electrons from a mole of atoms of the element in the gaseous state. It is measured in $kJ\,mol^{-1}$.

$$M(g) \rightarrow M^+(g) + e^- \quad \Delta H = \text{First ionisation energy, } kJ\,mol^{-1}$$

Further successive removals of the available electrons can be effected. For the removal of the second mole of electrons

$$M^+(g) \rightarrow M^{2+}(g) + e^- \quad \Delta H = \text{Second ionisation energy, } kJ\,mol^{-1}$$

The energy required to remove the second mole of electrons is greater than that required to remove the first, as there is now additional positive charge resisting the removal. Plotting the energy required to remove successive electrons against the number removed provides an interesting confirmation of the arrangement of the electrons in the energy levels just described. Figure 2.16 shows the curve obtained; as the energy values are large a logarithmic scale is used. The outermost electron, the $3s$ electron, requires little energy for removal. The next electron comes from a new quantum

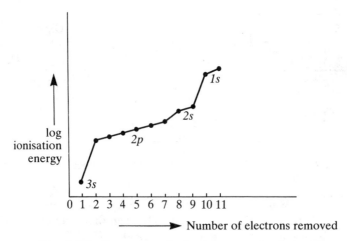

Fig. 2.16 Successive ionisation energies for sodium

level and so requires a substantial amount more of energy to remove it. The following five electrons all come from the 2p orbit and require only small successive increases in energy for their removal; a slightly greater increase is needed to remove the 2s electrons, followed by a considerable increase for the tenth and eleventh electrons, since they occupy a new quantum level.

The first four ionisation energies for sodium, magnesium and aluminium are given in Table 2.4 to illustrate the point that a big increase in the energy required occurs each time an electron is removed from a new quantum level.

Table 2.4 Ionisation energies for sodium, magnesium and aluminium

Element	Energy required to remove one mole of electrons, kJ			
	First	Second	Third	Fourth
Na	500	4600	6900	9500
Mg	740	1500	7700	10 500
Al	680	1800	2700	11 600

(ii) Convergence limits

One method of measuring the ionisation energy of an element is to study the frequency of the lines in the electromagnetic spectrum of the element. The lines in any one series will be seen to merge closer and closer together as the frequency of the radiation increases (Fig. 2.17).

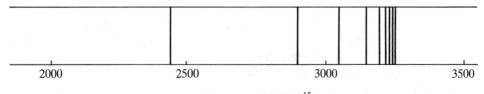

Frequency/Hz × 10^{15}

Fig. 2.17 Line spectrum for atomic hydrogen in the ultra-violet range (the Lyman series)

The frequency at which the lines finally merge is called the *convergence limit*. At this frequency an electron has gained sufficient energy to escape the atom altogether. If the series of lines chosen is that in which the electron originated from the lowest quantum level or *ground state* of the element ($n = 1$), the ionisation energy of the element can be found by using Planck's equation

$$E = h\nu \quad \text{(page 19)}$$

where E = ionisation energy, kJ mol^{-1}
$\quad h$ = Planck's constant
$\quad \nu$ = frequency at convergence limit, Hz

(iii) Electron impact method

This is used to measure the ionisation energy of an element in its gaseous form. A radio valve of the thermionic type containing the gaseous element, for example argon, at very low pressure is connected in a circuit as shown in Fig. 2.18.

The valve cathode is made to emit electrons by heating it. A potential difference is applied between the grid and the cathode to accelerate the electrons towards the grid; most of these fast-moving electrons pass through the spaces of the grid, and some will collide with argon atoms in the valve, the remainder being prevented from reaching the plate at the end of the valve by the application of a small negative potential. At first the electrons do not have enough energy to knock an electron out of the atoms with which they collide. The voltage between the cathode and the grid is then

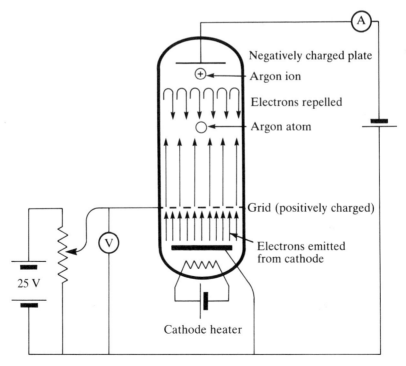

Fig. 2.18   Electron impact method for measuring ionisation energy

increased until the electrons are moving fast enough to have sufficient energy to ionise the atoms on collision.

$$Ar(g) + e^-_{fast} \rightarrow Ar^+(g) + 2e^-_{slow}$$

One of the slow electrons has been knocked out of the argon atom; the other is the colliding electron which has lost energy in the collision.

The occurrence of the positive argon ions are immediately detected by the milliammeter in the left-hand circuit shown in the diagram, since they are attracted to the negatively charged plate, causing a small current flow. The potential difference between the cathode and grid at which this current first flows is that required to give an electron sufficient energy to ionise an argon atom. From this voltage the ionisation energy of argon can be calculated, since the charge on an electron is 1.6×10^{-19} coulombs. One coulomb of charge passing through a potential difference of one volt requires one Joule of energy, therefore an electron passing through a potential difference of V volts (the grid potential) requires $1.6 \times 10^{-19} \times V$ Joules. To ionise one mole of argon atoms requires 6.02×10^{23} electrons (the Avogadro constant) and therefore $6.02 \times 1.6 \times 10^4 \times V$ Joules, or $96.3 \times V$ kJ. Hence the ionisation energy is given by multiplying the grid voltage by 96.3, the units being kJ mol^{-1}.

(d) Representation of atomic orbitals

As mentioned on page 18, the electron is no longer considered to occupy a specific limited orbit in the atom, but instead is regarded as existing in a more diffuse region known as an *atomic orbital*, the shape of which is of great significance in the bonding of atoms in chemical combination.

The orbitals of s electrons are considered to be spherical. That is, electrons in an s orbital are located somewhere in a given sphere with the nucleus at its centre. This can be represented by drawing a boundary round the nucleus depicting a spherical space beyond which the electron is not likely to be, or by showing the electron(s) as a cloud charge with areas of density shaded according to the degree of probability of the location of the electron.

Boundary of sphere
in which the electron is
located

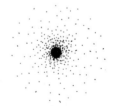

Probable position of electron
represented by density of
shading

Fig. 2.19 Alternative methods of representing the hydrogen atom

The s orbital being spherical, bonds made between atoms by s electrons are non-directional. p Orbitals are directional in nature; they are represented as three dumb-bell shapes at right angles to each other in three dimensions, the nucleus being at the origin of the axes. This orientation dictates the shapes of molecules in which the bonds between the atoms involve p electrons.

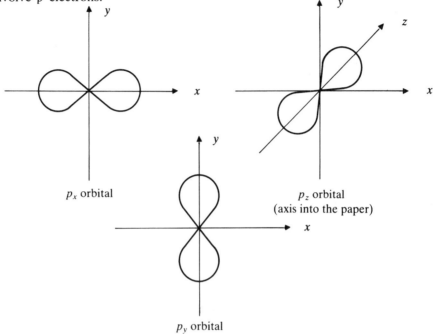

p_x orbital

p_z orbital
(axis into the paper)

p_y orbital

Fig. 2.20 p orbitals

The ten electrons in the d subshells occupy five orbitals oriented about three axes at right angles to one another, the nucleus again being at the origin of the axes. The shapes of the electron clouds are illustrated in Fig. 2.21.

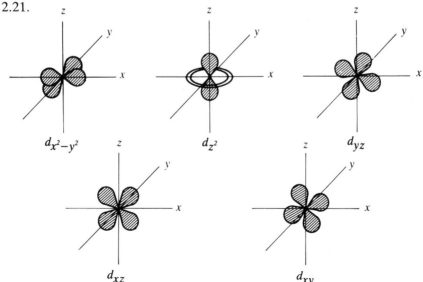

$d_{x^2-y^2}$

d_{z^2}

d_{yz}

d_{xz}

d_{xy}

Fig. 2.21 d orbitals

The $d_{x^2-y^2}$ orbital consists of four pear-shaped lobes oriented *along* the x and y axes. A second orbital d_{z^2} is oriented only along the z axis as shown in Fig. 2.21. The remaining three orbitals each consist of four pear-shaped lobes in the same plane oriented *between* the axes. Thus the four lobes of the d_{xy} orbital are in the same plane as the x and y axes, but occupy the spaces between them, while the lobes of the d_{xz} orbital fill the spaces between the x and y axes, and those of the d_{yz} orbital fill the spaces between the y and z axes.

The d orbitals are also directional in nature, and their disposition helps to explain the shape of molecules in which the central atom is covalently bonded to five or six atoms or groups of atoms, and of liganded ions (page 56).

The locations of the electrons, and the probable volumes of space (or orbitals) they occupy, are calculated from mathematical equations derived from a knowledge of the charges, magnetic fields, distances between nuclei and other measurable phenomena, and are the 'best' picture we have so far; further study may well modify views held at present.

(e) Group 0 and stable electronic structures

The elements helium, neon, argon and the other noble gases in Group 0 are the least reactive of all the elements. They exist as monatomic molecules which have the maximum number of electrons in their filled orbitals.

$$
\begin{array}{ll}
\text{He} & 1s^2 \\
\text{Ne} & 1s^2 2s^2 2p^6 \\
\text{Ar} & 1s^2 2s^2 2p^6 3s^2 3p^6 \\
\text{Kr} & 1s^2 2s^2 2p^6 3s^2 3p^6 3d^{10} 4s^2 4p^6
\end{array}
$$

These unreactive elements have high first ionisation energies since it is difficult to remove an electron from so stable an arrangement.

When chemical compounds are formed between reactive elements, the atoms tend to combine in such a way as to acquire, within the compound, an electronic structure which corresponds to the stable arrangement found in the inert gases. This is further discussed in Chapter 4.

Some compounds of noble gases have been made in recent years. Notably several xenon fluorides are known, including XeF_2, XeF_4 and XeF_6. The shapes of these molecules are shown in Fig. 2.22.

The noble gases are mainly of use because of their inert character. Helium, the second lightest element, is a safe substitute for hydrogen in balloons and airships; it can also replace nitrogen in deep sea diving bells

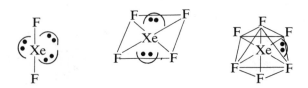

Xenon difluoride   Xenon tetrafluoride   Xenon hexafluoride
(linear)   (square planar)   (distorted octahedral)

Fig. 2.22   Shapes of xenon compounds

and suits since it is less soluble in the blood and less likely to cause 'the bends'. Argon provides an inert atmosphere for many metallurgical processes and is used to fill electric light bulbs. Neon is familiar in the discharge tubes used in advertising signs.

Questions

1. Briefly state the significance of (*a*) Geiger and Marsden's experiment with α rays and metal foils, and (*b*) the experiment in which Moseley bombarded elements with an electron beam.
2. Why does salt, spilt on a gas flame, impart a yellow colour to the flame?
3. Write the electron configuration for the element of atomic number 28. To what Group in the Periodic Table do you think this might belong?
4. Suggest reasons why the first ionisation energy of potassium (420 kJ mol^{-1}) is much lower than that of argon (1520 kJ mol^{-1}).
5. Outline, with the aid of a labelled diagram, the use of a mass-spectrometer in the determination of relative atomic mass.
   The mass spectrum of neon consist of three lines corresponding to relative mass/charge ratios of 20, 21 and 22 with relative intensities of $0.91:0.0026:0.088$ respectively. Explain the significance of these data and hence calculate the relative atomic mass of neon. (C)
6. The atomic spectrum of hydrogen is given by the following relationship,

$$\frac{1}{\lambda} = R_H\left(\frac{1}{n_1^2} - \frac{1}{n_2^2}\right)$$

(*a*) (i) What does λ represent?
   (ii) What do the terms n_1 and n_2 represent?
   (iii) What are the units of the constant R_H?
(*b*) The spectrum consists of a number of lines which may be divided into a number of series,
   (i) Why does the spectrum consist of lines?
   (ii) Why is there a small number of series in the spectrum?
   (iii) Explain why each series converges and in what direction it converges.
(*c*) What method is used to generate the light source for observing the atomic spectrum of hydrogen?
(*d*) Name the instrument used to resolve the hydrogen spectrum. (JMB)
7. (*a*) Explain the meaning of the term *first ionisation energy* of an atom.
   (*b*) The first, second, third and fourth ionisation energies of the elements X, Y, and Z are given below:

	First	*Second*	*Third*	*Fourth*
Ionisation energies in kJ mol^{-1}				
X	738	1450	7730	10550
Y	800	2427	3658	25024
Z	495	4563	6912	9540

Using this information, state, giving your reason, which element is most likely
  (i) to form an ionic univalent chloride
  (ii) to form a covalent chloride
  (iii) to have +2 as its common oxidation state.   (O)

3   Nuclear chemistry

3.1   The radioactive elements

Natural radioactivity was first observed by Becquerel in 1896 when he found that a photographic plate kept carefully wrapped in black paper in a dark cupboard had been 'fogged'; he discovered that the radiation which had penetrated the wrapping emanated from a crystal of a uranium compound kept in the same cupboard. He also established that the radiation would ionise gases, and that it was a property of uranium independent of the compound containing it. His discovery aroused great interest, and shortly after it was found that thorium had similar properties.

In 1898 M. and Mme. Curie isolated radium and polonium from pitchblende, after noticing that the radiation from the mineral was too strong to be accounted for by its uranium content alone. The extraction involved a long and tedious process of fractional crystallisation (page 110), resulting in the extraction of 0.2 gram of radium from one tonne of pitchblende.

Many elements are known to have weakly radioactive isotopes, e.g. potassium-40, but the main radioactive elements are those of atomic number 84 or above. They all spontaneously and continuously emit radiation of a penetrating nature at rates which are not affected by changes in temperature or pressure. The radiation ionises gases, affects photographic plates and causes certain compounds such as zinc sulphide to fluoresce; it has powerful physiological effects which may be harmful but can be harnessed for beneficial uses. Radioactivity is associated with the release of large amounts of energy.

3.2   Types of nuclear radiation

Radioactive elements emit three different types of radiation: alpha, beta and gamma rays. These differ in their penetrating ability and ionising power, and can be distinguished by the effects of a magnetic or electric field.

The effect of such a field is illustrated in Fig. 3.1. The sample of the radioactive element is placed in a deep hole in a block of lead; the emitted rays are subjected to a magnetic or electric field. The whole apparatus is enclosed under high vacuum.

α-Rays, which are the least penetrating, have a range of only a few centimetres in air, and can be stopped by a sheet of paper. They ionise strongly any gas they pass through. The way in which they are deflected in a magnetic or electric field indicates that they are composed of relatively heavy particles of positive charge. Measurements by Rutherford of the charge/mass ratio of α-particles led to the conclusion that they were helium ions with a double positive charge (i.e. helium nuclei $_2^4$He).

This was confirmed in 1909 by an experiment in which α-particles emitted by a sample of radon gas over a period of one week were collected; an electric discharge passed through the gas gave the characteristic spectrum of helium. (After emission α-particles rapidly pick up electrons from the surroundings to form neutral gas molecules.)

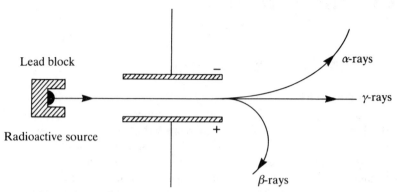

Fig. 3.1 Effect of electric field on α-, β- and γ-rays

β-rays do not ionise gases as strongly as α-rays, but they are much more penetrating: they may have a range of several metres in air and require a sheet of aluminium several millimetres thick to stop them. They are more strongly affected by magnetic or electrostatic fields than α-rays; the amount and direction of their deflection suggest that they are composed of particles of negligible mass and one unit of negative charge. Becquerel showed experimentally that β-rays were similar to cathode rays, and later workers confirmed that β-particles were fast-moving electrons.

Gamma rays are very penetrating, and even a sheet of lead several centimetres thick will fail to stop them. They ionise gases very weakly and are not affected by a magnetic or electrostatic field. γ-rays do not behave as particles, but are a form of electromagnetic radiation travelling with the velocity of light and of a wavelength which corresponds to that of very short X-rays.

Table 3.1 Summary of types of nuclear radiation

Emission	Composition	Electrical charge	Mass of particle	Relative penetrating and ionising power
α-rays	Helium nuclei ^4_2He	+2	4 units	Slightly penetrating, highly ionising
β-rays	Electrons	−1	1/1840 unit	Moderately penetrating and ionising
γ-rays	Electromagnetic radiation	No charge	—	Highly penetrating, slightly ionising

The collection of helium from a radioactive source provides a method of estimating the Avogadro constant L. The number of α-particles emitted from a sample of radium can be counted by means of a Geiger counter. The volume of helium gas collected, although very small, can be measured if the period of collection is long enough. For example, $0.043\ \text{cm}^3$ of helium at s.t.p. has been collected over a period of one year. In this time the number of α-particles emitted was 11.6×10^{17}. Hence $22\,400\ \text{cm}^3$ or 1 mole of monatomic helium gas contains $\dfrac{11.6 \times 10^{17} \times 22\,400}{0.043}$ or 6.04×10^{23} atoms; thus a value for the Avogadro constant has been found.

3.3   Radioactive disintegration

Radioactive disintegration causes changes that are quite unlike ordinary chemical changes; they are spontaneous, unaffected by physical conditions or chemical combination. The energy released by the dissociation of a substance emitting α-particles is at least a million times greater than that released in a chemical change of a comparable amount of any substance.

To account for these facts Rutherford and Soddy suggested in 1902 that the atoms of radioactive elements are unstable and break up in a radioactive change to form different atoms. This revolutionary idea was soon accepted since it satisfactorily accounted for all the known facts of radioactivity. The disintegration of an atom *is* accompanied by the emission of an α- or a β-particle, while the enormous release of energy is explained as the result of the very slight loss of mass that accompanies the emission of particles from the nucleus of the atom. This is in accordance with Einstein's theory that mass can be converted to energy. The minute loss of mass occurs when, for example, two protons and two neutrons combine in the nucleus to form an α-particle and is known as the *mass defect*. Rutherford and Soddy also suggested that the radioactive decay continues, at a definite rate, through a number of stages until a stable product is obtained (Fig. 3.2). The intermediate products depend on the nature of the particle emitted.

The loss of an α-particle reduces the mass number (A) (page 13) of the atom by four units and its atomic number (Z) by two units. For example, radium decays to radon by the loss of an α-particle:

$$^{226}_{88}\text{Ra} \rightarrow \,^{222}_{86}\text{Rn} + \,^{4}_{2}\text{He}$$

The new atom is now that of an isotope of the element occupying a position two places before the original element in the Periodic Table. Two of the electrons surrounding the nucleus disperse and the remainder rearrange to an electron configuration consistent with the new element.

The loss of a β-particle results from the emission of a high speed electron from a neutron. The neutron, having lost an electron, becomes a proton. The mass of the new atom is scarcely altered but there is now one more proton in the nucleus, so that the atomic number is increased by one and the new atom is that of an isotope of the next element in the Periodic Table. For example, thorium changes to protoactinium when it decays by β-emission.

$$^{231}_{90}\text{Th} \rightarrow \,^{231}_{91}\text{Pa} + \,^{0}_{-1}\text{e}$$

The way in which the changes occur is set out in the Group Displacement law, which can be summarised:

Table 3.2   Group displacement law

Particles lost	Change of mass number	Change of atomic number
α-particle ($^{4}_{2}\text{He}$)	-4	-2
β-particle ($^{0}_{-1}\text{e}$)	No change	$+1$

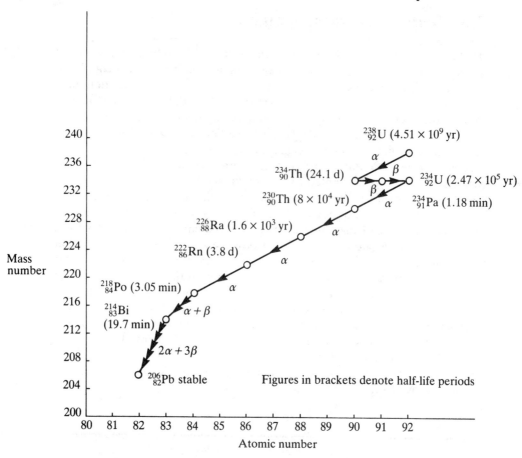

Fig. 3.2   Part of the decay series for uranium-238

3.4   Stability of radioactive atoms

The stability of an atom appears to depend on the relative number of protons and neutrons present in the nucleus. Information concerning stability is illustrated by plotting the number of neutrons against the number of protons present in the nucleus of all known nuclides. The result is shown in Fig. 3.3.

When this is done a continuous line can be drawn through those nuclides that are stable; the positions of the neutron/proton ratio for unstable nuclides lie in a shaded area on either side of the line.

From the diagram the following information can be deduced:

(a) The lighter stable nuclides mainly have equal numbers of neutrons and protons.

(b) The heavier stable nuclides have more neutrons than protons.

(c) Many stable nuclides have even numbers of protons and neutrons. For example, $^{16}_{8}O$, $^{28}_{14}Si$, $^{56}_{26}Fe$. Exceptions to this rule are $^{2}_{1}H$, $^{6}_{3}Li$, $^{10}_{5}B$ and $^{14}_{7}N$.

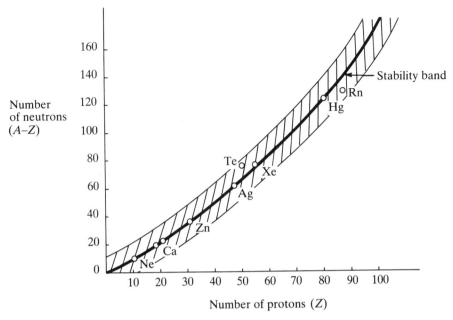

Fig. 3.3 Neutron/proton numbers for stable isotopes (only some isotopes are marked)

(d) Unstable nuclides disintegrate in such a way as to produce new isotopes nearer to the 'stability band'.

(e) A nuclide with a neutron/proton ratio above the line decays so as to gain an increase in atomic number, i.e. by beta emission which changes a neutron into a proton, thereby decreasing the neutron/proton ratio.

(f) An unstable nuclide with a neutron/proton ratio below the stability line disintegrates in such a way as to decrease its atomic number; in heavy nuclides this can occur by α-emission, which will increase the neutron/proton ratio (see series illustrated in Fig. 3.2).

3.5 Rate of radioactive decay

The intensity of radiation from a radioactive element decreases with time, since, as disintegration proceeds, fewer and fewer unstable atoms remain. The rate of radioactive decay is different for each radioactive isotope, and can be used to characterise it. The time taken for the radioactivity of a sample to fall to half its initial value is constant for any particular nuclide, and is known as its half-life, $t_{\frac{1}{2}}$. The half-life may be anything from millions of years to fractions of a second. The half-life of radium-226 is 1620 years, which means that of one gram of radium-226, half a gram will remain after 1620 years, one quarter of a gram will remain after 3240 years, one-eighth of a gram will still be radium-226 after 4860 years, and so on.

For short-lived nuclides the half-lives can be determined by plotting the activity of a sample against time. The activity, which is proportional to the

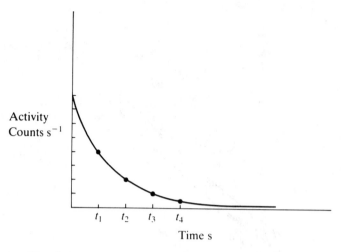

Fig. 3.4   Decay curve of protoactinium (half-life 1.18 min.)

number of radioactive atoms present in the sample, is measured by a Geiger counter—an instrument which is sensitive to both α- and β-particles. From the graph the time for the activity of the sample to fall to half its initial value can be read.

The rate of decay for a particular nuclide is also characterised by its *radioactive decay constant*.

Since the rate of decay is proportional to the number of atoms of the nuclide present, it follows first order kinetics and can be expressed in the relationship

$$-\frac{dN}{dt} = \lambda N$$

where λ is the radioactive decay constant and N is the number of atoms present at time t.

Rearranging,

$$-\frac{dN}{N} = \lambda \, dt$$

On integrating, this expression becomes

$-\ln N = \lambda t + C$ where C is the integration constant

Now if N_0 is the number of atoms present when $t = 0$, then

$$-\ln N_0 = C$$

Substituting in the integrated equation

$$-\ln N = \lambda t - \ln N_0$$

or

$$\ln \frac{N_0}{N} = \lambda t$$

Thus λ may be found by plotting $\ln \dfrac{\text{initial activity}}{\text{activity at time } t}$ against t (Fig. 3.5).

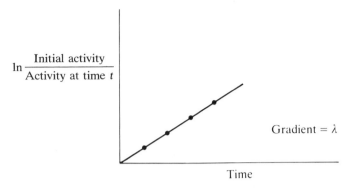

Fig. 3.5   Determination of radioactive decay constant

The half-life is used to characterise nuclides since it is independent of the initial concentration and therefore remains constant during the whole of the radioactive decay. It can be calculated from the radioactive decay constant, since at the half-life time the number of active atoms will have fallen to half the initial value, and

$$N = \frac{N_0}{2}, \qquad \ln 2 = \lambda t_{\frac{1}{2}}, \qquad \text{or } t_{\frac{1}{2}} = \frac{\ln 2}{\lambda}.$$

A nuclide with a large radioactive decay constant will have a short half-life.

Plainly these methods would not be suitable for determining the half-life of a nuclide such as uranium-238 which is 4.5×10^9 years! Other methods, outside the scope of this book, have to be used.

3.6   Nuclear reactions

(a) Transmutation of the elements

The alchemists' dream of turning base metals into gold is, in a sense, achieved in the transmutation of one element into another by bombarding suitable elements with fast-moving neutrons, protons, α-particles or deuterons (nuclei of deuterium atoms, ^2_1H)—although this has never resulted in obtaining quantities of gold from lead!

Rutherford observed in 1909 that subjecting nitrogen gas to bombardment by a stream of α-particles gave rise to a supply of fast-moving protons ^1_1H. Further examination showed that the other product of the collisions between nitrogen atoms and α-particles was an isotope of oxygen of mass number 17, ($^{17}_8\text{O}$.) The overall change can be represented* by the equation:

$$^{14}_7\text{N} \quad + \quad ^4_2\text{He} \quad \rightarrow \quad ^{17}_8\text{O} \quad + \quad ^1_1\text{H}$$

Nitrogen   Alpha   Oxygen   Proton
nucleus   particle   nucleus

---

* The change can also be represented in the following notation:

Initial nuclide (incoming particle, outgoing particle) final nuclide. Thus the effect of alpha particles on $^{14}_7\text{N}$ is denoted by

$$^{14}_7\text{N}(^4_2\text{He}, ^1_1\text{H})^{17}_8\text{O}$$

The effect of α-rays on lighter elements was studied further and, in 1930, the bombardment of beryllium by α-particles was found to give rise to a stream of fast-moving uncharged particles of a very penetrating nature. These were identified by Chadwick as neutrons, the existence of which had been predicted by Rutherford (page 12). The other product of the reaction was carbon-12:

$$_4^9Be + _2^4He \rightarrow _6^{12}C + _0^1n$$

Many different transmutations have now been carried out. In 1934 F. Joliot and his wife Irene Joliot-Curie, daughter of Marie Curie, produced the first artificial radioisotope by bombarding aluminium with a stream of α-particles; this resulted in the formation of an unstable isotope of phosphorus and was accompanied by the emission of neutrons:

$$_{13}^{27}Al + _2^4He \rightarrow _{15}^{30}P + _0^1n$$

Since 1934 at least one radioactive isotope of every element has been made. In addition, atoms with atomic number greater than 92 have been produced. These are atoms of new elements that do not occur naturally on earth, such as neptunium, plutonium, americum and lawrencium.

(b) Nuclear fission

Once neutrons had been identified the way to nuclear fission was opened up; the bombardment of uranium by neutrons produced a reaction of a different kind from the early nuclear reactions. During fission the uranium nucleus is split into large fragments of approximately equal mass; this disintegration causes the emission of further neutrons known as fission neutrons. The combined mass of the fission products is appreciably less than that of the original uranium atom and bombarding neutron (the mass defect). This disappearance of mass accounts for the tremendous release of energy that occurs during fission.

The fission of uranium-235 is brought about by collision with either fast or slow neutrons, and can be represented by the equation:

$$_{92}^{235}U + _0^1n \rightarrow _{56}^{141}Ba + _{36}^{92}Kr + 3_0^1n$$

However, products other than barium and krypton may also be formed, together with different numbers of neutrons.

The emitted neutrons cause fission of any further uranium-235 atoms they collide with. A chain reaction is set up which may proceed at a constant rate if only one of the emitted neutrons causes new fission, or may become multiplied to explosion rate if all three neutrons cause new fissions. In practice few neutrons cause new fissions as some of those emitted escape at the surface and others are absorbed by uranium-238 atoms (99.3% of natural uranium) or by other fragments present, without causing fission. For every system there is a critical mass which must be attained before even a steady chain reaction can start.

Uranium-238 behaves differently from uranium-235; it undergoes fission on collision with very fast neutrons but absorbs slow neutrons. The former event is comparatively rare, but the latter results in the formation of neptunium, which decays by β-emission to plutonium:

$$^{238}_{92}U + ^{1}_{0}n \rightarrow ^{239}_{92}U \rightarrow ^{239}_{93}Np + ^{0}_{-1}e$$
$$\text{slow}$$

$$^{239}_{93}Np \quad \rightarrow \quad ^{239}_{94}Pu \quad + \quad ^{0}_{-1}e$$
$$\text{Neptunium} \quad \text{Plutonium} \quad \beta\text{-particle}$$

Plutonium will also undergo fission with slow neutrons and is therefore a new and valuable nuclear fuel.

The reactor in a nuclear power station has a graphite core with a series of vertical channels. Sufficient rods of natural uranium enriched with uranium-235 are lowered into these channels to make the reaction critical, i.e. to start a steady chain reaction. The rate of the reaction is controlled, moderated or stopped by lowering rods of *neutron absorbers* such as cadmium or boron into vacant channels in the core. The heat from the reactor is used to produce steam for the turbines that drive the dynamos to produce electric power.

Another type of reactor is the so-called *fast breeder reactor* which contains a core of nearly pure uranium-235 surrounded by a sheath of uranium-238. The uranium-238 absorbs the fission neutrons escaping from the core and is converted into plutonium; thus as the uranium-235 core decays new fuel is produced.

Atomic bombs are operated by bringing together suddenly two masses of uranium-235 or plutonium-239, which together are greater than the critical mass. An uncontrolled chain reaction ensues and explosion follows.

The concentration of uranium-235 present in natural uranium is only 0.7%, making the separation of the pure isotope a lengthy and tedious process. It is effected by forming uranium hexafluoride and separating the gaseous compounds $^{235}UF_6$ and $^{238}UF_6$ by diffusion (page 74).

(c) Thermonuclear fusion

The fusion of light nuclei into a heavier nucleus can lead to a loss of mass and therefore the liberation of energy. This occurs in the hydrogen bomb and is believed to be the source of the Sun's energy, but so far attempts to harness the energy from a controlled reaction have not been successful. A typical reaction is that with heavy hydrogen (deuterium):

$$^{2}_{1}H + ^{2}_{1}H \rightarrow ^{3}_{2}He + ^{1}_{0}n$$

One of the difficulties is to raise the temperature sufficiently to start the reaction (several million degrees Celsius are probably required); the passing of large electric currents has been tried and the use of lasers may solve the problem. Once the reaction were started, enough heat would be generated to keep it going. In the hydrogen bomb an uncontrolled thermonuclear reaction is started by using an atomic fission bomb to trigger the fusion reaction. If a controlled reaction can be achieved a vast source of energy will be available since there is enough deuterium in the world's oceans to last for millions of years.

3.7   Use of radioactive isotopes

Radioisotopes are already used widely for a great variety of purposes, and new uses are continually being developed. In medicine γ-rays from cobalt-60 are used to treat cancer patients by destroying the malignant cells; small pellets of radioactive isotopes of short half-life, such as gold-198, have been implanted in tumours to destroy them; iodine-131 (half-life 8 days) is used to treat cancer of the thyroid gland. Medical instruments and bandages can be sterilised after packing by a brief exposure to γ-rays.

γ-rays from cobalt-60 can also be used to detect cracks and flaws in welds and castings. The ionising effect of radiation is used to remove the static electricity that builds up in the textile and paper-making industries constituting a fire risk. β-ray sources are used in the manufacture of plastics and paper to check the thickness of the material by measuring the amount of radiation which penetrates the sheet. A similar procedure can be used to control automatically the filling of powder and paste containers.

An interesting use of radioisotopes is as 'tracers'; for this purpose small amounts of the isotope are added at the start of the process to be followed which is then monitored by a radiation detector which registers the later whereabouts of the isotope. For example the distribution of sand and mud carried into estuaries by rivers can be followed; the rate of absorption of phosphorus in crops from phosphatic fertilisers is studied to obtain information about plant metabolism and about the efficacy of various fertilisers; the wear on bearings in machinery can be measured by adding radioactive metal to the bearings and observing the extent to which radioactive particles appear in the lubricating oil; thorough mixing of the ingredients of such diverse products as plastics and animal foods can be checked by adding a small amount of a radioisotope to one ingredient.

3.8   Carbon-14 dating

Carbon-14 is formed when cosmic ray neutrons collide with atmospheric nitrogen:

$$^{14}_{7}\text{N} + {}^{1}_{0}n \rightarrow {}^{14}_{6}\text{C} + {}^{1}_{1}\text{H}$$

The formation and decay of carbon-14, which has a half-life of about 5600 years, leads to a constant level of radioactivity in atmospheric carbon dioxide.

The carbon in plants and trees, which is taken up during photosynthesis from the atmosphere, also has a constant level of radioactivity. When the living material dies, and no more ingestion takes place, the decay of the carbon-14 reduces the radioactivity at a steady rate. The age of objects made of wood or linen cloth or other carbon-containing material can therefore be estimated by measuring the residual activity.

Questions

1. Write equations for the decay of $^{238}_{92}\text{U}$ to $^{241}_{91}\text{Pa}$ depicted in Fig. 3.2.

2. Explain, with examples, the differences between the transmutation of an element, nuclear fission and thermonuclear fusion.

3. (a) What is meant by the statement 'the half-life of the radioactive isotope of carbon of mass number 14 is 5730 years?

(b) Name two elements (other than carbon) which have naturally-occurring radioactive isotopes. For the two examples you quote, state whether α- or β-particles are emitted, and indicate which of the two isotopes has the longer half-life.   (O)

4. (a) What are α, β and γ emissions and how do they differ in their penetrating power and their behaviour in a magnetic field?

Explain the meaning of the two numbers before the symbol for uranium and identify P, Q, R and S in the following equations.

$$^{234}_{92}U \rightarrow \alpha + P$$

$$^{239}_{92}U \rightarrow \beta^- + Q$$

$$^{235}_{92}U \rightarrow \gamma + R$$

$$^{238}_{92}U + {}^{2}_{1}H \rightarrow {}^{239}_{92}U + S$$

(b) Deduce the nature of X in the following reaction.

$$^{235}_{92}U + {}^{1}_{0}n \rightarrow {}^{95}_{42}Mo + {}^{139}_{57}La + 2{}^{1}_{0}n + 7X$$

There is approximately 0.1 per cent less mass on the right-hand side of the equation than on the left-hand side. What is the significance of this?
                                                                            (JMB)

5. The ${}^{45}_{20}Ca$ isotope of calcium is radioactive: during decay each calcium atom emits one β-particle.

(a) Suggest an isotope of potassium which, on neutron bombardment, you might expect to give ${}^{45}_{20}Ca$. Write a balanced equation for the reaction you propose.

(b) Write a balanced equation for the decay of ${}^{45}_{20}Ca$.

(c) The half life of ${}^{45}_{20}Ca$ is 160 days. If a mass m of initially pure ${}^{45}_{20}Ca$ is taken, write down an expression for the mass of ${}^{45}_{20}Ca$ that will decay in 480 days.

(d) If the mass m of pure ${}^{45}_{20}Ca$ taken in (c) is 2.25 g, how many β-particles will be emitted during 480 days?

(e) From your knowledge of the natural occurrence and properties of calcium, suggest two possible uses of this isotope.   (C)

6. (a) How are the *atomic number* and *mass number* related to the structure of the nucleus of an atom?

(b) The following table gives the atomic numbers of certain elements together with the relative atomic masses of their stable isotopes.

Atomic number	Relative atomic masses
1	1, 2
5	10, 11
8	16, 18
12	24, 25, 26
17	35, 37
26	54, 57, 58
29	63, 65
40	87, 93
44	104, 107

(i) Plot a graph of atomic number against number of neutrons for each stable isotope, in order to establish a *band of stability*, with both upper and lower limits.

(ii) From the graph predict which of the following species are unstable

$$^{24}_{10}Ne, \quad ^{45}_{20}Ca, \quad ^{50}_{23}V, \quad ^{64}_{29}Cu, \quad ^{92}_{42}Mo$$

(iii) Having determined the unstable species, explain the *number* and *types* of emission required in order to attain stable structures. What would be the atomic number and atomic mass of the stable isotopes attained?

(c) (i) What are the two methods by which $^{27}_{13}Al$ could lose electrons? What would be the products in each case?

(ii) If $^{27}_{13}Al$ absorbed a neutron and subsequently underwent beta-emission, what would be the resulting isotope? Give its atomic number and atomic mass. (AEB 1978)

7. The table shows the mass numbers and the numbers of neutrons in the nucleus of four elements W, X, Y and Z

	W	X	Y	Z
Mass number	36	39	40	40
Neutrons in nucleus	18	20	21	22

(a)   (i) Write down the atomic numbers of the four elements.

(ii) Which of the four elements are isotopes of each other?

(iii) The electronic configuration of the element neon (atomic number 10) can be written in the form $1s^2 2s^2 2p^6$. Write down similarly the configurations of the elements W, X, Y and Z.

(b) The element Y is radioactive, decaying by emission of a β-particle, with a half-life of 1.3×10^9 years.

(i) What is a β-particle?

(ii) The decay of the nucleus of Y gives rise to nucleus V. What are the mass number and the atomic number of the nucleus V?

(iii) What is meant by the statement that 'the half-life is 1.3×10^9 years'?

(c) (i) In which group of the Periodic Table will the element X be placed?

(ii) Write down the formulae of the typical oxide and chloride of X.

(d) Given a supply of the hydroxide of the element of electronic configuration $1s^2 2s^2 2p^6 3s^1$, how would you prepare a pure sample of the chloride of the element? (L)

4   Bonding

4.1   Introduction

When elements combine to form compounds there is generally an increase in stability. Sodium burns in chlorine to produce the very stable white crystalline compound we use as table salt. Evidently the sodium, a dangerously active metal, and the chlorine, a poisonous corrosive gas, have changed their nature considerably, for in no respect does the product resemble the starting materials. Chlorine also forms a compound with carbon, the heavy inert liquid tetrachloromethane (carbon tetrachloride). However sodium chloride and tetrachloromethane are very different in many respects. These differences arise from the type of bonding between the atoms in the compounds. Sodium chloride is an ionic compound with electrovalent bonding whilst tetrachloromethane is a molecular compound with covalent bonding. These are the two main types of bonding; both result in the participating atoms acquiring more stable electronic configurations either by the transfer or by the sharing of electrons between atoms. The electrons involved in bonding are those which occupy the outermost orbital of the atom, the valency electrons.

4.2   Ionic or electrovalent bonding

Here valency electrons are transferred completely from one atom to another to form ions (page 3). The elements which most readily form electrovalent compounds are the more reactive metals, from Groups I and II, and the more reactive non-metals, from Groups VI and VII. The ions they form acquire an electronic structure similar to that of a noble gas from Group 0. Thus in the formation of sodium chloride, the sodium atom loses an electron to form a positive ion (cation) and this electron is gained by a chlorine atom to form a negative ion (anion). Both ions now have the structure of a noble gas.

Sketches

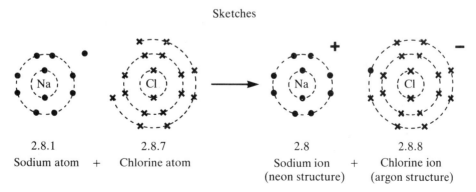

2.8.1	2.8.7	2.8	2.8.8
Sodium atom +	Chlorine atom	Sodium ion (neon structure) +	Chlorine ion (argon structure)

These electron transfers can conveniently be represented by 'dot-and-cross' diagrams in which only the outer, or valency, electrons are shown. Thus in the formation of calcium oxide the calcium atom loses two electrons to form a 2-valent positive ion, and the oxygen gains two electrons to form a

2-valent negative ion:

Calcium atom + oxygen atom   Calcium ion + oxide ion

$$Ca\!\!:\quad + \quad \overset{\times\times}{\underset{\times\times}{O}}\!\!\overset{\times}{} \quad\longrightarrow\quad Ca^{2+} \quad + \quad [\!:\!\overset{\times\times}{\underset{\times\times}{O}}\!\!\overset{\times}{}]^{2-}$$

2.8.8.2   2.6   2.8.8   2.8
                                 (Argon structure) (Neon structure)

The ions, once formed, being positively and negatively charged entities, are mutually attracted and assemble together in a crystalline structure held together by electrostatic forces. Therefore no separate molecules exist; the formula CaO simply gives the ratio of the atoms present. The arrangements of the ions in the crystal lattices are related to the sizes of the ions and the charges they carry, as in the sodium chloride lattice, Fig. 4.1.

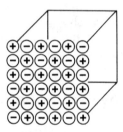

Fig. 4.1   Sodium chloride lattice

The strong forces between the ions give rise to a crystalline solid of high melting point, high boiling point and high latent heat of vaporisation. Ionic compounds also conduct electricity when molten or dissolved in water, the ions carrying the current (page 204).

Whereas positive ions of charge one or two are easily formed, the formation of those with three charges is less easy, as each electron removed leaves the ion with an additional positive charge that resists the removal of the next electron (page 23). Aluminium forms an Al^{3+} ion in some of its compounds (in others, notably aluminium chloride, the bonding is covalent (page 253)). The energy required for the formation of an ion with four positive charges is too great; Sn^{4+} and Pb^{4+} ions are quoted in tables of standard electrode potentials, but in fact Pb(IV) and Sn(IV) compounds are substantially covalent.

Not all cations attain a noble gas structure; notable exceptions are those formed from the d-block or transition elements of the first long period (described in Chapter 22).

The non-metal elements of Groups VI and VII readily form 2-valent and 1-valent negative ions, such as the O^{2-} ions in magnesium oxide and the Cl^- ions in sodium chloride. In Group V, nitrogen, and to a lesser extent, phosphorus form N^{3-} ions in nitrides and P^{3-} ions in phosphides, as in Mg_3N_2 and Ca_3P_2. Anions with a charge greater than three are unknown.

4.3 Covalent bonding

Atoms united by covalent bonding acquire noble gas structures by the sharing of a pair of electrons, one electron being contributed by each atom. For example, carbon in tetrachloromethane acquires a neon structure by sharing its four valency electrons with four chlorine atoms, each of which acquires an argon structure:

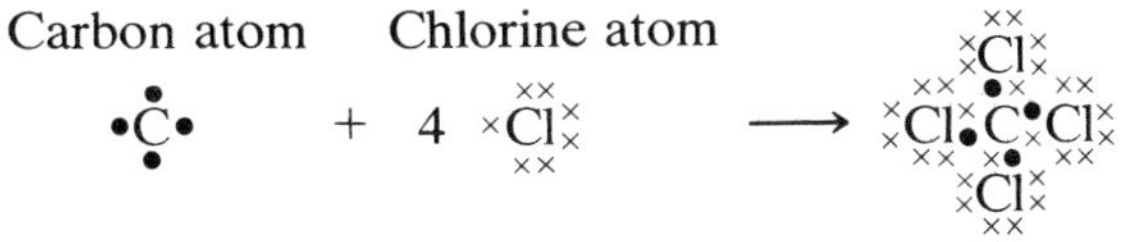

The shared pair of electrons makes a strong bond, and stable discrete molecules are formed. Since no electrons have been transferred from one atom to another the molecules are electrically neutral particles, and have little attraction for one another. Hence molecular compounds with covalent bonding are gases, liquids or low melting point solids which do not conduct electricity and dissolve readily in organic solvents.

Covalent bonds can also be formed in which four or six electrons are shared between two atoms. The sharing of two electrons between atoms as in a chlorine molecule is termed a *single bond* and is represented by a line joining the two atoms:

$$\text{:Cl:Cl:} \quad \text{or} \quad \text{Cl—Cl or Cl}_2$$

The sharing of four electrons is termed a *double bond*, and is represented by two lines between the atoms, as in a carbon dioxide molecule:

$$\text{:O::C::O:} \quad \text{or} \quad \text{O==C==O} \quad \text{or} \quad \text{CO}_2$$

Six shared electrons constitute a *triple bond*, as in a nitrogen molecule:

$$\text{:N:::N:} \quad \text{or} \quad \text{N≡N} \quad \text{or} \quad \text{N}_2$$

These double and triple bonds are an important feature of some carbon compounds. They are discussed more fully in Chapter 23.

Covalent bonding can also give rise to *macromolecules* in which the bonding is continuous from atom to atom throughout the structure, giving rise to *atomic crystals* such as diamond, graphite (page 83) or silicon dioxide (SiO_2).

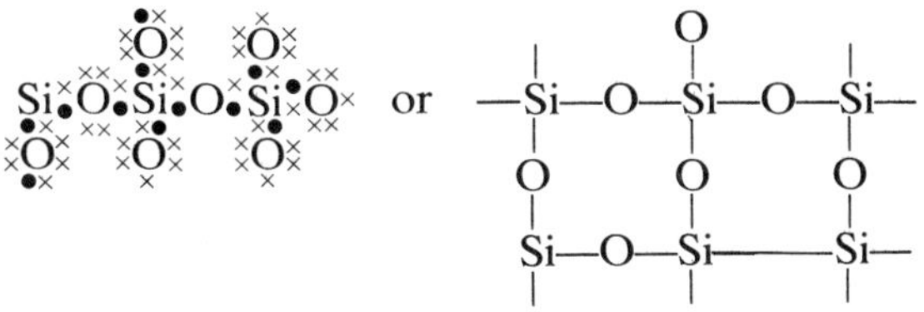

Fig. 4.2 Silicon dioxide (represented in two dimensions only)

The shared pair of electrons makes for very strong directional bonds between the atoms, and so substances composed of giant atomic crystal lattices are of characteristic shapes, have high melting and boiling points and are insoluble in water. Except for graphite, they do not readily conduct an electric current.

4.4 Intermediate bonding and polar covalent molecules

Although electrovalency and covalency are the extreme types of bonding, many bonds are intermediate in character. Compounds which potentially contain a small positive ion combined with a large negative ion tend to be partially covalent (Fajan's rule, 1923). This is because an electron is not so readily lost from a small atom where it is close to the nucleus, and because the electron cloud surrounding a large negative ion is readily distorted to give a higher electron density between the atoms and consequently a covalent character to the bond. Lithium chloride provides a good example of this effect (Fig. 4.3).

Li^+ Cl^- LiCl

Fig. 4.3 Distortion of the electron cloud by the small Li^+ ion

Similar examples of covalency are shown by compounds of beryllium (page 235) and aluminium (page 247).

The ability to distort the electron cloud or *polarising power* of a cation depends on both its size and charge; ions of small radius which carry a large charge, for example Al^{3+}, are the most effective polarisers. The most readily polarised anions are those with large ionic radii whose outer electrons are well shielded from the nucleus by completed shells of electrons e.g. the *polarisability* of the halide ion increases $Cl^-<Br^-<I^-$.

Pauling's concept of electronegativity is valuable in determining whether a compound will be electrovalent or covalent. *Electronegativity* can be defined as *the ability of an atom to attract electrons.* Pauling devised a scale of relative electronegativity which is shown qualitatively in Fig. 4.4. In general, the greater the difference in electronegativity between two atoms, the more likely is it that the bond between them will be ionic.

Increasing electronegativity →

```
        H
Li  Be  B  C  N  O  F  ↑
Na                 Cl     Increasing
K                  Br     electronegativity
Rb                 I
Cs
```

Fig. 4.4 Trends in electronegativity

It will be seen that the most electronegative element is fluorine, and that caesium is one of the least electronegative, making caesium fluoride, CsF, one of the most 'ionic' compounds. It will also be seen that electronegativity increases across the Periods of the Periodic Table and decreases down any Group.

Many wholly covalent compounds have a distinct polarity resulting from the unequal sharing of electrons. These polar covalent bonds result from the differences in electronegativity between the atoms. In the hydrogen chloride molecule the chlorine atom is more electronegative than the hydrogen atom and attracts the electrons more strongly, concentrating the negative charge at one end of the molecule.

Fig. 4.5 Representation of distribution of negative charge in the hydrogen chloride molecule

This uneven distribution of electric charge gives the molecule an electrical polarity or *dipole moment* which is proportional to the charges involved and the distance apart of the nuclei. The situation can be represented by the notation $^{\delta+}H—Cl^{\delta-}$, or by an arrow in the direction of the more electronegative element, as in the polarised bond between carbon and chlorine in chloromethane $H_3C\rightarrow—Cl$.

Whether a covalent compound has a dipole moment or not depends partly on the symmetry of the molecule. A chlorine molecule (Cl_2) has no dipole moment as the two chlorine atoms share the electrons equally. Chloroethane will be polar but tetrachloromethane will not, as the chlorine atoms are symmetrically disposed around the carbon atom in the latter.

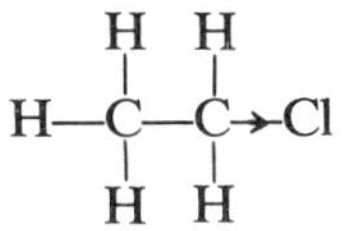

Chloroethane 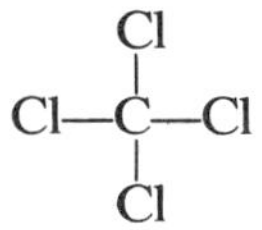   Tetrachloromethane

4.5   The shapes of some simple covalent molecules

Since the covalent bond is directional, molecules have definite shapes. These can be predicted by the electron repulsion theory of Sidgwick–Powell which assumes that bonding pairs of electrons in a molecule will take up positions as far apart as possible owing to their mutual repulsion.

Thus two pairs of bonding electrons will give a linear molecule, three pairs will result in a trigonal planar shape, and four pairs will result in a tetrahedral shape.

The presence of unbonded electrons in a molecule affects the shape. The lone pair of electrons on the nitrogen atom of the ammonia molecule exerts a repulsive force which gives the molecule a pyramidal shape with bond angles of 107° between the N—H bonds; the repulsive force of the lone pair

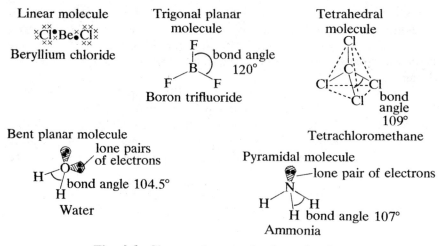

Fig. 4.6 Shapes of some simple molecules

is greater than that of a bonded pair, resulting in the reduction of the angle from that of a regular tetrahedron.

In the water molecule, H_2O, the oxygen atom carries two lone pairs which occupy orbitals as shown in Fig. 4.6. This results in a bent planar shape with an angle of 104.5° between the H—O bonds. When water freezes to ice the molecules effectively become tetrahedral, with two corners of the tetrahedron being occupied by lone pairs of electrons (page 229).

Trigonal bipyramidal and octahedral molecular shapes arise where five and six pairs of electrons form bonds. This situation is possible with atoms which have electrons in the s and p orbitals of the third quantum level and hence vacant $3d$ orbitals. They can expand the number of valency electrons to ten or twelve. Phosphorus $(1s^2 2s^2 2p^6 3s^2 3p^3)$ and sulphur $(1s^2 2s^2 2p^6 3s^2 3p^4)$ form phosphorus pentachloride and sulphur hexafluoride (Fig. 4.7).

Fig. 4.7 Trigonal bipyramidal and octahedral molecules

The presence of double or triple bonds in molecules also influences their shape. This is discussed in the next section and in Chapter 23.

4.6   Bond hybridisation and molecular orbitals

(a) Sigma (σ) and pi (π) bonds

When atoms combine to make covalently bonded compounds the orbitals of the electrons from each of the atoms forming the bonding pair overlap to give a new *molecular orbital* which both electrons now occupy. Considering first the spherical orbitals of *s* electrons and the dumbbell shaped orbitals of *p* electrons (page 27), the following bonds can be made:

(*i*) *Sigma (σ) bonds*
   (a) Formed by the overlap of two *s* orbitals:

   (b) Formed by the overlap of two *p* orbitals linearly oriented:

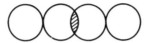

   (c) Formed by the overlap of one *s* orbital and a *p* orbital:

(*ii*) *Pi (π) bonds*
These are formed by the overlap of two *p* orbitals from adjacent atoms, laterally oriented:

π bonds form when the atoms are already linked by a σ bond. They are also formed by the overlap of a *p* and a *d* orbital:

p orbital   *d* orbital

The double overlap leads to a molecular orbital consisting of two separate volumes of electron density, one above and one below the σ bond.

   Molecular orbitals are thus formed when atoms approach so closely that the orbitals of the outer electrons of one atom overlap the orbitals of the outer electrons of another atom. The number of molecular orbitals formed equals the number of atomic orbitals from which they are derived. Each molecular orbital can contain up to two electrons which are differentiated by their spin quantum number (page 21). (This is in accordance with Pauli's Exclusion Law that no two electrons in any atom can have all four quantum numbers identical). The electrons enter the molecular orbital of lowest energy first, in accordance with Hund's rule (page 22).

   Molecular orbitals differ from atomic orbitals in that the electrons filling them are under the influence of two or more nuclei rather than a single nucleus. They are of two kinds of orbitals—*bonding orbitals* which are of lower energy than the atomic orbitals from which they are formed, and *antibonding* orbitals which are of higher energy than those from which they are formed. Bonding will not occur unless there is a net drop of energy (the bond energy); the greater the drop in energy, the stronger is the bond.

These concepts can be illustrated by considering the formation of the hydrogen molecule. Each hydrogen atom has a single electron in a 1s orbital. When two atoms approach closely enough for the orbitals to overlap, two molecular orbitals are formed. One of these orbitals has a high electron density betwen the atomic nuclei, binding them together to make a bonding orbital (σ orbital) of lower energy than the atomic orbitals. The other has only a very low electron density between the nuclei; this is the antibonding (σ^*) orbital, and is of higher energy than the contributing atomic orbitals. The electrons, following Hund's rule, enter the bonding orbital, while the antibonding orbital remains empty; thus a lowering of energy results and a strong bond is formed.

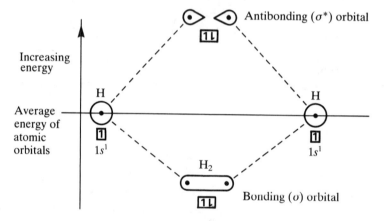

Fig. 4.8   Hydrogen molecule

Helium, which like hydrogen has electrons only in the 1s orbital, does not form molecules of formula He_2, and molecular orbital theory shows that bonding between two helium atoms is unlikely. Each helium atom has two electrons in the 1s orbital; if two atoms approach closely so that these orbitals overlap, then two molecular orbitals will be formed.

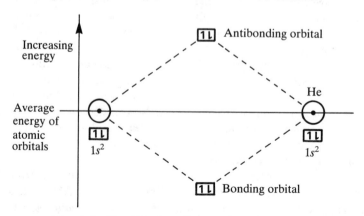

Fig. 4.9   Possible bonding orbitals in helium

The bonding orbital is at a lower energy level than the corresponding atomic orbitals, and is filled first with two electrons. In this case however there are two more electrons to be accommodated since each helium atom has *two* 1s electrons; these two must therefore go into the anti-bonding orbital. This orbital is of higher energy than the contributing atomic orbitals. The energy required to raise two electrons to the antibonding orbital is slightly greater than the drop in energy occurring when two electrons enter the bonding orbital. To form a helium molecule would therefore require an input of energy, the resulting molecule would be at a higher energy level than the separate atoms. Hence there is no tendency for helium to form diatomic molecules.

(b) sp^3 hybridisation

Methane is known to have four equal C—H bonds tetrahedrally disposed around the central carbon atom. Remembering that carbon $(1s^2 2s^2 2p^2)$ forms bonds by utilising the two 2s electrons and the two 2p electrons, the question arises as to how these bonds become identical and symmetrically disposed. It is thought that the 2s electrons become unpaired and that one of them is promoted to the vacant 2p orbital, giving a structure $1s^2 2s^1 2p_x^1 2p_y^1 2p_z^1$ (page 23). Following this the 2s and the three 2p orbitals hybridise to form four equivalent hybrid orbitals with unequal lobes tetrahedrally disposed and each occupied by one electron. Since they are formed from one s and three p electrons this is known as sp^3 *hybridisation.* Bonding with four hydrogen atoms, each contributing one s electron, now leads to the tetrahedral molecule of methane with four symmetrical molecular orbitals each occupied by two electrons, as predicted by the electron-repulsion theory (page 47).

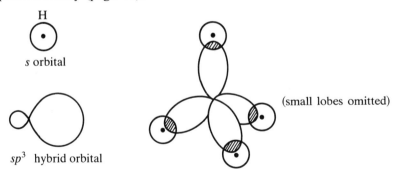

Fig. 4.10 σ bonds in methane

A similar model can be used to describe the shapes of the ammonia and water molecules.

(c) sp^2 hybridisation

In the formation of double bonds between carbon atoms, as in ethene, one of the $2s$ electrons of each carbon atom is again promoted to give the structure $1s^2 2s^1 2p_x^1 2p_y^1 2p_z^1$, but hybridisation occurs only between the $2s$ orbital and two of the $2p$ orbitals, making three identical orbitals in one plane at an angle of $120°$ to each other, with one unhybridised p orbital at right angles to these (Fig. 4.11).

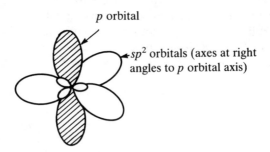

p orbital

sp^2 orbitals (axes at right angles to p orbital axis)

Fig. 4.11 sp^2 hybrid orbitals in ethene

Two of the sp^2 hybrid orbitals of each carbon atom form bonding orbitals with the s electrons of two hydrogen atoms, whilst the third hybrid orbitals form a σ-bond between the two carbon atoms; the unhybridised p orbitals overlap to form a π bond to complete the double bond. In the molecule the four hydrogen atoms all lie in the same plane with an angle of slightly less than $120°$ between the C—H bonds, while the cloud-charge resulting from the π bond is distributed above and below the plane of the molecule (Fig. 4.12). This cloud-charge is more easily polarised than that of the σ bond, and also inhibits free rotation about the axis between the carbon atoms.

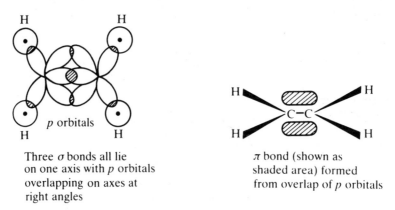

Three σ bonds all lie on one axis with p orbitals overlapping on axes at right angles

π bond (shown as shaded area) formed from overlap of p orbitals

Fig. 4.12   Square planar molecule of ethene

The concept of sp^2 hybridisation can be applied to explain the trigonal shape of such molecules as boron trifluoride (B; $1s^2 2s^2 2p^1$); the situation in this case being uncomplicated by a π bond.

The benzene molecule, like the ethene molecule, can be interpreted in terms of σ-bonds involving sp^2 and s orbitals with overlapping π bonds. The six carbons of the structure lie in a ring with bond angles of $120°$ to each other and each form one σ-bond with a hydrogen atom and two σ

bonds with adjacent carbon atoms:

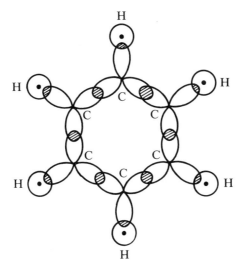

The unhybridised p orbitals lie at right angles to the three molecular orbitals of the bonds:

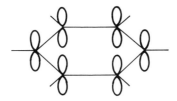

The p orbitals are close enough to overlap, resulting in π bonds forming as two electron clouds, one above the plane of the benzene ring, and one below:

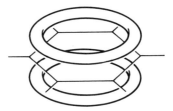

(d) *sp* hybridisation

sp hybridisation of an s orbital with a single p orbital gives rise to linear molecules such as the triple-bonded ethyne, HC≡CH. After promotion of one of the 2 s electrons, each carbon has a structure $1s^2 2s^1 2p_x^1 2p_y^1 2p_z^1$; hybridisation then gives rise to two sp orbitals and two unhybridised p orbitals all at right angles to one another. A σ bond is formed between each carbon atom and a hydrogen atom, with a further σ bond between the two carbon atoms. The electrons from the unhybridised p orbitals form two π bonds between the carbon atoms, completing the triple bond Fig. 4.13.

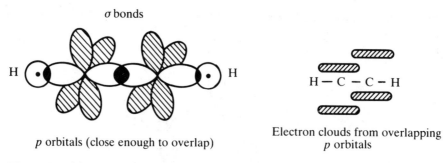

σ bonds

p orbitals (close enough to overlap)

Electron clouds from overlapping
p orbitals

Fig. 4.13 (a) Linear molecule of ethyne (b) π bonds shown as cloud-charge

The diatomic molecule of nitrogen also contains a triple bond. Each nitrogen atom has an electron configuration $1s^2 2s^2 2p^3$ and bonding takes place through the unpaired p electrons. When the nitrogen atoms approach closely the p_x orbitals overlap linearly, with the formation of a bond which draws the nuclei close together; the p_y orbitals now overlap laterally forming a π bond; the third bond, also a π bond, is formed by the lateral overlap of the p_z orbitals. (Fig. 4.14)

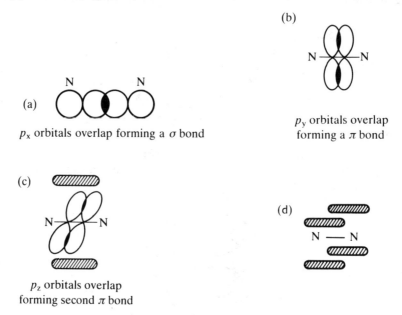

(a) p_x orbitals overlap forming a σ bond

(b) p_y orbitals overlap forming a π bond

(c) p_z orbitals overlap forming second π bond

(d) N — N

Fig. 4.14 Formation of nitrogen molecule

This leaves each nitrogen atom with an unbonded or lone pair of electrons.

4.7   Coordinate bonding and coordinate complexes

(a) The coordinate bond

A coordinate bond, or dative covalent bond, differs from an ordinary covalent bond in that the shared pair of electrons emanate from only one of the bonded atoms. The formation of a compound between ammonia and boron trifluoride provides an example. When boron combines with fluorine the boron atom does not acquire a complete noble gas structure. Boron (electronic structure 2,3 has three valency electrons, by bonding with three fluorine atoms it acquires an electronic structure 2,6 which falls short of a neon structure 2,8. This leaves the possibility of boron gaining two more electrons by further combination. Nitrogen (electronic structure 2,5) with five valency electrons acquires a neon structure by combining with three hydrogen atoms to form ammonia. Two of these electrons are unshared and can therefore be 'donated' to form a coordinate bond between the nitrogen and boron atoms:

Ammonia + Boron trifluoride $\longrightarrow$ Ammoniaborontrifluoride

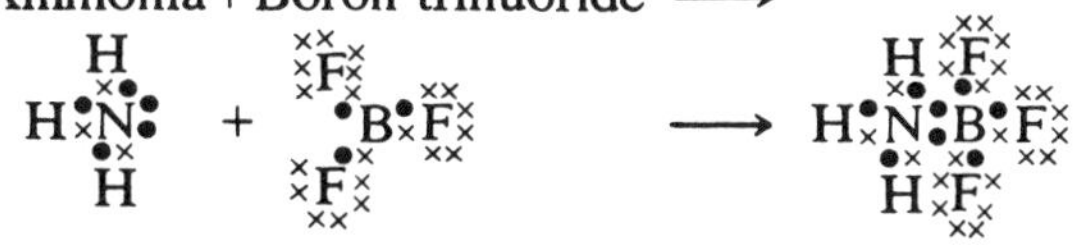

This can be represented by the notation:

$$H_3N \rightarrow BF_3$$

The coordinate bond occurs in a wide variety of compounds including complex ions which are formed in aqueous solution. Whereas the larger ions with low positive charges such as those of the alkali metals tend only to attract a loose sheath of water molecules to form hydrated ions of indeterminate composition, small ions with higher charges, such as Mg^{2+}, Al^{3+} and those of the transition metals, form hydrates of definite composition and considerable stability. The water molecules form coordinate bonds with the metal ion by the donation of a lone pair of electrons from the oxygen atom. For example, the copper(II) ion in solution forms the familiar blue hydrated ion $[Cu(H_2O)_4]^{2+*}$ in which the Cu^{2+} ion is surrounded by four water molecules each donating a pair of electrons:

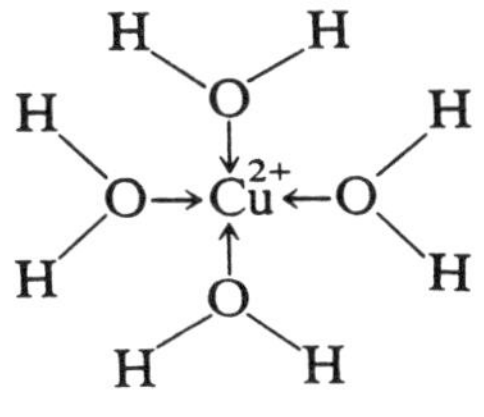

Addition of concentrated ammonia solution produces the characteristic royal blue colour; this results from the presence of $[Cu(NH_3)_4]^{2+}$ ions formed by the displacement of water molecules by ammonia molecules which can donate a lone pair of electrons from the nitrogen atoms:

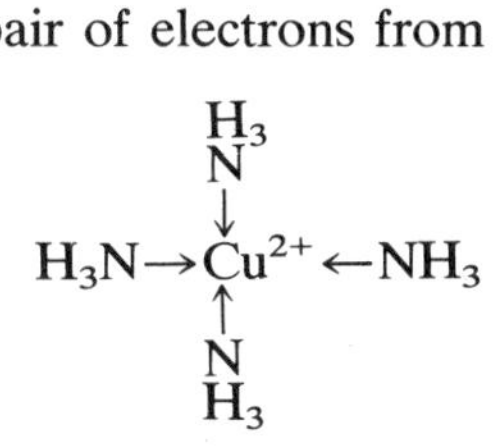

$$\underset{\underset{N}{\overset{H_3}{\uparrow}}}{H_3N \rightarrow Cu^{2+} \leftarrow NH_3}$$
$$\underset{H_3}{\overset{N}{\uparrow}}$$

* There is recent evidence that whereas $[Cu(H_2O)_4]^{2+}$ forms an ion in crystals of copper(II) compounds, the ion in solution is $[Cu(H_2O)_6]^{2+}$.

(b) Ligands

Molecules and ions which donate electrons to form coordinate bonds in this way are known as *ligands*; they include water, ammonia and carbon dioxide molecules, and chloride (Cl^-) and cyanide (CN^-) ions. The number of coordinate bonds formed by the metal ion is known as the *coordination number*, and is characteristic for the ion. The most common coordination numbers are 2, 4 and 6. Thus silver(I) ions show a coordination number of 2 in $[Ag(NH_3)_2]^+$, nickel(II) ions a coordination number of 4 in $[Ni(Cl^-)_4]^{2-}$ and iron(III) ions one of 6 in $[Fe(CN)_6]^{3-}$. The overall charge on the complex ion is given by the sum of the charges on the metal ion and those on the liganding ion.

The ligands mentioned so far each donate one pair of electrons; these are *monodentate* ligands. Other species are able to form more than one coordinate bond; for example ethane-1,2-diamine, $H_2NCH_2CH_2NH_2$, is *bidentate*, forming two liganding bonds by donating the lone pairs on each of the nitrogen atoms, as in the complex ion $[Cr(en)_3]^{3+}$ where en represents ethane-1,2-diamine:

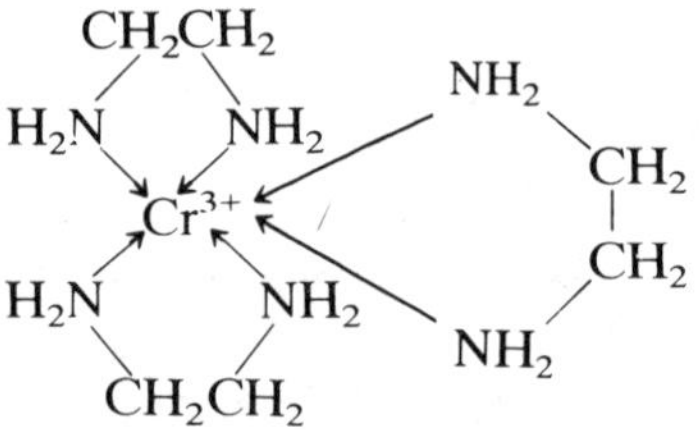

Such complexes are known as *chelates*; the name being derived from a Greek word meaning 'crab's claw'. Other bidentate groups include ethane-1,2-diol ($HO—CH_2CH_2—OH$), ethanedioate ions $\left(\begin{smallmatrix} COO \\ | \\ COO \end{smallmatrix}\right)^{2-}$ and aminoethanoic acid (H_2NCH_2COOH).

An important *polydentate* ligand is the hexadentate liganding ion derived from bis[di(carboxymethyl)amino] ethane, which is known for convenience as edta (from the traditional name ethylene diamine-tetra-acetic acid):

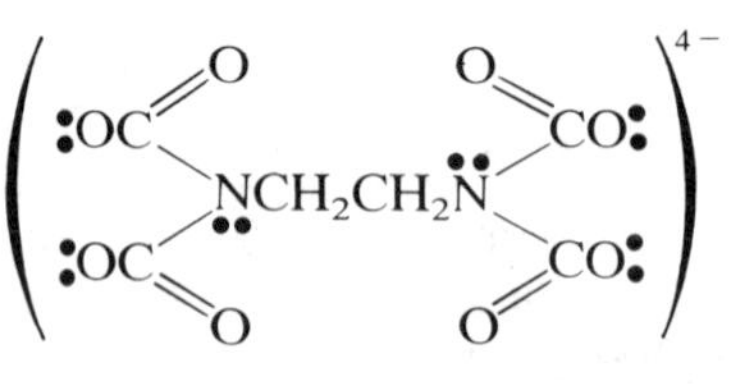

This has four negative charges and can form 6 coordinate bonds to give very stable complexes such as $[Cu(edta)]^{2-}$ or $[Ni(edta)]^{2-}$.

(c) Naming complex ions

The naming of complexes follows the systematic rules for nomenclature devised by IUPAC. The ligand is named first with a prefix such as di-, tri-, tetra-, penta- or hexa- to denote the number present. If the ligand is a neutral molecule its name is used unchanged except for water (aqua), ammonia (ammine) and carbon monoxide (carbonyl); for ligands which are negative ions the normal ending is changed to —o as in chloro, Cl^-, and cyano, CN^-. The oxidation state (page 198) of the central metal ion is denoted by using roman numerals in brackets after the ion. If the final complex is an anion the ending -ate is used for the metal ion. If the name already contains numbers, confusion is avoided by placing the ligand name in a bracket which is prefixed by bis, tris etc. as appropriate. Some examples are given in Table 4.1.

Table 4.1 Complex ions

Complex cations	Ligands
Tetraaquacopper(II) $[Cu(H_2O)_4]^{2+}$	H_2O
Hexaamminecobalt(III) $[Co(NH_3)_6]^{3+}$	NH_3
Tetraaquadichlorochromium(III) $[Cr(H_2O)_4(Cl)_2]^+$	H_2O and Cl^-
Tris(ethane-1,2-diamine) cobalt(III) $[Co(en)_3]^{3+}$	$H_2NCH_2CH_2NH_2$
Complex anions	Ligands
Hexacyanoferrate(III) $[Fe(CN)_6]^{3-}$	CN^-
Hexacyanoferrate(II) $[Fe(CN)_6]^{4-}$	CN^-
Diaquatetrachlorochromate(II) $[Cr(H_2O)_2(Cl)_4]^-$	H_2O and Cl^-

(d) Shapes of complex ions

The liganding species are arranged around the central metal ion in a *coordination sphere* of definite geometrical shape. Where there are two ligands the complex ion is linear with the ligands on opposite sides of the central metal ion. For complex ions of coordination number 4 two types of configuration are known. A large number of 4 coordinated bivalent metal ions have been found to be tetrahedral; these include tetraamminenickel(II), tetracyanocuprate(II) and tetrachlorocobaltate(II) ions.

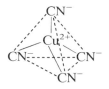

Fig. 4.15 Tetracyanocuprate(II) ion

Other 4 coordinated complexes have a square planar shape, as in the case of tetrachlorocuprate(II) ions.

Where there are six ligands an octahedral shape develops with one liganding species occupying each of the corners of an octahedron with the metal ion at the centre as, for example, in hexaamminechromium(III).

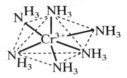

Fig. 4.16 Hexaamminechromium(III) ion

Many of these ions, especially those containing more than one liganding species, exhibit geometric and optical isomerism (page 332). Such isomers are described later in the text.

(e) The stability of complex ions

Complexes of the same central metal ion vary in stability according to the liganding species present; where more than one is available the ligand forming the more stable complex will displace the ligands of a less stable complex. Thus, while ammonia molecules displace the water molecules liganding a copper(II) ion, so edta ions will displace the ammonia molecules.

$$[Cu(H_2O)_4]^{2+} + 4NH_3 \rightleftharpoons [Cu(NH_3)_4]^{2+} + 4H_2O$$
$$[Cu(NH_3)_4]^{2+} + edta^{4-} \rightleftharpoons [Cu\ edta]^{2-} + 4NH_3$$

The displacements are not necessarily complete and proceed by a series of equilibrium reactions.

$$[Cu(H_2O)_4]^{2+} + NH_3 \rightleftharpoons [Cu(H_2O)_3NH_3]^{2+} + H_2O$$

The equilibrium constant, K, for the overall reaction is known as the *stability constant* of the complex ion. In general, the ligands of the complex ion with the higher stability constant will displace the ligands of the complex ion with the lower stability constant. This rule may be reversed where one liganding species is present in high concentration, as may be the case when water molecules are forming the ligands. Thus, if concentrated hydrochloric acid is added to a small quantity of copper(II) sulphate(VI) solution, the colour changes from blue through green to yellow as the chloride ions displace the water molecules from the $[Cu(H_2O)_4]^{2+}$ ion to form the $[Cu(Cl)_4]^{2-}$ ion which has a higher stability constant; but if this solution is then heavily diluted, the overwhelming concentration of water molecules now present results in the displacement of the chloride ion ligands and the reappearance of the blue $[Cu(H_2O)_4]^{2+}$ ions.

A more detailed study of stability constants and equilibrium constants in general will be found in Chapter 10.

A consequence of the great stability of some complex ions is the modification of the reactions of the central ion of the complex. For example, a solution of potassium hexacyanoferrate(III), $K_3Fe(CN)_6$, does not give a precipitate with sodium hydroxide solution, neither does it give a blood-red colouration with thiocyanate ions, (CNS⁻), since the metal ion is screened by the presence of the ligands. The central ion of a complex is also protected from oxidation and reduction.

(f) Methods of determining coordination number

There are several methods of determining the number of liganding groups in a complex ion. For highly coloured complexes, such as nickel(II) ions with edta, a colorimeter (page 101) can be used. Equimolecular solutions of Ni^{2+} ions and the sodium salt of edta are mixed in varying proportions; a one to one mole ratio of the ions produces the most intense colour, indicating complete reaction at this ratio.

For ligands soluble in organic liquids, a partition coefficient method (page 101) can be used. For example, measured quantities of solutions of ammonia and copper(II) ions of suitable concentrations are mixed so that the ammonia molecules are present in excess; the resulting solution is then shaken with tetrachloromethane which dissolves some of the free ammonia. The organic layer is separated and titrated with a suitable acid; application of the partition law then gives the total concentration of unliganded ammonia molecules in the aqueous and organic layers and hence the number of ammonia molecules associated with the Cu^{2+} ions can be found.

A third method uses as an indicator a liganded complex of lower stability than that of the complex ion being tested. The indicator ligand species must be a strongly coloured dye which forms a complex ion of a different colour with the metal ion in question; a small amount is added to a measured quantity of a solution of known concentration of the metal ion forming the complex. A solution of known concentration of the liganding ion under investigation is added. When sufficient has been added to react completely with the metal ions present, the indicator ligand, being of lower stability, is displaced and a colour change occurs.

(g) Carbonyls

Some metals, notably iron and nickel, form complexes with carbon monoxide. In these the metal has a oxidation state of zero, that is, it is not present as an ion but as a metal atom. Pentacarbonyliron(0) and tetracarbonylnickel(0) are both gases; the complexes are moderately stable but can be decomposed on heating.

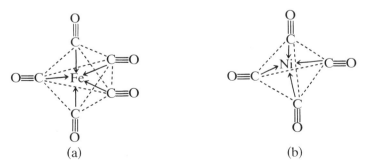

(a)   (b)

Fig. 4.17   (a) Bipyramidal molecule of pentacarbonyliron(0) (b) Tetrahedral molecule of tetracarbonylnickel(0)

(h) Coordination compounds in nature

There are many biologically important coordinate compounds, such as those enzymes which are composed of amino-acids liganded to metal ions. The haemoglobin and chlorophyll molecules are coordination compounds of iron and magnesium respectively.

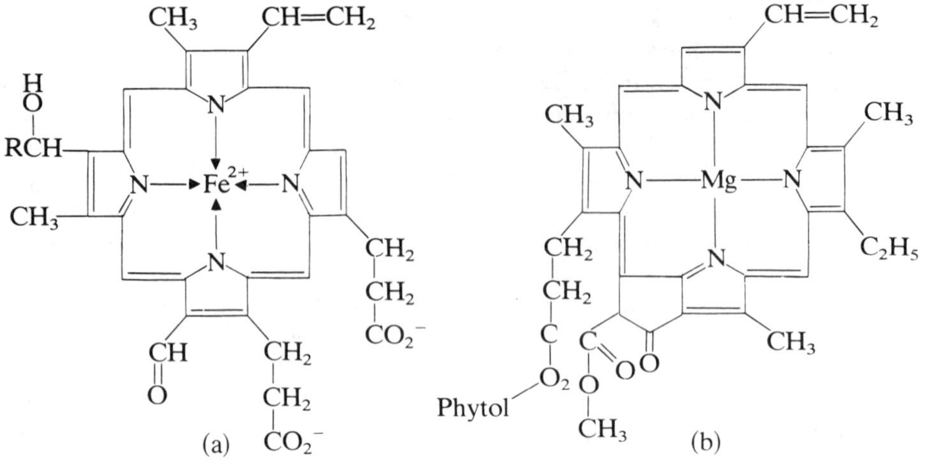

Fig. 4.18   (a) Heme unit of haemoglobin (b) Chlorophyll

4.8   Delocalisation of bonding electrons and resonance hybrids

There are some molecules and ions which cannot be accurately represented by a single, simple, structural diagram. A 'dot-and-cross' model reveals the possibility of two or more structures in which the atomic nuclei remain in the same relative position while the disposition of bonding electrons varies; such structures are known as *canonical* structures.

The actual molecule or ion exists as a blend of the canonical structures in which some of the electrons are delocalised i.e. they do not occupy bonding orbitals between any two particular atoms but spread their influence over the whole molecule or ion. Structures with delocalised electrons are known as *resonance hybrids*. They are of lower energy and therefore greater stability than any one of the canonical structures; the energy difference is known as resonance energy. For example, the nitrate ion is derived from nitric acid:

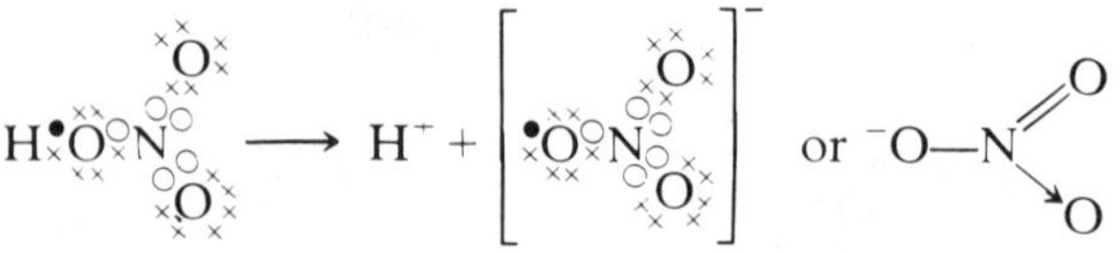

Plainly, each of the bonds could occur between the nitrogen atom and any one of the oxygen atoms, giving rise to three possible canonical structures:

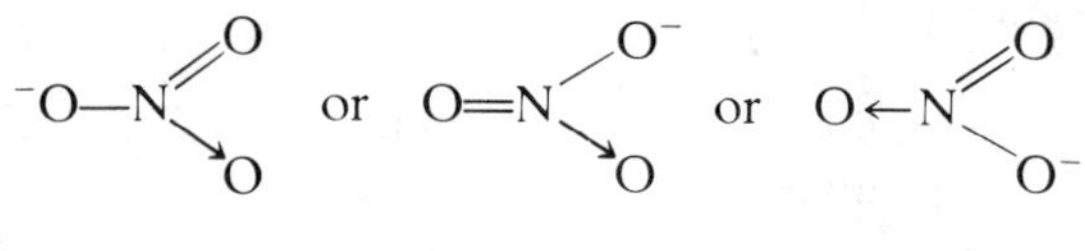

However, experimental work has shown that the nitrate ions exists in only one form, which is completely symmetrical, having three bonds of equal length at angles of 120° to one another. The explanation is that the nitrate ion is a resonance hybrid, with the electrons delocalised and shared equally. This is represented with the delocalised electrons shown as dotted lines:

$$\left[O\text{---}N \begin{array}{c} \nearrow O \\ \searrow O \end{array} \right]^{-}$$

Evidence to support the concept of delocalisation of electrons and the existence of resonance hybrids is derived from measurements of bond lengths and angles and of heats of formation of the structure. A more detailed consideration of these measurements is given on page 357 in connection with the structure of the benzene molecule.

4.9   Metallic bonding

Metals are giant crystalline substances in which the atoms are packed in various regular patterns, as described in Chapter 5. The outer electrons of the metallic atoms are released and delocalised, giving an ordered array of positive ions held together by a 'sea' of electrons, which are not firmly held by any particular atoms, but are free to form a network of mobile electrons which bind the cations together. This simple model accounts satisfactorily for most of the properties of metals, such as their good electrical conductivity, high melting point, boiling point, ductility and malleability.

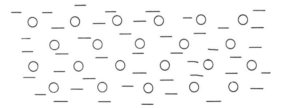

Fig. 4.19 Model for metallic bonding

4.10   Hydrogen bonding

There is another type of bonding which is intermolecular rather than interatomic. Bonds are formed between hydrogen atoms of one molecule and an electronegative atom of another molecule. These *hydrogen bonds* are electrostatic attractions, much weaker than normal bonds, having bond energies (page 131) of only 12.5–30 kJ mol^{-1}, as compared with 400 kJ mol^{-1} for an average covalent bond or 200 kJ mol^{-1} for a coordinate bond. They only occur between molecules in which the hydrogen is bonded to a strongly electronegative element such as fluorine, oxygen or nitrogen; thus they occur in liquid hydrogen fluoride, water or ammonia, but not in hydrogen bromide, hydrogen sulphide or phosphine.

In water, for example, the small, highly electronegative oxygen atom attracts the single electrons from the hydrogen atoms to such an extent as to produce a considerable dipole moment (page 47). Electrostatic attractions then develop between a hydrogen atom of one molecule and an oxygen atom of another molecule, making aggregates of molecules held together by hydrogen bonds.

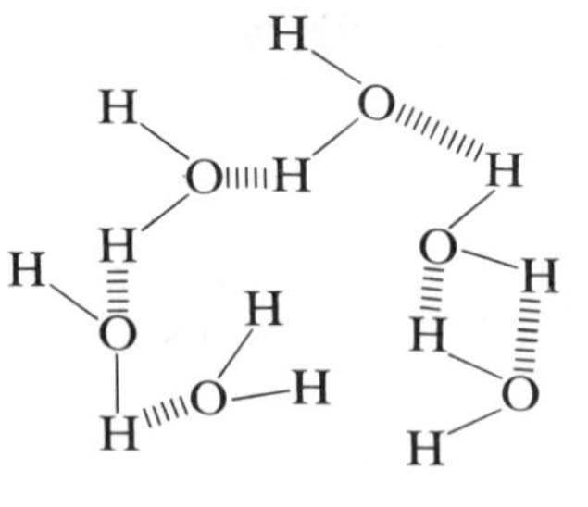

Fig. 4.20 Development of hydrogen bonds

Hydrogen bonds occur in many compounds containing —OH and —NH$_2$ groups, their presence accounts satisfactorily for the anomalously high boiling points of water, alcohols and carboxylic acids (among other effects which are dealt with separately as they arise in the text). They are also of great importance in biological systems; for example, they form the 'bridges' that maintain the spiral slope of DNA molecules.

Questions

1. Show by means of a dot-and-cross diagram the formation of the ionic compound sodium oxide.
2. Compare and contrast the properties of ionic compounds, molecular compounds and compounds of giant atomic structure. Explain the similarities or differences in terms of the bonding.
3. Sketch the shape of a silicon tetrachloride molecule, a phosphine molecule and a hydrogen sulphide molecule.
4. (*a*) Discuss the bonding in (i) calcium oxide, (ii) tetrachloromethane, (iii) ice, (iv) the molecule Al$_2$Cl$_6$.
   (*b*) What are the spatial arrangements of the atoms in (i) boron trichloride, (ii) ammonia, (iii) the gaseous compound SF$_6$?   (O)
5. What are the spatial arrangements of the atoms in the following molecular species and how may these be accounted for using the Sidgwick-Powell theory of electron pair repulsion: (i) BeCl$_2$; (ii) H$_2$O; (iii) BCl$_3$; NH$_3$? For each of these molecular species state whether or not it has a dipole moment. (Your answer should include simple diagrams wherever appropriate.)   (O)
6. The properties of substances are largely the result of the type and strength of bonding between the particles of which they are composed.
   Critically discuss this statement and in your answer refer specifically to the following substances: silica, graphite, sodium chloride, iodine and metallic copper.   (L)

7. Draw diagrams to illustrate the shape and symmetry of s and p orbitals. Write the electronic structure of (a) a carbon atom, (b) a chlorine atom, in terms of s and p electrons.

Use the electron repulsion theory to predict the shape of each of the following molecules:

   (i) phosphorus trichloride
   (ii) boron trichloride
   (iii) beryllium chloride ($BeCl_2$).   (C)

Part II
Physical Principles

5 Kinetic theory and states of matter

5.1 Kinetic theory

It is accepted that matter is made up of particles in continuous motion. The kinetic theory of matter was developed over a period of years towards the end of the nineteenth century by a number of physicists, including notably Maxwell and Boltzmann. It is of particular importance in accounting for the properties of gases, since the motion of the particles in liquids and solids is relatively restricted.

Briefly, it is considered that the molecules of a gas are in continuous rapid motion. They move in straight lines, changing direction only if they collide or hit the walls of the containing vessel. The molecules are regarded as being perfectly elastic, so that the collisions do not result in a change of the total kinetic energy of the gas. The collisions a particle undergoes may cause its individual kinetic energy to vary considerably, but the *average* kinetic energy of the molecules remains constant under constant conditions. Increase of temperature causes the molecules to move faster, increasing the average kinetic energy proportionally to the absolute temperature of the gas. The pressure of a gas arises from the collision of the molecules with the walls of the containing vessel; a given mass of gas in a container of constant volume will exert a greater pressure if the temperature is increased. A fundamental equation can be derived relating the volume V, pressure p, mass of one molecule m, number of molecules n and the root mean square velocity* of the molecules u:

$$pV = \tfrac{1}{3}mnu^2$$

From this equation all the gas laws given below can be derived, although they were originally worked out experimentally.

The distances between the molecules in a gas are large compared to their size, and so they develop little in the way of forces between them. In liquids the molecules are close enough together to exert a pronounced influence on each other; thus while the molecules of a liquid are also in rapid motion the nature of this motion is restricted by the forces between them. The motion of particles in solids is further reduced by the close proximity of neighbouring particles, and is substantially restricted to vibration.

5.2 Gases

(a) Boyle's law, Charles' law and the ideal gas equation

In 1662 Robert Boyle, experimenting with gases, formulated his law relating the pressure and volume of a fixed mass of gas. He found that, at constant

---

* The root mean square is a particular method of expressing an average value, and differs slightly from the arithmetic average value.

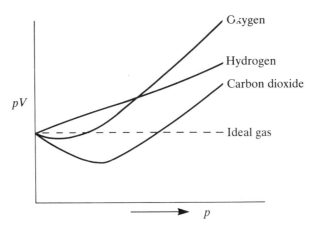

Fig. 5.1 pV plotted against p at about 320 K for some gases

temperature, the volume was inversely proportional to the pressure:

$$p \propto \frac{1}{V} \quad \text{or} \quad pV = \text{constant}$$

In fact the relationship holds only for 'ideal gases' and the deviation from it by real gases are quite large, particularly at high pressures and low temperatures (Fig. 5.1).

An equally important law was later deduced by Charles which related the volume of a fixed mass of gas to the absolute temperature. He found that the volume varied directly with the temperature at constant pressure:

$$V \propto T \quad \text{or} \quad \frac{V}{T} = \text{constant}$$

where V = volume of the fixed mass
T = absolute or Kelvin temperature

This law is subject to the same limitations as Boyle's law; nevertheless they both hold reasonably well at low pressures and raised temperatures, and provide a useful basis for quantitative experimental work with gases.

Combining the two laws leads to the equation

$$\frac{p_1 V_1}{T_1} = \frac{p_2 V_2}{T_2}$$

where, at T_1 p_1 = pressure of a fixed mass of gas, Nm^{-2}
V_1 = volume of the gas, m^3
and, at T_2 p_2 = pressure of the same mass of gas, Nm^{-2}

This equation allows volumes of gas collected in the laboratory to be compared even if they are collected at different temperatures and pressures, for the volumes can all be corrected to s.t.p: 273 K and one atmosphere pressure ($101\,325\ Nm^{-2}$). It is useful to remember in this context that the volume occupied by one mole of any gas at s.t.p. is $0.0224\ m^3$ or $22.4\ dm^3$ (22.4 litres).

It is evident from the equation that $\frac{pV}{T}$ has a constant value. This is usually expressed as $\frac{pV}{T} = k$, or as the ideal gas equation

$$pV = nRT$$

where for a fixed mass of gas containing n moles

$$p = \text{pressure } Nm^{-2}$$
$$V = \text{volume } m^3$$
$$T = \text{temperature } K$$
$$R = \text{the gas constant}$$

Note that the ideal gas equation can also be derived from the fundamental kinetic equation $pV = \frac{1}{3}nmu^2$. The kinetic energy of a molecule of mass m and velocity v is $\frac{1}{2}mv^2$, and the *average* kinetic energy of a gas molecule is $\frac{1}{2}mu^2$, where u is the root mean square velocity. The average kinetic energy is also proportional to the absolute temperature T, that is, $\frac{1}{2}mu^2$ varies as T. Then $\frac{1}{3}nmu^2$ also varies as T, or $\frac{1}{3}nmu^2 = kT$, where k is a constant or,

$$pV = kT$$
$$= RT \text{ for one mole of gas molecules.}$$

The value of R, the gas constant, can be found by substituting the values of p, V, n and T for one mole of gas at s.t.p:

$$R = \frac{pV}{nT} \qquad n = 1, p = 101\,325 \text{ Nm}^{-2}$$

$$V = 22.4 \times 10^{-3} \text{ m}^3\text{mol}^{-1} \ T = 273 \text{ K}$$

$$= \frac{101\,325 \text{ Nm}^{-2} \times 22.4 \times 10^{-3} \text{ m}^3}{273 \text{ K} \times 1 \text{ mol}}$$

$$= 8.31 \text{ Nm mol}^{-1} \text{ K}^{-1}$$

or $8.31 \text{ J mol}^{-1} \text{ K}^{-1}$ since one joule of energy results from the force of one newton moving through one metre.

(b) Dalton's law of partial pressures

When two or more gases are mixed they each contribute an independent partial pressure to the total pressure. In 1801 Dalton formulated a law, based on his experimental results, stating that when several gases occupy the same space, then each will contribute a partial pressure equal to that it would exert if it occupied the space alone. Thus for gases $A, B, C, \ldots$ occupying the same container, and not reacting with each other,

$$P_{\text{total}} = p_A + p_B + p_C \cdots$$

The partial pressure of a gas can be used as a measure of its concentration in a mixture since the concentrations of the gases are related to their partial pressures. Thus for a mixture of two gases A and B

$$\left(\frac{\text{moles of gas } A}{\text{moles of gas } A + \text{moles of gas } B}\right) = \text{mole fraction } A = \frac{p_A}{P_{\text{total}}}$$

$$\therefore \quad p_A = (\text{mole fraction } A) \times P_{\text{total}}$$

where p_A = the partial pressure of gas A.

P_{total} = the total pressure of the mixture.

The law of partial pressures is used to find the actual volume of a gas that has been collected over water. The gas is saturated with water vapour molecules so that the pressure at which it is collected (the atmospheric pressure at the time of collection) includes the water vapour pressure at the temperature of the container. This value can be found in tables of data and subtracted from the atmospheric pressure to give the pressure of the dry gas at that temperature:

$$p_{atmospheric} - p_{water\ vapour} = p_{gas}$$

The volume of the gas collected can then be correctly reduced to that at s.t.p.

(c) The vapour density of gases

The density of a gas can be expressed in $kg\ m^{-3}$ in the same way as any other substance, but a more convenient measure is the *relative density* or *vapour density* of the gas. This is defined as the mass of a given volume of gas compared with the mass of the same volume of hydrogen:

$$Vapour\ density\ of\ a\ gas\ or\ vapour = \frac{mass\ of\ any\ volume\ of\ the\ gas\ or\ vapour}{mass\ of\ the\ same\ volume\ of\ hydrogen}$$

all volumes being measured at the same temperature and pressure.

The vapour density of a gas can be related to its relative molecular mass by applying Avogadro's law, so that

$$Vapour\ density\ of\ gas\ or\ vapour = \frac{mass\ of\ n\ molecules\ of\ the\ gas}{mass\ of\ n\ molecules\ of\ hydrogen}$$

or
$$d = \frac{mass\ of\ 1\ molecule\ of\ gas\ or\ vapour}{mass\ of\ 1\ molecule\ of\ hydrogen}$$

Since hydrogen is diatomic, then

$$d = \frac{mass\ of\ 1\ molecule\ of\ gas\ or\ vapour}{2 \times mass\ of\ 1\ atom\ of\ hydrogen}$$

but,
$$\frac{mass\ of\ 1\ molecule\ of\ an\ element\ or\ compound}{mass\ of\ 1\ atom\ of\ hydrogen}$$

is the relative molecular mass

$$\therefore \quad d = \frac{relative\ molecular\ mass}{2*}$$

or
$$relative\ molecular\ mass = 2 \times vapour\ density$$

This enables the relative molecular mass of a gas or volatile liquid to be determined by measuring its vapour density. The original apparatus for

* strictly—2.016.

measuring vapour densities of gases was cumbersome, since the mass of even a large volume of gas is small. Regnault in 1845 used globes of 50 litres capacity which had to be evacuated of air and filled with the gas under test. Two such globes had to be balanced against each order to compensate for the buoyancy of the globes in the air. However, Lord Rayleigh, repeating these experiments in 1885, obtained results for the density of nitrogen so accurately that he was able to detect the presence of the previously undiscovered noble gases in atmospheric nitrogen.

The vapour densities of volatile liquids were obtained by Victor Meyer using an apparatus which measured the volume of air displaced by the rapid vaporisation of a small sample of the liquid at temperatures well above its boiling point, and by Dumas using the bulbs illustrated below.

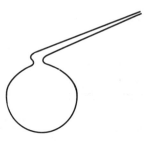

Fig. 5.2   Dumas bulb

Small Dumas bulbs for class use can conveniently be made from soft glass test tubes. The bulb is weighed to find the mass of the bulb plus the air it contains. It is then warmed slightly and the open tip placed in a beaker of the liquid whose vapour density is to be measured; as it cools a small quantity of liquid is drawn into the bulb. The bulb is then placed in a bath of liquid whose boiling point is well above that of the sample, which becomes vapourised, driving the air out of the bulb and filling it with vapour at atmospheric pressure. When all the liquid has vapourised the tip of the bulb is quickly sealed with a bunsen flame; the bulb is removed from the bath and allowed to cool before reweighing to find the mass of bulb plus vapour. To find the volume of the vapour the tip of the bulb is broken under water, which fills the vacuum left by the condensed vapour. The bulb plus water are weighed with the fragments of glass, and the mass of water found by subtracting the mass of bulb plus air. Since one gram of water has a volume of one cubic centimetre this gives the internal volume of the bulb, the volume occupied by the vapour. The mass of air in the bulb must be calculated; although it is negligible with respect to the mass of water it is comparable to the mass of the vapour.

An example serves to make the steps clear. Suppose the vapour density of trichloromethane was to be found. The following might be typical measurements:

Room temperature 16°C, atmospheric pressure 102.9 Nm^{-2}. Bath temperature 100°C (a water bath can be used since trichloromethane boils at

61°C)

$$
\begin{aligned}
&\text{Mass of bulb} + \text{air} && 10.320 \text{ g} \\
&\text{Mass of bulb} + \text{trichloromethane} && 10.428 \text{ g} \\
&\text{Mass of bulb} + \text{water} && 49.960 \text{ g} \\
&\text{Mass of water} = 49.960 - 10.320 = 39.64 \text{ g}
\end{aligned}
$$

∴ Internal volume of bulb = volume of water = 39.64 cm^3

$$
\text{Volume of air in bulb} = 39.64 \times \frac{102.9}{101.3} \times \frac{273}{289} \text{ at s.t.p.}
$$

$$
= 38.04 \text{ cm}^3
$$

Density of air at s.t.p. $= 0.0013 \text{ g cm}^{-3}$

∴ Mass of air in bulb $= 38.04 \times 0.0013 = 0.049 \text{ g}$

Mass of empty bulb $= 10.320 - 0.049 = 10.271 \text{ g}$:

∴ Mass of trichloromethane $= 10.428 - 10.271 = 0.157 \text{ g}$

$$
\text{Volume of trichloromethane at s.t.p.} = 39.64 \times \frac{102.9}{101.3} \times \frac{273}{373} = 29.47 \text{ cm}^3
$$

Density of hydrogen at s.t.p. $= 8.99 \times 10^{-5} \text{ g cm}^{-3}$

∴ vapour density of trichloromethane

$$
= \frac{\text{mass of } 29.47 \text{ cm}^3 \text{ of CHCl}_3}{\text{mass of } 29.47 \text{ cm}^3 \text{ of H}_2}
$$

$$
= \frac{0.157}{29.47 \times 8.99 \times 10^{-5}}
$$

$$
= 59.3
$$

and its relative molecular mass is 59.3×2 or 118.6 (accurately 59.3×2.016 or 119.5). Alternately, having found that 29.47 cm^3 of vapour has a mass of 0.157 g, the molar mass can be calculated from the knowledge that one mole of gas occupies $22\,400 \text{ cm}^3$. Thus

$$
\text{molar mass of trichloromethane} = \frac{0.157 \times 22\,400}{29.47}
$$

$$
= 119.3
$$

Nowadays gas syringes can be used to measure the vapour density of both gases and liquids (Fig. 5.3)

The mass of the gas can be obtained by direct weighing, or, if the gas is alkaline or acidic, by expelling the gas into suitable absorption tubes which are weighed before and after the operation. Note that if direct weighing is carried out, corrections must be made for the buoyancy in the air of the withdrawn plunger. If volatile liquids are used the syringe has to be encased in a heated jacket and the sample injected into the syringe by means of a hypodermic needle.

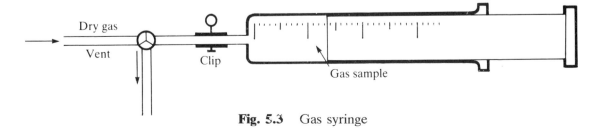

Fig. 5.3   Gas syringe

Certain compounds or elements undergo *thermal dissociation*; at raised temperatures the molecules break up into smaller units. Typical examples are phosphorus pentachloride, ammonium chloride, iodine and dinitrogen tetroxide;

$$PCl_5(g) \rightleftharpoons PCl_3(g) + Cl_2(g)$$
$$NH_4Cl(g) \rightleftharpoons NH_3(g) + HCl(g)$$
$$I_2(g) \rightleftharpoons 2I(g)$$
$$N_2O_4(g) \rightleftharpoons 2NO_2(g)$$

In these cases the measured vapour density will be lower than that expected from the molecular mass of the undissociated substance. When a weighed sample of the substance is heated the dissociation causes an increase in the number of particles present, which gives an increase in volume at constant pressure; since the mass is still the same the vapour density will be lower. The degree of dissociation varies with temperature, and if dissociation becomes complete, in any of the examples given above, then the vapour density will be halved. To calculate the degree of dissociation, the vapour density of the sample is measured and compared to the theoretical vapour density of the undissociated gas. At a given temperature, if a fraction (α) of one mole of, say, phosphorus pentachloride, is dissociated, then the amount of undissociated molecules will be $(1-\alpha)$ moles, and the quantity of each of the dissociated molecules will be α moles:

$$PCl_5(g) \rightleftharpoons PCl_3(g) + Cl_2(g)$$
$$(1-\alpha) \qquad \alpha \qquad \alpha$$

Thus the total number of moles present at this temperature will be $(1-\alpha) + \alpha + \alpha = (1+\alpha)$, and since, at constant pressure, the volume of a fixed mass of gas depends on the number of particles, then

$$\frac{\text{volume of dissociated sample}}{\text{volume of undissociated sample}} = \frac{1+\alpha}{1}$$

and since the vapour density is inversely proportional to the volume

$$\frac{\text{vapour density of dissociated sample}}{\text{vapour density of undissociated sample}} = \frac{1}{1+\alpha}$$

or in experimental terms

$$\frac{\text{measured vapour density of sample}}{\text{theoretical vapour density}} = \frac{1}{1+\alpha}$$

Some compounds undergo thermal dissociation without change in the actual number of particles present. For example, hydrogen iodide dissociates into hydrogen and iodine:

$$2HI(g) \rightleftharpoons H_2(g) + I_2(g)$$

In these cases the vapour density of the mixture remains the same, in spite of the dissociation.

In contrast, if the molecules in the vapour are associated, then the measured vapour density will be higher than that expected from the molecular formula. At certain temperatures the observed vapour density of ethanoic acid is 60, giving a relative molecular mass of 120, which corresponds to a formula $(CH_3COOH)_2$ rather than the accepted CH_3COOH, indicating that ethanoic acid molecules are associated in the vapour phase.

(d) Graham's law of diffusion

A characteristic of gases is that they will spread out, or *diffuse* to fill any volume into which they are introduced. A heavier gas will diffuse upwards to mix completely with a lighter gas, as shown when a gas jar of air is inverted over a gas jar containing bromine vapour; the upper jar quickly becomes coloured as the bromine vapour diffuses into it. Kinetically this is explained by the continuous motion of both air and bromine particles, which pass through the interface of the gases and continue to do so until the gases are completely mixed. Gases also diffuse through porous partitions, doing so at different rates. If a porous pot is connected to a manometer (Fig. 5.4) and a beaker of hydrogen is inverted over it, the mercury in the connecting arm of the manometer will be depressed as the pressure in the pot increases. The hydrogen, being the lighter molecule, diffuses into the pot more rapidly than the air diffuses out. The movement of the mercury can be reversed by replacing the beaker of hydrogen with one of carbon dioxide which is denser than air.

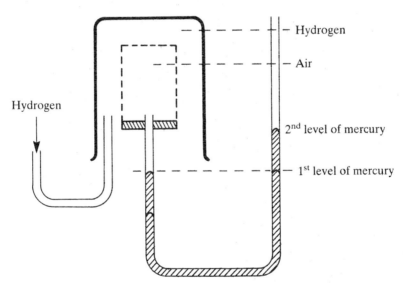

Fig. 5.4   Diffusion of hydrogen

This effect was first observed by Graham in 1846; his law states that the rate of diffusion of a gas, at constant pressure and temperature, is inversely proportional to the square root of its density. Thus for two gases A and B

$$\frac{\text{rate of diffusion of } A}{\text{rate of diffusion of } B} = \sqrt{\frac{\text{density } B}{\text{density } A}}$$

The rates are usually compared as rates of effusion, the time taken for equal volumes of gas to escape under slight pressure through a single small aperture. Since rate $\frac{1}{t}$, where $t = $ time of effusion, then

$$\frac{\text{time taken for a given volume of } A \text{ to escape}}{\text{time taken for a given volume of } B \text{ to escape}} = \frac{t_A}{t_B} = \sqrt{\frac{\text{density } A}{\text{density } B}}$$

The relative molecular mass of a gas can be determined by comparing its rate of effusion with that of a gas of known relative molecular mass. The apparatus shown in Fig. 5.5 is filled with water or some other suitable liquid in which neither gas is soluble.

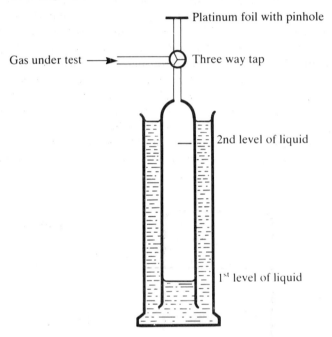

Fig. 5.5   Apparatus for measuring rate of effusion

Gas A is introduced into the central tube through the three-way tap and then allowed to escape through a pin-hole in the platinum foil covering the end of the tube. The time taken for the liquid to rise a measured distance up the tube is noted. The experiment is repeated, under the same conditions of temperature and pressure, using a gas B of known relative molecular mass. It follows from Avogadro's Law that the density of a gas is proportional to its relative molecular mass, so that:

$$\frac{t_A}{t_B} = \sqrt{\frac{\text{density gas } A}{\text{density gas } B}} = \sqrt{\frac{M_r(A)}{M_r(B)}}$$

Since $M_r(B)$, t_A and t_B are known then $M_r(A)$ can be calculated.

For many reasons, such as slight solubility of the gases in the liquid and variations of the effective pressure during the experiment, very accurate results are not obtained unless a much more sophisticated design of apparatus is used.

A major application of the laws of diffusion is the separation of the isotopes of uranium ^{253}U and ^{238}U. The uranium is converted to the gaseous hexafluoride. The resulting mixture of $^{235}UF_6$ and $^{238}UF_6$ is separated by a multistage diffusion process. The difference in density is so slight that only a partial separation is effected in one stage. The stages are arranged continuously so that the lighter fraction, richer in $^{235}UF_6$, is passed on whilst the other fraction is passed back. Several thousand stages are required to obtain a satisfactory separation.

(e) Distribution of molecular velocities

As the temperature of a gas increases the average kinetic energy, and hence the average velocity, of the molecules increases. The root mean square velocity of the molecules of a gas is proportional to the square root of the Kelvin temperature, which means that the average velocities are approximately doubled by a rise in temperature of 1000 K above room temperature. Most of the molecules have a velocity close to the average at a particular temperature, but some have velocities much higher or lower than the average. Maxwell and Boltzmann calculated the distribution of molecular velocities at a given temperature from the laws of probability, and it can be seen from the curves they constructed (Fig. 5.6) that a rise in temperature not only increases the average velocity but also the fraction having a higher velocity than the mean.

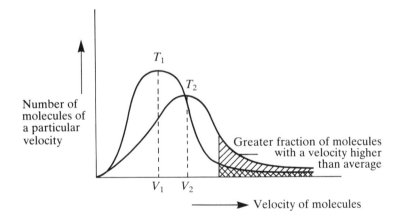

Fig. 5.6   Distribution curves for molecular velocities

v_1 = average velocity of molecules at temperature T_1
v_2 = average velocity of molecules at temperature T_2

(f) Deviations from ideal behaviour by real gases

As has been seen (page 66) Boyle's law relating pressure and volume at constant temperature applies only to a 'perfect' gas. In fact, no such gas exists, and the elementary kinetic theory is an over-simplification. To regard the molecules as points undergoing perfectly elastic collisions and having no influence on one another may be approximately true at low pressures and raised temperatures, when the size of the molecules is small compared to the space they occupy and the distance between them is relatively large.

However at increased pressure and lower temperatures the molecules become crowded together and their size becomes significant in comparison with the space they occupy, perfectly elastic collisions no longer take place and weak cohesive forces develop between them. These forces were first recognised by van der Waals, and are known as *van der Waals forces*. He worked out a corrected form of the ideal gas equation which depends on constants, different for each gas, derived from the *critical constants* of the particular gas.

(g) Critical constants

Gases can be liquefied if they are compressed and cooled below certain temperatures. For each gas there is a *critical temperature* above which the gas cannot be liquefied no matter how high the pressure applied to it. The pressure that will liquefy a gas at its critical temperature is called the *critical pressure*, and the volume occupied by 1 mole of a gas at its critical temperature and pressure is known as its *molar critical volume*.

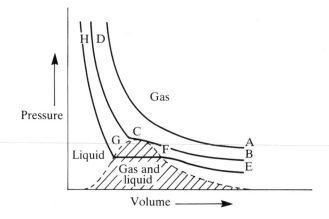

Fig. 5.7   Isothermals for a real gas

Figure 5.7 shows constant temperature curves (isothermals) relating pressure and volume for a quantity of a real gas (carbon dioxide).

The curve A shows the change of volume with pressure at a temperature above the critical temperature. The gas does not liquefy at any pressure (although it may become as dense as a liquid at very high pressures).

The curve BCD, at the critical temperature, shows a change of slope at C, the critical point. At pressures greater than that at C, the gas liquefies, and the isothermal soon runs practically parallel to the vertical axis, since liquids are little affected by pressure.

The curve EFGH shows the behaviour of the gas well below the critical temperature. The portion EF represents compression of the gas. At F liquid begins to form, and the volume reduces while the pressure remains constant until all the gas is converted to liquid at G; the portion GH again runs substantially parallel to the axis.

It should be noted that in the region of the critical point the physical properties of gas and liquid are very similar.

5.3 Liquids

(a) van der Waals forces

The weak forces of cohesion known as van der Waals forces restrict the freedom of movement of the molecules in liquids since these molecules are much closer together than those in gases. They are thought to arise from dipole-dipole attractions. Even in completely non-polar molecules the electron cloud surrounding the molecule may be momentarily displaced, producing a slight dipole moment in the molecule. This may induce an opposite dipole in an adjacent molecule so that the two are momentarily attracted. These dipole-dipole forces are thought to be transient, making and breaking continuously as the molecules move in the liquid. They are far weaker than any normal bonds and weaker even than hydrogen bonds.

The strength of the van der Waals forces increase with increasing numbers of electrons in the molecule, and vary with the shape of the molecule. For example, the boiling points of the straight chain alkanes rise steadily as the mass of the molecules increases, but that this is as much due to increasing van der Waals forces as to increasing mass can be seen by comparing the isomers of pentane, C_5H_{12} (page 345). These each have a relative molecular mass of 72, but the straight chain molecule, pentane, boils at 36°C, whereas its isomer, 2,2-dimethylpropane, which has an approximately spherical molecule, boils at only 9°C (Fig. 5.8). Evidently van der Waals forces are greater when the molecules can come into more extensive contact.

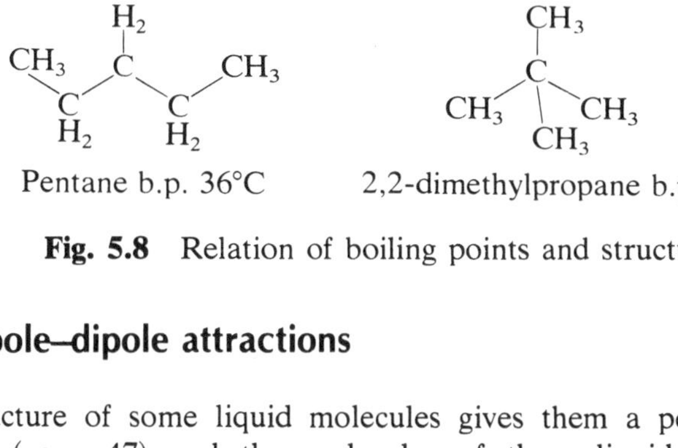

Pentane b.p. 36°C   2,2-dimethylpropane b.p. 9°C

Fig. 5.8 Relation of boiling points and structures

(b) Dipole–dipole attractions

The structure of some liquid molecules gives them a permanent dipole moment (page 47) and the molecules of these liquids have stronger cohesive forces between them than do non-polar molecules. Propanone, of

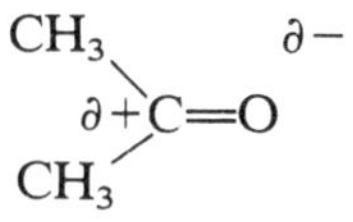

relative molecular mass 58, boils at 56°C. The presence of the dipole-dipole bonds can be detected by taking a volume of propanone and adding to it an equal volume of tetrachloromethane. A drop in temperature of a few degrees will be observed on mixing the liquids, indicating the absorption of energy. This energy must have been used to break bonds. Since propanone and tetrachloromethane do not react, and the latter is perfectly symmetrical, these bonds must be the dipole-dipole attraction between the propanone molecules.

(c) Hydrogen bonding in liquids other than water

In Chapter 4 it was seen that the unusually high boiling point of water can be explained in terms of hydrogen bonding between the molecules. Hydrogen bonds are also formed in liquids whose molecules contain suitable groupings of oxygen and hydrogen atoms. The boiling points of the alcohols are higher than might be expected from their relative molecular masses, and it is interesting to observe that if the oxygen atom of ethanol C_2H_5OH (b.p. 78°C) is replaced by a less electronegative sulphur atom to give the corresponding thiol, C_2H_5SH, the boiling point is reduced to 37°C in spite of the greater mass of the thiol molecule. Similarly if the hydrogen atom of the hydroxyl group in ethanol is replaced by an ethanoate group, thus removing the possibility of hydrogen bonding, then the ester formed has a boiling point of only 77°C in spite of the approximate doubling of the molecular mass (Fig. 5.9).

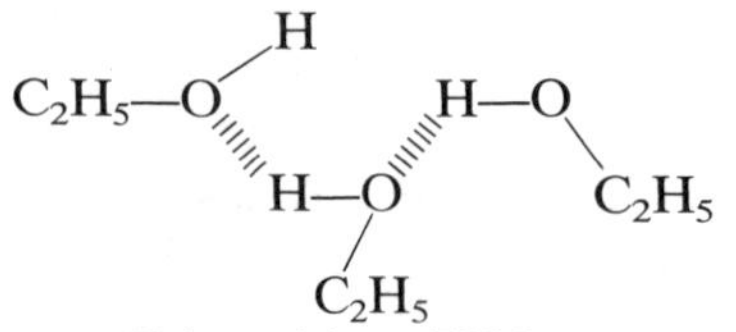

Ethanol b.p. 78°C
Aggregate formed by hydrogen bonds

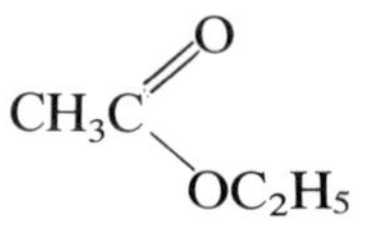

Ethyl ethanoate b.p. 77°C
No hydrogen bonding

C_2H_5SH
Ethyl mercaptan b.p. 37°C
No hydrogen bonding

Fig. 5.9   Boiling points of ethanol and derivatives

It is perhaps worth pointing out that hydrogen atoms in non-polar molecules or groups do not form hydrogen bonds. Thus while the H atom of the —OH group in ethanol will form such bonds, the H atoms in the —CH₃ group do not. Similarly, the H atoms in methane, CH_4, do not make hydrogen bonds with other molecules, but the H atom in the polar molecule trichloromethane, CCl_3H, is sufficiently activated to do so.

The dimerisation of ethanoic acid (page 101) can be explained in terms of hydrogen bonding, and there is evidence from X-ray diffraction studies to support the idea.

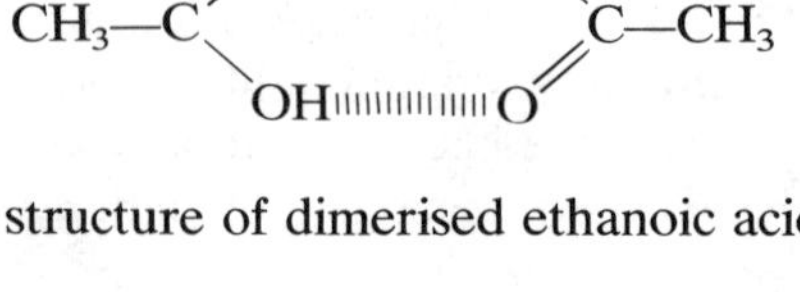

Fig. 5.10   Ring structure of dimerised ethanoic acid molecules

Measurement of the relative molecular mass of ethanoic acid dissolved in water indicates that here the molecules exist singly, not as dimers. Presumably they form hydrogen bonds preferentially with the water molecules.

The simple amines develop some hydrogen bonding, but this not as marked as in the alcohols because nitrogen atoms are less electronegative than oxygen atoms. A comparison of the boiling points of butane, butanol and aminobutane serves to illustrate this.

Butane	C_4H_{10}	b.p. 0.5°C
Butan-1-ol	C_4H_9OH	b.p. 117°C
Aminobutane	$C_4H_9NH_2$	b.p. 78°C

5.4   Solids

(a) Crystal lattices

The particles—ions, atoms or molecules—in solids have little freedom of movement, and exist in an orderly array which determines the external shape of the crystal.

The main crystal systems are shown in Fig. 5.11.

The arrangement of the particles in a regular system is called a *space lattice* and is made up of *unit cells*. A unit cell is defined as the smallest portion of a space lattice which can generate the complete space lattice by moving a distance equal to its own dimensions in various directions. The simplest unit cells are based on cubes.

(b) The stacking of atoms in metal crystals

Metals have been pictured as ions held in a sea of electrons (page 61), and since these ions are approximately spherical the structure of the metal crystal can be regarded as the stacking of spheres. Three different arrangements are possible. In *hexagonal close packing* (Fig. 5.13) each sphere is surrounded by six others in one plane, with three more above and three below, making twelve 'nearest neighbours' in all, so it is said to have a *coordination number* of 12. An equally efficient arrangement, also leading to the closest packing possible, is the *face-centred cubic close packing system*. Again the spheres have twelve 'nearest neighbours' and a coordination number of 12. A third arrangement leads to a less close-packed system in which each sphere has eight 'nearest neighbours' and a coordination number of eight; this is the *body-centred cubic system*.

No definite correlation between the type of structure and the position of the metal in the Periodic Table has yet been demonstrated. (A better understanding of the structures can be gained by making or studying models).

(c) Ionic crystals

Unlike metal crystals, ionic crystals, being composed of cations and anions, are not made up of identical units. The type of lattice structure arising

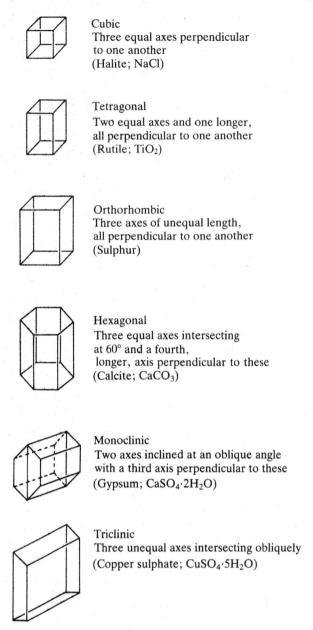

Cubic
Three equal axes perpendicular
to one another
(Halite; NaCl)

Tetragonal
Two equal axes and one longer,
all perpendicular to one another
(Rutile; TiO$_2$)

Orthorhombic
Three axes of unequal length,
all perpendicular to one another
(Sulphur)

Hexagonal
Three equal axes intersecting
at 60° and a fourth,
 longer, axis perpendicular to these
(Calcite; CaCO$_3$)

Monoclinic
Two axes inclined at an oblique angle
with a third axis perpendicular to these
(Gypsum; CaSO$_4$·2H$_2$O)

Triclinic
Three unequal axes intersecting obliquely
(Copper sulphate; CuSO$_4$·5H$_2$O)

Fig. 5.11   Basic crystal systems

Fig. 5.12   Unit cell for simple cubic system

Hexagonal close packing
Coordination number 12
(Mg Ca Zn)

Face centred
cubic close packing
Coordination number 12
(Cu Ag Au Ca Al)

Body centred cubic
Coordination number 8
(Na K Ba Fe V)

Fig. 5.13   Metal crystal systems

depends on both the ratio of cations to anions in the formula and the ratio of the radius of the cation to that of the anion ($r_c/r_a = radius\ ratio$).

In binary compounds with a cation : anion ratio of 1 : 1, the structure varies with the relative sizes of the ions since this determines the number of anions that can be packed around the cation, and vice versa. A large cation such as that of caesium (ionic radius 0.169 nm) can accommodate eight chloride ions (radius 0.181 nm) around it. Thus caesium chloride, CsCl, forms a crystal in which eight chloride ions surround each caesium ion and eight caesium ions surround each chloride ion. The cations and anions in the structure each have a coordination number of 8, the lattice being composed of interlocking simple cubes of caesium ions and chloride ions.

The sodium ion (radius 0.095 nm), being much smaller than that of caesium, can accommodate only six chloride ions around it, giving rise to a lattice structure in which six chloride ions surround each sodium ion octahedrally while six sodium ions surround each chloride ion. Thus the sodium chloride lattice, with a coordination number 6.6, has a cubic structure as shown in Fig. 5.14.

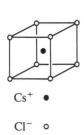
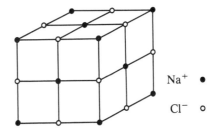

Cs⁺ •

Cl⁻ ∘

Na⁺ •

Cl⁻ ∘

Unit cell of caesium chloride

Coordination 8 . 8

Unit cell of sodium chloride

Coordination 6 . 6

Fig. 5.14   Unit cells

The radius ratio for the caesium chloride structure is $\dfrac{0.169}{0.181}$, or 0.93, and

that for sodium chloride is $\dfrac{0.095}{0.181}$, or 0.52, and it has been shown that crystals having a radius ratio of cation to anion exceeding 0.73 exhibit a caesium chloride type of structure; if the radius ratio lies between 0.73 and 0.41 the lattice is of the sodium chloride type. Below a radius ratio of 0.41 the lattice structure is tetrahedral and the compounds mainly covalent, as in zinc sulphide.

For ternary compounds such as $Ca^{2+}(F^-)_2$ the ratio of cation to anion is $2:1$. The ionic radius of Ca^{2+} is 0.99 nm and that of F^- is 0.136 nm, giving a radius ratio of 0.73, corresponding to a caesium chloride structure. In this, each calcium ion is surrounded by eight fluoride ions, but, since there are only half the number of calcium ions, each fluoride ion is surrounded by only four calcium ions, giving a crystal with coordination number 8.4. The lattice is simple cubic, but only every other cube of fluoride ions has a calcium ion at its centre (Fig. 5.15).

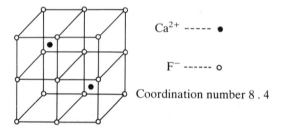

Ca^{2+} ----- ●

F^- ------ ○

Coordination number 8 . 4

Fig. 5.15   Calcium fluoride structure, coordination number 8 . 4

Some examples of compounds having the caesium chloride and sodium chloride structure are given in Table 5.1.

Table 5.1   Lattice types

Lattice type	Compounds	Radius ratio	Coordination numbers
Caesium chloride	Cs^+Br^- Cs^+I^- Rb^+Cl^- $Ca^{2+}(F^-)_2$	>0.73	8 . 8 8 . 8 8 . 8 8 . 4
Sodium chloride	Li^+F^- Na^+Br^- K^+I^-	0.73–0.41	6 . 6 6 . 6 6 . 6

(d) Covalent or atomic crystals

Atomic crystals can be regarded as macro-molecules in which the atoms are covalently bonded into a regular lattice. In diamond, the carbon atoms are each surrounded tetrahedrally by four other carbon atoms (Fig. 5.16).

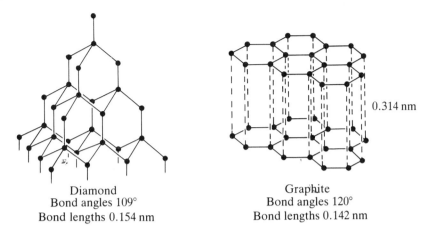

Diamond
Bond angles 109°
Bond lengths 0.154 nm

Graphite
Bond angles 120°
Bond lengths 0.142 nm

0.314 nm

Fig. 5.16 Diamond and graphite

The strength of the covalent bonds explains the hardness and lack of reactivity of diamond, while the directional nature of the bonding gives the crystalline shape. Silicon, germanium and tin have similar structures, but are much softer as the bonds are not so strong.

The structure of graphite differs from that of diamond in that each atom is bonded to only three other carbon atoms, forming a layer structure of regular hexagons. The bond length between the carbons is shortened, indicating that the fourth bonding electron from each carbon is delocalised, giving some double bond character to the structure as in the benzene molecule (page 357). The layers are held together by weak van der Waals forces. The slippery feel of graphite and its value as a lubricant can be explained by the gliding of these layers over one another.

Diamond and graphite, being different forms of the same element, are known as *allotropes*, and are said to be *polymorphic* (page 91). Many elements exhibit allotropy, for example tin has three allotropes (including one which has the same structure as diamond). The term polymorphic is also extended to compounds which may exist in two or more different crystalline forms. Calcium carbonate crystallises as calcite or as aragonite, both having the same formula, $CaCO_3$.

Compounds which form atomic crystals include silicon dioxide, SiO_2, (page 45) and the two forms of zinc sulphide. The latter are both built up

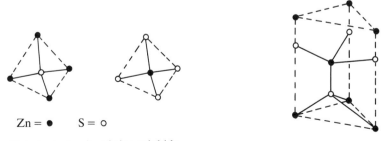

Zn = ●   S = ○

Tetrahedral units of zinc sulphide

Fig. 5.17 Structures of zinc sulphide

from tetrahedral units, with each zinc atom surrounded by four sulphur atoms and each sulphur atom surrounded by four zinc atoms. In zinc blende the resulting lattice is of the diamond type, but a slight difference in stacking the units results in another, but very similar, crystalline form for wurtzite (Fig. 5.17). The difference is best studied from models of the structures.

(e) Molecular crystals

Molecular crystals, composed of molecules held together only by weak van der Waals forces, are soft and of low melting point. For example elemental iodine contains diatomic molecules with strong covalent forces between the atoms but little attraction between the separate molecules which form a lustrous black crystalline solid, which is readily vaporised at room temperature. Other non-metal elements which form molecular crystals include phosphorus and sulphur. Solid, covalently bonded, organic substances such as naphthalene and benzoic acid also form low melting point covalent crystals.

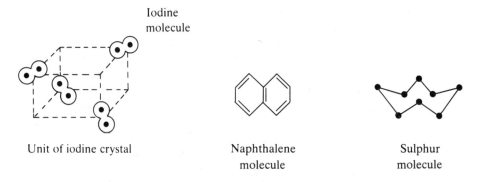

Iodine
molecule

Unit of iodine crystal

Naphthalene
molecule

Sulphur
molecule

Fig. 5.18   Covalent crystals

(f) X-ray crystallography

One of the tools used to determine the disposition of the ions, atoms or molecules in a crystal is the diffraction of X-rays. If a beam of white light falls on a grid of fine lines etched on glass, a *diffraction grating*, the path of the rays is altered according to their wavelength, giving rise to a succession of coloured spectra. If monochromatic light is used, this *diffraction* of the rays results in a series of bright lines on a dark background.

Now crystal lattices are of such dimensions that they can act as diffraction gratings for the electromagnetic radiation of short wave length known as X-rays. For example, X-rays of a wave length of the order of 5.85×10^{-2} nm will be diffracted by sodium chloride crystals, where the distance between the ions is about 2.82×10^{-1} nm, and if such a beam of X-rays is passed through a thin crystal of sodium chloride a characteristic pattern of spots can be recorded on a photographic plate. The interpretation of the pattern is complicated and involves detailed mathematical treatment.

In practice the diffraction of X-rays from the successive planes of particles in a crystal is used.

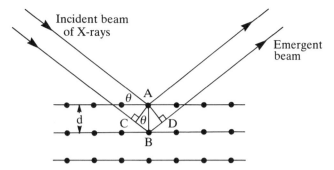

Fig. 5.19   Diffraction of X-rays by crystal lattice

The rays diffracted from successive planes obey the laws of reflection, and if the extra distance CB + BD travelled by the ray deflected from the second plane is equal to a whole number of wave lengths, then the waves will reinforce and a bright spot will appear.

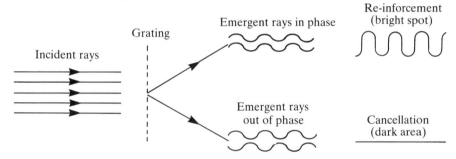

Fig. 5.20   Diffraction patterns

Geometry (Fig. 5.19) shows that $CB = BD = AB \sin \theta$, hence if $CB + BD$ is equal to a whole number of wavelengths, n, then

$$n\lambda = CB + BD$$
$$= 2AB \sin \theta$$

$AB = d$, the distance between two planes of particles

$$\therefore \quad n\lambda = 2d \sin \theta$$

This is known as the *Bragg equation*; by choosing suitable values for λ (wave length of the X-ray) and θ (the angle of incidence of the X-ray), the distance between planes of particles in a crystal can be measured.

Inter-nuclear distances can also be found, and from these the radius of the ions. Pauling drew up a table of ionic radii by fixing the radius of the oxide ion at 0.146 nm and that of the fluoride ion at 0.133 nm, thus allowing the remainder to be found by difference. For example, the internuclear distance, x, in sodium fluoride is 0.228 nm:

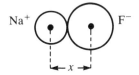

Hence the ionic radius of $Na^+ = (0.228 - 0.133) = 0.095$ nm. Note that the values found may differ slightly with the location of a particular ion.

The distance between the nuclei of atoms in a molecule (the covalent radius) and the inter-nuclear distance between atoms in neighbouring molecules (the van der Waals radius) can also be measured.

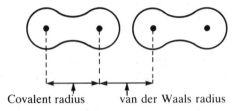

Covalent radius   van der Waals radius

Fig. 5.21.   Covalent and van der Waals radii

(g) Determination of the Avogadro constant

The formula mass of potassium chloride, i.e. the mass of a mole of K^+Cl^- ion pairs, is 74.5 g. The measured density of potassium chloride is 1.989 g cm^{-3}, so the volume occupied by a mole of K^+Cl^- ion pairs is $\dfrac{74.5}{1.989}$ cm^3. The internuclear distance between K^+ ions and Cl^- ions in potassium chloride, measured by X-ray analysis, is 0.3145 nm, or 0.3145×10^{-7} cm, so that the volume of the small cube formed in isolation by four K^+Cl^- ion pairs would be $(0.3145)^3 \times 10^{-21}$ cm^3.

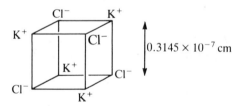

Fig. 5.22   Potassium chloride lattice

In the three-dimensional array of the potassium chloride crystal each ion forms a corner of eight adjacent cubes, giving the same number of ions as cubes, or one ion pair to two cubes. The volume of one ion pair in the structure is therefore $2 \times (0.3145)^3 \times 10^{-21}$ cm^3. By definition the number of ion pairs in one mole is equal to the Avogadro constant, L, hence

$$L \times 2(0.3145)^3 \times 10^{-21} \text{ cm}^3 = \frac{74.5}{1.989} \text{ cm}^3$$

$$L = \frac{74.5}{1.989 \times 2(0.3145)^3} \times 10^{21}$$

$$= 6.02 \times 10^{23}$$

This is one of the most accurate methods of determining the Avogadro constant.

Questions

1. State Graham's law of diffusion.

   A fluoride of phosphorus is found to diffuse in the gaseous state more slowly by a factor of 2.12 than nitrogen under the same conditions. Calculate the relative molecular mass of this fluoride, and given that its molecule contains one atom of phosphorus, write down its formula.   (O)

2. (a) Sketch a graph, with appropriately labelled axes, illustrating the distribution of molecular velocities in a gas. Describe how this distribution changes with an increase in temperature.

   (b) Describe the differences between molecular motion in gases and liquids. Show how the distribution of molecular velocities in a liquid can be used to account for the observed variation of the vapour pressure of a liquid with the temperature.   (JMB)

3. (a) Outline how you would determine the relative molecular mass of a volatile liquid. In what circumstances will this method produce abnormal results for the relative molecular mass?

   (b) A compound of phosphorus and fluorine contains 24.6 per cent by mass of phosphorus. 1.00 g of this compound has a volume of 194.5 cm^3 at a pressure of 1 atm and a temperature of 25°C.

   (i) Deduce the molecular formula for the compound.

   (ii) What is the shape of a molecule of the compound   (JMB)

4. (a) Books of data often include details of the following characteristics of the elements: (i) atomic radius, (ii) covalent radius, and (iii) van der Waals radius. Explain these terms, by use of labelled diagrams, or otherwise.

   (b) Draw diagrams to show the structures of graphite and diamond. Indicate the approximate distances between carbon atoms in these structures and comment on their relative magnitude. Explain the different electrical properties of these two forms of carbon.

   (c) Give one example of a molecular crystal and discuss the structure in terms of inter-molecular forces.   (L)

5. X-ray diffraction is one of the most powerful techniques for the elucidation of the structure of crystalline solids. Give a brief outline of the principles on which this technique is based.

   Crystalline solids may be classified, according to the nature of their structural units, as ionic molecular or atomic crystals. For each of these classes give an appropriate example, state how the particles of the substances chosen are held together in the solid state and discuss briefly the relationship between the structures and properties of these materials.   (L)

6. (a) In what ways does the structure of a gas differ from the structure of a liquid?

   (b) What is an *ideal gas*? Explain why real gases do not obey the equation $PV = nRT$

   (c) What is the difference between a gas and a vapour?

   (d) Carbon dioxide has a *critical temperature* of 31°C and a *critical pressure* of 7.3 MPa (73 atm).

   Define the terms in italics. Sketch, and fully label, graphs of pressure/volume for carbon dioxide at 15°C, 31°C and 45°C.

(*e*) Define the term *partial pressure*. 200 cm^3 of hydrogen at 100 kPa (1 atm) and 20°C, and 150 cm^3 of helium at 200 kPa (2 atm) and 20°C were mixed in a total volume of 500 cm^3. Calculate the partial pressure at 20°C of each gas in the mixture.   (AEB 1976)

6 Physical equilibrium

6.1 Introduction

There are a number of physical equilibria that are important in the study of chemistry. The relationship between solid, liquid and vapour states of single substances, the effect of cooling on the composition of solutions, and the conditions under which polymorphic forms of substances exist all need to be taken into account. The conditions necessary for successful fractional distillation are deduced from study of the vapour pressure/temperature equilibria of mixed liquids; the partitioning of a solute between two liquids is a valuable method of extracting and purifying some organic compounds.

In discussing these equilibria the term *phase* is used to denote the various parts of the system such as solid, liquid or gas, which are in contact with each other but separated by distinct boundaries. A separate phase must be homogeneous i.e. of uniform composition; it must also be physically different and physically separate from other parts of the same system. Thus a single gas in contact with a liquid is a two phase system, as is a solid in contact with a liquid, or two liquids that do not mix. However, two liquids that mix completely, or any mixture of gases, or solids that completely dissolve in a liquid, form single phase systems. A homogeneous mixture of solids such as an alloy constitutes a single phase, but a mixture of solids of variable and unspecified composition has as many phases as there are solids present; similarly a mixture of allotropes is a multiphase system.

6.2 Single component systems

(a) Phase equilibrium diagrams

The relationship between the solid, liquid and vapour phases of a single substance is illustrated by constructing a *phase equilibrium diagram*; the experimentally determined vapour pressure of the substance is plotted against the temperature of measurement, and a further series of determinations gives a curve showing the effect of applied pressure on the melting point of the solid (or freezing point of the liquid). These curves are represented pictorially in the same diagram although the pressure curves are not usually on the same scale as the range is too great. A generalised form of such a diagram is shown in Fig. 6.1. The line BO represents the conditions of temperature and pressure under which the solid substance is in equilibrium with its vapour; at any point along this line an increase in temperature results in a change of solid directly into vapour. Theoretically if OB was plotted to its extreme it would end at absolute zero (0 K); but in most cases the vapour pressure would become vanishingly small before this.

The line OA is the saturated vapour pressure/temperature curve for the liquid substance in equilibrium with its vapour; it ends abruptly at A which corresponds to the critical point, beyond which liquid and vapour are indistinguishable (page 76).

At O, which is known as the *triple point*, the solid and liquid are in

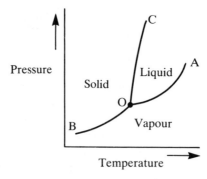

Fig. 6.1 Generalised phase equilibrium diagram

equilibrium under their own vapour pressure; this may be marginally different from the data book melting point at which solid and liquid are in equilibrium under a pressure of $101\,325\ Nm^{-2}$.

The line OC (which has no definite end) represents the effect of increasing pressure on the melting point of the solid. From the diagram it can be deduced that at any point in the area BOC the substance can only exist as a solid, in the area COA only liquid exists and in the area below AOB only vapour.

(b) Phase diagram for carbon dioxide

The triple point of carbon dioxide is at a temperature of about 216 K, and it can be seen from Fig. 6.2 that the vapour pressure of the solid is then much greater than atmospheric pressure; in fact the vapour pressure of the solid reaches one atmosphere at 195 K which is well below the melting point. For this reason solid carbon dioxide sublimes at normal pressures. This is of great advantage in the handling of solid carbon dioxide as a portable refrigerant since it disperses as vapour without the formation of any liquid. Liquid carbon dioxide does not in fact form until the pressure is about five atmospheres, and it can be seen from the slope of the line OC that the temperature of the change from solid to liquid increases slightly as the pressure is increased above five atmospheres.

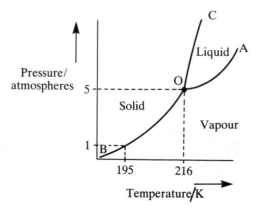

Fig. 6.2 Phase diagram for carbon dioxide

(c) Phase diagram for ice/water/water vapour

The triple point at which ice, water and water vapour are all in equilibrium occurs at a temperature 273.01 K, only marginally above the melting point of ice, and at a pressure 613 Nm^{-2}, far below atmospheric pressure.

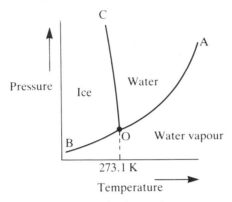

Fig. 6.3 Phase diagram for ice/water/water vapour (not to scale)

Ice, therefore, does not sublime under normal atmospheric conditions although it does have an appreciable vapour pressure. The line OC slopes towards the vertical axis since increased pressure lowers the melting point of ice; this unusual feature is consistent with the open lattice structure of ice (page 229), which is easily crushed under pressure.

(d) Allotropy or polymorphism

Elements that can exist in two or more forms are said to exhibit allotropy, the different forms being called *allotropes*. These may result from a different arrangement of atoms in the molecule, as in oxygen (O_2) and ozone (O_3), or a different arrangement of atoms in a giant molecular structure, as in the allotropes of carbon, diamond and graphite. Allotropic forms of molecular crystals such as those of sulphur and phosphorus arise from the different stacking of the unit molecules in the crystal.

Where the allotropes are of different crystalline forms they are termed polymorphs. Polymorphic substances include compounds as well as elements; examples are calcium carbonate, which exists as calcite and aragonite, and silica, which exists as quartz, tridymite and cristobalite.

Polymorphic substances are of two kinds. Enantiotropic polymorphs change form reversibly at a definite *transition temperature*; the different forms are stable within set temperature ranges and change into the other forms on passing through the transition temperature. Monotropic polymorphs have one stable form. There is no transition temperature and the unstable form(s) can exist at all temperatures but will always revert to the stable form in time. The reversion is sometimes very slow, allowing the less stable polymorph to exist for a considerable time in a *metastable* form—indeed, diamond is metastable to graphite.

The difference between enantiotropy and monotropy is illustrated by pressure/temperature diagrams for sulphur and phosphorus.

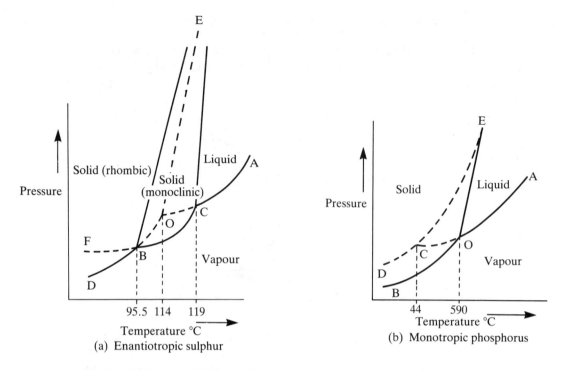

Fig. 6.4 Phase equilibrium diagrams illustrating allotropy (not to scale)

Figure 6.4(a) shows that rhombic and monoclinic sulphur are each stable over a different range of temperature; at B, which is the lowest triple point, called the transition temperature, both forms have the same vapour pressure and are in equilibrium.

If rhombic sulphur is slowly heated the vapour pressure rises along the curve DB; at B it changes to the monoclinic form, the vapour pressure-temperature equilibrium being given by the curve BC. C is the triple point at which monoclinic sulphur is in equilibrium with liquid and vapour, and occurs at approximately the normal melting point of monoclinic sulphur. CA is the vapour pressure-temperature curve for liquid sulphur in equilibrium with sulphur vapour.

If the system is cooled slowly the liquid solidifies to monoclinic sulphur which may persist unchanged in this crystalline form below the transition temperature in a metastable state; its vapour curve in this state is given by the dotted line BF. The metastable monoclinic crystals existing in the area above DBE may revert at any time to the stable rhombic form.

If, on the other hand, rhombic sulphur is heated rapidly it will reach its melting point (114°C) without passing through the allotropic monoclinic state. In this case the vapour pressure of the metastable rhombic sulphur follows the curve BO to the triple point O at which rhombic sulphur is in equilibrium with liquid sulphur and sulphur vapour; the vapour pressure curve after this point is OCA. The slopes of the lines OE and CE indicate that increasing pressure raises the melting points of both rhombic and monoclinic sulphur, and that of the line BE shows that the transition temperature is also raised by increased pressure.

In the case of red and white phosphorus the allotropy is monotropic. There is no definite transition temperature and a direct change of form can only occur from the metastable to the stable state.

The vapour pressure-temperature curve of red phosphorus is shown in Fig. 6.4(b) by the line BO. When the temperature of red phosphorus is raised slowly the vapour pressure increases steadily until the triple point O is reached; here red phosphorus, liquid phosphorus and phosphorus vapour are in equilibrium at the melting point of red phosphorus (590°C). The equilibrium curve continues along the line OA which is the vapour pressure-temperature curve for liquid phosphorus. Now if the liquid is cooled very rapidly the vapour pressure curve follows the line AOC and white phosphorus is formed; C is a second triple point at which white phosphorus, liquid phosphorus and vapour are in equilibrium at the melting point of white phosphorus (44°C).

The vapour pressure curve DC for white phosphorus lies above that for red phosphorus, as is always the case with the unstable form of a monotropic element. Thus red and white phosphorus can both exist in the area BOE, but the white will always revert eventually to the red.

6.3 Eutectics

When a pure liquid is cooled it normally starts to solidify at a fixed characteristic temperature; the temperature remains constant until all the liquid has changed to solid and then continues to fall at a rate depending on the external cooling.

The presence of a solute, either solid or liquid, lowers the freezing point and renders it less definite. For example, if a dilute solution of sodium chloride is cooled, crystals of pure ice separate when the temperature reaches 273 K. Further cooling results in more ice being formed whilst the solution becomes more and more concentrated.

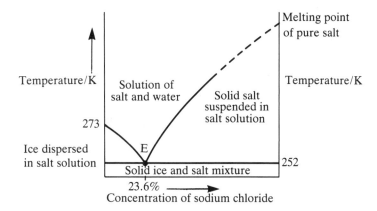

Fig. 6.5 Temperature/composition diagram for salt and water solution

When the concentration of salt in the solution reaches 23.6% the mixture solidifies without change of concentration at a constant temperature of 252 K. This point is called the *eutectic point*, and the 23.6% salt solution is known as the *eutectic mixture*. Similarly, a concentrated salt solution deposits pure salt on cooling until the concentration of the solution falls to 23.6%, when the eutectic mixture solidifies at constant temperature.

Although eutectic mixtures are of fixed composition, they do not, except by chance, correspond to definite chemical compounds, nor, where water is the solvent, to definite hydrates of the salts.

Eutectic curves are important in metallurgical studies; liquid mixtures of molten metals such as tin and lead change composition on cooling as one or other of the metals solidifies alone until a solid of fixed composition is deposited at the eutectic point. For tin and lead the composition of the eutectic mixture is 38% lead and 62% tin. The freezing point of tin is 505 K, that of lead is 600 K and that of the eutectic mixture is 456 K. From these data a freezing point diagram can be constructed, choosing axes of suitable scale, as in Fig. 6.6.

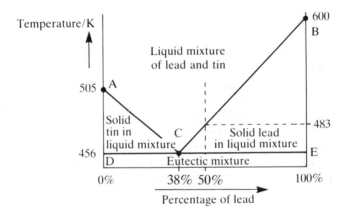

Fig. 6.6 Temperature/composition diagram for tin and lead*

The freezing points of tin and lead are marked on parallel vertical axes (points A and B). A line is then drawn at the eutectic temperature (DE) and a point C marks the composition of the eutectic mixture. Joining AC and BC completes the diagram. Now if a liquid mixture of 50% tin and 50% lead is cooled, it can be seen that pure lead will separate as the temperature reaches 483 K; deposition of lead will continue until the concentration of lead falls to 38%. Since this is the eutectic mixture solidification will continue without change of composition at a constant temperature of 456 K until the whole mass is solidified. These changes can also be represented on a cooling curve (Fig. 6.7).

* Lines derived from practical data would, in fact, be slightly curved.

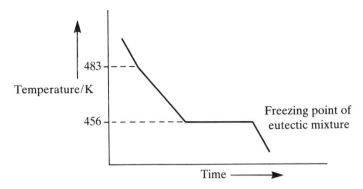

Fig. 6.7   Cooling curve for a mixture of 50% tin and 50% lead

6.4   Vapour pressure of miscible liquids

(a) Ideal mixtures and Raoult's law

In a mixture of perfectly compatible liquids the intermolecular forces between the components are virtually identical with the forces between the molecules of the individual components. Thus the behaviour of each component is unaffected by the presence of the others.

The vapour pressures of such ideal solutions are governed by Raoult's law, which he first expressed in 1886. He found experimentally that the partial vapour pressure of a component in a solution is equal to the vapour pressure of the pure component multiplied by the mole fraction of it present in the solution, all measurements being taken at the same temperature. Thus for a two-component system of compatible liquids such as hexane and heptane:

Partial pressure of hexane, $p_{(hexane)}$

$$= \frac{\text{moles hexane}}{\text{moles (hexane + heptane)}} \times \text{v.p. hexane}$$

and partial presure of heptane, $p_{(heptane)}$

$$= \frac{\text{moles heptane}}{\text{moles (hexane + heptane)}} \times \text{v.p. heptane}$$

and further, from Dalton's law:

$$P_{(total)} = p_{(hexane)} + p_{(heptane)}$$

A vapour pressure-composition diagram for an ideal mixture at a constant temperature is shown in Fig. 6.8.

Vapour pressure-composition curves are important in the consideration of the conditions necessary for separating liquids by distillation. For this purpose they are generally transposed into boiling point-composition (constant pressure) diagrams as in Fig. 6.9, the lines sloping the opposite way to those of vapour pressure diagrams since a high vapour pressure corresponds to a low boiling point. For various boiling points the composition of the liquid and of the vapour in equilibrium with the liquid are determined, giving two curves as in Fig. 6.9. Since for mixtures suitable for separation by

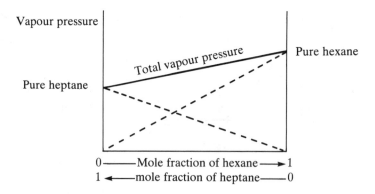

Fig. 6.8   Vapour pressure-composition diagram

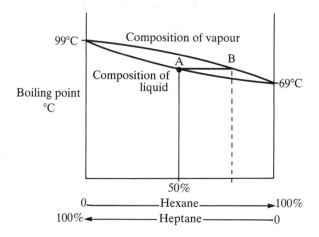

Fig. 6.9   Boiling point/composition diagram

distillation the vapour at any intermediate temperature is richer in the lighter component than the liquid, the vapour composition curve lies to the right of the liquid composition curve.

Suppose a mixture of 50% hexane and 50% heptane is distilled. The boiling point of the mixture is given by the point A in the diagram; at that temperature the composition of the vapour is given by B, and so any distillate collected will be richer in the lower boiling point component. If this distillate is redistilled further enrichment of the lighter component can be effected, and repeated separate redistillation of both distillates and residues would eventually give rise to substantially pure samples of the two liquids.

(b) Fractional distillation

In practice the lengthy and inefficient process of successive redistillations suggested above is avoided in the laboratory by fitting the distilling flask with a *fractionating column*, a long tube filled with glass beads, short lengths of glass tube or other devices giving a large surface area on which liquid and vapour can intermingle.

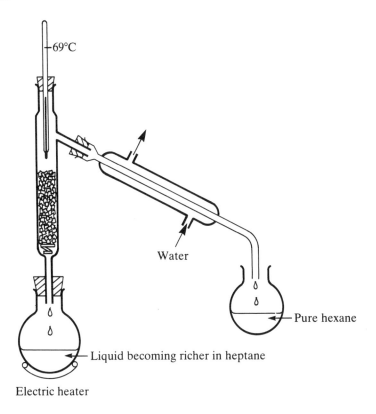

Fig. 6.10 Laboratory fractionating column

As the flask is heated the mixture vaporises, the vapours rising up the column. A liquid richer in the higher boiling component condenses on the packing and trickles back into the flask. Ascending vapours richer in the lower boiling point component pass through the descending liquid so that liquid and vapour are in momentary equilibrium at a number of points in the column; each equilibrium stage results in further separation of the two components, so that eventually only the vapour of the lighter liquid reaches the top of the column. The heating is controlled so that the top of the column is at the boiling point of the lighter liquid, which distils over while the heavier liquid continually returns to the distilling flask. The efficiency of a fractionating column is improved by arranging for controlled cooling at the top of the column and insulating the body to minimise heat loss, thus ensuring an even flow of condensate down the packing.

Fractional distillation is extensively employed in the petroleum industry to separate the large number of similar liquid hydrocarbons which make up crude oil. A continuous process, rather than the batch process described above, is used, and the fractionating column may be many feet in diameter and a hundred feet high, sometimes arranged so that condensate can be drawn off at intermediate points in the column, giving 'fractions' of different boiling ranges (page 479).

Fractional distillation is also used to separate oxygen and nitrogen in air liquefaction (page 475).

(c) Non-ideal solutions and azeotropes

Many liquids which are completely miscible form mixtures which deviate from ideal behaviour; these are liquids which develop intermolecular forces which are different from those between the molecules of the individual liquids. For example, when trichloromethane is mixed with ethyl ethanoate a slight rise of temperature is observed, indicating that bonds have been formed. These are hydrogen bonds between the molecules of the two liquids. These intermolecular attractions reduce the tendency of the

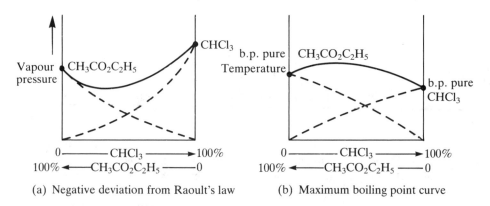

molecules to escape the liquid mixture, resulting in a lower vapour pressure and consequently a maximum boiling point for the liquid higher than that of either component.

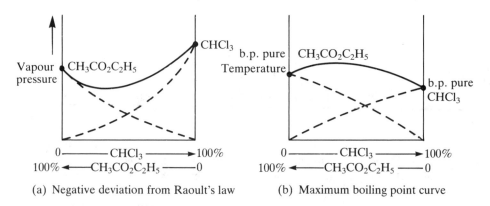

(a) Negative deviation from Raoult's law   (b) Maximum boiling point curve

Fig. 6.11 Vapour pressure-composition and boiling point-composition diagrams for trichloromethane and ethyl ethanoate mixtures

The lowering of the vapour pressure is referred to as a negative deviation from Raoult's law; other solutions giving rise to maximum boiling point mixtures are nitric acid and water, hydrochloric acid and water, and propane and trichloromethane. Positive deviations from Raoult's law are seen in liquid mixtures where the new intermolecular forces are less strong than those existing between the molecules of the individual liquids. For example, when cyclohexane is added to propanone a slight drop of temperature is observed; this results from the breaking of some of the dipole-dipole attractions that occur in propanone (page 77) by the interference of the cyclohexane. The resulting mixture therefore exhibits a vapour pressure slightly higher than might be expected from the vapour pressures of the individual liquids, and in turn the boiling point curve reaches a minimum point. Other binary mixtures giving a minimum boiling point solution are ethanol and water, and propanol and water.

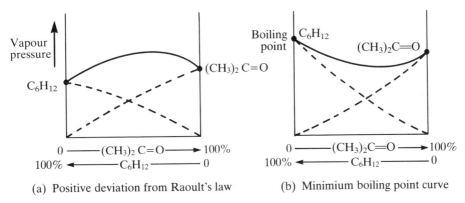

(a) Positive deviation from Raoult's law   (b) Minimum boiling point curve

Fig. 6.12 Vapour pressure-composition and boiling point-composition diagrams for propanone and cyclohexane mixtures

Mixtures which show substantial deviations from Raoult's law can never be completely separated by distillation, since at the maximum or minimum boiling point a mixture of constant composition distils. For example, if dilute nitric acid is fractionally distilled, the distillate will be mainly water until the concentration of nitric acid reaches 68%. At this point a constant boiling mixture is formed; the liquid and vapour both have the same composition and so all the remaining solution distils at constant temperature (121°C) without change of concentration (Fig. 6.13(a)).

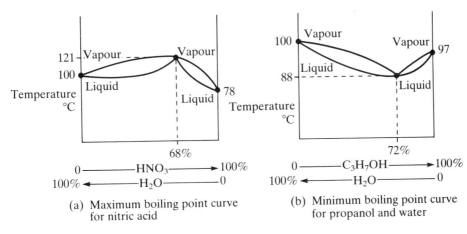

(a) Maximum boiling point curve   (b) Minimum boiling point curve
    for nitric acid   for propanol and water

Fig. 6.13 Maximum and minimum boiling point curves

An example of a minimum boiling point constant boiling mixture is shown in Fig. 6.13(b), which illustrates the formation of a constant boiling mixture containing 72% of propanol in water. Pure propanol can only be obtained from aqueous solutions containing more than 72% of propanol at the start; it is left as a residue on distilling off the constant boiling mixture.

Constant boiling point mixtures are known as *azeotropes*; those of the mineral acids are useful in preparing standard solutions, for the azeotropic mixture needs only to be diluted to the required concentration.

(d) Steam distillation

The vapour pressure over two immiscible liquids is normally that of the upper layer only. However, if the mixture is agitated so that some of each liquid is at the liquid/vapour boundary then the vapour pressure will be the sum of the vapour pressures that each liquid exerts individually at that temperature.

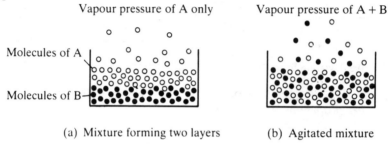

(a) Mixture forming two layers   (b) Agitated mixture

Fig. 6.14 Vapour pressure of immiscible liquids

Provided that there is enough of each liquid in the agitated mixture to give rise to its saturated vapour pressure then the total vapour pressure is independent of the amounts of each liquid present. The result is that an agitated mixture of two immiscible liquids will boil at a temperature lower than that of either of the two constituents.

Organic compounds which are spoiled by heating to their distillation temperature can be purified and separated from non-volatile products by *steam-distillation*. For example, phenylamine, b.p. 184°C, prepared from nitrobenzene, is placed in a flask with water; the mixture is heated and at the same time agitated by passing steam through it, and a mixture of phenylamine and water distils into the receiver and separates into two layers. Phenylamine is in fact slightly soluble in water, and to counter this salt is added to the aqueous layer of the distillate before separation.

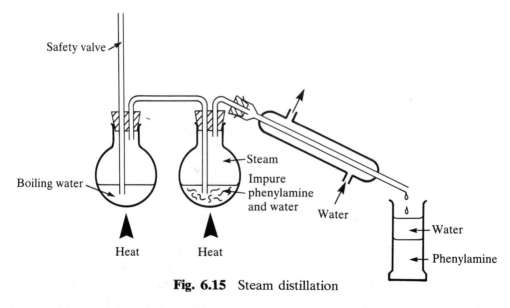

Fig. 6.15 Steam distillation

In steam distillation the mass of the required product in the distillate can be found from the expression

$$\frac{\text{mass of product}}{\text{mass of water}} = \frac{\text{(relative molecular mass} \times \text{vapour pressure) of product}}{\text{(relative molecular mass} \times \text{vapour pressure) of water}}$$

where the vapour pressures are measured at the common boiling point.

6.5   Partition law

(a) Partition coefficients

When a small quantity of iodine is shaken with a mixture of water and benzene, some iodine will dissolve in both liquids; if the mixture is allowed to separate into two layers and left until equilibrium conditions are established the iodine concentration in each layer can be determined by titration with sodium thiosulphate (page 301). Repeated determinations using different masses of iodine, water and benzene show that at a fixed temperature the ratio of the concentration of iodine in the two layers is constant:

$$\frac{\text{concentration of iodine in water}}{\text{concentration of iodine in benzene}} = \text{a constant}$$

The constant is called the *partition coefficient* or *distribution constant* for the solute between two particular solvents at a given temperature. The *partition* or *distribution law* states that at a fixed temperature a solute distributes itself between two immiscible solvents in such a way that the ratio of the concentration of solute in each layer is constant.

This law, like so many others, is modified if the solute is associated or dissociated in one or other of the liquids. If the solute forms associated molecules in one of the layers, as, for example, ethanoic acid dimerises in benzene (page 78), then the expression becomes

$$\frac{\text{concentration of unassociated solute in liquid } A}{\sqrt{\text{concentration of associated solute in liquid } B}} = \text{a constant}$$

or, for ethanoic acid partitioned between benzene and water

$$\frac{\text{concentration of ethanoic acid in water}}{\sqrt{\text{concentration of ethanoic acid in benzene}}} = \text{a constant}$$

Experimental results can be used to determine whether or not a compound is associated in a particular solvent. For example, analysis of the concentration of a substance A shaken with water and tetrachloromethane and allowed to come to equilibrium gave the following results:

Concentration of A in CCl$_4$ C_T mol l^{-1}	Concentration of A in water C_W mol l^{-1}	$\dfrac{C_T}{C_W}$	$\dfrac{\sqrt{C_T}}{C_W}$
0.106	0.051	2.0	6.38
0.500	0.110	4.5	6.42
2.600	0.250	10.4	6.45
8.500	0.450	18.8	6.48

It can be seen that a constant value for the expression is obtained if the square root of the concentration of A in tetrachloromethane is taken, hence A must form dimers in tetrachloromethane.

(b) Solvent extraction

In the preparation of organic compounds the product is frequently obtained dissolved in water together with by-products such as inorganic salts, so that to recover the product by evaporation of the water is not feasible. These substances are extracted by shaking the aqueous mixture with an organic solvent which is immiscible with water; the organic layer can be separated from the aqueous layer in a separating funnel and the product recovered by evaporating the solvent. Ethoxyethane or some other low boiling point liquid is usually chosen.

The solvent extraction is made more efficient by using the solvent in two or three small batches rather then all at one time. That more of the product is recovered can be seen from the following example:

Suppose 1.0 gram of a compound X dissolved in 100 cm^3 of water is to be extracted with 100 cm^3 of ethoxyethane. Given that the distribution constant of X between ethoxyethane and water is 5, then after shaking the aqueous solution with 100 cm^3 of ethoxyethane:

$$\frac{\text{concentration of } X \text{ in ethoxyethane}}{\text{concentration of } X \text{ in water}} = \frac{5}{1}$$

and if W_1 g of X dissolve in the ethoxyethane this becomes

$$\frac{W_1/100}{(1-W_1)/100} = \frac{5}{1} \quad \text{or} \quad \frac{W_1}{(1-W_1)} = 5 \quad \text{and} \quad W_1 = \frac{5}{6} \quad \text{or} \quad 0.83 \text{ g}$$

Now suppose the ethoxyethane were used in two batches of 50 cm^3 each and that the first extraction gives a concentration of W_2 g in 50 cm^3. Then

$$\frac{W_2/50}{(1-W_2)/100} = \frac{5}{1}; \quad \frac{2W_2}{(1-W_2)} = 5; \quad W_2 = \frac{5}{7} \quad \text{or} \quad 0.71 \text{ g}$$

There is left in the water 0.29 g; if the aqueous layer is shaken with the second 50 cm^3 of ethoxyethane to extract W_3 g:

$$\frac{W_3/50}{(0.29-W_3)/100} = \frac{5}{1}; \quad \frac{2W_3}{(0.29-W_3)} = 5; \quad W_3 = \frac{1.45}{7} \quad \text{or} \quad 0.21 \text{ g}$$

The total extracted by using two batches of ethoxyethane is therefore (0.71+0.21) or 0.92 g, as against 0.83 g using the same amount in a single batch.

6.6 Chromatography

(a) General

Chromatography has become a major method of separating closely related substances. The name, literally 'colour writing', was used to describe the results of early experiments all concerned with the separation of coloured substances; nowadays the name has been extended to describe similar methods applied to the separation of colourless substances, metal ions and gases.

(b) Paper partition chromatography

This method depends on the distribution of the substances to be separated between paper and a moving solvent. In fact it is the water molecules strongly adsorbed on the fibres of pure cellulose paper which act as the *stationary or standing phase* while the solvent is the *moving phase*. For example, to analyse the colours used in food colouring extracts, a spot of each colour is placed at a marked spot on a strip of chromatographic paper. Control spots of recognised permitted colours may be added as references. The colours are absorbed by the water molecules adhering to the cellulose fibres; when they are touch dry the paper is made into a cylinder which is stood in a solvent in a gas jar closed with a lid. The solvent rises up the paper by capillary action, carrying the various colours to heights which depend on the partition coefficient of each between the adsorbed water and the solvent.

The solvents may include ethanol, propanone, propanol, butanol, ethanoic acid, ammonia solution or special mixtures of these and are carefully chosen to give a good separation of the components. The components of the mixture can be identified by observing which of the separated colours rises to the same height as the corresponding pure colour. More accurately, the *retention factor* or R_f *value* of each separate colour can be measured for a specific solvent at a set temperature. This is the ratio of the distance moved by the spot to the distance moved by the solvent front in a given time.

Paper chromatography is used to separate amino-acids and lower molecular mass carbohydrates, which are all colourless. To identify the positions reached by the various components of the mixture the chromatogram is dried and then developed by the use of a suitable reagent which combines with the acids or the sugars to give coloured spots.

The method described above uses an ascending liquid, but the same results are obtained by suspending the paper in a trough so that the solvent travels downwards—*descending chromatography*. Mixtures of amino-acids

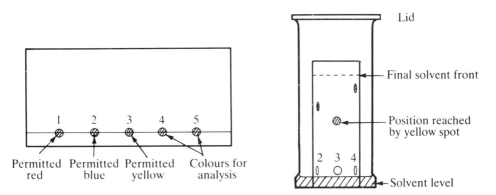

Fig. 6.16   Paper chromatography

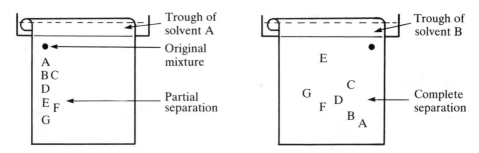

Fig. 6.17 Separation of a mixture of amino acids by two-way chromatography

may be so complicated that *two-way paper chromatography* is needed to separate them. A spot of the dissolved mixture is placed at one corner of the chromatographic paper, which is placed in a solvent which at least partially separates the components. The paper is dried and turned through 90° and placed in a second solvent to complete the separation. Considerable skill is needed in the interpretation of these chromatograms.

(c) Thin layer chromatography

Effective separations can be made by using thin layer chromatography. Here, the standing phase is made by forming a very thin layer of adsorptive material such as kieselguhr, silica gel, alumina or even 'Polycell' on glass slides; the layer is dried thoroughly and used in much the same way as chromatographic paper. It is advantageous in that very small quantities can be used and a very rapid separation obtained.

(d) Elution chromatography

Here, as in thin layer chromatography, the components of a mixture are separated by selective adsorption of the compounds on a standing phase made of materials such as alumina, silica gel, chalk or kielselguhr. The finely divided material is packed into a long column fitted with a tap (a burette can be used). The mixture is dissolved in a suitable solvent, the *eluting liquid*. A small quantity of the solution is introduced into the tube, followed by pure solvent which elutes, or washes, the components down the column. Those which are least strongly adsorbed by the material in the column travel most quickly, and can be collected separately as they pass out at the bottom of the tube, and recovered by evaporating the solvent. This method was used by a Russian botanist, Tswett, who was one of the earliest workers in this field; he resolved chlorophyll into its constituents by using an alcoholic extract and a tube packed with finely ground chalk.

(e) Gas chromatography

This is a development of the elution method described above; it is used to analyse mixtures of gases or low boiling point liquids that are very difficult to distinguish by other means.

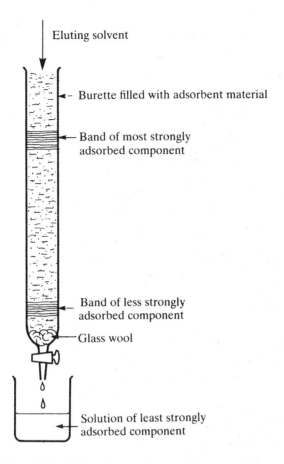

Fig. 6.18 Elution or column chromatography

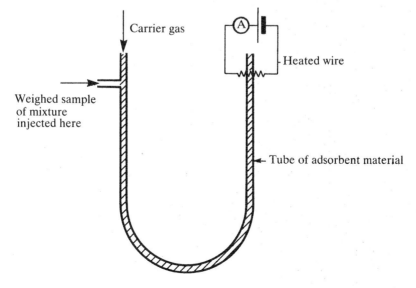

Fig. 6.19 Apparatus for gas chromatography

The standing phase is composed of adsorbent material such as charcoal or those already mentioned enclosed in a long tube. These materials are sometimes coated with a thin layer of adsorbed oil, such as dibutyl phthalate or silicone oils. A small quantity of the gaseous mixture is injected into the column at a convenient point and is adsorbed by the material in the column. An inert gas such as nitrogen is then passed through the tube. As the *carrier gas* passes over the adsorbed components of the mixture it displaces them and sweeps them further along the tube. Obviously the rate of displacement will depend on how strongly the components of the mixture are adsorbed and therefore separated. Various methods are employed to detect the emergence of each separate component; these include the use of a heated wire carrying a current placed across the top of the column. The resistance of the wire is altered by the passage over it of a gas other than the carrier gas. The change of current is observed on an ammeter in the circuit. The apparatus is calibrated before use by passing through it pure samples of the gases expected in the mixture.

Questions

1. Show *on one diagram*:
(i) the variation of the vapour pressure of ice and water with temperature;
(ii) the effect of pressure on the melting point of ice.

Mark on your diagram the triple point and the critical point and briefly explain their significance.   (O)

2. Construct from the following data the freezing-point diagram for mixtures of zinc and cadmium:

zinc, m.p. 419°C; cadmium, m.p. 321°C;
eutectic temperature 263°C:
composition of eutectic mixture, 83% cadmium by mass.

Explain what happens when a melt containing equal masses of zinc and cadmium is cooled, and sketch a cooling curve for the process.

What is the physical nature of a mixture represented by a point on the freezing-point diagram with coordinates T, 350°C; 20% cadmium by mass?   (C)

3. Discuss the origin of the various possible types of curve arising when the vapour pressure of a mixture of two miscible volatile solvents is plotted against the composition expressed in mole fraction. You should refer to the ideal case and also to two non-ideal cases.

For each case mentioned above, draw also the boiling point/composition curve. In the ideal case, show how the curve may be used to explain the process of fractional distillation; and for one non-ideal case, show how the curve may be used to explain the existence of an azeotropic mixture.   (JMB)

4. (*a*) At atmospheric pressure tetrachloromethane boils at 77°C, and tin(IV) chloride at 114°C. A liquid mixture of these two compounds can be separated by fractional distillation.

(i) Draw rough graphs to show (1) how you would expect the total vapour pressure of such a mixture to depend on its composition at a given temperature, (2) the relation between the boiling point at a fixed pressure and the composition of the liquid and vapour phases.

(ii) On the basis of the diagram in your answer to (i) (2), explain the process of fractional distillation.

(b) Sulphur can exist as rhombic sulphur, monoclinic sulphur, liquid sulphur and sulphur vapour, according to the temperature and pressure. Sketch a pressure-temperature diagram to illustrate this statement, and state clearly the significance of the lines on your diagram.   (O)

5. 100 cm^3 of a solution of iodine in tetrachloromethane was vigorously shaken with 900 cm^3 of water and the resulting mixture allowed to settle.

Separate portions of the mixture (containing aqueous and non-aqueous layers) were subjected to the following operations:

(a) water was added;

(b) iodine was added;

(c) the mixture was warmed whereby some of the tetrachloromethane evaporated, and then cooled to room temperature;

(d) aqueous potassium iodide solution was added;

(e) a very small quantity of sodium thiosulphate was added;

(f) a large quantity of sodium thiosulphate was added.

Suggest what effect, if any, each of these operations would have had on the concentration of iodine in the two layers and on the *ratio* of these concentrations. Give reasons for your conclusions.   (L)

6. Ethanol (b.p. 78.5°C) and water form a constant boiling mixture having a boiling point of 78.2°C and a composition of 95.6% ethanol.

(a) Define the term *constant boiling mixture.*

(b) Sketch, and label fully, the boiling point/composition diagram for ethanol and water.

(c) An ethanol/water mixture shows positive deviations from Raoult's Law. Explain and account for this and state the law.

(d) What intermolecular change takes place when ethanol is added to water?

(e) State qualitatively the result of distilling (i) a mixture containing 75% ethanol, (ii) a mixture containing 97.5% ethanol.

(f) State with reasons which one of the following pairs of substances most closely obeys Raoult's Law:

(i) $C_2H_5NH_2$ and $C_6H_5NH_2$,

(ii) CH_3COCH_3 and $CH_3COC_2H_5$,

(iii) C_6H_6 and $C_6H_5CH_3$.   (AEB 1976)

7   Solubility and colligative properties

7.1   Introduction

Factors affecting the solubilities of solids and gases in liquids are discussed in Section 7.2 below, together with the construction and use of solubility curves.

The lowering of vapour pressure, elevation of the boiling point, depression of the freezing point and magnitude of the osmotic pressure of a solution all depend on the relative number of non-volatile particles present in the solution. These effects are known collectively as the colligative properties of the solution, and are discussed in Sections 7.3–7.7. The laws governing the colligative properties apply only for dilute solutions, and it is assumed here that such solutions are being considered.

7.2   Solubility and solubility curves

(a) Concentration of solutions

A solution may be defined as a perfectly homogeneous mixture. This description can be applied to solutions of gases in gases, gases in liquids, liquids in liquids, gases in solids, solids in liquids or solids in solids. In this section the systems considered will be confined to solutions of solids and gases in liquids, and the main solvent considered will be water.

The solvent is the substance in excess and the solute is the substance dissolved in it. A standard solution is one in which the concentration of solute is precisely defined. This may be done in several ways:

Percentage by mass. This can be expressed either as the mass in grams of solute dissolved in 100 g of solvent or as the mass of solute dissolved in 100 g of solution.

Grams per litre. This is sometimes called the mass concentration; it is the number of grams of solute in one litre of solution, $(g\,l^{-1})$ or $(g\,dm^{-3})$; other units such as $kg\,m^{-3}$ can be used.

Moles per litre (molar solution). This has already been described in Chapter 1. The concentration is expressed as $mol\,dm^{-3}$.

Mole fraction. The concentrations of solute or solvent can be expressed as mole fractions; the ratios of moles of one or the other to the total number of moles of the two together. For example, if 0.5 mole of a solute A is dissolved in 1.5 moles of a solvent B, then:

$$\text{Mole fraction of } A = \frac{0.5}{(0.5+1.5)} \text{ or } 0.25$$

$$\text{Mole fraction of } B = \frac{1.5}{(0.5+1.5)} \text{ or } 0.75$$

The sum of the mole fractions of A and B is of course unity.

(b) Saturated and supersaturated solutions

Most solids become more soluble as the temperature increases (notable exceptions are anhydrous sodium sulphate and sodium carbonate-10-water), so it is necessary to quote solubilities at a given temperature. A saturated solution is one in which no more of a particular solute can be dissolved at that temperature; if further amounts of solute are added the concentration of dissolved solid is not increased and some solid remains undissolved. An equilibrium is established between the saturated solution and the undissolved solute; solid is deposited from the solution at the same rate as undissolved solid passes into solution. A saturated solution is therefore described as one which is in equilibrium with undissolved solute at the temperature of the solution. If the temperature is raised, solid dissolves until the solution is saturated at the higher temperature; similarly if the solution is cooled, crystals are deposited until the concentration of the solution is reduced to that of a saturated solution at the lower temperature.

Under certain conditions a solution may become *supersaturated*; in such a solution the amount of solid dissolved is greater than that required to saturate the solution at the existing temperature. These solutions are unstable, and if disturbed by the addition of a 'seed' crystal of the solute, or even by a speck of dust, they will deposit crystals until the concentration reaches that of a saturated solution. This effect can be illustrated by warming sodium thiosulphate-5-water crystals ($Na_2S_2O_3 \cdot 5H_2O$) in a test tube; the crystals dissolve in their water of crystallisation, and if cooled to room temperature without shaking the tube, they remain dissolved in a supersaturated solution which stays clear until 'seeded', when the whole recrystallises with considerable evolution of heat.

(c) Measurement of solubility and solubility curves

The solubility of a solid in a liquid is defined as the maximum quantity in grams of the solid that will dissolve in one hundred grams of solvent, at a given temperature, in the presence of undissolved solid.

In the laboratory, the solubility may be measured by stirring excess of the solid in the solute, in a water bath kept at the required temperature, until no more dissolves. After allowing the mixture to settle a sample of the saturated solution is withdrawn and weighed. The mass of solute is found by evaporation and weighing, or by volumetric analysis e.g. chlorides can be estimated with standard silver nitrate solution. The mass of solvent in the sample is found by difference, and the results quoted in grams of solute per hundred grams of solvent ($g(100g)^{-1}$).

The variation of solubility with temperature is shown by plotting one against the other; the graphs are known as solubility curves (Fig. 7.1).

Solubility curves give information concerning the conditions for crystallisation of solids from solutions. For example, a solution of potassium nitrate containing 90 g of solid dissolved in 100 g of water will be unsaturated at 80°C; if the solution is cooled to 60°C (X on graph) it will become saturated; further cooling to, say, 40°C will cause deposition of 40 g of crystals, leaving a solution saturated with 50 g potassium nitrate in 100 g of water (Y on graph).

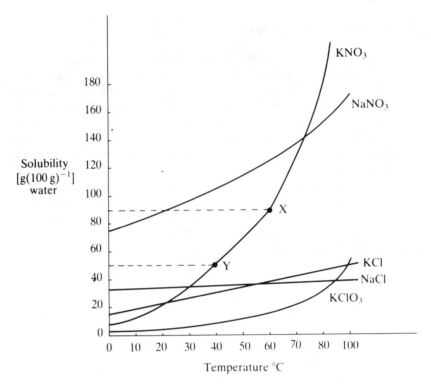

Fig. 7.1   Some typical solubility curves

(d) Fractional crystallisation

Solid compounds can be separated from mixtures by dissolving them in a solvent in which the solubilities of the compounds vary differently with temperature. For example, a mixture of 20 g of potassium chloride with 20 g of potassium chlorate will dissolve completely in 100 g of water at 80°C; if this solution is now cooled to 30°C, some 15 g of pure chlorate crystals will be deposited while the potassium chloride remains in a solution which is still not saturated with respect to potassium chloride at this temperature. This technique is used to purify the crude products from preparations of organic compounds by recrystallisation; the impurities are present only in small concentrations and remain dissolved in the solvent when the pure product crystallises from the solution.

   If the solubilities of the two solids are very similar then a repeated process of *fractional crystallisation* must be employed. A solution is prepared of such concentration that, on cooling, about half of the solids crystallises from the solution, yielding a mixture of crystals slightly richer in the less soluble constituent and leaving the mother liquor richer in the more soluble solid. Both crystals and mother liquor are then further treated, each successive crystallisation giving rise to a better separation of the solids until a pure sample of each is obtained.

(e) Solutions of gases in water

The solubility of a gas in a liquid can be measured in terms of the absorption *coefficient*, which is defined as the volume of gas, measured at s.t.p., which is dissolved by unit volume of the liquid, in contact with the gas at the specified temperature and pressure.

Solutions of gases in water fall into two classes: very soluble gases, which are those which react with water, such as hydrogen chloride or ammonia, and those gases which dissolve without reaction and are only sparingly soluble, such as nitrogen or oxygen. The absorption coefficients for the solution of some gases in water at 0°C and 1 atmosphere pressure are:

Gas	Ammonia	Hydrogen chloride	Carbon dioxide	Oxygen	Nitrogen
Absorption coefficient	1300	520	1.70	0.050	0.024

The effect of pressure on the solution of gases is given by *Henry's law*, which states that the *mass* of gas that dissolves in a given volume of a liquid at a given temperature is directly proportional to the pressure of the gas. In terms of volume this means that the *volume* of gas dissolved is independent of pressure, since if the pressure is doubled the same volume will be occupied by twice the mass. This is important in the manufacture of 'fizzy' drinks and soda-water, in which carbon dioxide is dissolved at a pressure of eight atmospheres; the mass of carbon dioxide dissolved is therefore eight times the amount dissolved at atmospheric pressure, and bubbles are expelled as soon as the bottle is opened.

The solubilities of most gases decrease with increase of temperature. The expulsion of gases from water on boiling is however mainly the result of the displacement of gases by steam; the solubility of oxygen at 100°C is still about one-third that at room temperature, but all the dissolved oxygen can be removed from water by brisk boiling for several minutes.

7.3   Lowering of vapour pressure

All liquids exposed to the air exert a vapour pressure which results from the escape of molecules of liquid into the air above. Normally some molecules escape completely from the liquid by diffusion, and the liquid slowly evaporates; the rate of evaporation is greatly increased by the action of draughts or by a rise in temperature. If, however, the liquid is in a closed container an equilibrium results as the rate of movement of the molecules from the liquid as vapour is balanced by the rate at which vapour molecules re-enter the liquid; the vapour pressure of a given liquid at equilibrium at a set temperature is constant, and is known as the *saturation vapour pressure* of the liquid.

The addition of a non-volatile solute to a liquid results in the lowering of the vapour pressure; this occurs because the presence of the solute molecules reduces the number of solvent molecules that can reach the surface of the liquid and therefore the number that can escape as vapour.

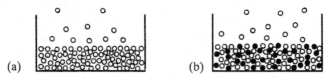

o   Solvent molecules

●   Solute particles

Fig. 7.2   Vapour pressure above (a) pure solvent (b) solution

The relationship of the vapour pressure to the concentration of the solute is defined by Raoult's law. (page 95).

The relative lowering of the vapour pressure is the difference between the vapour pressure of the pure solvent (p_0) and the vapour pressure of the solution (p) divided by the vapour pressure of the pure solvent; and the mole fraction of the solute is the ratio of moles of solute (n) to the total moles of solute plus solvent ($n + N$). That is,

$$\frac{p_0 - p}{p_0} = \frac{n}{n + N}$$

For a dilute solution n is small compared to N and can be ignored in the denominator, giving:

$$\frac{p_0 - p}{p_0} = \frac{n}{N}$$

That is to say, the relative lowering of the vapour pressure is proportional to the mole quantity of dissolved solid in a given mass of solvent; for a particular solvent it will be the same for equimolar quantities of any solute provided that the solute is neither associated nor dissociated in the solvent.

The relative molecular mass of compounds that do not associate or dissociate in the solvent can be found from measurements of the vapour pressure of solutions. For a mass m_1 of a compound whose relative molecular mass is M_1, dissolved in m_2 grams of a solvent of relative mass M_2:

$$n = \frac{m_1}{M_1} \quad \text{and} \quad N = \frac{m_2}{M_2}$$

$$\frac{p_0 - p}{p_0} = \frac{\dfrac{m_1}{M_1}}{\dfrac{m_2}{M_2}} \quad \text{and} \quad M_1 = m_1 \times \frac{M_2}{m_2} \times \frac{p_0}{p_0 - p}$$

In practice this method is rarely used since measurements of vapour pressure are difficult to carry out accurately; the other colligative effects are usually employed as they are more easily measured.

7.4 Elevation of the boiling point

The vapour pressure of a liquid heated in an open vessel increases until it reaches atmospheric pressure, at which point the liquid is said to boil. The temperature remains constant and any further addition of energy by heating goes to convert the liquid into vapour. The normal boiling point of a liquid is therefore defined as the temperature at which its vapour pressure reaches normal atmospheric pressure, $101\,325\,\mathrm{Nm}^{-2}$.

Now since the vapour pressure of a solution is reduced below that of the pure solvent by the addition of a non-volatile solute, it is evident that a higher temperature will be required for the vapour pressure to reach atmospheric i.e. the boiling point of the solution will be higher than that of the pure solvent (Fig. 7.3).

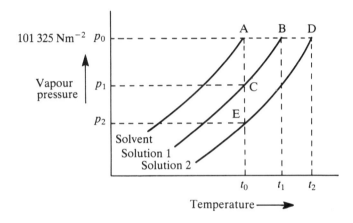

Fig. 7.3 Elevation of boiling point of solutions.

It can be seen from Fig. 7.3 that t_0 K is the boiling point of pure solvent, t_1 K the boiling point of solution 1 and t_2 K the boiling point of solution 2. Now for a dilute solution the vapour pressure/temperature curves approximate to straight lines, and therefore ABC and ADE are similar triangles. Hence

$$\frac{AB}{AD}=\frac{AC}{AE} \quad \text{or} \quad \frac{t_1-t_0}{t_2-t_0}=\frac{p_0-p_1}{p_0-p_2}$$

or $\quad\dfrac{\text{elevation of b.p. of solution 1}}{\text{elevation of b.p. of solution 2}}=\dfrac{\text{lowering of v.p. of solution 1}}{\text{lowering of v.p. of solution 2}}$

That is to say that the elevation of the boiling point is proportional to the lowering of the vapour pressure. Since the latter is proportional to the concentration of dissolved solid then so also is the elevation of the boiling point; again one mole of any solute dissolved in a constant mass of a given solvent will raise the boiling point by a characteristic amount, provided the solute is neither associated nor dissociated. The elevation of the boiling point caused by the presence of one mole of solute in 100 g of solvent is called the *molar elevation constant* of the solvent; the terms boiling point constant and ebullioscopic constant are also used. However, in most cases one mole of a compound would not dissolve in 100 g of solvent, and even if it did the solution would be too concentrated to obey Raoult's law; for this reason the boiling point constant is frequently quoted for one kilogram of solvent (or, for water, 1000 cm³). Some values of boiling point constants are given in Table 7.1.

Table 7.1   Molar elevation constants

Solvent	per 100 g	per kg
Water	5.20 K	0.520 K
Benzene	27.0 K	2.70 K
Trichloromethane	36.6 K	3.66 K

These figures are used to determine relative molecular masses; for example, if m_1 grams of a compound dissolved in m_2 grams of water is found to raise the boiling point by T K, then:

$$\frac{T}{0.52} = \frac{m_1/m_2}{M/1000} = \frac{m_1}{m_2} \times \frac{1000}{M}$$

and

$$M = \frac{1000 m_1}{m_2} \times \frac{0.52}{T}$$

where M is the relative molecular mass of the compound.

Measurements are made by placing the solvent in a tube with a side arm to which a condenser is attached to prevent loss of solvent during heating and boiling. Successive additions of small weighed quantities of the solute are made; the boiling point at the new concentration being noted each time. Anti-bumping granules are added to ensure smooth boiling without superheating of the liquid above its boiling point; elaborate precautions have to be taken to ensure steady heating and to avoid fluctuations caused by draughts. A much simplified version of the apparatus used is shown in Fig. 7.4.

Since the elevation of the boiling point is small, a thermometer that will register very small changes of temperature has to be used. Such a thermometer was designed by Beckmann; it has a very narrow capillary tube to contain the mercury thread, attached to a large thermometric bulb for mercury at its lower end and to a reservoir bulb at its upper end. The capillary tube spans a scale of about six degrees, calibrated to read to 0.01 K; the quantity of mercury in the lower bulb is adjusted by use of the reservoir so that the mercury thread is on the scale within the required range of temperature, for example, between 270 K and 276 K or 370 K and 376 K.

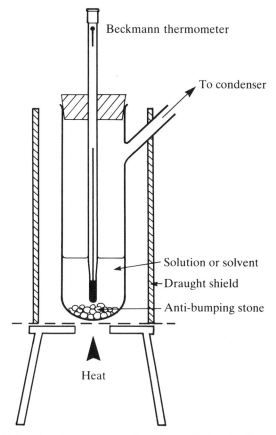

Fig. 7.4 Simple apparatus for determining boiling points

7.5 Depression of the freezing point

Another colligative property brought about by the addition of a non-volatile solute to a liquid is that the freezing point of the solution is depressed. This is only true of solutions sufficiently dilute to deposit pure solvent on cooling. This is illustrated in Fig. 7.5, showing the relation of the freezing point and vapour pressure change with temperature.

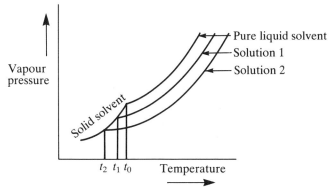

Fig. 7.5 Depression of the freezing point

t_0 is the freezing point of the pure solvent, t_1 and t_2 are the freezing points of solutions after adding successive amounts of solids. The depressions $(t_1 - t_0)$ and $(t_2 - t_0)$ are again proportional to the amount of solute added. One mole of any compound dissolved in 100 g of solvent, provided it is neither associated nor dissociated in the solvent, produces a depression of freezing point which is characteristic for that liquid. This is the *freezing point constant* or *molar depression constant* for the liquid, formerly known as the cryoscopic constant of the liquid. Some typical values are:

Table 7.2 Molar depression constants

Solvent	*Molar depression constant*	
	per 100 g	*per kg*
Water	18.6 K	1.86 K
Benzene	50.0 K	5.00 K
Ethanoic acid	39.0 K	3.90 K

The measurement of freezing point depression is relatively easy; the major care being to avoid supercooling taking the liquid below its freezing point without the appearance of crystals. Vigorous stirring at the temperature just above the expected freezing point is used to induce crystallisation. The apparatus is sketched in Fig. 7.6.

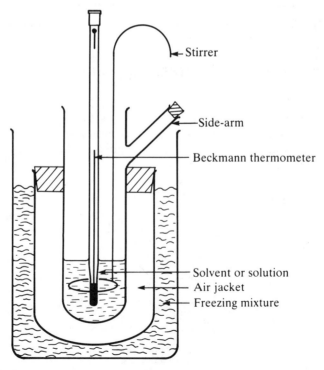

Fig. 7.6 Apparatus for measurement of freezing point depressions

The pure solvent is placed in the inner tube which is fitted with a Beckmann thermometer, adjusted for the required temperature range. The tube is surrounded by an air jacket to ensure even cooling and the whole is placed in a beaker of freezing mixture such as ice and salt.

When the freezing point of the pure solvent has been measured, a weighed quantity of solute is added through the side arm of the inner tube; great care must be taken to ensure that all the solute dissolves. The freezing point of the solution is then determined, and, if required, further determinations can be made after the addition of more solute.

To calculate the relative molecular mass of a compound the molar depression constant is used. Suppose m_1 grams of a compound whose relative molecular mass M is to be found is dissolved in m_2 grams of water; the molar depression constant for water is 18.6 K per 100 grams and the depression measured will be proportional to this,

$$\therefore \quad \frac{18.6}{T} = \frac{M/100}{m_1/m_2} \quad \text{or} \quad \frac{M}{100} \times \frac{m_2}{m_1}$$

$$M = \frac{18.6}{T} \times \frac{m_1}{m_2} \times 100$$

where T is the measured depression of freezing point.

7.6   Osmotic pressure

(a) Osmosis

It is well known that the particles, ions or molecules, of a crystal of a soluble compound placed in water will be distributed evenly throughout the solution by a slow process of diffusion; in the same way, if a layer of a saturated solution of, say, copper sulphate is carefully placed under a layer of pure water, the two layers will gradually merge by diffusion of both solute particles and solvent particles until the whole is one solution of even concentration.

If as in Fig. 7.7(c) some pure solvent is separated from a solution by a membrane which will permit the passage of solvent molecules but prevent the passage of solute molecules then there will be an overall flow of solvent molecules into the solution layer. If two solutions of different concentrations are used, then the flow of solvent molecules will be from the less to the more concentrated solution, and the movement will continue until the solutions are of equal concentration. Of course, solvent molecules move in both

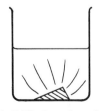

(a) Diffusion of a soluble solid into a liquid

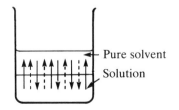

(b) Diffusion of both solvent and solute particles

Pure solvent
Solution

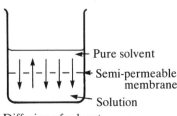

(c) Diffusion of solvent particles only

Pure solvent
Semi-permeable membrane
Solution

Fig. 7.7 Diffusion and osmosis

directions, but the rate of flow into the more concentrated solution is greater.

This process is called *osmosis*, and the dividing membranes are called *semi-permeable membranes*. Some natural semi-permeable membranes include animal bladders, the skins of fruits such as plums, and the membranes under the shells of eggs. Cellophane and parchment both act as semi-permeable membranes, and very strong semi-permeable systems can be made by precipitating salts such as copper(II) hexacyanoferrate(II) in the pores of a porous pot. The properties of the latter can be demonstrated by dropping a copper sulphate crystal into a solution of potassium hexacyanoferrate(II); the two salts react forming a layer of $Cu_2[Fe(CN)_6]$ on the surface of the crystal; water molecules pass by osmosis through this layer towards the more concentrated layer of copper sulphate solution within; the membrane swells and bursts allowing a bubble of concentrated copper sulphate solution to emerge, and this reacts, swells and burst in its turn to produce a 'crystal growth'. The well-known crystal gardens made with sodium silicate solution 'grow' by a similar process.

Osmosis is also demonstrated by the swelling of dried prunes or peas placed in water, or by the bursting of red blood corpuscles placed in a solution of lower osmotic pressure relative to themselves. In the latter case all the solution becomes coloured red, but if blood corpuscles are placed in a solution of relatively higher osmotic pressure they shrink and settle to the bottom of the container.

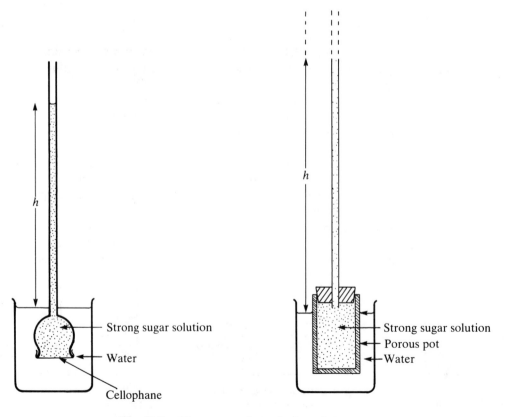

Fig. 7.8. Demonstration of osmotic pressure

The development of osmotic pressure can be shown by fixing a piece of cellophane across the mouth of a thistle funnel filled with a concentrated sugar solution. This is inverted in a beaker of water (Fig. 7.8).

Water passes into the thistle funnel and the solution rises up the stem. The flow will continue until the hydrostatic pressure of the column of solution above the level of the water in the beaker, h, is equal to the osmotic pressure between the solution and the water in the beaker. A better measure of h is obtained by using a semi-permeable membrane made by depositing copper(II) hexacyanoferrate(II) in the pores of a porous pot; the apparatus is stronger and more leak-proof, and it will be found that the solution in the tube rises to a remarkably high level.

Osmotic pressure is often spoken of carelessly as though it were a property of a solution similar to its boiling point or freezing point, but it should be remembered that a solution does not exert an osmotic pressure in isolation. To exert osmotic pressure, a solution must be in contact with the pure solvent, or another solution of different concentration, through a semi-permeable membrane. The strict definition of the osmotic pressure of a solution is the pressure required to prevent osmosis when the solution is separated from pure solvent by a semi-permeable membrane.

(b) Measurement of osmotic pressure

Methods of measuring the osmotic pressure of a solution are based on determinations of the pressure required to prevent osmosis.

The pressures involved are quite high (of the order of five to ten atmospheres) and the apparatus used must be strong and the connections and joints able to withstand such pressures. A simple apparatus was designed by Pfeffer, who did much of the early work on osmosis. The semi-permeable membrane is of the porous pot type, fitted with a mercury manometer whose

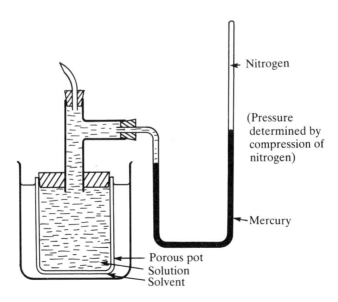

Fig. 7.9 Pfeffer's apparatus

closed end is filled with nitrogen, thus enabling the pressure to be determined without significant volume change. The porous pot and the tubes
connecting it to the manometer are filled with the solution through an inlet
tube which is then sealed off with a flame. The pot is then placed in a beaker
of pure solvent and the whole kept at constant temperature.

The maximum pressure shown by the compression of nitrogen in the
manometer, the pressure required to prevent further osmosis, is recorded.

A modern osometer is shown in Fig. 7.10.

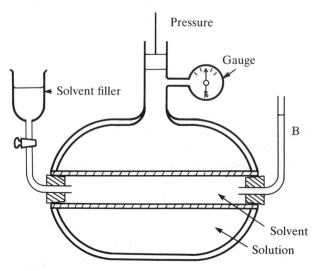

Fig. 7.10   Berkeley' and Hartley's apparatus.

In this apparatus the semi-permeable membrane is formed in a tube of
porous clay by depositing copper(II) hexacyanoferrate(II) in the pores; the
tube is filled with solvent and placed within another cylinder containing
solution. This outer cylinder is fitted with a device for applying pressure to
prevent the flow of solvent into solution by osmosis. The pressure required
to maintain a constant solvent level in tube B is measured.

(c) Effect of concentration and temperature on osmotic pressure

The osmotic pressure, π, of a solution of a compound that is not associated or
dissociated in the solvent is proportional to the concentration of the solute.

$$\pi \propto concentration \qquad\qquad\text{(i)}$$

For a fixed amount of solute this may be expressed in terms of the volume of
solution:

$$\pi \propto \frac{1}{volume\ of\ solution} \qquad\qquad\text{(ii)}$$

The osmotic pressure of a solution is increased by a rise in temperature and
is in fact directly proportional to the absolute temperature:

$$\pi \propto T \qquad\qquad\text{(iii)}$$

Now if equations (ii) and (iii) are combined, we get

$$\pi \propto \frac{T}{V}, \quad \text{or} \quad \pi = \text{constant} \times \frac{T}{V}$$

For a given mass of solute, therefore,

$$\pi = k \times \frac{T}{V}$$

an expression plainly analogous to the relationship between the pressure, temperature and volume of gases expressed in the gas laws (page 66). For a solution containing one mole of solute in 22.4 litres of solution at a temperature 273 K the osmotic pressure is 101 325 Nm^{-2}, and these figures give a value for k of 8.314 $J\,K^{-1}\,mol^{-1}$ which is numerically the same as the gas constant R. Hence $\pi V = RT$.

Thus the osmotic pressure exerted by a dilute solution separated from a solvent by a semi-permeable membrane is numerically the same as the pressure an identical mass of solute would exert if it were a gas contained in a vessel of volume equal to that of the solution.

(d) Calculation of relative molecular mass from osmotic pressure

The equation $\pi V = RT$ is used to calculate the relative molecular masses of compounds that are not associated or dissociated in solution. For example, the osmotic pressure of a solution containing 12.28 grams of a protein dissolved in one litre of water at 25°C was found to be 963 Nm^{-2}. Here the volume containing one mole of the protein

$$V = \frac{1000 \times M}{12.28}\,cm^3, \quad \text{or} \quad \frac{10^{-3}\,M}{12.28}\,m^3$$

where M is the relative molecular mass of the protein.

Hence

$$963 \times \frac{10^{-3}\,M}{12.28} = 8.314 \times 298$$

$$M = \frac{8.314 \times 298 \times 12.28}{963 \times 10^{-3}}$$

$$= 31.6 \times 10^3$$

The following example illustrates an alternative method of calculation.

The osmotic pressure of a solution containing 5.1 g of a sugar dissolved in one litre of water was measured and found to be 68 100 Nm^{-2} at 20°C.

then 68 100 Nm^{-2} is the osmotic pressure at 273 K of 1 litre
of solution containing $5.1 \times \dfrac{293}{273}$ grams

and 68 100 Nm^{-2} is the osmotic pressure at 273 K of 22.4 litres
of solution containing $5.1 \times \dfrac{293}{273} \times 22.4$ grams

and 101 325 Nm^{-2} is the osmotic pressure at 273 K of 22.4 litres

of solution containing $5.1 \times \dfrac{293}{273} \times 22.4 \times \dfrac{101\ 325}{68\ 100}$

or 182.4 grams.

but 101 325 Nm^{-2} is the osmotic pressure developed by one mole of solute in 22.4 litres of solution at 273 K. Hence the relative molecular mass of the sugar is 182.4.

(e) Importance of osmotic pressure measurements

The advantage of the relatively high osmotic pressures developed by solutions of quite low concentrations is that a method is provided of determining the relative molecular masses of big molecules such as those of plastics, proteins or carbohydrates. Solutions of these compounds, of reasonable concentration by mass, contain relatively few particles; for example, a solution of ten grams of a protein of relative molecular mass in the range of 5000 to 10 000 in one litre of water will have a molar concentration of only 1/500 to 1/1000 mol l^{-1}, which will give rise to a very small depression of the freezing point. The same solution will however exert an osmotic pressure of reasonable size.

Great care must be taken in preparing solutions for medication, either by injection into the bloodstream or for administration in the form of eye or nose drops. The solutions must not cause shrinkage or swelling of the body tissue cells by osmosis through the cell walls. To ensure this, tissue cells are placed in the medicinal solution and observed under a microscope; the concentration of the medicinal solution is adjusted until no osmosis occurs; at this point the solution is said to be *isotonic* with the cell fluids. It may not be exactly iso-osmotic with the cell fluid because cell walls are not perfect semi-permeable membranes but have some selective effect on the process of osmosis. *Iso-osmotic* solutions are those which placed separately in contact with pure solvent through a semi-permeable membrane develop identical osmotic pressures.

(f) Effects of dissociation and association of solute molecules

The colligative properties depend for their magnitude on the number of particles of non-volatile solute present in the solution. For this reason the property measured, be it lowering of vapour pressure, elevation of boiling point, depression of freezing point or development of osmotic pressure, will be greater if the compound is dissociated in solution, thus providing more particles, and less if the compound is associated, thus reducing the number of particles present.

This will result in abnormal values for relative molecular masses calculated from measurements of the colligative properties. The relative molecular mass calculated for a substance such as sodium chloride, which is dissociated in solution, will be much lower than the theoretical value, while that calculated for one which associates, such as ethanoic acid in benzene (page 101) will be much higher than the theoretical value.

The degree of dissociation of partially dissociated compounds can be calculated by comparing the apparent relative molecular mass, determined by, for example, depression of the freezing point, with the known theoretical relative molecular mass.

Consider a compound AB partially dissociated in solution with a degree of dissociation α:

$$AB \rightleftharpoons A^+ + B^-$$
$$(1-\alpha) \quad \alpha \quad \alpha$$

The total number of particles from unit mass of AB will be greater than would be expected from the undissociated compound by the factor $(1 - \alpha + 2\alpha)$, or $(1+\alpha)$. The observed depression of the freezing point will be greater than the expected depression by this factor, and the apparent and theoretical relative molecular masses, being inversely proportional to the freezing point depressions, will be related:

$$\frac{\text{calculated apparent relative molecular mass}}{\text{known theoretical relative molecular mass}} = \frac{1}{1+\alpha}$$

Hence α can be found. The degree of association for a compound can be calculated in the same way.

If the dissociation is complete, the relative molecular mass found from measurements of the colligative properties will be half the theoretical value, and, by the same reasoning, the experimental value will be double the theoretical for molecules which are completely associated in solution.

Questions

1. Explain what is meant by the term 'colligative property'. Name three such properties and describe the practical determination of one such property. Explain how such a measurement can be used to determine relative molecular mass. Give two instances when the method selected would give unexpected results.

   A 0.10 M solution of $HgCl_2$ in water freezes at $-0.186°C$ whereas a 0.10 M solution of $Hg(NO_3)_2$ freezes at $-0.558°C$. Deduce the structural states of these two compounds. Given that the melting point of mercury(II) chloride is 280°C what is the predominant type of bonding in this compound?

   (The freezing point depression constant for water is 1.86 K for 1 mole of solute in 1 litre of solution). (JMB)

2. (a) Assuming that the vapour pressure of water at any constant temperature is reduced on adding an involatile solute such as sugar by an amount proportional to the concentration of the solute, show with the aid of a diagram that the elevation of the boiling point of such a solution is proportional to the concentration of the solute.

   (b) Give a labelled sketch of the apparatus you would use to determine the depression of the freezing point of a solution of sugar.

   (c) The freezing point of pure benzene is 5.533°C. The freezing point of a solution of 6.40 g of naphthalene ($C_{10}H_8$) in 1000 g of benzene is 5.277°C, while that of a solution of 15.25 g of benzoic acid, $C_7H_6O_2$, in

1000 g of benzene is 5.175°C. What conclusions can you draw from this information? (O)

3. (*a*) State with reasons, why measurement of the elevation of boiling point could be used with an aqueous solution, to obtain the molar mass of urea (carbamide), but not ethanoic acid.

(*b*) A solution of 2.8 g of cadmium iodide (CdI_2) in 20 g of water boiled at a temperature which was 0.20 K higher than the boiling point of pure water under the same conditions of pressure. Calculate the molar mass of the solute, and comment on the result. (The boiling point elevation constant for water is 0.52 K kg/mol). (AEB 1978)

4. When 1.44 g of a monobasic carboxylic acid were dissolved in 100 g of water, the solution froze at −0.39°C. The same mass of acid displaced from Victor Meyer's apparatus 253 cm^3 of air, measured at 23°C and 740 mmHg (corrected for the vapour pressure of water). What values do these data give for the relative molecular mass of the acid? What qualitative explanation can you give for these values?

(Cryoscopic (freezing point) constant for water = 1.86°C per 100 g.)
(C)

5. (*a*) State Raoult's law as applied to dilute solutions.

(*b*) In the same diagram sketch the two curves which show the effect of temperature on the vapour pressures of (i) a pure solvent and (ii) a solution of a non-volatile solute in the same solvent.

Account for the difference in these curves and use them to explain the difference in boiling points between the solvent and the solution.

(*c*) Explain why the vapour pressure above a solvent rises with an increase in temperature.

(*d*) A solution was made by dissolving 8.4 g of a non-volatile organic compound X (relative molecular mass = 168) in 69 g of ethanol.

Calculate

(i) the mole fraction of X in the solution,

(ii) the vapour pressure of the solution at 20°C given that the vapour pressure of pure ethanol at this temperature is 26.6 kPA,

(iii) the boiling point of the solution, given that ethanol boils at 78°C and the boiling point constant for ethanol is 1.15 K kg mol^{-1}.

(AEB 1980)

6. (*a*) What is meant by the terms *colligative property* and *osmosis*? State how the osmotic pressure of a system is related to (i) the concentration of the solute, (ii) temperature and (iii) the relative molecular mass of the solute.

(*b*) Describe a method used to give a fairly accurate determination of the relative molecular mass of a solute using osmosis. What kind of solutes are best suited for such a determination?

(*c*) An aqueous solution of a substance *X* at 27°C exhibited an osmotic pressure of 779 kPa (7.69 atmospheres). What would be the freezing point of the same solution at standard atmospheric pressure? The molecular depression constant for water is 1.86 K kg/mol. (*X* is known not to associate or dissociate in aqueous solution). (AEB 1978)

7. A solution of 2.00 g of a polymer in 1.00 dm^3 of water was found to have an osmotic pressure of 273 N m^{-2} at 0°C. Calculate to two significant figures the relative molecular mass (molecular weight) of the polymer.
(O)

8. Freezing point measurements indicate that the relative molecular mass of ethanoic (acetic) acid in benzene is very different from what it is in water. What are the approximate values of these relative molecular masses and how is the difference explained?   (O)

Part III
Chemical Principles

8 Thermochemistry

8.1 Introduction

When carrying out reactions in the laboratory, such as the dissolving of magnesium ribbon in acid, it is often noticeable that the reaction mixture becomes hot. In other cases, such as the dissolving of ammonium nitrate in water, the mixture becomes cold. In either case the system has undergone an energy change, heat having been evolved or absorbed. Such energy changes during chemical reactions are of vital importance as they enable us to live and also bring the comforts of civilisation. Thus plants convert energy from the sun by photosynthesis, providing food which we convert into body tissue and energy by chemical processes; we burn fossil fuels to heat our homes, and to drive our internal combustion engines and the turbines that produce our electricity. The study of energy changes is therefore an important part of chemistry, especially of applied chemistry. To do this, it is necessary to measure the heat changes that occur during reactions; standard conditions and precise definitions are needed, and some of these are given below.

The heat content of a system is known as the *enthalpy* of the system. The change of enthalpy is given by the heat of reaction at constant pressure, and is symbolised as ΔH (Delta H), where Δ signifies 'change of' and H is measured in kJ. Thus, for a reaction

$$\text{Reactants} \rightarrow \text{Products}$$

$$\Delta H = H_{\text{Products}} - H_{\text{Reactants}}$$

If the reaction is *exothermic* and heat is evolved, ΔH is given a negative sign since the chemical system has lost energy to the surroundings. If the reaction is *endothermic* and heat is absorbed, then ΔH is given a positive sign since the chemical system has gained energy from the surroundings.

The enthalpy change that occurs as a result of a reaction is called the *enthalpy of reaction*, defined as 'the heat change when a reaction takes place between the masses of reactants indicated by the equation for the reaction'.

Under standard conditions the equation represents the reaction of mole quantities reacting at a pressure of 1 atmosphere and at 298 K, so that both products and reactants are in their normal physical state. The change is represented* as $\Delta H_{298}^{\ominus}$. If reactions will not take place at this temperature they are conducted at a suitable temperature and a value for ΔH at 298 K is calculated. Solutions must be of a concentration of 1 mole dm^{-3}, and solids which exist in more than one form are assumed to be in their most stable state. Thus for the combustion of carbon the equation

$$C(\text{graphite}) + O_2(g) \rightarrow CO_2(g) \qquad \Delta H_{298}^{\ominus} = -393.4 \text{ kJ mol}^{-1}$$

means that if 1 mole (12 g) of graphite reacts with 1 mole (32 g) of oxygen gas at 298 K and 1 atmosphere, then 1 mole (44 g) of carbon dioxide will be formed with the evolution of 393.4 kJ.

The enthalpy changes for some different types of reaction are defined:

*It may also be represented $\Delta H_m^{\ominus}$ (298 K).

128

(a) Enthalpy of formation

The *standard enthalpy of formation* $\Delta H^{\ominus}_{f298}$ of a compound represents the heat change when 1 mole of the compound is formed from its elements under the standard conditions. The standard enthalpy of formation of water is $-285.5\,\text{kJ mol}^{-1}$:

$$H_2(g) + \tfrac{1}{2}O_2(g) \rightarrow H_2O(l) \qquad \Delta H^{\ominus}_{f298} = -285.5\,\text{kJ mol}^{-1}$$

(As elements appear in these equations in their normal physical states they are assigned zero enthalpy.)

(b) Enthalpy of combustion

The *standard enthalpy of combustion*, $\Delta H^{\ominus}_{(combustion)298}$, of an element or compound is the enthalpy change when 1 mole of the substance is completely burned in oxygen. The enthalpy of combustion of methane is $-890\,\text{kJ mol}^{-1}$:

$$CH_4(g) + 2O_2(g) \rightarrow CO_2(g) + 2H_2O(l) \qquad \Delta H^{\ominus}_{(combustion)298} = -890\,\text{kJ mol}^{-1}$$

(c) Enthalpy of atomisation

The *standard enthalpy of atomisation* $\Delta H^{\ominus}_{at.298}$ of an element is that energy required to form one mole of gaseous atoms from the element in its standard state. Thus the enthalpy of atomisation of hydrogen is $218\,\text{kJ mol}^{-1}$:

$$\tfrac{1}{2}H_2(g) \rightarrow H(g) \qquad \Delta H^{\ominus}_{at.298} = +218\,\text{kJ mol}^{-1}$$

(d) Enthalpy of hydrogenation

The *standard enthalpy of hydrogenation* $\Delta H^{\ominus}_{hydrogenation298}$ is the enthalpy change when 1 mole of an unsaturated compound is fully saturated by reaction with gaseous hydrogen. The enthalpy of hydrogenation of ethene is $-137.5\,\text{kJ mol}^{-1}$:

$$C_2H_4(g) + H_2(g) \rightarrow C_2H_6(g) \qquad \Delta H^{\ominus}_{hydrogenation\,298} = -137.5\,\text{kJ mol}^{-1}$$

The enthalpy changes that occur in many other reactions, such as dissociation, solution, neutralisation, crystallisation or sublimation, can be defined similarly. In discussion it is not necessary to refer continually to the standard conditions, but if conditions are not standard, this should be stated.

8.2   Hess's law

Hess's law states that the heat evolved or absorbed during a chemical change is independent of the route by which the change is accomplished. For example, a solution of ammonium chloride can be made from ammonia and hydrogen chloride gas by two methods:
(a) the two gases can be reacted together to form solid ammonium chloride:

$$NH_3(g) + HCl(g) \rightarrow NH_4Cl(s) \qquad \Delta H = -176\,\text{kJ mol}^{-1}$$

and the solid ammonium chloride can then be dissolved in water:

$$NH_4Cl(s) \rightarrow NH_4Cl(aq) \qquad \Delta H = +16.3 \text{ kJ mol}^{-1}$$

The overall enthalpy change is $-176.1 + 16.3 = -159.8 \text{ kJ mol}^{-1}$
(b) each gas can be separately dissolved in water to give solutions of ammonia and hydrochloric acid:

$$NH_3(g) + aq \rightarrow NH_3(aq) \qquad \Delta H = -35.2 \text{ kJ mol}^{-1}$$
$$HCl(g) + aq \rightarrow HCl(aq) \qquad \Delta H = -72.4 \text{ kJ mol}^{-1}$$

the two solutions are now mixed to give ammonium chloride solution:

$$NH_3(aq) + HCl(aq) \rightarrow NH_4Cl(aq) \qquad \Delta H = -52.2 \text{ kJ mol}^{-1}$$

In this case the total enthalpy change is $-(35.2 + 72.4 + 52.2) = -159.8 \text{ kJ mol}^{-1}$ which is the same as by route (a).

In fact Hess's law follows directly from the law of conservation of energy, because if a greater amount of heat could be obtained from an alternative route for the same chemical change we could in theory use this to 'create' energy.

Hess's law is applied to determine enthalpies of formation or reaction indirectly in cases where it is difficult or impossible to measure the change directly. For example, it is impossible to measure the enthalpy of formation of hydrocarbons simply by reacting carbon and hydrogen together in the correct proportions; however, the enthalpies of combustion of carbon, hydrogen and the product can all be measured. Thus, to find the enthalpy of formation of methane ΔH_{f298} the following procedure is followed:
The equation for the formation of methane is:

$$C(\text{graphite}) + 2H_2(g) \rightarrow CH_4(g) \tag{1}$$

The standard enthalpies of combustion of carbon, hydrogen and methane are found from a book of data:

$$CH_4(g) + 2O_2(g) \rightarrow CO_2(g) + 2H_2O(1) \qquad \Delta H^{\ominus}_{298} = -890.3 \text{ kJ mol}^{-1} \tag{2}$$

$$C(\text{graphite}) + O_2(g) \rightarrow CO_2(g) \qquad \Delta H^{\ominus}_{298} = -393.5 \text{ kJ mol}^{-1} \tag{3}$$

$$H_2(g) + \tfrac{1}{2}O_2(g) \rightarrow H_2O(1) \qquad \Delta H^{\ominus}_{298} = -285.5 \text{ kJ mol}^{-1} \tag{4}$$

Now by Hess's law the enthalpy change will be the same whether we make $(CO_2 + 2H_2O)$ by using steps (1) and (2) or by steps (3) and (4)×2, hence:

$$\Delta H^{\ominus}_{f298}(CH_4) + \Delta H^{\ominus}_{c298}(CH_4) = \Delta H^{\ominus}_{c298}(C) + 2 \times \Delta H^{\ominus}_{c298}(H_2)$$

or

$$\Delta H^{\ominus}_{f298} = -393.5 - 2 \times 285.5 + 890.3$$
$$= -74.2 \text{ kJ mol}^{-1}$$

Enthalpies of hydrogenation are also difficult to determine directly, particularly for partial hydrogenation, but again knowledge of the enthalpies of combustion of products and reactants can be used. To find the enthalpy change involved in converting one mole of ethyne to one mole of ethene, the enthalpies of combustion of ethyne, hydrogen and ethene are required:

$$C_2H_2(g) + H_2(g) \rightarrow C_2H_4(g) \qquad \Delta H^{\ominus}_{h298} = x \text{ kJ mol}^{-1} \qquad (1)$$

$$C_2H_2(g) + 2\tfrac{1}{2}O_2(g) \rightarrow 2CO_2(g) + H_2O(l) \qquad \Delta H^{\ominus}_{c298} = -1310 \text{ kJ mol}^{-1} \qquad (2)$$

$$H_2(g) + \tfrac{1}{2}(O_2)(g) \rightarrow H_2O(l) \qquad \Delta H^{\ominus}_{c298} = -285 \text{ kJ mol}^{-1} \qquad (3)$$

$$C_2H_4(g) + 3O_2(g) \rightarrow 2CO_2(g) + 2H_2O(l) \qquad \Delta H^{\ominus}_{c298} = -1393 \text{ kJ mol}^{-1} \qquad (4)$$

By Hess's law the enthalpy changes for the production of two moles of carbon dioxide and two moles of water will be the same whether route (1) and (4) or route (2) and (3) are followed. Hence:

$$\Delta H^{\ominus}_{h298}(C_2H_2) + \Delta H^{\ominus}_{c298}(C_2H_4) = \Delta H^{\ominus}_{c298}(C_2H_2) + \Delta H^{\ominus}_{c298}(H_2)$$
$$\Delta H^{\ominus}_{h298}(C_2H_2) = -1310 - 285 + 1393$$
$$= -202 \text{ kJ mol}^{-1}$$

Thus the hydrogenation of ethyne to ethene is exothermic and the heat evolved is 202 kJ mol^{-1}.

8.3 Bond energies

(a) Average standard bond energies

When a chemical change takes place energy has to be supplied to break the bonds between the atoms of the reactants, and is released when new bonds form between these atoms to make the products. Bonds between any particular pair of atoms are associated with an individual amount of energy known as the bond energy. This can be shown by plotting the enthalpies of combustion of a homologous series of alcohols (Fig. 8.1). These substances are all of similar nature and physical structure, each differing from the next by the addition of one —CH$_2$— unit (page 344).

Table 8.1 Alcohols

Structure	Name	Relative molecular mass	$\Delta H^{\ominus}_{c298}$ kJ mol^{-1}
CH$_3$CH$_2$OH	Ethanol	46	−1367
CH$_3$CH$_2$CH$_2$OH	Propanol	60	−2017
CH$_3$CH$_2$CH$_2$CH$_2$OH	Butanol	74	−2675
CH$_3$CH$_2$CH$_2$CH$_2$CH$_2$OH	Pentanol	88	−3323
C$_2$H$_5$CH$_2$CH$_2$CH$_2$CH$_2$OH	Hexanol	102	−3976

The constant slope of the graph indicates a regular increment to the energy change contributed in each case by the conversion of the —CH$_2$— unit to carbon dioxide and water:

$$-CH_2- + 1\tfrac{1}{2}O_2 \rightarrow CO_2 + H_2O$$

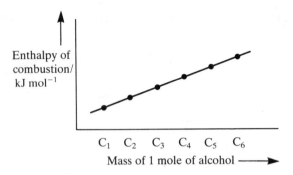

Fig. 8.1 Heats of combustion of alcohols

and this can be considered to imply a constant contribution from each individual bond.

Values for bond energies are mainly found by indirect methods. For example, the total energy required to break a molecule of methane into its constituent atoms in the gaseous phase can be determined.

$$CH_4(g) \rightarrow C(g) + 4H(g) \qquad \Delta H^\ominus_{298} = 1662 \text{ kJ mol}^{-1}$$

Assuming that all four bonds in the methane molecule are equal then the bond energy of a C—H bond must be $1662/4$ kJ mol$^{-1} = 415$ kJ mol^{-1}.

The enthalpy of atomisation of two consecutive higher alkanes can yield information about both C—H and C—C bonds in these molecules. The enthalpy of atomisation of propane is 3996 kJ mol^{-1} and that of butane is 5168 kJ mol^{-1}.

$$
\begin{array}{ccc}
H & H & H \\
\backslash & | & / \\
H-C-C-C-H \\
| & | & | \\
H & H & H
\end{array}
\quad (g) \longrightarrow 3C(g) + 8H(g) \quad \Delta H^\ominus_{298} = +3996 \text{ kJ mol}^{-1}
$$

$$
\begin{array}{cccc}
H & H & H & H \\
| & | & | & | \\
H-C-C-C-C-H \\
| & | & | & | \\
H & H & H & H
\end{array}
\quad (g) \longrightarrow 4C(g) + 10H(g) \quad \Delta H^\ominus_{298} = +5168 \text{ kJ mol}^{-1}
$$

The enthalpy of atomisation of propane results from the breaking of 2 (C—C) bonds and 8 (C—H) bonds, while that of butane results from the breaking of 3 (C—C) bonds and 10 (C—H) bonds. Representing the bond energies as E(C—C) kJ and E(C—H) kJ, then:

$$2 \text{ E(C—C)} + 8 \text{ E(C—H)} = 3996 \qquad\qquad (1)$$

$$3 \text{ E(C—C)} + 10 \text{ E(C—H)} = 5168 \qquad\qquad (2)$$

Solving the simultaneous equations (1) and (2)

$$\text{E(C—H)} = 413 \text{ kJ mol}^{-1} \qquad \text{and} \qquad \text{E(C—C)} = 346 \text{ kJ mol}^{-1}.$$

The bond energy for any particular bond varies slightly according to the molecule it is in, and the bond energy terms quoted in data books are average values. Some of these values are quoted at the end of the chapter.

(b) Use of bond energies

Bond energies can be used to calculate theoretical values for enthalpies of formation and enthalpies of reaction. The agreement between theoretical and experimental values is very close in most cases; where there is a discrepancy this can usually be attributed to the existence of resonance structures such as, for example, those proposed for polyatomic ions (page 60) or benzene (page 357).

To calculate the enthalpy of formation a structural formula must be drawn and the bond energies of the various bonds added together; these must be subtracted from the enthalpies of atomisation of the elements comprising the compound. For example, the enthalpy of formation of ethanol is calculated:

$$2C(\text{graphite}) + \tfrac{1}{2}O_2(g) + 3H_2(g) \longrightarrow H\!-\!\underset{\underset{H}{|}}{\overset{\overset{H}{|}}{C}}\!-\!\underset{\underset{H}{|}}{\overset{\overset{H}{|}}{C}}\!-\!O\!-\!H \quad (l)$$

Enthalpy of atomisation, $kJ\,mol^{-1}$			Bond energies $kJ\,mol^{-1}$	
2C (graphite)	2×714	1428	C—C	346
$\tfrac{1}{2}O_2(g)$		247	C—O	358
$3H_2(g)$	6×216	1296	O—H	463
			$5 \times$ (C—H)	2065
			(5×413)	
		2971		3232

$$\Delta H_{f298}C_2H_5OH = 2971 - 3232 = -261 \text{ kJ mol}^{-1}$$

Enthalpies of reaction can be calculated similarly, and this is useful where the formation of an intermediate compound is involved. The hydrogenation of ethyne to ethene provides a good example, since experimentally it would be difficult to avoid the formation of at least some ethane.

$$H\!-\!C\!\equiv\!C\!-\!H(g) + H_2(g) \longrightarrow \overset{H}{\underset{H}{\diagdown}}C\!=\!C\overset{H}{\underset{H}{\diagup}} \quad (g)$$

Energy required to break bonds, $kJ\,mol^{-1}$		Energy released in making new bonds, $kJ\,mol^{-1}$	
$2 \times$ C—H $= 2 \times 413$	826	$4 \times$ C—H $= 4 \times 413$	-1652
$1 \times$ C$\equiv$C	835	$1 \times$ C$=$C	-611
$1 \times$ H—H	435		
	2096		-2263

The overall release of energy is (2096—2263) or -167 kJ, or the enthalpy of hydrogenation of ethyne to ethene is -167 kJ mol^{-1}.

8.4   Lattice energies

The forces holding oppositely charged ions into a crystal lattice are electro-static. The *lattice energy* is defined as the change of enthalpy (heat change) that occurs when 1 mole of a solid crystalline substance is formed from its gaseous ions. The gaseous ions must be separated by such large distances that the forces between them are zero.

The lattice energy can be found by applying Hess's law to an energy cycle known as the Born-Haber cycle. For example, the lattice energy of sodium chloride can be calculated from the following data, all of which can be measured experimentally.

(i) Conversion of sodium metal to gaseous atoms (atomisation):

$$Na(s) \rightarrow Na(g) \qquad \Delta H^{\ominus}_{298} = +108.3 \text{ kJ mol}^{-1}$$

(ii) Conversion of gaseous sodium atoms to ions (ionisation):

$$Na(g) \rightarrow Na^+(g) + e^- \qquad \Delta H^{\ominus}_{298} = +500 \text{ kJ mol}^{-1}$$

(iii) Atomisation of chlorine:

$$\tfrac{1}{2}Cl_2(g) \rightarrow Cl(g) \qquad \Delta H^{\ominus}_{298} = +121 \text{ kJ mol}^{-1}$$

(iv) The gaining of an electron by chlorine to form chloride ions (this is the *electron affinity* of the element and is defined as the energy change when a gaseous atom accepts an electron to form a negative ion):

$$Cl(g) + e^- \rightarrow Cl^-(g) \qquad \Delta H^{\ominus}_{298} = -364 \text{ kJ mol}^{-1}$$

(v) The standard enthalpy of formation of sodium chloride Na^+Cl^-:

$$Na(s) + \tfrac{1}{2}Cl_2(g) \rightarrow Na^+Cl^-(s) \qquad \Delta H^{\ominus}_{298} = -411 \text{ kJ mol}^{-1}$$

The Born-Haber cycle is then constructed, remembering that elements in their normal physical state are assigned zero enthalpy. It can be seen from the diagram that the only term unknown is the lattice energy, and that this can be found by equating the figures on the left hand side with those on the right hand side:

$$411.0 + 108.3 + 500.0 + 121.1 - 364 = 776.4 \text{ kJ}$$

Thus the lattice energy of sodium chloride is $-776.4 \text{ kJ mol}^{-1}$.

The calculation of the lattice energy of magnesium oxide serves to illustrate some further points. The data required are:

(i) Atomisation of magnesium:

$$Mg(s) \rightarrow Mg(g) \qquad \Delta H^{\ominus}_{298 \text{ K}} = 146 \text{ kJ mol}^{-1}$$

(ii) Ionisation of magnesium:

$$Mg(g) \rightarrow Mg^+(g) + e^- \qquad \Delta H^{\ominus}_{298} = +736 \text{ kJ mol}^{-1}$$
$$Mg^+(g) \rightarrow Mg^{2+}(g) + e^- \qquad \Delta H^{\ominus}_{298} = +1448 \text{ kJ mol}^{-1}$$

Overall

$$Mg(g) \rightarrow Mg^{2+}(g) + 2e^- \qquad \Delta H^{\ominus}_{298} = +2184 \text{ kJ mol}^{-1}$$

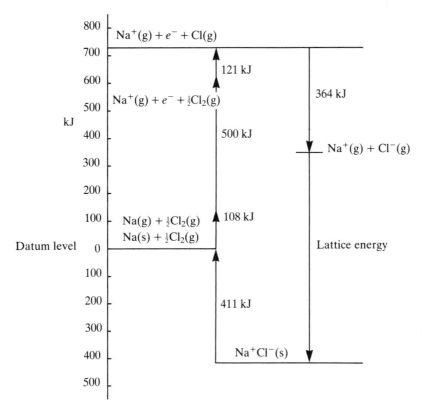

Fig. 8.2 Born-Haber cycle for sodium chloride

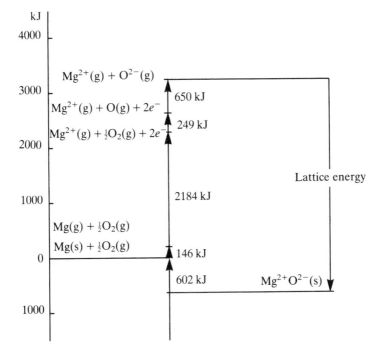

Fig. 8.3 Born-Haber cycle for magnesium oxide

(iii) Atomisation of oxygen:

$$\tfrac{1}{2}O_2 \rightarrow O(g) \qquad \Delta H^{\ominus}_{298} = 249.2 \text{ kJ mol}^{-1}$$

(iv) Electron affinity of oxygen:

$$O(g) + e^- \rightarrow O^-(g) \qquad \Delta H^{\ominus}_{298} = -141.4 \text{ kJ mol}^{-1}$$

$$O^-(g) + e^- \rightarrow O^{2-}(g) \qquad \Delta H^{\ominus}_{298} = +790.8 \text{ kJ mol}^{-1}$$

Overall electron affinity is $+649.9$ kJ mol^{-1} (note that oxygen accepts one electron with the release of energy, but the acceptance of a second electron requires the input of energy).

(v) Enthalpy of formation of magnesium oxide:

$$Mg(s) + \tfrac{1}{2}O_2(g) \rightarrow MgO(s) \qquad \Delta H^{\ominus}_{f298} = -601.7 \text{ kJ mol}^{-1}$$

Construction of the energy cycle in this case includes a positive value for electron affinity (Fig. 8.3).

$$\text{Lattice energy} = -(601.7 + 146 + 736.0 + 1448.0 + 249.2 + 649.4)$$
$$= -3830.3 \text{ kJ mol}^{-1}$$

This very high lattice energy reflects the great stability of magnesium oxide (page 239).

Lattice energies can also be calculated theoretically. If the ions are regarded as perfect spheres the normal equation for calculating the electrostatic forces can be applied. This requires knowledge of the charges on the ions (which depend on the compounds used) and measuring the distance between the ions (which can be done by X-ray analysis). The forces depend on the magnitude of the charges, the distance between them and the permittivity of the medium (in this case, vacuum).

For a completely ionic crystal the experimentally determined lattice energy agrees closely with the theoretically calculated value; any great discrepancy between the two figures can be taken as evidence that the crystal is, at least to some extent, covalent. Some theoretically and experimentally determined values of lattice energies are given in Table 8.2.

Table 8.2

Compound	Born–Haber cycle experimental value for lattice energy kJ mol^{-1}	Electrostatically calculated theoretical value for lattice energy kJ mol^{-1}
Sodium chloride Na$^+$Cl$^-$	776.4	766.1
Potassium iodide K$^+$I$^-$	631.8	630.9
Silver chloride AgCl	916	768.5
Zinc sulphide ZnS	3615	3427

The agreement for the two alkali metal salts is close, as would be expected for two such typically ionic salts. The values found theoretically and experimentally for the lattice energies of silver chloride and zinc sulphide differ quite markedly, and this is consistent with the covalent bonding of these compounds (pages 309 and 335).

The enthalpy of solution of a wholly ionic compound depends on the difference between the lattice energy of the crystal and the *hydration energy* of the ions. The latter is the energy released when one mole of gaseous ions are hydrated. For soluble salts it is of approximately the same magnitude as the lattice energy, so the energy required for breaking the lattice is balanced by the energy released in hydrating the ions.

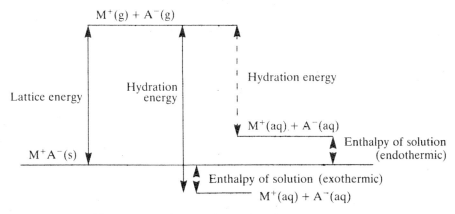

Fig. 8.4 Enthalpy of solution for a metal salt

If the hydration energy is greater than the lattice energy then ΔH is negative and the dissolution is exothermic; if the lattice energy is the greater then ΔH is positive and the dissolution is endothermic.

8.5   Enthalpies of neutralisation

The enthalpy of neutralisation of an acid by a base is defined as the enthalpy change when one mole of hydrogen ions from an acid is neutralised by one mole of hydroxyl ions from a base in dilute aqueous solution. For the neutralisation of strong acids by strong bases this figure is remarkably constant:

$$NaOH(aq) + HCl(aq) \rightarrow NaCl(aq) + H_2O(l)$$

$$\Delta H^{\ominus}_{298} = -57.36 \text{ kJ mol}^{-1}$$

$$KOH(aq) + HCl(aq) \rightarrow KCl(aq) + H_2O(l)$$

$$\Delta H^{\ominus}_{298} = -57.33 \text{ kJ mol}^{-1}$$

$$NaOH(aq) + HNO_3(aq) \rightarrow NaNO_3(aq) + H_2O(l)$$

$$\Delta H^{\ominus}_{298} = -57.33 \text{ kJ mol}^{-1}$$

$$\tfrac{1}{2}Ca(OH)_2(aq) + HNO_3(aq) \rightarrow \tfrac{1}{2}Ca(NO_3)_2(aq) + H_2O(l)$$

$$\Delta H^{\ominus}_{298} = -57.36 \text{ kJ mol}^{-1}$$

If it is assumed that all these acids and bases are fully ionised in dilute solution, each equation can be written ionically:

$$H^+(aq) + OH^-(aq) \rightarrow H_2O(l)$$

which is simply the equation for the formation of liquid water from aqueous ions; this explains the constancy of the values for enthalpy of neutralisation.

The enthalpy of neutralisation of a weak acid by a strong base is usually somewhat lower.

$$CH_3COOH(aq) + NaOH(aq) \rightarrow CH_3COONa(aq) + H_2O(l)$$
$$\Delta H^\ominus_{298} = -56.1 \text{ kJ mol}^{-1}$$

This is because weak acids are not fully ionised even in dilute solution and therefore some energy is absorbed in the ionisation of the acid as it reacts:

$$CH_3COOH(aq) \rightleftharpoons CH_3COO^-(aq) + H^+(aq) \quad \Delta H \text{ positive}$$

A similar situation leads to a lower enthalpy of neutralisation for weak bases neutralised by strong acids. The evolution of heat when a weak acid is neutralised by a weak base is quite small.

8.6   Calorimetry

Enthalpies of combustion of elements or compounds can be measured by the use of a bomb calorimeter. This is a cylindrical steel 'bomb' strong enough to withstand high pressure internally (Fig. 8.5).

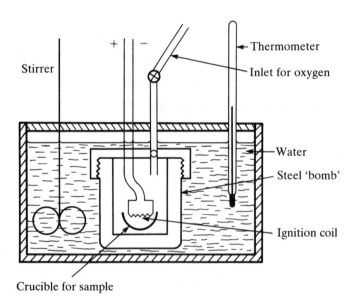

Fig. 8.5   Bomb calorimeter

The weighed sample is placed in a crucible in the bomb, which is closed with a tightly fitting screw lid. Oxygen is passed in at high pressure and the prepared bomb is placed in a calorimeter containing a measured quantity of water. The sample is ignited by passing a current through the ignition wire and the rise in temperature of the water is noted on a sensitive thermometer. The heat evolved warms the water. 4.18 J raises the temperature of 1 gram of water by 1°C so the amount of heat evolved is given by:

The mass (W) of water × rise in temperature (t) × 4.18 J.

This is evolved by m grams of sample, so the molar enthalpy of combustion of the substance is given by

$$\frac{W \times t \times 4.18 \times \text{mass of one mole of substance}}{1000 \times m} \text{ kJ}$$

In practice many corrections must be made, including those for the specific heat capacity of the calorimeter, bomb, stirrer and thermometer. Also a correction has to be made because measurements recorded at constant volume give only the internal energy change. Normally reactions are carried out in open vessels at constant pressure, which means that the recorded energy change is reduced by the amount of work that the system has to do against atmospheric pressure if there is a change of volume. In most cases the difference is quite small; where there is no change of volume then the internal energy change is equal to the enthalpy change.

Enthalpies of combustion of alcohols can be measured in the laboratory by burning the alcohol in a spirit lamp as shown in Fig. 8.6. The mass of alcohol burned is measured by weighing the spirit lamp before and after combustion. The heat evolved is calculated from the rise of temperature in the calorimeter, exactly as for the bomb calorimeter, and once again the specific heat capacities of calorimeter, stirrer and thermometer must be taken into account. The apparatus can be adapted to measure the enthalpies

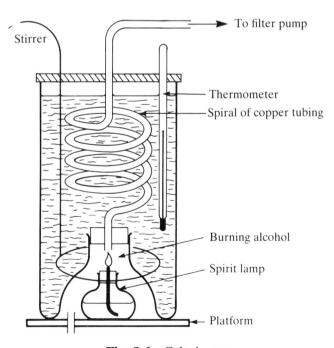

Fig. 8.6   Calorimeter

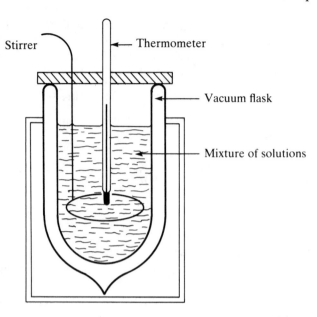

Fig. 8.7   Calorimeter for determining enthalpies of neutralisation

of combustion of gases which can be burned at a jet. In this case the volume burned can be measured by a meter. A major source of error in either case is the incomplete combustion of the alcohol or gas, and care must be taken to make sure that the flame is perfectly non-luminous. Even so, some incomplete combustion yielding carbon monoxide rather than carbon dioxide is bound to occur.

To measure enthalpies of neutralisation the apparatus shown in Fig. 8.7 is used. The accuracy of the results obtained is increased if reasonably large volumes of solutions are used. Convenient quantities are 500 cm³ of 2 M solutions of both acid and base, so that the heat recorded will be that for 1 mole of reactants. The temperature rise is noted and the heat evolved calculated as before from the expression

$$\frac{W \times t \times 4.18 \times \text{mass of one mole}}{1000 \times \text{mass of sample}} \text{ kJ}$$

If it is assumed that 1000 cm³ of solution has the same heat capacity as 1000 g of water (W) and one mole of reactants (sample) is used, this simplifies to $4.18 \times t$ kJ, where t is the rise in temperature. The specific heat capacities of the stirrer, thermometer and flask must also be taken into account.

In all thermochemical measurements it is the usual practice to construct a cooling curve to compensate for loss of heat, even by heavily insulated calorimeters. The temperature of the reactants is recorded for a period of time before the experiment begins. The reactants are then mixed, or the sample is ignited, and the temperature recorded for several minutes after the reaction has taken place. The cooling curve is then extrapolated back to the time of the reaction and the corrected temperature rise read from the graph.

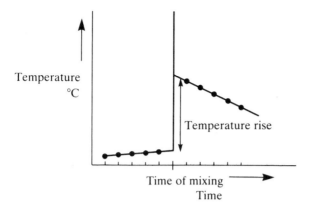

Fig. 8.8 Cooling Curve

8.7 Entropy and free energy

In general, given the correct conditions, reactions that proceed spontaneously are those which are exothermic. Thus the enthalpy of the system decreases. However, a number of endothermic processes (in which heat is absorbed) take place spontaneously under suitable conditions. For example, salts such as ammonium nitrate dissolve readily with the absorption of heat. Also, the formation of water gas, the combination of carbon and sulphur to make carbon disulphide and the decomposition of dinitrogen tetraoxide are endothermic.

$$C(s) + H_2O(g) \rightleftharpoons CO(g) + H_2(g) \qquad \Delta H^{\ominus}_{298} = +181 \text{ kJ mol}^{-1}$$

$$C(s) + 2S(s) \quad \rightleftharpoons CS_2(l) \qquad \Delta H^{\ominus}_{298} = +19 \text{ kJ mol}^{-1}$$

$$N_2O_4(g) \quad \rightleftharpoons 2NO_2(g) \qquad \Delta H^{\ominus}_{298} = +57 \text{ kJ mol}^{-1}$$

From these examples it is clear that some other factor besides the change of enthalpy must be considered before the course of the reaction can be predicted.

Consider two gases which do not react with each other; if these are placed in adjacent compartments of a container at the same temperature and pressure, and the partition wall is removed, they will mingle by diffusion until they are completely mixed. No reaction has taken place, no heat has been evolved or absorbed, yet the system has undergone a change that is not spontaneously reversible. The system has become less ordered or more random. We say that the *entropy* of the system has increased, and further, that increase of entropy will result from a spontaneous change.

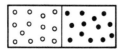

(a) Ordered state with
    each gas separate

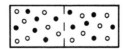

(b) Gases mixed to
    give a less ordered state

Fig. 8.9 Order and disorder

The study of entropy is part of rather advanced thermodynamics, but a qualitative pictorial consideration of the concept can be useful. Since we are considering entropy as indicative of the degree of disorder of a system, it follows that at absolute zero (0 K) a crystalline solid will be in a state of perfect order or zero entropy. With increasing temperature entropy increases as the atoms, ions or molecules become increasingly mobile and disordered on passing from solid to liquid to gas. At 298 K and atmospheric pressure each substance has an entropy value known as its molar entropy, S_m, which is measured in joules per mole per kelvin, $J\,mol^{-1}\,K^{-1}$. These values are listed in data books, and can be equated in a similar way to enthalpies of reaction. For example, the standard molar entropies ($S^{\ominus}_{298}$) for ethene, hydrogen and ethane are 219.5, 130.6 and 229.5 $J\,mol^{-1}\,K^{-1}$ respectively. Hence the change of entropy for the hydrogenation of ethene to ethane is $229.5 - (219.5 + 130.6)$, or $-120.6\,J$. That is:

$$C_2H_4(g) + H_2(g) \rightarrow C_2H_6(g) \qquad \Delta S^{\ominus}_{298} = -120.6\,J\,mol^{-1}\,K^{-1}$$

The process of vaporisation at boiling point (the change of liquid molecules to more disordered gaseous molecules), involves an increase in entropy. If the molar enthalpies of vaporisation of liquids are plotted against their boiling points, a linear relationship is shown (provided the liquids do not have any extensive intermolecular bonding).

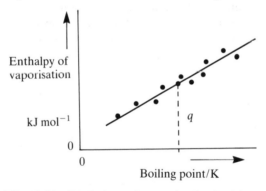

Fig. 8.10 Enthalpy of vaporisation/boiling point curve

The slope of this line is $\dfrac{q\,kJ\,mol^{-1}}{T\,K}$, where q is the quantity of heat absorbed or evolved in the entropy change at T K. Hence we can say that the change of entropy on vaporisation, ΔS, at constant temperature and pressure, is given by $\dfrac{q}{T}\,kJ\,mol^{-1}\,K^{-1}$. Hence $\Delta S = \dfrac{q}{T}$, or $q = T\Delta S$.

It is found in practice and can be shown thermodynamically that only part of the heat evolved during a reaction can be turned into useful work. A distinction is therefore made between the enthalpy, H, of a substance and its free energy G. As with enthalpy values, absolute values for standard free energies of individual substances cannot be measured, but there are a number of methods by which the free energy change, ΔG, for a specific reaction may be determined. This free energy change is defined as the amount of energy derived from the reaction that can be used to do work; it can be regarded as the difference between the free energies of the products and of the reactants. The change of enthalpy during a reaction and the change of free energy are given by the equations

and
$$\Delta H = H_{\text{products}} - H_{\text{reactants}}$$

$$\Delta G = G_{\text{products}} - G_{\text{reactants}}$$

There may be many factors contributing to the difference between the change of enthalpy and the change of free energy during a reaction, but for a process taking place at constant temperature and pressure (as in an open flask or test-tube) the difference between ΔG and ΔH will be the amount of energy absorbed or evolved in the change of entropy, or

$$\Delta H - \Delta G = T\Delta S$$

or
$$\Delta G = \Delta H - T\Delta S$$

The conditions for change to take place, therefore, are that there should be a decrease of free energy and an increase of total entropy. Hence ΔG must be negative if a reaction is to take place, that is,

$$\Delta H - T\Delta S < 0.$$

The equation $\Delta G = \Delta H - T\Delta S$ can be used to predict the probability of a chemical change, but it does not give any indication of the rate of the probable reaction, which may be so slow as to be negligible. Also, we have been considering the entropy change of a totally isolated system, which is in fact an entirely theoretical concept, since the entropy change of the surroundings must also be a factor. However, the equation is a useful guide, as can be seen from the following examples:

(i)   Reactions that are both exothermic and result in an increase of entropy take place readily, since all factors lead to a decrease of free energy. For example, the burning of carbon:

$$C(s) + O_2(g) \rightarrow CO_2(g) \begin{cases} \Delta H = -393.5 \text{ kJ mol}^{-1} \\ \Delta S = +3.4 \text{ J mol}^{-1} \text{ K}^{-1} \end{cases}$$

and
$$\Delta H - T\Delta S < 0$$

(ii) Reactions that are exothermic but result in a decrease of entropy take place provided that the entropy change does not make ΔG positive. For example, the hydrogenation of ethene to ethane:

$$C_2H_4(g) + H_2(g) \rightarrow C_2H_6(g) \quad \begin{matrix} \Delta H = -136 \text{ kJ mol}^{-1} \\ \Delta S = -120.6 \text{ J mol}^{-1} \text{ K}^{-1} \end{matrix}$$

Here, even at room temperature

$$\Delta H - T\Delta S = -136 + (298 \times 120.6 \times 10^{-3}) \quad \text{or} \quad -100 \text{ kJ}$$

The reaction is therefore feasible and takes place readily under conditions of moderate temperature and pressure.

(iiii) Reactions that are endothermic, and therefore not favoured energetically, take place if the gain in entropy is sufficient to make $T\Delta S$ larger than ΔH. This may not occur under normal conditions, but the reaction may take place at raised temperature. For example, the reaction between steam and coke is endothermic:

$$C(s) + H_2O(g) \rightarrow CO(g) + H_2(g) \qquad \Delta H = +131 \text{ kJ mol}^{-1}$$

However the gain in entropy resulting from the reaction is considerable

$$C(s) + H_2O(g) \rightarrow CO(g) + H_2(g) \qquad \Delta S = +135 \text{ J mol}^{-1}\text{K}^{-1}$$

and at high temperature the reaction proceeds. For example, at 1500 K

$$\Delta H - T\Delta S = 131 - (1500 \times 135 \times 10^{-3}) \quad \text{or} \quad -61.5 \text{ kJ}$$

This is of great importance in the production of water gas (page 260).

It must be emphasised, however, that in the examples above little has been said about the total entropy, which includes that of the surroundings; it is the increase in total entropy that leads to change:

$$S_{TOTAL} = S_{SURROUNDINGS} + S_{SYSTEM}$$

A proper appreciation of entropy and its meaning can only be gained by first studying thermodynamics.

Table 8.3 Bond energy terms

Bond	Bond energy kJ mol^{-1}
C—C	346
C=C	611
C≡C	835
C—H	415
C—O	358
C=O	803
O—H	463
N—H	389
F—H	554
Cl—H	431
O=O	497

Questions

1. Define the terms *enthalpy change of reaction* and *enthalpy of neutralisation*.

   Explain why the quantity of heat evolved when one mole of hydrochloric acid, HCl, is neutralised by one mole of sodium hydroxide, NaOH, is the same as the quantity of heat evolved when one mole of nitric acid, HNO$_3$, is used instead of the hydrochloric acid.   (AEB 1977)

2. (a) Outline the Born-Haber cycle for the formation of sodium chloride from its elements, giving, for each step, an equation and the name and sign of the energy change. Show how the lattice energy may be calculated from the cycle.

(b) Describe and discuss the crystal structures of sodium chloride and of caesium chloride. Indicate briefly how these structures were determined.

(JMB)

3. (a) State Hess's law and explain what is meant by the *enthalpy change* of a reaction.

(b) Calculate the *enthalpy change* of the reaction

$$CO(g) + 2H_2(g) \rightarrow CH_3OH(l)$$

at 298 K, from the following data at 298 K:

$$CO(g) + \tfrac{1}{2}O_2(g) \rightarrow CO_2(g); \quad \Delta H = -283.0 \text{ kJ mol}^{-1};$$
$$H_2(g) + \tfrac{1}{2}O_2(g) \rightarrow H_2O(l); \quad \Delta H = -285.8 \text{ kJ mol}^{-1};$$
$$CH_3OH(l) + 1\tfrac{1}{2}O_2(g) \rightarrow CO_2(g) + 2H_2O(l); \quad \Delta H = -715.0 \text{ kJ mol}^{-1}.$$

(O)

4. The following table lists the enthalpies of combustion at 298 K of several monohydride alcohols.

Alcohol	$\Delta H^{\ominus}_{c298}$ (kJ mol^{-1})
Methanol	−715
Ethanol	−1367
Propan-1-ol	−2017
Butan-1-ol	−2675

(a) Draw a plot of $\Delta H^{\ominus}_{c298}$ against the relative molecular mass of the alcohols.

(b) From your graph estimate a value for the enthalpy of combustion of pentan-1-ol.

(c) Given $\Delta H^{\ominus}_{f298}(H_2O) = -286 \text{ kJ mol}^{-1}$ and $\Delta H^{\ominus}_{f298}(CO_2) = -394 \text{ kJ mol}^{-1}$, calculate the standard enthalpy of formation at 298 K of ethanol.

(d) The enthalpy of combustion at 298 K of ethane-1,2-diol is −1180 kJ mol^{-1}. What would you expect the corresponding value for propane-1,3-diol to be?

(e) Write down the balanced equation for the combustion of
  (A) methane gas,
  (B) methanol vapour

(ii) Briefly compare and contrast the entropy changes associated with the two reactions.

(L)

5. Explain what is meant by the term 'entropy'. Briefly discuss the significance of entropy in the context of chemistry.

Each of the following reactions is accompanied by a large entropy change, the direction of which is indicated. Examine each reaction and suggest reasons for the entropy change.

(a) $PCl_5(s) \rightarrow PCl_3 + Cl_2(g)$ $\quad \Delta S$ is positive

(b) $Li^+(g) + aq \rightarrow Li^+(aq)$ $\quad \Delta S$ is negative

(c) $KNO_3(s) + aq \rightarrow K^+(aq) + NO_3^-(aq)$ $\quad \Delta S$ is positive

(L)

9 Rates of reaction

9.1 Introduction

You will have noticed by now that some reactions take place much faster than others. Ethanol is rapidly converted to chloroethane by the action of phosphorus pentachloride, but is only slowly oxidised to ethanoic acid by refluxing with potassium dichromate at a raised temperature (Chapter 25). Knowledge of the rate at which a reaction proceeds is vital to manufacturers designing new processes, and is taken into account by chemical engineers designing plant for the production of all kinds of products such as paints, plastics, synthetic fibres or medicines. Rates of reaction are also studied by workers researching into how and why reactions take place.

9.2 Factors affecting rates of reaction

Since the particles—molecules or ions—must collide to react, there are several factors that directly affect reaction rate.

(a) Temperature

A rise in temperature greatly increases the rate of reaction. The average kinetic energy of the particles is increased; a higher proportion of them have an energy greater than the activation energy required for reaction, and the number of effective collisions is higher.

(b) Concentration

Increased concentration of the reactants increases the rate of reaction because collisions between the particles are more frequent.

(c) Pressure

An increase in pressure increases the rate of reaction between gases because the molecules are brought closer together and collide more frequently. This is, in effect, an increase in concentration.

(d) State of reactants

In reactions involving solids the rate of reaction is increased if the solid is finely divided, thus exposing a greater surface to attack. Fine iron powder will dissolve in cold dilute hydrochloric acid while coarse iron filings do not dissolve until the acid is heated.

(e) Catalysts

Catalysts can dramatically increase the rate of a chemical reaction. They are defined as substances which will alter the rate of a chemical reaction while remaining chemically unchanged at the end of the reaction. Only a small amount of a catalyst is required, and it can affect the rate of reaction for a long time. Catalysts are specific in their action; a catalyst that increases the rate of one reaction may have no effect on another.

The mechanism of catalysis is not fully understood, but there appear to be two main types. Some catalysts increase the rate of a reaction by the formation of intermediate compounds with the regeneration of the catalyst at the end of the reaction. Thus if A + B react slowly to give D, the presence of a catalyst C might increase the rate by allowing the reaction to take place in two fast stages:

$$\left.\begin{array}{l} A+C \rightarrow AC \\ AC+B \rightarrow D+C \end{array}\right\} \text{ fast reactions}$$

The catalyst C is continuously regenerated and so only a small amount is needed to speed up the reaction. The reactants in this type of reaction are usually all in the same phase and this is *homogeneous* catalysis. The speeding up of the thermal decomposition of potassium chlorate by the addition of manganese dioxide is thought to be an example of this.

In the other form of catalysis the reactants and catalyst are not in the same phase; for example, the rate of reaction between gases is often catalysed by finely divided solids. This is *heterogeneous* catalysis. The molecules of the reacting gases are adsorbed on to the surface of the catalyst, bringing them into closer contact with one another and making them more reactive by weakening the bonds within the molecules. This effect is observed in the reaction between nitrogen and hydrogen to make ammonia, for which the catalyst is finely divided iron (page 475).

$$N_2(g) + 3H_2(g) \rightleftharpoons 2NH_3(g)$$

9.3 Rate equations

The effect of the concentration of the reactants was first studied by Guldberg and Waage in 1864, for simple one-step reactions. They formulated a law of mass action which stated that the rate of a reaction, at constant temperature, was proportional to the product of the active masses of the reacting substances. Modern rate laws have been developed from their work. For simplicity, the active mass of a reactant A can be taken as the concentration of A in mole dm^{-3} and is written [A]. Thus for the simple reaction:

$$A+B \rightarrow C$$

the rate of reaction, at constant temperature, is proportional to the molar concentration of A multiplied by the molar concentration of B:

$$\text{Rate of reaction} \propto [A] \times [B]$$

or

$$\text{Rate of reaction} = k[A][B]$$

This is the *rate equation* or *rate expression*, and k is called the *velocity constant* or *rate constant* for the reaction.

9.4   Order of reaction

The way in which the rate of a reaction depends on the concentration of the reactants is expressed in terms of the *orders* of the reaction. For the reaction:

$$A + B \rightarrow C$$

the rate may be found to depend simply on the concentration of A, and not on the concentration of B at all. In this case:

$$\text{Rate of reaction} = k[A]$$

and we say that the reaction is first order with respect to A, zero order with respect to B, and first order overall.

In another case it may be found that the rate depends on the concentration of both A and B, so that:

$$\text{Rate of reaction} = k[A][B]$$

and the reaction is first order with respect to A, first order with respect to B and second order overall.

In a more complicated case the rate of reaction may be found to depend on the concentration of A and the square of the concentration of B. Then the rate expression becomes:

$$\text{Rate of reaction} = k[A][B]^2$$

and the reaction is first order with respect to A, second order with respect to B, and third order overall.

So the overall order of a reaction is the sum of the indices in the rate expression. All this has to be found experimentally because the orders of a reaction do not necessarily coincide with the figures used in a balanced equation. For example, potassium peroxodisulphate reacts with potassium iodide to form iodine. Ionically the equation is:

$$S_2O_8{}^{2-} + 2I^- \rightarrow I_2 + 2SO_4{}^{2-}$$

The rate equation is found to be:

$$\text{Rate of reaction} = k[S_2O_8{}^{2-}][I^-]$$

So the reaction is first order with respect to both peroxodisulphate and iodide ions, although from the equation you might think that it was second order (involving two ions) for the iodide. Determination of the orders of a reaction can be used to find out *how* it takes place.

9.5   Finding the rate of reaction

The rate of a reaction at a given moment is usually expressed as the amount of reaction taking place in unit time, and to find it the change of concentration of one component of the reaction mixture is measured. Such characteristics as change of colour, volume of gas evolved, or change of pressure may be used. The rate found should be quoted in terms of the change used. For example: 'Rate of reaction with respect to concentration of iodine' or 'Rate of reaction with respect to oxygen evolved.'

Consider the decomposition of hydrogen peroxide, catalysed by manganese dioxide (manganese(IV) oxide):

$$2H_2O_2(aq) \rightarrow 2H_2O(l) + O_2(g)$$

The oxygen evolved can be collected in a syringe and its volume measured at set time intervals (say fifteen seconds). A graph of the results can be plotted.

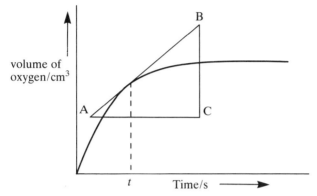

Fig. 9.1 Rate of decomposition of hydrogen peroxide with respect to oxygen evolved

When the concentration of hydrogen peroxide is high, at the start, the reaction is fast and the slope steep. As the concentration falls the reaction slows down, and the curve flattens out when all the hydrogen peroxide is decomposed. The rate, at a particular time (t), is given by the slope of the curve, and this can be found by drawing a tangent AB to the curve at t.

$$\text{Slope} = \frac{BC}{AC} = \frac{\text{volume of oxygen collected}}{\text{time}}$$

$$= \text{Rate of decomposition of } H_2O_2$$

The units in this case are $cm^3 s^{-1}$ of oxygen evolved. The volume of oxygen evolved is proportional to the amount of hydrogen peroxide decomposed, and the rate could have been determined by plotting the concentration of hydrogen peroxide remaining undecomposed.

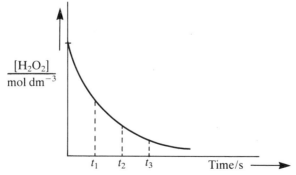

Fig. 9.2 Rate of reaction with respect to concentration of hydrogen peroxide

Since the concentration of hydrogen peroxide is decreasing the curve will be of the same shape as that plotted for the evolution of oxygen, but inverted; the units are now mol $dm^{-3}s^{-1}$.

9.6 Finding the order of reaction

To find the order of a reaction several successive rates are measured by drawing tangents at times $t_1, t_2, t_3 \cdots$ as in Fig. 9.3, These rates are plotted against the concentration of hydrogen peroxide at the times $t_1, t_2, t_3 \cdots$. If a straight line results, then the reaction is first order.

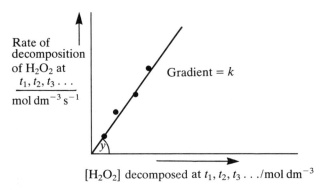

Fig. 9.3   First-order reaction

For second order reactions, where the rate depends on the concentrations of two reacting molecules, then, when only one species is involved, a straight line is obtained when the rates at $t_1, t_2, t_3 \cdots$ are plotted against the squares of the concentrations at these times. For example, the rate of decomposition of hydrogen iodide into hydrogen and iodine depends on $[HI]^2$.

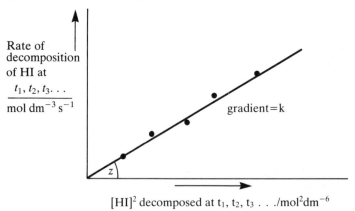

Fig. 9.4   Second-order reaction

In practice, rather than finding the rate at different points during the course of a single experiment, a series of experiments may be carried out using a different concentration each time and measuring the initial rate of reaction.

The examples given above deal with reactions involving only one reactant.

Where there is more than one reactant, the rate for each in turn must be measured.

To find the order of a reaction such as the hydrolysis of 1-bromobutane in the presence of potassium hydroxide:

$$C_4H_9Br + KOH \rightarrow C_4H_9OH + KBr$$

two sets of experiments have to be performed. The rate of reaction is measured several times using different initial concentrations of bromobutane while keeping the concentration of potassium hydroxide constant. Plotting the initial rates of reaction found against the concentrations of bromobutane used gives the order of reaction with respect to bromobutane. A second set of measurements is then made keeping the bromobutane concentration constant while varying the concentration of potassium hydroxide, thus determining the order of the reaction with respect to potassium hydroxide. The overall order will be the sum of the two orders found.

These methods can be applied to simple first and second order reactions. In many actual cases the overall rate equation is more complicated and may even involve fractional indices; these are of great practical importance in industrial chemistry but need not be considered at this stage.

9.7   Finding the velocity constant

When the rate of reaction is known, k, the velocity constant, can be calculated by substituting the known values in the rate equation. Thus, for a first-order reaction such as the decomposition of hydrogen peroxide

$$Rate = k[H_2O_2]$$

or

$$k = \frac{rate}{[H_2O_2]}$$

Since the rate is measured in $mol\,dm^{-3}s^{-1}$ and the concentration in $mol\,dm^{-3}$, the units for k are s^{-1} for first order reactions.

For a second-order reaction such as the decomposition of hydrogen iodide the units of k are $dm^3mol^{-1}s^{-1}$, since

$$k = \frac{rate}{[HI]^2}$$

The velocity constants are also given by the gradients of the lines obtained by plotting rates against concentrations, as in Fig. 9.3 and Fig. 9.4.

A more direct method of finding the velocity constant is to express the rate of reaction as $\frac{dx}{dt}$ where

$$x = \text{moles of reactant changed at time } t$$
$$a = \text{initial concentration of reactant}$$
$$(a - x) = \text{concentration at time } t. \text{ Then}$$

$$\text{rate of reaction} = \frac{dx}{dt} = k(a - x)$$

or
$$\frac{dx}{(a-x)} = k\,dt$$

Integration of this expression gives

$$-\ln(a-x) = kt + \text{constant} \qquad (1)$$

When $t=0$ then $x=0$ and

$$-\ln a = \text{constant}$$

substituting in equation (1)

$$-\ln(a-x) = kt - \ln a$$

or

$$\ln\frac{a}{(a-x)} = kt$$

Hence the velocity constant k may be found by plotting values of $\ln\dfrac{a}{(a-x)}$ against time, as in Fig. 9.5.

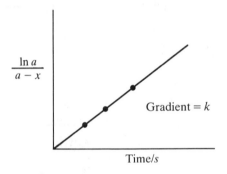

Fig. 9.5 Velocity constant for a first order reaction

For a second-order reaction the rate depends on the square of the concentration with a single reactant, or on the product of the concentrations with two reactants. Assuming the simplest case where the concentrations of two reactants are equal, then

$$\frac{dx}{dt} = k(a-x)^2$$

or

$$\frac{dx}{(a-x)^2} = k\,dt$$

on integrating this becomes

$$\frac{1}{(a-x)} = kt + \text{constant}$$

when $t=0$ then $x=0$, and the constant is $\dfrac{1}{a}$, giving

$$\frac{1}{(a-x)} = kt + \frac{1}{a}, \quad \text{or} \quad \frac{x}{a(a-x)} = kt$$

Thus the velocity constant for a second order reaction is given by the slope of the line obtained by plotting $\dfrac{x}{a(a-x)}$ against time:

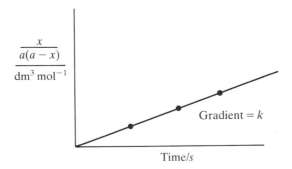

Fig. 9.6   Velocity constant for a second-order reaction

9.8   Half-life and order of reaction

Another way of determining the order of a reaction is to find the half-life time. The half-life of a reaction is the time taken for the initial concentration of a reactant to fall to half its value. For a first-order reaction the half-life is independent of the initial concentration. Thus the time taken for the concentration of the reactant to fall to half its original value is the same as the time taken to fall from a half to a quarter of its value, and from a quarter to an eighth, and so on.

The reaction between bromine and propanone is a first-order reaction and can be followed by measuring the falling concentration of bromine. A curve as shown in Fig. 9.7 will be obtained by plotting the concentration of bromine against time.

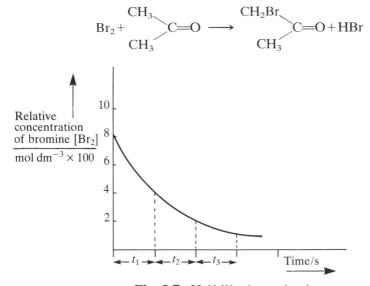

Fig. 9.7   Half-life determination

Since the reaction is first-order then $t_1 = t_2 = t_3$ etc. This is of great importance in radiochemistry since all radioactive decay follows first-order kinetics (page 35).

The half-life is readily calculated from the equation obtained in the determination of the velocity constant. At the half-life time, $t_{\frac{1}{2}}$, the initial concentration has fallen to half its value and $(a - x) = \dfrac{a}{2}$. The equation $\ln \dfrac{a}{(a-x)} = kt$ then becomes

$$\ln 2 = kt_{\frac{1}{2}}, \quad \text{or} \quad t_{\frac{1}{2}} = \frac{\ln 2}{k}$$

For a second-order reaction between equal concentrations of reactants the time for the concentration to fall to half the original value does depend on the initial concentration and is inversely proportional to it. The time taken can be calculated from the equation

$$\frac{x}{a(a-x)} = kt$$

If the initial concentration is halved then the concentration x at this time is $\dfrac{a}{2}$ and $t_{\frac{1}{2}} = \dfrac{1}{ka}$.

Thus first- and second-order reactions can be distinguished by considering their half-life times.

9.9 Rate-determining step and the mechanism of a reaction

If you think about the reaction between bromine and propanone it may seem a little odd that it is first order since there are two reactants. The explanation may be that the rate depends upon the concentration of only one of the reactants; experiments show that the rate is proportional to the concentration of propanone but is independent of the concentration of bromine. This leads to the idea that the reaction takes place in stages: a slow stage, or bottle-neck, involving propanone only, followed by fast reactions involving bromine. The slow stage controls the rate of reaction and is called the *rate-determining* step.

Possibly the propanone molecule undergoes a structural change from a *ketonic* form to an *enolic* form:

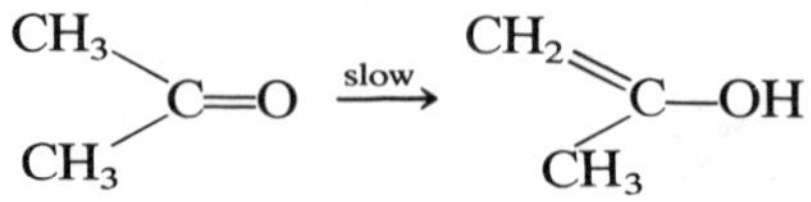

followed by a rapid reaction with bromine and a return to the *ketonic* form:

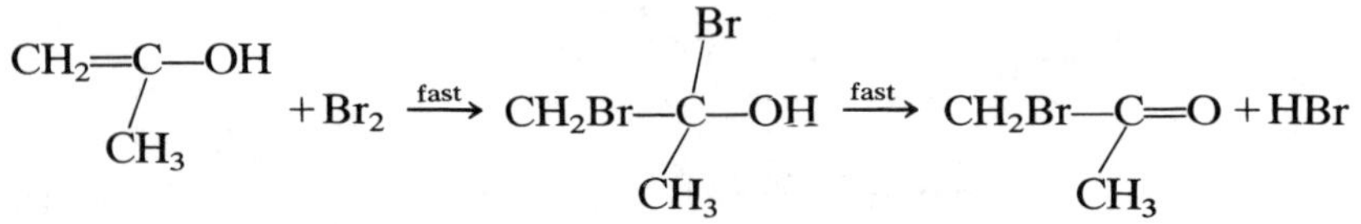

Reactions 2 and 3 are so rapid that they do not affect the overall rate at all.

Study of the thermal decomposition of hydrogen iodide and dinitrogen pentoxide provides further examples of how knowledge of the rate and order of a reaction can be used to determine the mechanism of a reaction.

The decomposition of hydrogen iodide

$$2HI \rightarrow H_2 + I_2$$

has been found to be second order; the mechanism is thought to be a slow reaction involving the collision of two hydrogen iodide molecules

$$(1) \quad HI + HI \xrightarrow{\text{slow}} \begin{array}{c} H\!-\!I \\ \vdots \\ H\!-\!I \end{array}$$

followed by two fast reactions:

$$(2) \quad \begin{array}{c} H\!-\!I \\ \vdots \\ H\!-\!I \end{array} \xrightarrow{\text{fast}} H_2 + 2I$$

$$(3) \quad 2I \xrightarrow{\text{fast}} I_2$$

Since reaction (1) is slow it is the rate-determining step.

In contrast the decomposition of dinitrogen pentoxide

$$2N_2O_5 \rightarrow 4NO_2 + O_2$$

is found to be first order, indicating that the rate-determining step involves one molecule only. The steps suggested are:

$$(1) \quad N_2O_5 \xrightarrow{\text{slow}} NO_2 + NO_3$$

followed by fast reactions summarised as:

$$(2) \quad 2NO_2 + 2NO_3 \xrightarrow{\text{fast}} 4NO_2 + O_2$$

These are simple cases, but sufficient to indicate how the study of rates of reaction can be used to find out how reactions take place.

9.10 Molecularity

The molecularity of a reaction is given by the number of molecules taking part in the rate-determining step, and must be a whole number. Thus, in the simple cases considered above, the decomposition of hydrogen iodide is bimolecular and the decomposition of dinitrogen pentoxide is unimolecular. In more complex situations the order is not necessarily the same as the molecularity.

9.11   Activation energy

Kinetic studies of reactions in the gaseous phase show that there is an 'energy barrier' to be overcome before reaction takes place. A given sample of gas at a fixed temperature will contain not only molecules with average kinetic energy but some with very low energy, and also a proportion with much higher than average energy. It can be shown that only those molecules having energy above a certain minimum value will react on collision; this minimum value is known as the activation energy. Thus the reaction paths for exothermic and endothermic reactions may be represented by the diagrams in Fig. 9.8.

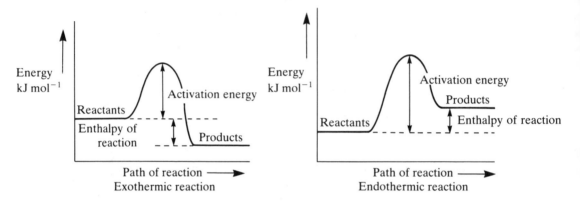

Fig. 9.8   Activation energy

For reactions in the gaseous phase the number of molecules having energy higher than the activation energy E can be derived from Maxwell's distribution of velocity law (page 75), expressed in a simplified form as:

$$n = n_0 e^{-E/RT}$$

where E = the activation energy
R = the gas constant
T = temperature $/K$
n_0 = total number of molecules
n = number of molecules with energy greater than E

If reactions take place only between molecules having an energy greater than E then both the rate of reaction and the velocity constant will be proportional to n at a given concentration, hence

$$\text{Rate of reaction} \propto k \propto A e^{-E/RT}$$

where A is a constant. n_0 is fixed for a given concentration.

Then on taking logarithms a relationship between the velocity constant k and the activation energy E can be established:

$$\log k = C - \frac{E}{2.3R} \times \frac{1}{T}$$

where C is a constant and $R = 8.3 \, \text{JK}^{-1}\text{mol}^{-1}$.

This can be used to find activation energies. The rate of reaction k is measured at various temperatures and $\log k$ plotted against $\dfrac{1}{T}$. The slope of the line will be given by $\dfrac{E}{2.3R}$. Thus if the angle of the slope is θ then $\tan \theta = \dfrac{-E}{2.3R}$ and $E = -2.3R\tan\theta$ J mol^{-1}.

The effect of some catalysts may well be that, by forming intermediate compounds, they reduce the activation energy necessary for the overall reaction.

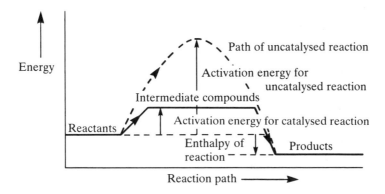

Fig. 9.9 Activation energies for catalysed and uncatalysed reactions

Less activation energy may be required for the two stages of the reaction involving the catalyst than is required if the reaction takes place without the catalyst.

9.12 Methods of following the course of a reaction

(a) Titration

The course of a reaction such as the hydrolysis of an ester can be followed by estimating the amount of acid formed:

$$CH_3COOC_2H_5 + H_2O \rightleftharpoons CH_3COOH + C_2H_5OH$$

The reaction mixture is made up with known volumes of ester and water and a known concentration of a mineral acid as catalyst to increase the rate of reaction. Small samples of known volume are withdrawn at set intervals and diluted to prevent further reaction. These are titrated with sodium hydroxide solution of known molarity to determine the concentration of ethanoic acid. The volume of sodium hydroxide solution used to neutralise the mineral acid present has to be deducted each time. In other experiments, iodine concentrations can be found by titration with standard sodium thiosulphate solution, or potassium manganate(VII) concentrations with ethanedioic acid.

(b) Clock reactions

The reaction between potassium peroxodisulphate(VI) and potassium iodide produces iodine quite slowly:

$$S_2O_8{}^{2-} + 2I^- \rightarrow I_2 + 2SO_4{}^{2-}$$

The iodine produced can be detected by adding some starch solution which will turn blue. To measure the rate at which the iodine is produced, mixtures of potassium peroxodisulphate of varying concentrations are made up containing starch solution and equal portions of a standard solution of sodium thiosulphate. To each mixture, in turn, an equal amount of standard potassium iodide is added, and the time for the iodine to be released and turn the starch blue is measured. The thiosulphate reacts instantly with the iodine produced in the peroxodisulphate reaction and so delays the appearance of iodine in the reaction mixture until enough has been produced to react with all the thiosulphate. This gives the rate of reaction with respect to iodine produced, and the effect of the concentration of potassium peroxodisulphate can be found.

(c) Collecting the gas evolved in a syringe

The volume of gas produced at set time intervals by reactions such as the decomposition of hydrogen peroxide or the solution of magnesium in acid

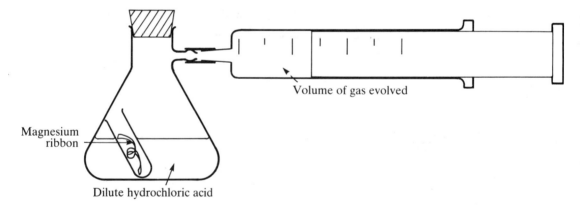

Fig. 9.10   Measurement of gas evolved

can be measured to give the rate of the reaction with respect to cm^3 of gas collected.

(d) Colorimetric measurements

The change in concentration of reactants or products such as potassium manganate(VII), or iodine, that are strongly coloured can be followed in a colorimeter. Light is passed through a sample of the reaction mixture and allowed to fall on a photoelectric cell connected to a microammeter. The darker the solution, the less light it transmits, and the lower the reading on the microammeter.

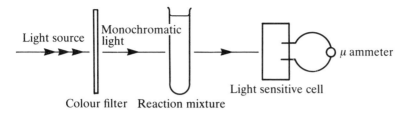

Fig. 9.11 Colorimetric measurement

The change of concentration is followed by taking readings with the microammeter. The instrument has to be calibrated so that the concentration of the coloured reactant is related to the meter reading.

(e) Conductivity measurements

For reactions in solution where there is a change in the number of ions present, the change in the conductivity of the solution (page 209) can be used to follow the course of the reaction. Potassium bromide reacts with potassium bromate in acid solution:

$$5Br^- + BrO_3^- + 6H^+ \rightarrow 3Br_2 + 3H_2O$$

The number of ions is reduced from twelve on the left hand side of the equation to zero on the right side; the conductivity will steadily decrease as the reaction proceeds.

(f) Use of a polarimeter

The rate of inversion of cane-sugar (sucrose) can be conveniently studied by measuring the change in optical activity with a polarimeter. (page 369). Natural sucrose is dextrorotatory, and is hydrolysed by acid to a mixture of glucose and fructose. Glucose is also dextrorotatory but fructose is laevorotatory, giving the overall effect that the rotation of the plane of polarisation of monochromatic light steadily decreases as the hydrolysis proceeds.

Questions

1. What do you understand by (a) Rate of reaction (b) Rate equation (c) Order of reaction (d) Velocity constant?
2. In the reaction $R + Q \rightarrow T$, the reaction is found to be first order with respect to R, second order with respect to Q.
   (a) Write the rate equation for the reaction.
   (b) Give the overall order of the reaction.
3. A substance X has an initial concentration of 0.60 mol dm^{-3}. After 30 seconds the concentration is 0.30 mol dm^{-3}. After a further 30 seconds it is 0.15 mol dm^{-3}. What is the order of the reaction?
4. The overall reaction $2A + B \rightarrow C + D$ is found to occur by two elementary steps, via an intermediate species X.

Thus

$$A + B \rightarrow X \qquad \qquad (1)$$
$$X + A \rightarrow C + D \qquad \qquad (2)$$

The rate of the overall reaction is given by

$$Rate = k[A]^2[B]$$

(a) (i) Define molecularity.
    (ii) What is the molecularity of step (1)?
    (iii) What is the molecularity of step (2)?
(b) (i) Define order of reaction.
    (ii) What is the overall order of the reaction?
(c) If the concentration of only one of the reactants can be doubled, which would give the greater increase in the overall rate?
(d) Give the reasons for your answer to (c).
(e) Suggest two other ways by which the rate of the reaction might be increased.   (JMB)

5. Two gases, X and Y, react according to the equation

$$X(g) + 2Y(g) \rightarrow XY_2(g)$$

The progress of the reaction was followed by measuring the variation of the total pressure, p, of the mixture with time, t, at 300 K and the initial rate of increase of the concentration of XY_2 was obtained.

	Initial concentration/mol litre^{-1}		*Initial rate of increase of XY_2/mol litre^{-1} s^{-1}*
Experiment	of X	of Y	
1	0.10	0.10	0.0001
2	0.10	0.20	0.0004
3	0.10	0.30	0.0009
4	0.20	0.10	0.0001
5	0.30	0.10	0.0001

(a) What is the order of the reaction with respect to X and Y?
(b) Write a *rate equation* for the reaction.
(c) Using the results of one of the experiments, calculate the numerical value for the rate constant, k, and give its units.
(d) Working from the overall rate equation, a student thought that the rate equation must be

$$rate = k' = [X][Y]^2$$

Why is it not possible to deduce the rate equation from the overall equation?   (C)

6. The reaction between iodine and propanone (acetone) in aqueous solution

$$CH_3COCH_3 + I_2 \rightarrow CH_2ICOCH_3 + HI$$

is catalysed by hydrogen ions.

(a) Describe briefly the experiments you would perform to determine the order of the reaction with respect to iodine, to propanone and to hydrogen ions.

(b) It is found that the rate of this reaction is proportional to the first power of the propanone concentration and to the first power of the hydrogen ion concentration, but is independent of the iodine concentration. What can you deduce from this?

(c) Explain why the usual effect of an increase of temperature on a reaction is to increase the rate at which it takes place.   (O)

7. In the presence of hydrochloric acid, N-chloracetanilide (A) is changed to its isomer, 4-chloroacetanilide (B):

$$\text{C}_6\text{H}_5-\text{N(Cl)}-\text{CO·CH}_3 \longrightarrow \text{Cl}-\text{C}_6\text{H}_4-\text{NHCOCH}_3$$

The progress of the reaction can be followed because A liberates iodine from potassium iodide solution whereas B does not react. The iodine can be estimated by titration with standard sodium thiosulphate solution. The table shows the volume, x, of sodium thiosulphate solution needed at various times in the course of a particular experiment, in each case for a fixed volume of reaction mixture. (x measures the amount of A left at each titration.)

t(min)	0	15	30	45	60	75
x(cm^3)	24.5	18.1	13.3	9.7	7.1	5.2

(a) Using suitable scales, plot a graph of x against t.

(b) From your graph, read off the time taken
   (i) for half of the original A
   (ii) for three quarters of the original A, to have been transformed into B.

What does a comparison of these two figures tell us about the order of the reaction?

(c) Describe in outline how you would investigate the effect of changing acid concentration on the rate of the reaction.

(d) Discuss briefly the effect that an increase in temperature would have on the rate of the reaction.

(e) Does the fact that a reaction is of the first order necessarily mean that it is unimolecular? Explain.   (L)

8. (a) The following data represent the activity of the radioactive isotope $^{223}_{87}\text{Fr}$, the longest lived isotope of francium.

Time in min	0	5	10	15	20	25	30	35	40
Activity in counts/min	680	575	495	425	355	300	265	230	210

(i) Use these data to plot a graph.

(ii) From the graph determine the time for the count to drop
    (1) to half the first value;
    (2) from three-quarters to three-eights of the first value;
    (3) from two-thirds to one third of the first value.

(iii) Determine the half-life of the radioactive isotope $^{223}_{87}Fr$.

(iv) Of what order is this decay process? Give your reasoning.

(v) Calculate the rate constant for the radioactive decay.

(b) The reaction between an aqueous solution of hydrogen peroxide and an aqueous solution of an iodide, in the presence of an acid, can be represented by the equation:

$$H_2O_2(aq) + 2I^-(aq) + 2H^+(aq) \rightleftharpoons 2H_2O(1) + I_2(aq)$$

Explain why, using the above information, it is wrong to state that the forward reaction is 5.   (AEB 1978)

0   Chemical equilibrium

10.1   Reversible reactions and equilibrium constants

(a) Introduction

In the previous chapter we discussed reactions that 'go to completion' at a measurable rate. A great many reactions do not result in complete transformation of reactants into products, but come to a state of equilibrium in which the proportions of reactants and products remain constant.

This occurs because the reactions are reversible; the products, when formed, react together reforming the original reactants. Since the reactions are proceeding both forwards and backwards the equations are written with a double arrow:

$$A + B \rightleftharpoons C + D$$

The mixture is in fact in a state of dynamic equilibrium, with a continuous change of molecules of A and B into molecules of C and D, and vice versa.

The rate of the left to right or forward reaction can be described as

$$k'[A][B]$$

and the rate of the right to left or backward reaction is given by

$$k''[C][D]$$

where k' and k'' are the rate constants (Chapter 9) and [A], [B], [C] and [D] are the concentrations of A, B, C and D in $mol\,dm^{-3}$, and are taken as the active masses of the components involved.

Since the mixture is in equilibrium, the ratio of the reactants and products being constant, the rate of the forward reaction is equal to the rate of the backward reaction:

$$k'[A][B] = k''[C][D]$$

or

$$\frac{k'}{k''} = K = \frac{[C][D]}{[A][B]}$$

The new constant K is called the equilibrium constant for that reaction at that temperature, and is defined as the product of the concentrations of the products divided by the product of the concentrations of the reactants in the equilibrium mixture. If the concentrations are measured in $mol\,dm^{-3}$ then the equilibrium constant is written as K_c, while if the concentrations are measured in terms of partial pressures (as in reactions between gases) then it is written as K_p.

When the balanced equation for the reactions involves more than one molecule of any reactant or product, then a factor for each molecule must appear in the expression for the equilibrium constant. So that if two molecules of a compound are involved then the concentration figure for that compound is squared, and if three are involved the figure is cubed.

For example, sulphur dioxide and oxygen react reversibly at raised temperature to form sulphur trioxide:

$$2SO_2 + O_2 \rightleftharpoons 2SO_3$$

The equilibrium constant for the reaction is given by the expression:

$$K_c = \frac{[SO_3]^2}{[SO_2]^2[O_2]} \tag{a}$$

The reaction between nitrogen and hydrogen to produce ammonia involves three molecules of hydrogen and two of ammonia:

$$N_2 + 3H_2 \rightleftharpoons 2NH_3$$

giving an equilibrium constant

$$K_c = \frac{[NH_3]^2}{[N_2][H_2]^3} \tag{b}$$

Since both these reactions are in the gaseous phase the concentrations of the components at equilibrium can be expressed in terms of their partial pressures (page 68). Thus for (a)

$$K_p = \frac{P_{SO_3}{}^2}{P_{SO_2}{}^2 \times P_{O_2}}$$

and for (b)

$$K_p = \frac{P_{NH_3}{}^2}{P_{N_2} \times P_{H_2}{}^3}$$

when P_{SO_3}, $P_{SO_2} \cdots$ etc. are the partial pressures of the gases in the equilibrium mixtures. K_p will be expressed in terms of atmospheres and will not have the same numerical value as K_c although the two do of course bear a mathematical relationship to one another.

The value of the equilibrium constant indicates the extent to which the forward reaction has taken place; if the reaction has proceeded far to the right then the equilibrium mixture will contain a high concentration of the products and K will be large; if the forward reaction is very small and a low concentration of product is formed then K will be small.

(b) Factors affecting equilibrium constants

(i) Le Chatelier's principle

The principle of Le Chatelier describes the effect of external factors on equilibrium mixtures, stating that 'if a system in equilibrium is subjected to a constraint, then the equilibrium will tend to alter in such a way as to annul the effect of the constraint'. These 'constraints' can be changes of concentration, pressure, or temperature.

(ii) Changes in concentration

The position of equilibrium is altered by changing the concentrations of the components in the equilibrium mixture at constant temperature. A familiar example is the addition of acid to a solution of potassium chromate (yellow) which results in the formation of potassium dichromate (orange).

$$2CrO_4^{2-} + 2H^+ \rightleftharpoons Cr_2O_7^{2-} + H_2O$$

$$K_c = \frac{[Cr_2O_7^{2-}][H_2O]}{[CrO_4^{2-}]^2[H^+]^2}$$

Addition of acid increases the concentration of H^+ ions so, since K_c remains constant, the equilibrium shifts to the right to increase the concentrations of $Cr_2O_7^{2-}$ ions and H_2O molecules. Similarly removal of H^+ ions by the addition of alkali results in the movement of the equilibrium to the left to increase $[CrO_4^{2-}]$ and $[H^+]$, thus K_c again remains constant.

(iii) Pressure

In gaseous reactions such as the conversion of a mixture of nitrogen and hydrogen to ammonia the effect of pressure on the composition of the equilibrium reaction is considerable:

$$N_2 + 3H_2 \rightleftharpoons 2NH_3$$

The four molecules of reactants on the left hand side of the equation are reduced to two molecules of product. Thus there is a considerable decrease in volume when ammonia is formed; an increase of pressure applied to the system will therefore cause the reaction to proceed to the right, reducing the volume of the mixture and tending to annul the effect of the applied pressure. This is important in the production of ammonia by the Haber process (page 475).

In reactions such as the formation of hydrogen iodide by the reaction between hydrogen and iodine:

$$H_2 + I_2 \rightleftharpoons 2HI$$

where there are an equal number of molecules on both sides of the equation, an increase of pressure make no difference to the position of equilibrium since it effectively increases the concentrations of both reactants and products equally.

(iv) Temperature

Changes of temperature affect both the composition of the equilibrium mixture and the equilibrium constant. The production of ammonia from hydrogen and nitrogen is exothermic:

$$N_2 + 3H_2 \underset{\text{endothermic}}{\overset{\text{exothermic}}{\rightleftharpoons}} 2NH_3$$

$$K_c = \frac{[NH_3]^2}{[N_2][H_2]^3}$$

Thus the forward reaction in which heat is given out will be favoured, on Le Chatelier's principle, by keeping the system at a low temperature. The dissociation of ammonia proceeds more rapidly when the temperature is raised. This results in a new equilibrium being established when the temperature is changed, accompanied by a new value for the equilibrium constant.

For an equilibrium such as the one above, K will be greater at lower temperatures since this will favour the exothermic reaction, and the concentration of ammonia will be greater at low temperatures (although the rate of reaction may then be impossibly low).

(v) Catalysts

A catalyst cannot alter the composition of an equilibrium mixture since its only function is to increase the rate of reaction. The position of equilibrium and the value of the equilibrium constant thus remain the same, but equilibrium conditions will be reached sooner in the presence of a catalyst. This is of special importance in the industrial application of reactions such as the synthesis of ammonia described in (iii) and (iv) above.

(c) Determination of equilibrium constants

The equilibrium constant of a reversible reaction has to be found by determining the concentrations of the substances present in the equilibrium mixture.

   The reaction between ethanoic acid and ethanol to form ethyl ethanoate and water is reversible:

$$CH_3COOH(l) + C_2H_5OH(l) \rightleftharpoons CH_3COOC_2H_5(l) + H_2O(l)$$

Suppose a moles of ethanoic acid and b moles of ethanol are sealed in a tube and left for about a week to come to equilibrium at a constant temperature. At the end of the week the amount of ethanoic acid remaining, x moles, can be measured by breaking the tube under water and rapidly titrating with a standard solution of sodium hydroxide. From this the concentrations of all the components of the equilibrium mixture, of total volume V dm^3, can be worked out with the help of the balanced equation. The number of moles of ethanoic acid that reacted is $(a-x)$ so, as an equal number of moles of ethanol must have reacted, then the number of moles of ethanol in the equilibrium mixture is $(b-(a-x))$. Similarly since each mole of ethanoic acid gave rise to one each of ester and water then the number of moles of each of these present at equilibrium is $(a-x)$. Thus we have:

$$CH_3COOH + C_2H_5OH \rightleftharpoons CH_3COOC_2H_5 + H_2O$$

| concentrations at equilibrium | $\dfrac{x}{V}$ | $\dfrac{(b-(a-x))}{V}$ | $\dfrac{(a-x)}{V}$ | $\dfrac{(a-x)}{V}$ |

and

$$K_c = \frac{[CH_3COOC_2H_5][H_2O]}{[CH_3COOH][C_2H_5OH]}$$

$$= \frac{(a-x)^2}{x(b-a+x)}$$

In this case, where there are equal numbers of molecules on either side of the equation, the volume V cancels and does not appear in the final equation. In other cases, such as the dissociation of phosphorus penta-chloride, this does not happen. Suppose that a moles of phosphorus pentachloride are sealed in a tube of volume V dm^3 at a temperature T K, and that when equilibrium is established x moles of chlorine are found to be present in the mixture:

$$PCl_5 \rightleftharpoons PCl_3 + Cl_2$$

Concentrations at equilibrium $\left.\right\}$ $\quad \dfrac{(a-x)}{V} \quad \dfrac{x}{V} \quad \dfrac{x}{V}$

$$K_c = \frac{[Cl_2][PCl_3]}{[PCl_5]}$$

$$= \frac{\left(\dfrac{x}{V}\right)^2}{\dfrac{a-x}{V}}$$

$$= \frac{x^2}{(a-x)V} \text{ mol dm}^{-3}$$

The presence of V in this equation emphasises the need to work in concentrations of reactants rather than in masses or moles of reactants, and the importance of stating the equation for the reaction concerned.

Once the equilibrium constant is found, it can be used to predict the equilibrium concentrations produced by other mixtures of reactants.

(d) Heterogeneous equilibrium

In the examples quoted so far the reactants and products have all been in the same phase, either all gases or all ions in solution. Equilibrium reactions such as the thermal dissociation of calcium carbonate involve both solids and gases; these are heterogeneous equilibria. If calcium carbonate is raised to a constant high temperature in an enclosed container an equilibrium is set up:

$$CaCO_3(s) \rightleftharpoons CaO(s) + CO_2(g)$$

and the equilibrium constant is given by:

$$K_p = \frac{P_{CO_2} \times P_{CaO}}{P_{CaCO_3}}$$

The vapour pressure exerted by solids is very small and independent of the amount of solid present, so that, provided excess calcium carbonate is present:

$$K_p = P_{CO_2}$$

At a given temperature the pressure of carbon dioxide is thus constant and independent of the amounts of calcium carbonate or calcium oxide present. This pressure, which depends only on the temperature of the equilibrium, is called the *dissociation pressure*. Above 900°C the pressure exceeds atmospheric pressure, so that heating calcium carbonate in the open above this temperature results in a rapid escape of carbon dioxide and the complete decomposition of the carbonate to give quicklime (page 242).

10.2   Solubility of sparingly soluble salts

(a) Solubility products

Most electrolytes that are commonly regarded as insoluble do, in fact, dissolve to a certain extent. If silver chloride is shaken with distilled water a very small quantity dissolves until a saturated, but very dilute, solution is formed. An equilibrium is established between undissolved solid and ions in solution:

$$AgCl(s) \rightleftharpoons Ag^+(aq) + Cl^-(aq)$$

Since the solution is very dilute the equilibrium can be described by:

$$K = \frac{[Ag^+][Cl^-]}{[AgCl]}$$

where $[Ag^+]$, $[Cl^-]$ and $[AgCl]$ are the concentrations in $mol\,dm^{-3}$ of the silver ions, chloride ions and undissolved silver chloride. Provided the solution is in contact with excess solid silver chloride then $[AgCl]$ remains constant and the *solubility product* (Ks) can be expressed in terms of the concentrations of the ions:

$$K_{s(AgCl)} = [Ag^+][Cl^-]\,mol^2\,dm^{-6}$$

For electrolytes where the numbers of positive and negative ions are not equal the concentrations are raised to the appropriate powers as in the expressions for equilibrium constants. Thus lead chloride dissolves to produce two chloride ions for each lead ion:

$$PbCl_2(s) \rightleftharpoons Pb^{2+}(aq) + 2Cl^-(aq)$$

and

$$K_{s(PbCl_2)} = [Pb^{2+}][Cl^-]^2\,mol^3\,dm^{-9}$$

Since the solutions are very dilute the values for K_s are low. Some typical values are given below.

Table 10.1   Solubility products at 298 K

Compound	K_s
$Ca(OH)_2$	4×10^{-6}
$BaSO_4$	1×10^{-10}
$AgCl$	1×10^{-10}
$Mg(OH)_2$	2×10^{-11}
Ag_2CrO_4	3×10^{-12}
$Fe(OH)_2$	1×10^{-15}
$Cu(OH)_2$	2×10^{-19}
$Fe(OH)_3$	8×10^{-4}

Solubility products are used to predict whether a sparingly soluble electrolyte will dissolve or be precipitated from solutions under various conditions. In general, where the product of the ionic concentrations exceeds the solubility product, i.e. $[A^+][B^-] > K_s$, then some solid will be precipitated, while if $[A^+][B^-] < K_s$, then some solid will be dissolved.

(b) Common ion effect

A sparingly soluble salt, such as silver ethanoate, can be precipitated from its saturated solution by the addition of sodium ethanoate. The solubility product of silver ethanoate is derived from the equilibrium equation:

$$AgEt(s) \rightleftharpoons Ag^+(aq) + Et^-(aq) \text{ (Where Et}^- \text{ represents CH}_3COO^-)$$

$$K_{s(AgEt)} = [Ag^+][Et^-]$$

Addition of the completely soluble and fully ionized sodium ethanoate increases the concentration of ethanoate ions, so that $[Ag^+][Et^-] > K_s$; therefore silver ethanoate is precipitated until the equilibrium is restored. This is of great importance in the recovery of silver in processes involving the use of silver solutions.

Sodium chloride is similarly purified from traces of the deliquescent magnesium and calcium chlorides that are present in crude common salt. The impure product is made into a saturated solution and hydrogen chloride gas bubbled through it. The resulting concentration of chloride ions causes precipitation of sodium chloride from its saturated solution but leaves the impurities, which are present only in small quantities, in solution.

(c) Calculation of solubility products from solubilities

The solubility product can be calculated if the solubility of the salt is either measured or looked up in a data book. For example, the solubility of silver chloride is 0.0015 g dm^{-3}, which is $1.04 \times 10^{-5} \text{ mol dm}^{-3}$. Assuming that the silver chloride is fully ionised in its saturated solution:

$$AgCl(s) \rightleftharpoons Ag^+(aq) + Cl^-(aq)$$

then

$$[Ag^+] = [Cl^-] = 1.04 \times 10^{-5} \text{ mol dm}^{-3}$$

and

$$K_s = [Ag^+][Cl^-] = (1.04)^2 \times 10^{-10} \text{ mol}^2 \text{ dm}^{-6}$$
$$= 1.08 \times 10^{-10} \text{ mol}^2 \text{ dm}^{-6}$$

The solubility of silver sulphide is $5 \times 10^{-17} \text{ mol dm}^{-3}$ and the equilibrium equation is $Ag_2S(s) \rightleftharpoons 2Ag^+(aq) + S^{2-}(aq)$ making

$$[Ag^+] = 2 \times 5 \times 10^{-17}$$

and

$$[S^{2-}] = 5 \times 10^{-17}$$

hence

$$K_s = [Ag^+]^2[S^{2-}]$$
$$= [10]^2[5] \times 10^{-51}$$
$$= 5 \times 10^{-49} \, mol^3 \, dm^{-9}$$

(d) Calculating solubility from solubility product

The solubility product of magnesium hydroxide is $2 \times 10^{-11} \, mol^3 \, dm^{-9}$

$$Mg(OH)_2(s) \rightleftharpoons Mg^{2+}(aq) + 2OH^-(aq)$$

hence

$$K_{s\,Mg(OH)_2} = [Mg^{2+}][OH^-]^2$$

From the equation it can be seen that $[OH^-] = 2[Mg^{2+}]$ hence

$$K_s = 4[Mg^{2+}]^3 = 2 \times 10^{-11}$$
$$[Mg^{2+}] = \sqrt[3]{5} \times 10^{-4}$$
$$= 1.7 \times 10^{-4} \, mol \, dm^{-3}$$

Each mole of magnesium ions comes from one mole of magnesium hydroxide so the solubility of magnesium hydroxide is $(1.7 \times 10^{-4} \times 58) \, g \, dm^{-3}$ which is $9.8 \times 10^{-3} \, g \, dm^{-3}$.

(e) Calculation of solubility in the presence of a common ion

The addition of a common ion can reduce the solubility of a sparingly soluble salt dramatically; the effect can be calculated from a knowledge of the solubility product.

The solubility of silver chloride is $1.50 \times 10^{-3} \, g \, dm^{-3}$, and the solubility product is $1.0 \times 10^{-10} \, mol^2 \, dm^{-6}$. To find the solubility of silver chloride in say one litre of 0.001 M potassium chloride solution we assume that the chloride ion concentration contributed by the silver chloride itself is negligible, and that the effective concentration is the 10^{-3} mole contributed by the potassium chloride:

$$K_{s\,AgCl} = [Ag^+][Cl^-]$$
$$\therefore \quad 1.0 \times 10^{-10} = [Ag^+][1 \times 10^{-3}]$$
$$[Ag^+] = 1 \times 10^{-7} \, mol \, dm^{-3}$$

Since each mole of silver ions comes from one mole of silver chloride, then the concentration of silver chloride is:

$$1 \times 10^{-7} \times 143.5 \, g \, dm^{-3} \simeq 1.4 \times 10^{-5} \, g \, dm^{-3}$$

which is one hundred times less than the solubility in pure water!

Questions

1. State *Le Chatelier's principle*.

$$N_2O_4 \rightleftharpoons 2NO_2; \quad \Delta H, \text{ positive.}$$

How is the composition of the equilibrium mixture for the above gaseous equilibrium affected by (a) change in temperature, (b) change in pressure, (c) the presence of a catalyst?

At 30°C and 1 atm pressure, 30% of a sample of dinitrogen tetraoxide, N_2O_4, is dissociated. Calculate K_p for the above equilibrium. What pressure would be necessary at 30°C to increase the dissociation to 40%?   (C)

2. (a) Write an equation to show the relationship between the equilibrium constant, K_p, and the partial pressures, $P_{N_2O_4}$, and P_{NO_2}, of the reactants in the following gaseous equilibrium:

$$N_2O_4 \rightleftharpoons 2NO_2; \qquad \Delta H(298 \text{ K}) = 54 \text{ kJ mol}^{-1}.$$

(b) State the effect, if any, on the above equilibrium of (i) increasing the pressure, (ii) raising the temperature. Give reasons for your answers.
(c) It was found that one dm^3 of the gaseous mixture weighed 2.777 g at 50.0°C and under a pressure of 1.01×10^5 Nm^{-2} (=1 atmosphere). Calculate:
(i)   the fraction of the N_2O_4 that is dissociated;
(ii) the percentage of NO_2 molecules in the mixture;
(iii) the value of K_p.
(One mole of a gas occupies 22.4 dm^3 at s.t.p.)   (O)

3. The solubility product of magnesium hydroxide, $Mg(OH)_2$, is 1.25×10^{-10} $mol^3 l^{-3}$
(a) Calculate the solubility (in grammes per litre) of magnesium hydroxide in water.
(b) Suppose you are given 1 litre of a 0.01 M aqueous ammonia solution. What is the maximum quantity (in moles) of magnesium hydroxide that can be dissolved in it?

$$(K_{\text{dissociation}} = 1.8 \times 10^{-5} \text{mol l}^{-1} \text{ for } NH_3(aq))$$

(c) How does the solubility product of $Mg(OH)_2$ compare with the solubility products of the other alkaline earth metal hydroxides? (Mg = 24.3, H = 1.00, O = 16.0)

4. (a) In the hydrolysis of ethyl ethanoate (ethyl acetate) state how the *position of equilibrium*, and the *rate of attainment of equilibrium* are affected by (i) temperature, (ii) the use, as reagents for the hydrolysis, of (1) water, (2) dilute sulphuric acid.
(b) For the equilibrium between propanoic acid and ethanol

$$C_2H_5OH(l) + C_2H_5COOH(1) \rightleftharpoons C_2H_5COOC_2H_5(1) + H_2O(1)$$

the equilibrium constant in terms of concentration, K_c, is 7.5 at 50°C.
(i) Write an expression for K_c for this reaction.
(ii) What mass of ethanol must be mixed with 74 g of propanoic acid at 50°C in order to obtain 80 g of ethyl propanoate in the equilibrium mixture?   (AEB 1977)

5. (a) (i) Write an equation for the equilibrium which may be presumed to exist between solid silver chloride and dissolved silver chloride in a saturated solution.
(ii) Write an expression for the solubility product of silver chloride and state the conditions under which this applies.

(iii) The solubility of silver chloride in water at 25°C is 1×10^{-5} mol/l. Calculate the solubility product of silver chloride and state the units in which it is expressed.

(b) The solubility product of lead(II) bromide at 25°C is 3.9×10^{-5} mol^3/l^3. Calculate the solubility in g/l at 25°C of lead(II)bromide.

(c) *Outline* a practicable method for determining the solubility of a named sparingly soluble ionic compound in water at 25°C.   (AEB 1977)

11 Acids and bases

11.1 Theory and definitions

Acids were defined in 1884 by Arrhenius as substances that dissolved in water to give hydrogen ions as the only positive ions.

$$H_2SO_4(aq) \rightarrow H^+(aq) + HSO_4^-(aq)$$

$$HCl(aq) \rightarrow H^+(aq) + Cl^-(aq)$$

More recent experimental evidence shows that the hydrogen ion is hydrated and the solution of an acid in water results in a reaction which produces an *oxonium* ion H_3O^+ which is a hydrated proton ($H^+ \cdot H_2O$):

$$HCl(g) + H_2O(l) \rightarrow H_3O^+(aq) + Cl^-(aq)$$

In a similar way bases were defined as substances that reacted with acids to form a salt and water only, or, in ionic terms, as substances that reacted with H^+ ions to form water:

$$CuO(s) + H_2SO_4(aq) \rightarrow CuSO_4(aq) + H_2O(l)$$

or

$$CuO(s) + 2H^+(aq) \rightarrow Cu^{2+}(aq) + H_2O(l)$$

Soluble bases, or alkalis, dissolve in water to form OH^- ions, so the definition applies equally to them:

$$Na^+ OH^-(aq) + H^+Cl^-(aq) \rightarrow Na^+Cl^-(aq) + H_2O(l)$$

or

$$OH^-(aq) + H^+(aq) \rightarrow H_2O(l)$$

Hydrogen ions are in fact protons, since the loss of an electron by a hydrogen atom leaves only the positive nucleus (page 226), so in all these reactions the acids have given up a proton to a base.

This is the basis of the definition of acids suggested by Brønsted and Lowry in 1922, which is widely used as it has the advantage of applying also to acids and bases in non-aqueous solvents.

Acids are defined as substances, molecules or ions, that can give up a proton to a base (i.e. proton donors).

Bases are defined as substances, molecules or ions, that can accept protons from an acid (i.e. proton acceptors).

The relationship between acids and bases can be summarised as:

$$\text{Acid} \rightleftharpoons \text{Base} + H^+$$
$$\text{(proton donor)} \qquad \text{(proton acceptor)}$$

or, since the proton is considered to be hydrated:

$$\text{Acid} + H_2O \rightleftharpoons \text{Base} + H_3O^+$$

For strong acids the equilibrium lies well to the right, the acid being fully dissociated, while for weak acids the equilibrium lies to the left:

$$CH_3COOH + H_2O \rightleftharpoons CH_3COO^- + H_3O^+$$

The acid and the base it gives rise to are known as *conjugate pairs*; the ethanoate ion is the *conjugate base* of ethanoic acid. Some examples will help to clarify this. (Conjugate pairs are underlined).

$$\text{Acid (1)} + \text{Base (2)} \rightleftharpoons \text{Base (1)} + \text{Acid (2)}$$

$$\underline{H_2SO_4} + H_2O \rightleftharpoons \underline{HSO_4^-} + H_3O^+$$

$$\underline{HSO_4^-} + H_2O \rightleftharpoons \underline{SO_4^{2-}} + H_3O^+$$

$$\underline{HCl} + H_2O \rightleftharpoons \underline{Cl^-} + H_3O^+$$

$$\underline{NH_4^+} + H_2O \rightleftharpoons \underline{NH_3} + H_3O^+$$

$$\underline{H_2O} + H_2O \rightleftharpoons \underline{OH^-} + H_3O^+$$

Note that both the hydrogen sulphate ion and the ammonium ion are regarded as acids, also that water can behave as both an acid or a base. In all cases Acid (1) is conjugate with Base (1).

Although these reactions are correctly written with H_3O^+ ions, it is the usual practice to use the simple H^+ ion and to represent, for example, the dissociation of ethanoic acid as:

$$CH_3COOH \rightleftharpoons CH_3COO^- + H^+$$

and the ionisation of water as:

$$H_2O \rightleftharpoons H^+ + OH^-$$

11.2   Dissociation constants

The extent to which an acid or a base is dissociated is called the *degree of dissociation* or *degree of ionisation*; it is expressed as the fraction or percentage of the molecules that are dissociated.

A useful relationship between the concentration, degree of ionisation and the ionisation constant of a weak acid or base in dilute solution at a given temperature can be derived. For a solution of ethanoic acid of concentration c mol dm^{-3} with a fractional degree of ionisation α, the concentrations of acid and ions at equilibrium will be:

$$CH_3COOH \rightleftharpoons CH_3COO^- + H^+$$
$$c(1-\alpha) \qquad\qquad c\alpha \qquad\quad c\alpha$$

Applying the equilibrium law, the *dissociation constant*, or ionisation constant, K, will be given by:

$$K = \frac{[CH_3COO^-][H^+]}{[CH_3COOH]} = \frac{c\alpha \times c\alpha}{c(1-\alpha)} = \frac{c\alpha^2}{(1-\alpha)}$$

If the degree of ionisation is very small then $(1-\alpha) \simeq 1$, and $c\alpha^2 = K$ or $\alpha = \sqrt{\dfrac{K}{c}}$. This is expressed in Ostwald's dilution law, which states that 'for a weak binary electrolyte, with a small degree of ionisation, the degree of ionisation is proportional to the square root of the reciprocal of the concentration.'

The degree of ionisation can be measured by various methods such as conductivity (page 212) or depression of the freezing point (page 123), and the value of K_a, the acid dissociation constant, can be found. Some values for weak monobasic acids are given below. Since the values for K_a are small, the strengths of weak acids are more readily compared by using pK_a where

$$pK_a = -\log K_a$$

Table 11.1 Dissociation constants for weak acids

Acid	K_a mol dm^{-3}	pK_a
Nitric(III) acid (HNO$_2$)	5×10^{-4}	3.3
Methanoic acid (HCOOH)	1.6×10^{-4}	3.8
Benzoic acid (C$_6$H$_5$COOH)	6.3×10^{-5}	4.2
Ethanoic acid (CH$_3$COOH)	1.7×10^{-5}	4.7
Chloric(I) acid (HClO)	3.7×10^{-8}	7.4
Hydrocyanic acid (HCN)	4.9×10^{-10}	9.3
Phenol (C$_6$H$_5$OH)	1.0×10^{-10}	10.0

The weaker the acid, the lower the acid dissociation constant for the same concentration at the same temperature. This concept does not apply to strong acids, which are considered to be fully ionised at most concentrations, nor does it hold for concentrated solutions where other factors have to be taken into account. Similarly, dissociation constants can be found for weak bases in dilute solution.

Table 11.2 Dissociation constants for weak bases

Base	K_b mol dm^{-3}
Ammonia (NH$_3$)	1.8×10^{-5}
Hydrazine (N$_2$H$_4$)	1.6×10^{-6}
Phenylamine (C$_6$H$_5$NH$_2$)	4.6×10^{-10}

11.3 pH values

The pH scale covers a range of solutions from those with a hydrogen ion concentration of 10^{-1} mol dm^{-3} (0.1 M) to those with a hydrogen ion concentration of 10^{-14} mol dm^{-3} which are strongly alkaline and have a hydroxide ion concentration of about 10^{-1} mol dm^{-3}.

The scale is derived from a consideration of the equilibrium resulting from the very small degree of ionisation of water into H$^+$ and OH$^-$ ions.

$$H_2O(l) \rightleftharpoons H^+(aq) + OH^-(aq)$$

The equilibrium constant K for this ionisation of water is given by:

$$K = \frac{[H^+][OH^-]}{[H_2O]}$$

Since the ionisation is so small the concentration of water molecules $[H_2O]$ can be taken as constant, so that the ionic product of water, $K_w = [H^+][OH^-]$. Conductivity measurements show that the value of K_w is 10^{-14} at 25°C, so that for pure distilled water, or neutral solutions, $[H^+][OH^-] = 10^{-14}$, and $[H^+] = [OH^-] = 10^{-7}$ mol dm^{-3}.

For acid solutions $[H^+]$ will be greater than 10^{-7} and $[OH^-]$ less than 10^{-7}, while for alkaline solutions $[H^+]$ will be less than $[10^{-7}]$ and $[OH^-]$ greater than 10^{-7}.

The concentration of hydrogen ions present in solution does not make a convenient scale, and in 1909 Sørensen suggested using the logarithm of the reciprocal value of the hydrogen ion concentration. This he called the pH scale, where p stands for 'potenz', meaning strength. Thus:

$$pH = \log \frac{1}{[H^+]} = -\log[H^+]$$

From this it follows that an acid solution having a concentration of 10^{-1} moles of hydrogen ions per dm^3 (0.1 M) has a pH of 1, while a neutral solution in which $[H^+] = 10^{-7}$ has a pH of 7.

In a solution of 0.1 M potassium hydroxide the concentration of OH$^-$ ions is 10^{-1} mol dm^{-3}. Hence, using the equilibrium equation, the concentration of H$^+$ ions is given by:

$$K_w = 10^{-14} = [H^+][OH^-] = [H^+][10^{-1}]$$

hence

$$[H^+] = 10^{-13}$$

and the pH of the solution is 13.

Some worked examples illustrate the uses of the pH scale.

(a) Find the pH of a solution made by mixing 25 cm^3 of 0.1 M hydrochloric acid with 15 cm^3 of 0.1 M sodium hydroxide.

Since the solutions are equimolar and

$$NaOH(aq) + HCl(aq) \rightarrow NaCl(aq) + H_2O(l)$$

then 15 cm^3 of sodium hydroxide will neutralise 15 cm^3 of hydrochloric acid, leaving 10 cm^3 of 0.1 M hydrochloric acid unneutralised in a total volume of 40 cm^3. Now 10 cm^3 of 0.1 M HCl contain $\dfrac{10 \times 0.1}{1000}$ moles H$^+$ ions. These are now present in 40 cm^3 solution, so that the concentration of the solution with respect to H$^+$ ions is:

$$\frac{10 \times 0.1}{1000} \times \frac{1000}{40} = 0.025 \text{ mol dm}^{-3}, \text{ and } pH = -\log[H^+] = -\log 0.025$$

$$= 1.6$$

Thus the solution has a pH of 1.6.

(b) What is the pH of a solution of 0.05 M sodium hydroxide? 0.05 M sodium hydroxide has a concentration of OH$^-$ ions of 5×10^{-2} mol dm^{-3}, and, since $K_w = 10^{-14} = [H^+][OH^-]$, the concentration of H^+ ions in this solution is given by $[H^+] = \dfrac{1 \times 10^{-14}}{5 \times 10^{-2}} = 0.2 \times 10^{-12}$

$$\therefore \quad pH = -\log(2 \times 10^{-13}) = 12.7$$

Thus the pH of 0.05 M sodium hydroxide is 12.7.

(c) To find the pH of a solution of a weak acid. For example, given that the acid dissociation constant for ethanoic acid is 1.8×10^{-5}, find the pH of a solution of 0.01 M ethanoic acid.

From Ostwald's dilution law the degree of dissociation (α) of ethanoic acid in a 0.01 M solution is given by:

$$\alpha = \sqrt{\frac{1.8 \times 10^{-5}}{10^{-2}}} = \sqrt{18 \times 10^{-4}} = 4.2 \times 10^{-2}$$

Hence the concentration of H^+ ions in a 0.01 M solution of ethanoic acid is $10^{-2} \times 4.2 \times 10^{-2}$ or 4.2×10^{-4}, and $pH = -\log[H^+] = -\log(4.2 \times 10^{-4}) = 3.38$. This is using the approximate form of equation for Ostwald's dilution law. If the unapproximated equation is used we have:

$$\alpha^2 = \frac{K}{c} \times (1 - \alpha) = \frac{1.8 \times 10^{-5}(1 - \alpha)}{10^{-2}}$$

or

$$\alpha^2 + 1.8 \times 10^{-3}\alpha - 1.8 \times 10^{-3} = 0$$

and

$$\alpha = 4.1 \times 10^{-2}$$

and

$$pH = 3.39$$

The difference in pH is not great and for most practical purposes the approximate expression of Ostwald's dilution law will be adequate.

11.4 Acid-base indicators

An indicator is a chemical compound which changes colour according to the hydrogen ion concentration of the aqueous solution it is dissolved in. Originally such indicators were extracted from vegetables, flower petals and lichens, the best known of these being litmus which is extracted from certain lichens abundant in the Mediterranean countries. Nowadays synthetic organic substances are used. Many of them are compounds with large molecules that have different coloured isomers in dynamic equilibrium. More readily understood are those that are weak acids which are feebly ionised in aqueous solution. The undissociated acid (HIn) is coloured and is in equilibrium with hydrogen ions and indicator ions (In$^-$) of a different colour:

$$HIn \quad \rightleftharpoons \quad H^+ + In^-$$
$$\text{Colour A} \qquad \text{Colour B}$$

Plainly the presence of excess hydrogen ions giving an acid solution will drive the equilibrium to the left giving colour A, while the removal of hydrogen ions by the addition of alkali will drive the equilibrium to the right giving colour B. The pH at which the colour change occurs is different for various indicators; some typical values are given below.

Table 11.3 Indicators

Indicator	Acid colour	pH change point	Alkali colour	pK value
Methyl orange	Pink	3.1–4.4	Orange	3.6
Methyl red	Pink	4.4–6.3	Yellow	5.0
Bromothymol blue	Yellow	6.0–7.6	Blue	7.1
Litmus	Red	5.0–8.0	Blue	7.0
Phenolphthalein	Colourless	8.3–10.0	Magenta	9.6

Most of the indicators change colour over a range of pH. At the midpoint of the change the concentration of the undissociated indicator is equal to the concentration of indicator ions:

$$HIn \rightleftharpoons H^+ + In^-$$

$$K_{HIn} = \frac{[H^+][In^-]}{[HIn]}$$

when

$$[HIn] = [In^-]$$

$$K_{HIn} = [H^+]$$

Thus at the mid point of the colour change the equilibrium constant for the indicator is equal to the hydrogen ion concentration of the solution it is in. The pH value at which an indicator is midway between its colour change is known as the pK value for the indicator.

Single indicators are used to find the neutralisation point when acid and alkaline solutions are titrated against one another as described below. Mixtures of indicators can be made up which will give successive colour changes to match a range of pH values. These are the *universal indicators* used in the laboratory to find the pH of a particular solution.

11.5 pH changes during titrations and choice of indicator

The pH changes that occur during titrations of acids with alkalis can be followed by using a *universal indicator* solution or a pH meter (page 201). The results, when plotted, reveal certain characteristics which depend on whether strong or weak acids or alkalis are involved. The following curves are typical of those obtained when solutions of 0.1 M alkalis are titrated against 25 cm³ portions of 0.1 M acids. The pH of the original acid solution is recorded; small quantities of alkali are added from a burette and the pH recorded after each addition. The pH is then plotted against the volume of alkali added. When 25 cm^3 of alkali has been added, equivalent amounts of acid and alkali are present.

(a) Titrating a strong acid with a strong alkali

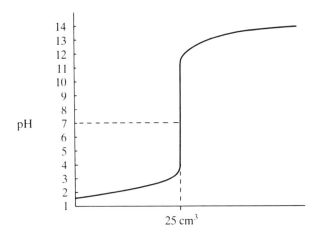

Volume of 0.1 M NaOH added to 25 cm^3 0.1 M HCl

Fig. 11.1   Titration curve (a)

The solution stays strongly acidic up to the addition of 24.9 cm^3 of alkali but the pH changes rapidly on the addition of one more drop, giving a sharp 'end-point' which exactly determines the equivalence point. For these titrations any indicator can be used since the pH changes from 3 to 10 on the addition of only one drop of alkali.

(b) Titrating a strong acid with a weak alkali

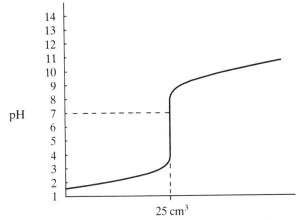

Volume of 0.1   M ammonia solution added to 25 cm^3 0.1   M HCl

Fig. 11.2   Titration curve (b)

The strongly acidic hydrochloric acid solution maintains its pH again until 24.9 cm^3 of ammonia solution have been added, but the addition of one drop more of ammonia changes the pH only from approximately 3 to 8 because ammonia is a weak alkali. A sharp end-point to indicate when the equivalence point is reached can be obtained only by the use of an indicator which changes colour in this range; for example, methylorange.

(c) Titrating a weak acid with a strong alkali

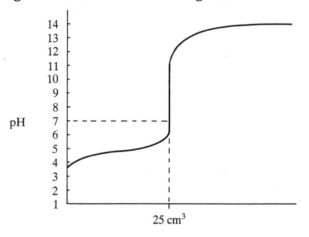

Volume of 0.1 M NaOH added to 25 cm^3 0.1 M ethanoic acid

Fig. 11.3   Titration curve (c)

Here the pH of the weakly acidic solution increases slowly as the sodium hydroxide is added until, at the equivalence point, it changes rapidly from approximately 6 to 12 on the addition of one drop of excess alkali. To obtain a sharp end-point here the indicator used must have a colour change in the range of 8–11; phenolphthalein may be used.

(d) Titrating a weak acid with a weak alkali

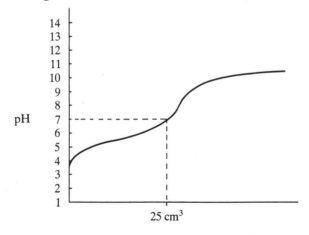

Volume of 0.1 M ammonia solution added to 25 cm^3 of 0.1 M ethanoic acid

Fig. 11.4   Titration curve (d)

The change of pH is not sharp at any point during this titration; for this reason weak acids are not used to standardise weak alkalis or vice versa.

To sum up, suitable indicators for various titrations are as follows.

Table 11.3   Use of indicators

Titration	Indicator
Strong acid/strong alkali	Most are suitable
Strong acid/weak alkali	Methyl orange, methyl red, Bromothymol blue
Weak acid/strong alkali	Phenolphthalein

The pH curves for the titrations of alkalis against dibasic or tribasic acids have two or three equivalence points corresponding to the respective completions of the formations of the acid salts and the normal salt. For example, if phosphoric acid(V) (H_3PO_4) is used rapid changes of pH will occur at the points at which sufficient $0.1\,M$ sodium hydroxide has been added to form the salts NaH_2PO_4, Na_2HPO_4 and Na_3PO_4.

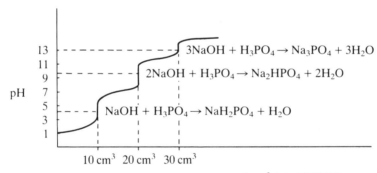

Volume of $0.1\ \ M$ NaOH added to $10\,cm^3\,0.1\ \ M\,H_3PO_4$

Fig. 11.5   Titration curve for phosphoric(V) acid

11.6   Hydrolysis of salts

Salts of strong acids and strong bases such as sodium sulphate dissolve in water to form neutral solutions. Salts of strong acids with weak bases, such as copper sulphate, however, form acidic solutions while those, like sodium ethanoate, of strong bases and weak acids form alkaline solutions. These differences arise from the hydrolysis of the ions formed when the salt dissolves in water. A solution of sodium ethanoate contains sodium and ethanoate ions from the fully dissociated salt together with a small number of hydrogen and hydroxyl ions from the weakly ionised water:

$$CH_3COONa(aq) \rightarrow CH_3COO^-(aq) + Na^+(aq)$$

$$H_2O(aq) \rightleftharpoons H^+(aq) + OH^-(aq) \tag{1}$$

A further equilibrium is set up between the hydrogen and ethanoate ions which results in the withdrawal of hydrogen ions to form ethanoic acid:

$$CH_3COO^-(aq) + H^+(aq) \rightleftharpoons CH_3COOH(aq) \qquad (2)$$

The withdrawal of the H^+ ions upsets equilibrium (1) giving rise to further ionisation of water and the formation of more undissociated ethanoic acid as in equation (2). As the process continues the concentration of OH^- ions increases steadily and the solution becomes alkaline:

$$\left.\begin{array}{l} CH_3COONa \rightarrow CH_3COO^- + Na^+ \\ \qquad\qquad\qquad + \\ H_2O \rightleftharpoons H^+ \quad + OH^- \end{array}\right\} \text{fully ionised}$$
$$\Updownarrow$$
$$CH_3COOH$$

A similar argument can explain the acidity of the salt of a strong acid with a weak base. Ammonium chloride is fully dissociated into ions which come into equilibrium with the hydrogen and hydroxide ions arising from the water. The hydroxide ions and the ammonium ions combine to form undissociated ammonium hydroxide so that an excess of hydrogen ions are left in the solution making it acidic:

$$\left.\begin{array}{l} NH_4Cl \rightarrow NH_4^+ + Cl^- \\ H_2O \rightleftharpoons OH^- + H^+ \end{array}\right\} \text{fully ionised}$$
$$\Updownarrow$$
$$NH_4OH \text{ (weak base)}$$

11.7   Buffer solutions

These very useful solutions are designed to be of constant pH which will not change even if small quantities of acids or alkalis are added or the solutions are diluted. Acid buffer solutions contain a weak acid together with the sodium salt of a weak acid. If sodium ethanoate is added to a solution of ethanoic acid the excess ethanoate ions from the fully dissociated sodium ethanoate will suppress the ionisation of the ethanoic acid. After equilibrium is reached the solution will have a pH that is related to the concentrations of ethanoic acid and ethanoate ions:

$$CH_3COONa(aq) \rightarrow CH_3COO^-(aq) + Na^+(aq)$$

$$CH_3COOH(aq) \rightleftharpoons CH_3COO^-(aq) + H^+(aq)$$

This pH will remain constant because any H^+ ions introduced by the addition of acid will be removed by reaction with the excess ethanoate ions to form undissociated ethanoic acid. If alkali is added the pH will again remain unchanged because the OH^- ions introduced will react with the excess ethanoic acid ions:

$$CH_3COOH(aq) + OH^-(aq) \rightarrow CH_3COO^-(aq) + H_2O$$

Dilution will affect the concentration of both acid and ethanoate ions but not the ratio of one to another, and so the equilibrium will remain undisturbed and the pH unchanged. The pH can be calculated from the acid dissociation constant of ethanoic acid, K_a, and the concentrations of solutions used.

$$K_a = \frac{[H^+][CH_3COO^-]}{[CH_3COOH]}$$

$$\therefore \quad [H^+] = \frac{K_a[CH_3COOH]}{[CH_3COO^-]}$$

Since the dissociation of the acid is very small $[CH_3COOH]$ is very nearly equal to the initial concentration of the acid used. The concentration of ethanoate ions, $[CH_3COO^-]$, will be very nearly equal to the initial concentration of sodium ethanoate added, as the amount of ethanoate ions resulting from the ionisation of the acid will also be very small. Thus:

$$[H^+] = \frac{K_a \times [ACID]}{[SALT]}$$

where $[ACID]$ and $[SALT]$ are the initial concentrations of ethanoic acid and sodium ethanoate used. Hence

$$\log[H^+] = \log K_a + \log \frac{[ACID]}{[SALT]}$$

but

$$pH = -\log[H^+]$$

$$\therefore \quad pH = -\log K_a - \log \frac{[ACID]}{[SALT]}$$

so that, given K_a and the concentrations $[ACID]$ and $[SALT]$, the pH of the solution can be calculated.

Similarly, given K_a, the ratio $[ACID]/[SALT]$ to provide a solution of a required pH can be calculated. Suppose a solution of pH 4.5 is to be made by mixing solutions of 0.1 M ethanoic acid and 0.1 M sodium ethanoate. K_a for ethanoic acid is 1.8×10^{-5}

$$pH = -\log K_a - \log \frac{[ACID]}{[SALT]}$$

$$\therefore \quad 4.5 = -\log 1.8 \times 10^{-5} - \log \frac{[ACID]}{[SALT]}$$

$$\therefore \quad \log \frac{[ACID]}{[SALT]} = -\log 1.8 \times 10^{-5} - 4.5$$

$$= 4.75 - 4.5 = 0.25$$

the antilog of 0.25 is 1.78

$$\therefore \quad \frac{[ACID]}{[SALT]} = 1.78$$

Since the solutions are both 0.1 M then 17.8 cm³ of ethanoic acid must be mixed with 10 cm³ of sodium ethanoate.

Buffer solutions will keep indefinitely and are of great practical use. Many fermentation processes depend on enzymes which will function only at constant pH. The pH of the blood is critical, so solutions for injection are made up in buffered solutions of the correct pH. Other uses include the buffering of solutions for electroplating. A range of buffer solutions can be made up and used to determine the pH of soils by colour matching the filtrates after adding universal indicator.

11.8   Lewis acids and bases

In 1923 G.N.Lewis suggested defining acids and bases in terms of electron-pair acceptors and donators. A Lewis acid is then 'a molecule or ion capable of accepting a pair of electrons in the formation of a covalent bond', while a Lewis base is a 'molecule or ion capable of donating a pair of electrons in the formation of a covalent bond'. Thus on the Lewis theory the hydrogen ion is acidic because it accepts a pair of electrons from a water molecule in forming the oxonium ion:

$$H^+ + \overset{\bullet\bullet}{O}H_2 \longrightarrow [H{\leftarrow}OH_2]^+, \text{ or } H_3O^+$$

The Lewis definition extends the concept of acids to substances that do not contain hydrogen and cannot (as they can in the Brønsted-Lowry theory) donate a proton. For example, boron trifluoride and aluminium chloride are Lewis acids (page 248). In the formation of boron trifluoride-ammonia, boron trifluoride accepts a pair of electrons from ammonia:

$$\begin{array}{ccc} F & H \\ | & | \\ F{-}B + \overset{\bullet}{\underset{\bullet}{:}}N{-}H & \longrightarrow & F_3B{\leftarrow}NH_3 \\ | & | \\ F & H \end{array}$$

Similarly sulphur trioxide, usually regarded as an acid anhydride, is a Lewis acid. In the reaction between sulphur trioxide and calcium oxide, the trioxide accepts a pair of electrons from the oxide ion to form a sulphate ion:

$$O{=}\overset{O}{\underset{O}{\overset{||}{S}}} + \overset{\bullet\bullet}{\underset{\bullet\bullet}{:}}\overset{\bullet\bullet}{O}{:}^{2-} \longrightarrow \left[O{=}\overset{O}{\underset{O}{\overset{||}{S}}}{\leftarrow}O \right]^{2-} , \text{ or } SO_4^{2-}$$

Besides extending the definition of acids to a wider range of substances, the Lewis theory brings the acid/base concept closer to the electronic theory of oxidation and reduction, since oxidising agents *accept* electrons *donated* by reducing agents.

For most elementary work the Brønsted-Lowry theory of acids is sufficient.

Questions

1. Distinguish between the terms 'strong' and 'concentrated' as applied to acids.

2. Give the names and formulae of the conjugate bases of the following acids: HNO_3, H_2O, C_2H_5COOH, NH_4^+, HCl.

3. The following ions can all behave as either acids or bases. Give (a) the name and formula of the acids from which they are derived (b) the name and formula of the conjugate bases they can give rise to: HSO_4^-, $H_2PO_4^-$, HCO_3^-.

4. Calculate the pH of the following solutions (a) $0.01\,M\,HCl$ (b) $0.2\,M.\,CH_3COOH$ (c) $0.04\,M\,NaOH$. (The acid dissociation constant for ethanoic acid is $1.8 \times 10^{-5}\,mol\,dm^{-3}$.)

5. Calculate the pH of a solution resulting from the addition of (a) $24.9\,cm^3$ of $0.1\,M\,NaOH$ to $25\,cm^3$ of $0.1\,M\,HCl$ (b) $25.1\,cm^3$ of $0.1\,M\,NaOH$ to $25\,cm^3$ of $0.1\,M\,HCl$. Comment on the significance of the two figures.

6. Explain why it would not be suitable to try to find the concentration of a solution of ammonia by titration with a solution of ethanoic acid of concentration $0.1\,M$.

7. According to modern definitions an aqueous solution of hydrogen chloride is regarded as an acid whilst an aqueous solution of ammonia is regarded as a base. Explain these definitions with respect to the molecules mentioned.

How does water fit into this definition?   (AEB 1980)

8. This question is concerned with acids and bases.

(a) In each of the following reactions, one of the reactants acts as an acid. Identify the species which you consider to be the acid.

(i) $H_2O + NH_3 \rightarrow NH_4^+ + OH^-$

(ii) $CH_3CO_2H + HClO_4 \rightarrow CH_3CO_2H_2^+ + ClO_4^-$

(iii) $HCO_3^- + HSO_4^- \rightarrow H_2O + CO_2 + SO_4^{2-}$

(iv) $HNO_3 + 2H_2SO_4 \rightarrow NO_2^+ + H_3O^+ + 2HSO_4^-$

(v) $H_3O^+ + OH^- \rightarrow 2H_2O$

(b) A $0.1\,M$ solution of hydrofluoric acid has a pH of 2.15.

(i) What is the concentration of H^+ in this solution?

(ii) $0.1\,M$ solutions of the other hydrogen halides have a pH close to 1. What is the reason for the different acidity of hydrofluoric acid?

                                                              (L)

9. (a) What deductions may be made from the following figures, which were obtained when $20\,cm^3$ of a solution of phosphoric(V) acid, H_3PO_4, was titrated with a standard solution of sodium hydroxide under the conditions stated?

(i) $10\,cm^3$ of the alkali was required for neutralisation with methyl orange as an indicator.

(ii) $20\,cm^3$ of the alkali was needed for neutrality when phenolphthalein was used as the indicator.

(iii) $30\,cm^3$ of the alkali was required when titration (ii) was repeated in the presence of an excess of aqueous barium chloride.

(b) Calculate the relative atomic mass of an s-block metal, X, which forms an insoluble carbonate of formula XCO_3, from the following data:

$50\,cm^3$ of $0.100\,M\,Na_2CO_3(aq)$ was added to a hot aqueous solution containing $0.500\,g$ of the anhydrous chloride of X. This mixture was filtered and the filtrate (including the washings of the precipitate) was found to neutralise $51.9\,cm^3$ of $0.100\,M$ hydrochloric acid.   (O)

10. In a titration, 0.37 g of a weak organic acid (of relative molecular mass 74) was neutralised by 50 cm^3 of 0.1 M potassium hydroxide.

(a) Calculate the basicity of the weak organic acid. The pH of this acid in 0.05 M solution is 2.3.

(b) Calculate

(i)   the concentration of hydrogen ions, $[H^+(aq)]$,

(ii) the degree of dissociation, α.

(iii) the dissociation constant, K_a, of the acid.

(c) What would be the pH of the 0.05 M aqueous solution if the organic acid were a strong acid?   (C)

11. (a) Write an expression for the acid dissociation constant (K_a) of ethanoic(acetic) acid in the following equilibrium:

$$CH_3COOH(aq) \rightleftharpoons H^+(aq) + CH_3COO^-(aq)$$

(b) State the relationship between pK_a and K_a.

(c) Given that the approximate value of K_a for ethanoic acid is 1.8×10^{-5} mol litre^{-1}, calculate (i) the pK_a and (ii) the percentage ionisation of the acid in a 1 M aqueous solution.

(d) The pK_a values of $(CH_3)_3CCOOH$ and Cl_3CCOOH at 20°C are 5.05 and 0.65, respectively.

(i) Which is the stronger acid?

(ii) Explain your answer.   (JMB)

2 Electrochemical cells and redox reactions

12.1 Electrode potentials

If a rod of metal is placed in a solution of ions of the same metal an equilibrium is established:

$$M(s) \rightleftharpoons M^{n+}(aq) + ne^-$$

where n may be one, two or three.

The position of the equilibrium depends on the ease with which the metal loses electrons and the energy of hydration of the ions. For a reactive metal, such as zinc, the equilibrium lies well to the right. The electrons which are released accumulate in the metal, which becomes negatively charged and attracts a layer of positive ions forming an electrical double layer known as the Helmholtz double layer (Fig. 12.1).

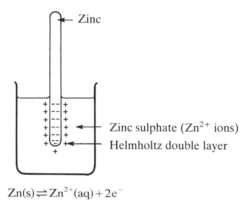

$$Zn(s) \rightleftharpoons Zn^{2+}(aq) + 2e^-$$

Fig. 12.1 Helmholtz double layer

This results in a potential difference between the metal and its solution known as the *electrode potential* of the metal; the electrode potential differs in magnitude with the tendency of the metal to lose electrons and form hydrated ions. For zinc this is plainly a negative electrode potential. For a less reactive metal such as copper the equilibrium between metal and metal ions lies more to the left:

$$Cu(s) \rightleftharpoons Cu^{2+}(aq) + 2e^-$$

There is a greater tendency for copper ions to be deposited from the solution than for copper to dissolve, and the pole acquires a positive electrode potential.

12.2 Electrochemical cells

That zinc forms hydrated ions more readily than copper is easily demonstrated by placing some metallic zinc in a solution of copper ions (copper

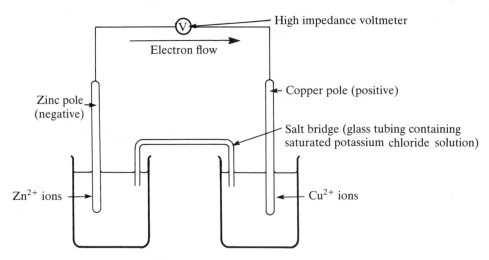

Fig. 12.2   Electrochemical cell

sulphate solution). The zinc dissolves to form zinc ions, losing two electrons in the process; the electrons are transferred to the copper ions, which are deposited as copper metal:

$$Zn(s) + Cu^{2+}(aq) \rightarrow Zn^{2+}(aq) + Cu(s)$$

The reaction is accompanied by a considerable release of energy in the form of heat. Some of this energy can be obtained as electrical energy if the electrons are transferred from the zinc to the copper through an external circuit, which is completed by connecting the solutions they are immersed in through a salt bridge or a porous partition, forming an *electrochemical cell* (Fig. 12.2).

In this cell zinc is dissolved from the zinc pole while copper is deposited on the copper pole; the external flow of electrons is from zinc to copper, that is, from the metal with the more negative electrode potential to the metal with the less negative electrode potential. In the solutions the electrons are transported by the ions present and pass from copper to zinc. (Conventionally, the *current* is said to flow from the positive copper pole to the negative zinc pole externally, and from the zinc to the copper in the cell solutions).*

If the cell is made up of a strip of magnesium immersed in a solution of magnesium ions and connected to a strip of zinc immersed in zinc ions, then the electron flow is from magnesium to zinc, since magnesium has a greater tendency than zinc to lose electrons, and so zinc becomes the positive pole. Similarly, if a copper half-cell is connected to a half-cell made by immersing a strip of silver in a solution of silver ions, the copper becomes the negative pole and the silver the positive pole (Fig. 12.3).

---

* Since an anode is defined as the electrode at which positive electricity enters the cell, the zinc, or negative electrode, is the anode of an electrochemical cell while the copper, or positive electrode, is the cathode. Thus the signs of the anode and cathode in an electrochemical cell are opposite to the signs of the anode and cathode in electrolysis.

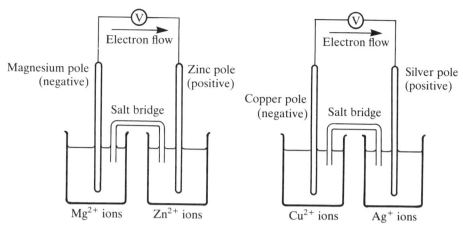

Fig. 12.3   Further examples of electrochemical cells

The potential difference between the two metal electrodes is measured by a high-impedance voltmeter (one that allows virtually no current to pass). This is the electromotive force, e.m.f., of the cell; it is measured in volts and given the symbol E. The value of E varies with temperature, concentration of the solution and the state of the metal used. The e.m.f. under standard conditions is that when the metal is in bar form and the concentration of the solution is $1\ mol\ dm^{-3}$ at a temperature of $25°C$. Under these conditions the e.m.f. of the zinc/copper cell is 1.1 volts. The cell is conventionally represented:

$$Zn(s)\,|\,Zn^{2+}(aq)\,\vdots\,Cu^{2+}(aq)\,|\,Cu(s) \qquad E = +1.1\ volts$$

The solid black lines represent the boundaries between the metals and their ions, and the dashed line* represents the salt bridge, porous partition or other device for connecting the half-cells. The sign of the e.m.f. of the cell represents the polarity of the right-hand electrode. The cell could be written in reverse order, with the zinc as the right-hand pole, in which case the value of E would be negative:

$$Cu(s)\,|\,Cu^{2+}(aq)\,\vdots\,Zn^{2+}(aq)\,|\,Zn(s) \qquad E = -1.1\ volts.$$

This particular cell is well known as the Daniell cell.

12.3 The hydrogen electrode and standard electrode potentials

It is not possible to measure the absolute value of a single electrode potential, since this necessarily implies the use of a second electrode to measure it against. To overcome this difficulty and to establish a series of *standard electrode potentials*, a relative scale of measurement has been adopted using a standard hydrogen electrode which is arbitrarily assigned an electrode potential of zero.

The hydrogen electrode is maintained by a stream of hydrogen gas bubbled over a platinum electrode, coated with platinum black (finely

* Other conventions are to use a single solid line | or double solid lines ‖.

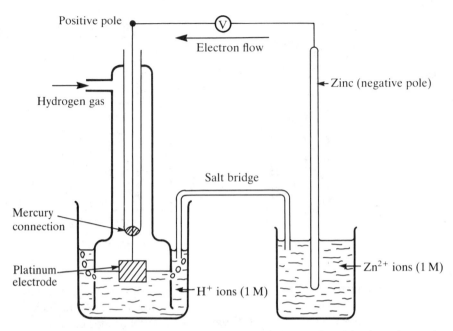

Fig. 12.4   Cell coupling standard hydrogen electrode with zinc electrode

divided platinum), immersed in a solution of H^+ ions. An equilibrium is set up between the gas adsorbed on the electrode and the hydrogen ions in the solution. The electrode potential of this electrode is fixed as zero under standard conditions of temperature (25°C) and pressure (1 atmosphere) using a concentration of hydrogen ions of one mole per litre.

Connecting other electrodes in circuit with the hydrogen electrode, the potential difference can be measured and a scale of standard electrode potentials established. It is conventional in the cell diagram to write the hydrogen electrode on the left in order to establish the polarity of the electrode of the half-cell on the right. If a zinc electrode is used the cell diagram is:

$$\text{Pt } H_2(g) \,|\, 2H^+(aq) \,\vdots\, Zn^{2+}(aq) \,|\, Zn(s) \qquad E^\ominus = -0.76 \text{ volts}$$

Thus the standard electrode potential for zinc is -0.76 volts. The zinc is the negative pole of the cell because it loses electrons more readily than hydrogen. Zinc is therefore a more powerful reducing agent (provider of electrons) than hydrogen in aqueous solution.

To find the overall reaction for the cell the half reactions should be written in order, starting with the more electronegative system:

$$Zn^{2+}(aq) + 2e^- \rightleftharpoons Zn(s) \qquad E^\ominus = -0.76 \text{ V} \qquad (1)$$

$$H^+(aq) + e^- \rightleftharpoons \tfrac{1}{2}H_2(g) \qquad E^\ominus = 0.00 \text{ V} \qquad (2)$$

Reaction (1) takes place from right to left since the zinc loses electrons more readily, while reaction (2) takes place from left to right since the H^+ ion collects the electrons and is reduced to hydrogen; combining (1) and (2) the cell reaction becomes:

$$Zn(s) + 2H^+(aq) \rightarrow Zn^{2+}(aq) + H_2(g)$$

which is consistent with the observed reaction when zinc metal is placed in

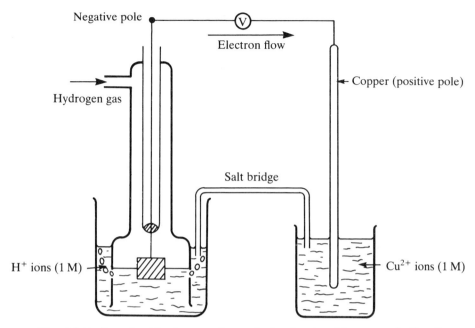

Negative pole

Electron flow

Copper (positive pole)

Hydrogen gas

Salt bridge

H^+ ions (1 M)

Cu^{2+} ions (1 M)

Fig. 12.5 Standard hydrogen electrode in circuit with copper half cell

acid solution. The equation is balanced by equating the number of electrons transferred in each half reaction.

When the hydrogen electrode is connected to a copper electrode, the current flows in the opposite direction. The electron flow in the external circuit is from the hydrogen electrode to the copper, since copper does not lose electrons as readily as hydrogen. In this case, the copper is the positive pole and the cell diagram therefore becomes:

$$Pt\ H_2(g)\ |\ 2H^+(aq)\ \vdots\ Cu^{2+}(aq)\ |\ Cu(s) \qquad E^\ominus = +0.34\ V$$

The positive e.m.f. of the cell once more gives the polarity of the right hand electrode.

Some values of standard electrode potentials are given below; fuller tables can be found in data books. Electrode systems may be written as Li^+/Li or $Li^+(aq) + e^- \rightarrow Li(s)$.

Table 12.1 The electrochemical series

Electrode	$E^\ominus$ volts
Ca^{2+}/Ca	−2.87
Na^+/Na	−2.71
Mg^{2+}/Mg	−2.37
Al^{3+}/Al	−1.67
Zn^{2+}/Zn	−0.76
Pb^{2+}/Pb	−0.13
$H^+/\frac{1}{2}H_2Pt$	0.00
Cu^{2+}/Cu	+0.34
Ag^+/Ag	+0.80

The order in which the elements occur in the table is known as the electrochemical series; the order of the metals corresponds in general to their reactivity.

12.4   Calculating the e.m.f. of cells from tables

Since the standard electrode potentials are all measured relative to the hydrogen electrode, the e.m.f. obtainable by combining any pair of electrodes can be calculated from the tables. Consider a cell made from zinc and copper electrodes. The electrode potentials given for each are -0.76 and $+0.34$ volts. Writing the cell diagrams so that the hydrogen electrodes link the two cells we have:

$$Zn(s) \,|\, Zn^{2+}(aq) \,\vdots\, 2H^+(aq) \,|\, H_2(g)Pt + Pt\, H_2(g) \,|\, 2H^+(aq) \,\vdots\, Cu^{2+}(aq) \,|\, Cu(s)$$

The value for the left hand cell is reversed giving an overall e.m.f. of $+0.76 + 0.34 = +1.10$ volts; the sign indicating that the copper electrode is the positive pole. This can be confirmed by measurement.

Thus in calculating the electromotive force of any given cell the following steps must be involved:

(a) Write the cell diagram for the system chosen.
(b) Look up the electrode potentials for each of the half-cells.
(c) Reverse the sign of the left-hand cell (since the system is reversed in the diagram).
(d) Add the revised values to obtain the e.m.f. of the cell and the polarity of the right-hand electrode.

Some worked examples will help to make this clear.
(i) Calculate the e.m.f. of a cell composed of lead and magnesium half-cells. The systems involved are

$$Mg^{2+} \,|\, Mg \qquad E^{\ominus} = -2.37\text{ V}$$

and

$$Pb^{2+} \,|\, Pb \qquad E^{\ominus} = -0.13\text{ V}$$

The cell diagram can be written:

$$Pb(s) \,|\, Pb^{2+}(aq) \,\vdots\, Mg^{2+}(aq) \,|\, Mg(s)$$

Reversing the sign of the lead electrode and adding the revised values we have:

$$0.13 - 2.37 = -2.24$$

indicating that the e.m.f. of the cell at 25°C is 2.24 volts and that magnesium is the negative pole.
(ii) Calculate the e.m.f. of a cell using magnesium and silver half-cells. The system involved this time are

$$Mg^{2+} \,|\, Mg \qquad E^{\ominus} = -2.37\text{ V}$$

and

$$Ag^+ \,|\, Ag \qquad E^{\ominus} = +0.80\text{ V}$$

The cell diagram can be written:

$$Mg(s) \mid Mg^{2+}(aq) \mid Ag^{+}(aq) \mid Ag(s)$$

Reversing the sign of the left-hand electrode and adding the revised values

$$2.37 + 0.80 = 3.17$$

giving a potential difference of $+3.17$ V, indicating that the cell has an e.m.f. of 3.17 V and that silver is the positive electrode.

12.5 The calomel electrode

The hydrogen electrode provides a valuable standard for comparing electrode potentials, but it is difficult to set up and to maintain under standard conditions; experimentally it is more convenient to use a reference electrode whose potential compared with the hydrogen electrode is known and which can be easily maintained at constant value.

One such electrode is the calomel electrode; this is set up as shown in Fig. 12.6. it has the advantage that it is easy to handle and maintains a steady electrode potential (0.334 volts) that is not altered by usage or storage.

The standard potential of an electrode system can now be measured by connecting it in circuit with the calomel electrode and reading the e.m.f. on the high impedance voltmeter. Since the calomel electrode has a standard potential of $+0.334$ volts this value must be added to the value found. For example, when coupled with a zinc electrode a reading of -1.094 volts is found; adding 0.334 to this gives the standard electrode potential for a zinc half-cell as -0.76, which corresponds to the value found using a hydrogen electrode.

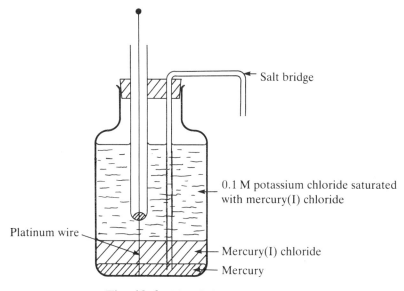

Fig. 12.6 The Calomel electrode

12.6   Concentration cells and the Nernst equation

So far all the cells considered have been standard cells in which the concentrations of the ions have been 1 mole dm^{-3} (1 M). If the concentration of the ions in the half-cells is altered the potential of the electrode is altered and consequently the e.m.f. of the cell. For a cell made up of copper and silver electrode systems the cell diagram is:

$$Cu(s) \,|\, Cu^{2+}(aq) \,\vdots\, Ag^{+}(aq) \,|\, Ag(s) \qquad E^{\ominus} = +0.46 \text{volts}$$

The half-reactions for the cells are:

$$Cu^{2+}(aq) + 2e^{-} \rightleftharpoons Cu(s) \quad +0.34 \text{ V}$$

$$Ag^{+}(aq) + e^{-} \rightleftharpoons Ag(s) \quad +0.80 \text{ V}$$

Suppose that the concentration of ions in the copper half-cell is reduced; by Le Chatelier's principle (page 164) the equilibrium will shift to the left, resulting in an increase in the supply of electrons to the copper electrode; the electrode becomes more negative (less than 0.34) and the e.m.f. of the cell increases. On the other hand, if the concentration of copper ions is kept constant and the concentration of silver ions is progressively reduced, the e.m.f. of the cell will steadily decrease. In this case the equilibrium again shifts to the left and the supply of electrons to the silver pole increases, making it more negative and bringing its electrode potential closer in value to that of the copper pole. Reverse effects can be obtained by increasing the concentrations of the copper or the silver ions.

A relationship between the concentrations of the ions on the half-cell and its electrode potential, known as the Nernst equation, has been derived:

$$E = E^{\ominus} + \frac{RT}{zF} \ln [\text{ion}]$$

where E is the new electrode potential/V
     $E^{\ominus}$ is the standard electrode potential/V
     R = the gas constant (8.313 J K^{-1})
     T = temperature/K
     F = the Faraday constant (96500 coulombs)
     z = charge on the metal ion in the cell
  [ion] = concentration of metal ion in the cell/mol dm^{-3}

Converting to logarithms to the base 10 this becomes:

$$E = E^{\ominus} + \frac{2.3 \, RT}{zF} \log [\text{ion}]$$

The value of $\dfrac{2.3 \, RT}{F}$ at 298 K is approximately 0.06 and the Nernst equation is frequently quoted as:

$$E = E^{\ominus} + \frac{0.06}{z} \log [\text{ion}]$$

The change of electrode potential with concentration of ions makes it possible to obtain small potential differences by connecting half-cells using the same metal but containing different concentrations of ions. Such an arrangement is called a concentration cell. For example, an e.m.f. of 0.059 volts can be obtained from the following arrangement:

$$Cu(s) \,|\, Cu^{2+}(0.01\ M) \,\vdots\, Cu^{2+}(aq)(1.0\ M) \,|\, Cu(s)$$
$$\text{negative pole} \qquad\qquad \text{positive pole}$$

$$E_{cell} = +0.059\ \text{volt}$$

Copper will dissolve from the negative pole as copper ions and will be deposited from the solution on to the positive electrode until the concentrations of each half-cell are equal, at which point the potential difference will vanish.

The Nernst equation can be used to find the concentration of ions in very dilute solutions. For example, a silver half-cell could be set up using a saturated solution of silver chloride as the cell solution, and its electrode potential measured. The standard electrode potential can be found in tables of data, and the charge on the silver ion is unity. These values substituted in the Nernst equation enable the concentration of silver ions to be calculated and hence the solubility of silver chloride.

12.7   Standard electrode potentials of other redox systems

During reactions such as:

$$I_2(aq) + 2e^- \rightleftharpoons 2I^-(aq)$$
$$Fe^{3+}(aq) + e^- \rightleftharpoons Fe^{2+}(aq)$$

electrons are transferred; systems involving such redox reactions (page 197) can be made into half-cells by inserting an inert electrode, usually platinum, into a solution of the appropriate ions.

If a cell such as that shown in Fig. 12.7 is set up, an e.m.f. of 0.23 V will be recorded.

The left hand electrode is represented as $I_2(aq), 2I^-(aq) \,|\, Pt$ and the right hand electrode as $Fe^{3+}(aq), Fe^{2+}(aq) \,|\, Pt$, the convention being that the more

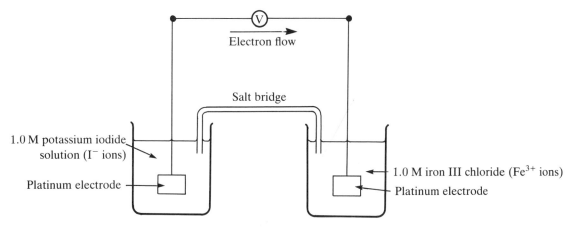

Fig. 12.7   E.M.F. from redox reaction

reduced species is written next to the platinum electrode. Electrons are transferred to the iron(III) chloride solution by the external circuit. The resultant formation of iodine in the left hand cell can be demonstrated by adding starch solution; the presence of Fe^{2+} ions in the other half cell is shown by adding potassium hexacyanoferrate(III) solution (page 331). When the cell is working the overall reaction is:

$$2Fe^{3+}(aq) + 2I^-(aq) \rightarrow 2Fe^{2+}(aq) + I_2(aq)$$

The electrode potentials for each half cell can be measured as for other electrodes by connecting them in circuits with hydrogen electrodes or other reference electrodes. The values are $+0.54$ volts for the I_2, $2I^-$ electrode and $+0.77$ volts for the Fe^{3+}, Fe^{2+} electrode. The cell diagram can be written:

$$Pt \mid 2I^-(aq), I_2(aq) \mathrel{\vdots} Fe^{3+}(aq)Fe^{2+}(aq) \mid Pt$$

The calculated e.m.f. is $(-0.54 + 0.76)$ volts, or 0.23 volts which corresponds to the measured value and shows the Fe^{3+}/Fe^{2+} electrode to be the positive pole.

The *standard redox potentials* for these and other electrodes are given below together with some of the metal/metal ion electrodes for comparison.

Table 12.2   Standard redox potentials

Electrode system	Standard redox potential $E^{\ominus}$ volts	Increasing reducing power	Increasing oxidising power
$Zn^{2+}(aq) \mid Zn(s)$	-0.76	↑	
$Fe^{2+}(aq) \mid Fe(s)$	-0.44		
$Sn^{2+}(aq) \mid Sn(s)$	-0.14		
$2H^+(aq), H_2(g) \mid Pt$	0.00		
$Sn^{4+}(aq), Sn^{2+}(aq) \mid Pt$	$+0.15$		
$Cu^{2+}(aq) \mid Cu(s)$	$+0.34$		
$I_2(aq), 2I^-(aq) \mid Pt$	$+0.54$		
$Fe^{3+}(aq), Fe^{2+}(aq) \mid Pt$	$+0.77$		
$Br_2(aq), 2Br^-(aq) \mid Pt$	$+1.09$		
$[Cr_2O_7^-(aq) + 14H^+(aq)], [2Cr^{3+}(aq) + 7H_2O(l)] \mid Pt$	$+1.33$		
$Cl_2(aq), 2Cl^-(aq) \mid Pt$	$+1.36$		
$[MnO_4^-(aq) + 8H^+(aq)], [Mn^{2+}(aq) + 4H_2O(l)] \mid Pt$	$+1.51$		↓

The order of the electrode potentials corresponds to the oxidising and reducing tendencies of the systems (Table 12.3).

12.8   Oxidation and reduction

The early simple definition of oxidation as the addition of oxygen to an element or compound:

$$2Mg(s) + O_2(g) \rightarrow 2MgO(s) \tag{1}$$

$$C_2H_5OH(l) + 3O_2(g) \rightarrow 2CO_2(g) + 3H_2O(l) \tag{2}$$

leads to the idea that reduction is the removal of oxygen (metal ores are reduced to metals):

$$2FeO(s) + C(s) \rightarrow 2Fe(s) + CO_2(g) \qquad (3)$$

and it is evident from the above equation that while iron oxide has been reduced to iron, carbon has been oxidised to carbon dioxide. Thus the two processes are complementary, and every oxidation is accompanied by a reduction. Considering the reaction between hydrogen and oxygen:

$$2H_2(g) + O_2(g) \rightarrow 2H_2O(l) \qquad (4)$$

it can be said that oxygen has oxidised hydrogen to water, or conversely that hydrogen has reduced oxygen to water. This extends the definitions of oxidation to include the removal of hydrogen from a compound and of reduction to include the addition of hydrogen to an element or compound.

$$2H_2S(g) + O_2(g) \rightarrow 2S(s) + 2H_2O(l) \qquad (5)$$

The removal of hydrogen from hydrogen sulphide can be effected by using chlorine, so that an oxidation process can be carried out without the use of oxygen:

$$H_2S(g) + Cl_2(g) \rightarrow S(s) + 2HCl(g) \qquad (6)$$

Here chlorine is the *oxidising agent*, while carbon in equation (3) is the *reducing agent*.

The idea of oxidation and reduction can be extended to the reactions of ions in solution, where they are explained in terms of electron transfer. For example, iron(II) oxide is oxidised to iron(III) oxide:

$$4FeO(s) + O_2(g) \rightarrow 2Fe_2O_3(s) \qquad (7)$$

Here the iron ions have been changed from Fe^{2+} to Fe^{3+}. A solution of iron(II) ions made by dissolving iron(II) sulphate in water can be oxidised by chlorine:

$$2Fe^{2+}(aq) + Cl_2(aq) \rightarrow 2Fe^{3+}(aq) + 2Cl^-(aq) \qquad (8)$$

This involves the loss of an electron from each Fe^{2+} ion. Thus we arrive at a universal definition of *oxidation as a loss of electrons*. Also, chlorine has been reduced to chloride ions by the gain of electrons, giving the definition of *reduction as a gain of electrons*.

Similarly, displacement reactions between metals and metal ions involve electron transfers, and are examples of redox reactions:

$$Zn(s) + Cu^{2+}(aq) \rightarrow Zn^{2+}(aq) + Cu(s) \qquad (9)$$

$$Cu(s) + 2Ag^+(aq) \rightarrow Cu^{2+}(aq) + 2Ag(s) \qquad (10)$$

Half-reactions can be written for *redox reactions*, showing the electrons involved. Thus for (9) the half-reaction concerning zinc is:

$$Zn(s) \rightarrow Zn^{2+}(aq) + 2e^-(aq)$$

while that for copper is:

$$Cu^{2+}(aq) + 2e^-(aq) \rightarrow Cu(s)$$

Zinc has been oxidised by the loss of electrons and the copper(II) ions have

Table 12.3 Oxidising and reducing agents

Oxidising agents	Reducing agents
$I_2(aq) + 2e^- \rightarrow 2I^-(aq)$ $Br_2(aq) + 2e^- \rightarrow 2Br^-(aq)$ $Cr_2O_7^{2-}(aq) + 14H^+(aq) + 6e^-$ $\rightarrow 2Cr^{3+}(aq) + 7H_2O(l)$ $Cl_2(aq) + 2e^- \rightarrow 2Cl^-(aq)$ $MnO_4^-(aq) + 8H^+(aq) + 5e^-$ $\rightarrow Mn^{2+}(aq) + 4H_2O(l)$ $S_2O_8^{2-}(aq) + 2e^- \rightarrow 2SO_4^{2-}(aq)$	$Zn(s) \rightarrow Zn^{2+}(aq) + 2e^-$ $H_2(g) \rightarrow 2H^+(aq) + 2e^-$ $H_2S(aq) \rightarrow 2H^+(aq) + S(s) + 2e^-$ $Sn^{2+}(aq) \rightarrow Sn^{4+}(aq) + 2e^-$ $Fe^{2+}(aq) \rightarrow Fe^{3+}(aq) + e^-$

been reduced by the gain of two electrons. Oxidising agents are defined as *electron acceptors* and reducing agents as *electron donors*.

Half-reactions for some typical oxidising and reducing agents are included in Table 12.3. The stronger the oxidising agent, the greater is its tendency to accept electrons.

The use of half-reactions makes the balancing of complex equations a much easier task. For example, iron(II) ions can be oxidised to iron(III) ions by the use of acidified potassium manganate(VII) solution. From the half-reactions, it can be seen that to change Fe^{2+} to Fe^{3+} involves the transfer of one electron, while the reduction of the manganate(VII) ion to the manganese(II) ion involves five electrons. Therefore in the balanced equation one manganate(VII) ion and five iron(II) ions must appear together with sufficient hydrogen ions to convert the oxygen, released in the breakdown of the manganate(VII) ion, to water.

$$MnO_4^-(aq) + 5Fe^{2+}(aq) + 8H^+(aq) \rightarrow Mn^{2+}(aq) + 5Fe^{3+}(aq) + 4H_2O(l)$$

The oxidation of iodide ions to iodine by potassium peroxodisulphate(VI) provides another example. Each peroxodisulphate ion, $S_2O_8^{2-}$, accepts two electrons on reacting to form two sulphate ions, SO_4^{2-}, and each iodide ion loses one electron when oxidised to iodine. Therefore two iodide ions are oxidised by each peroxodisulphate ion, and the equation may be written:

$$2I^-(aq) + S_2O_8^{2-}(aq) \rightarrow I_2(aq) + 2SO_4^{2-}(aq)$$

Even more complicated equations can be balanced quite readily if the transfer of electrons in the half-reactions is considered.

12.9 Oxidation numbers and nomenclature

The assignment of oxidation numbers to elements in a compound is a useful concept. The oxidation number is governed by the number of electrons gained, lost or shared by the element in bonding with other elements present. This is simple if the bonding is ionic, and is applied to covalently bonded elements by adopting the convention that the electron would pass to the element of greater electronegativity. Some elements, which enter only into ionic bonding, have invariable oxidation numbers. Others, including the transition metals, form a range of compounds in which they exhibit different oxidation numbers. To find the oxidation number of such an element in a molecule or a complex ion the following rules must be applied:

(a) The oxidation number of any element in its normal or ground state is zero e.g. S in S_8 or H in H_2.

(b) The oxidation number of an element present as an ion is the same as the charge on the ion; thus in sodium chloride, Na^+Cl^-, the oxidation number of sodium is +1 and of chlorine is −1, and in copper(II) chloride, $Cu^{2+}(Cl^-)_2$, the oxidation number of copper is +2 and of chlorine is −1.

(c) The oxidation number of the more electronegative element in a covalent molecule is negative and that of the less electronegative is positive, the values depending on the number of electrons shared by the element in question. Thus in ammonia, NH_3, the oxidation number of nitrogen is −3 and that of hydrogen is +1.

(d) The algebraic sum of the oxidation numbers of all the atoms present in any compound is zero; likewise the algebraic sum of the oxidation numbers of the atoms in a complex ion is equal to the charge on the ion. Thus in the sulphate ion, SO_4^{2-}, the charge on the ion is −2, the sum of the oxidation numbers of four oxygen atoms is 4(−2) or −8, and hence the oxidation number of sulphur in the ion is +6.

(e) Certain elements have fixed oxidation numbers:

(i) Group I metals invariably have an oxidation number +1.

(ii) Group II metals invariably have an oxidation number +2.

(iii) Hydrogen has an oxidation number +1 (except in metallic hydrides (−1)).

(iv) Oxygen has an oxidation number −2 (except in peroxides (−1) and difluorine monoxide F_2O (+2)).

(v) Fluorine always has an oxidation number −1.

Applying these rules it can be seen that the oxidation number of manganese in $KMnO_4$ is +7, since that of potassium is +1 and of four oxygen atoms is −8, giving a sum of zero. Also, chlorine in the chlorate(V) ion has an oxidation number of +5, giving a charge of −1 on the ion, ClO_3^-.

It is from this concept of oxidation number that the IUPAC nomenclature arises. The numbers in roman figures in the name of a compound indicate the oxidation number of the simple ion in a compound and the oxidation number of the central atom in a molecule or complex ion. Thus copper(II) sulphate(VI) indicates that copper is present as an ion with two positive charges (oxidation number +2) and that the sulphur atom has an oxidation number of +6. Similarly the name copper(I) oxide clearly denotes the formula of red copper oxide as Cu_2O, and copper(II) oxide indicates that black copper oxide is CuO.

The concept of oxidation number can also be used in the balancing of equations. If it is known that aqueous acidified potassium manganate(VII) is reduced to manganese ions by the action of hydrogen sulphide, which is itself oxidised to sulphur, then the oxidation number of manganese has been reduced from +7 to +2 while that of sulphur has been raised from −2 to 0. Hence the reacting ratios of hydrogen sulphide and manganate(VII) ions must be 5:2, and the equation can be written:

$$5H_2S(aq) + 2MnO_4^-(aq) \rightarrow 2Mn^{2+}(aq) + 5S(s)$$

This does not account for the oxygen, so hydrogen ions (from the acidification) must be added to the equation to balance it:

$$5H_2S(aq) + 2MnO_4^-(aq) + 6H^+(aq) \rightarrow 2Mn^{2+}(aq) + 5S(s) + 8H_2O(l)$$

It can also be seen that oxidation of an element leads to an increase in oxidation number and reduction to a decrease in oxidation number. An oxidation number chart for an element with variable oxidation number, such as that given for nitrogen (page 275), provides a useful summary of the types of compounds the element will form.

12.10 Use of standard redox potentials to predict the course of reactions

In general, it has been shown that electrode systems high in the electrochemical series are more powerful reducing agents than electrode systems with a more positive potential, and an equation for the overall reaction can be written. A knowledge of the standard electrode potentials of redox half reactions can therefore be used to predict the direction of a reaction and to determine if one component is able to oxidise or reduce another. For example, in the last section it was shown that iron(III) ions oxidise iodide ions to iodine, so it might be expected that a similar reaction would take place between iron(III) ions and chloride ions. However, chlorine is a more powerful oxidising agent than iodine, and a study of the electrode potentials gives the following figures:

$$I_2(aq), 2I^-(aq) \mid Pt \qquad E^{\ominus} + 0.54 \text{ volts} \qquad (1)$$

$$Fe^{3+}(aq), Fe^{2+}(aq) \mid Pt \qquad E^{\ominus} + 0.77 \text{ volts} \qquad (2)$$

$$Cl_2(aq), 2Cl^-(aq) \mid Pt \qquad E^{\ominus} + 1.36 \text{ volts} \qquad (3)$$

The appropriate half-reactions for (1) and (2), written in the order in which they occur in the table, are:

$$I_2(aq) + 2e^- \rightleftharpoons 2I^-(aq)$$

$$Fe^{3+}(aq) + e^- \rightleftharpoons Fe^{2+}(aq)$$

The iodide ions in the more negative electrode system have the greater reducing power, so the first reaction moves to the left, supplying electrons; the Fe^{3+} ions in the more positive electrode system have a greater oxidising power, so the second reaction moves to the right, taking electrons. The overall reaction is therefore:

$$2Fe^{3+}(aq) + 2I^-(aq) \rightarrow 2Fe^{2+}(aq) + I_2(aq)$$

and the iodine ions have been oxidised, as expected. The half-cell reactions for (2) and (3) are:

$$Fe^{3+}(aq) + e^- \rightleftharpoons Fe^{2+}(aq)$$

$$Cl_2(aq) + 2e^- \rightleftharpoons 2Cl^-(aq)$$

In this case the chlorine half-cell is the more positive and has the greater oxidising power, so the overall reaction is from chlorine to chloride, and iron(II) ions are oxidised to iron(III) ions by chlorine:

$$Cl_2(aq) + 2Fe^{2+}(aq) \rightarrow 2Cl^-(aq) + 2Fe^{3+}(aq)$$

Thus to predict the overall reaction the half reactions must be written in the correct order; it then follows that the first reaction proceeds from right to left and the second from left to right, and the two can be combined into one equation.

The general rule is that a system higher in the electrochemical series will be oxidised by a system lower in the series. However, although a correct prediction of the direction of a possible reaction can be made, it does not necessarily follow that such a reaction will take place. The prediction does not include any information concerning the rate of reaction, which may be so slow that the reaction, although possible, does not actually take place. The second limitation is that reactions between electrode systems that have only a small difference in electrode potential can readily be reversed by altering the conditions. Reactions between electrodes with a potential difference greater than ± 0.4 V will normally follow the predicted direction, but those with potentials closer than this may be affected by the conditions of the reaction.

Potassium manganate(VI) (page 327) can be made by mixing manganese(IV) oxide and potassium manganate(VII) in very alkaline conditions, although the standard electrode potentials are:

$$2MnO_4^-(aq) + 2e^- \rightleftharpoons 2MnO_4^{2-}(aq) \quad E^\ominus = +0.56 \text{ V}$$

$$MnO_4^{2-}(aq) + 2H_2O(l) + 2e^- \rightleftharpoons MnO_2(s) + 4OH^-(aq) \quad E^\ominus = +0.60 \text{ V}$$

From these the expected reaction would be the disproportionation of the manganate(VI) ion to give manganese dioxide and manganate(VII) ions; this reaction does take place if acid is added to the reaction mixture. However, provided the equilibrium of the lower reaction is upset by the presence of excess OH^- ions, manganate(VI) ions are formed.

12.11   Potentiometric titrations and the pH meter

A standard hydrogen electrode made up with a concentration of hydrogen ions of one mole per litre is assigned an electrode potential of zero; as with other cells the potential of this electrode changes with a change of concentration of ions in the half-cell. For a concentration of H^+ ions greater than one mole per litre the potential becomes more positive, while with a lower concentration of ions the potential becomes more negative. Thus the change of pH during an acid/alkali titration can be followed by immersing a hydrogen electrode in the solution to be titrated and connecting it to a reference electrode such as the calomel electrode. As the alkali is added the hydrogen ion concentration changes, causing a change in potential difference between the two electrodes which can be measured on a high impedance voltmeter. Since pH $= -\log[H^+]$ the variation in pH during the titration can be followed.

A more convenient instrument, the pH meter, has been developed from this arrangement. The hydrogen electrode is replaced by a glass electrode consisting of a thin-walled glass bulb into which is sealed a platinum wire immersed in a solution of hydrogen ions of constant concentration (a buffer solution). When such a bulb is immersed in an acid solution a potential drop develops across the thin glass membrane which is related to the hydrogen ion concentration, and therefore to the pH, of the acid solution. Such an electrode can be connected through the platinum wire to a silver electrode, and the resulting e.m.f. measured. In practice the silver electrode and the glass electrode are incorporated into one unit, or probe, which can be plugged into a high impedance voltmeter calibrated to read pH directly. Titrations of solutions of dark colours, where an indicator cannot be used, can be carried out using a pH meter. The pH probe and meter can also be used to measure the pH of solutions without contaminating them.

Questions

1. A cell is set up between copper and silver:

$$Cu(s) \,|\, Cu^{2+}(aq) \,\vdots\, Ag^+(aq) \,|\, Ag(s)$$

the reaction being:

$$Cu(s) + 2Ag^+(aq) \rightleftharpoons Cu^{2+}(aq) + 2Ag(s)$$

(i) Calculate the standard e.m.f. of this cell.
(ii) Explain how this e.m.f. would change if the concentration of $Cu^{2+}(aq)$ were increased.   (C)

2. Consider the following standard electrode potentials, all of which refer to aqueous solutions.

$$Cr^{3+}, Cr^{2+} \,|\, Pt \qquad\qquad\qquad\qquad E^{\ominus} = -0.41 \text{ V}$$
$$(Cr_2O_7^{2-} + 14H^+), (2Cr^{3+} + 7H_2O)/Pt \quad E^{\ominus} = +1.33 \text{ V}$$
$$Ce^{4+}, Ce^{3+} \,|\, Pt \qquad\qquad\qquad\qquad E^{\ominus} = +1.70 \text{ V}$$

(i) Write down the equations for the half-cell reactions corresponding to these three standard electrode potentials.
(ii) Find the e.m.f. of the following cells, all concentrations being unit.

$$Pt \,|\, Cr^{2+}, Cr^{3+} \,\vdots\, Ce^{4+}, Ce^{3+} \,|\, Pt$$
$$Pt \,|\, (2Cr^{3+} + 7H_2O), (Cr_2O_7^{2-} + 14H^+) \,\vdots\, Ce^{4+}, Ce^{3+} \,|\, Pt$$

(iii) In view of the e.m.f. so calculated, what would be expected to happen when a solution containing an equal number of moles of Ce^{3+} and Ce^{4+} ions was added to an acidified solution of Cr^{3+} ions? Write the equation for the reaction.   (L)

3. (*a*) How does a knowledge of the relevant standard electrode potentials enable a comment to be made on the likelihood that a particular redox reaction will occur?

(*b*) The standard electrode potentials, in volts, for a number of half-reactions in aqueous acid are as follows:

$$Fe^{3+} + e^- \rightleftharpoons Fe^{2+} \qquad E^{\ominus} = +0.77 \text{ V}$$
$$Ag^+ + e^- \rightleftharpoons Ag \qquad E^{\ominus} = +0.80 \text{ V}$$
$$\tfrac{1}{2}I_2 + e^- \rightleftharpoons I^- \qquad E^{\ominus} = +0.54 \text{ V}$$
$$\tfrac{1}{2}Br_2 + e^- \rightleftharpoons Br^- \qquad E^{\ominus} = +1.07 \text{ V}$$

Comment on the following:

(i) If 10 cm^3 of 1.0 M iron(III) chloride solution is added to 10 cm^3 of 1.0 M potassium iodide solution, iodine is liberated. However if 100 cm^3 of 1.0 M sodium fluoride solution is added to the iron(III) chloride solution before it is added to the iodide solution, no iodine is liberated.

(ii) No iodine is liberated when 1.0 M silver nitrate solution is added to 1.0 M sodium iodide solution.

(c) Would you expect 1.0 M iron(III) chloride solution to liberate bromine from 1.0 M sodium bromide solution?

(d) What would you expect to be the outcome of adding silver powder to 1.0 M iron(III) nitrate solution? (O)

4. (a) The following aqueous equations represent oxidation/reduction reactions occurring in aqueous solution. By considering each reaction in terms of electron transfer explain which species is *oxidised* and which is *reduced*.

(i) $2FeCl_2 + Cl_2 \rightarrow 2FeCl_3$

(ii) $I_2 + 2Na_2S_2O_3 \rightarrow 2NaI + Na_2S_4O_6$

State, and account for, the colour change observed in each case.

(b) By considering the ionic half equations:

$$Ce^{4+} + e^- \rightarrow Ce^{3+}$$
$$C_2O_4^{2-} \rightarrow 2CO_2 + 2e^-$$
$$Fe^{2+} \rightarrow Fe^{3+} + e^-$$

calculate the volume of cerium(IV) sulphate (of concentration 0.2 mol/l) required to oxidise 25 cm^3 of iron(II) diethanoate (iron(II) oxalate) solution of concentration 0.6 mol/l.

(c) Given the following standard electrode potentials at $25°C$

$$Pb^{2+} + 2e^- \rightleftharpoons Pb \qquad E^{\ominus} = -0.126 \text{ V}$$
$$Zn^{2+} + 2e^- \rightleftharpoons Zn \qquad E^{\ominus} = -0.763 \text{ V}$$

calculate the e.m.f. of the cell:

$$Zn(s) \,|\, Zn^{2+}(aq, 1 \text{ M}) \,\vdots\, Pb^{2+}(aq, 1 \text{ M}) \,|\, Pb(s)$$

(AEB 1976)

13   Electrolysis

13.1   Introduction

Electrolysis, which was developed soon after Volta's invention of the electrochemical cell in 1800, is one of the modern chemist's most useful tools. Humphry Davy used electrolysis in the early years of the nineteenth century to prepare the very reactive metals of Groups I and II, which had so far defied isolation. Later, after the further development of electric power, the extraction of aluminium by electrolysis made this most useful metal cheap and readily available.

Electrolysis is also employed to purify copper, for electroplating, and in the manufacture of chlorine and sodium hydroxide from brine. In the laboratory it can be used to measure the solubility of very sparingly soluble salts such as silver chloride, and to determine Avogadro's constant.

Electrolysis is the decomposition of a molten or dissolved *electrolyte* by the passage of an electric current between two electrodes connected to a battery or other source of direct current. The chemical changes which result occur at the electrodes; the electrode connected to the positive terminal of the battery is the *anode* and that connected to the negative terminal is the *cathode*.

An electrolyte is a substance which conducts electricity when fused or dissolved in a polarising solvent (usually water) and is thereby decomposed. Electrolytes are salts, basic oxides, hydroxides or acids, all of which provide ions to transport electrons through the liquid from cathode to anode. The battery or other source of electricity, required to provide the electrons, can be regarded as an 'electron pump'.

The ions are attracted to the electrode of opposite charge to themselves; those travelling to the anode (anions or negative ions), give up electrons while those travelling to the cathode, (cations or positive ions), collect electrons. The ions are said to be *discharged* at the electrodes.

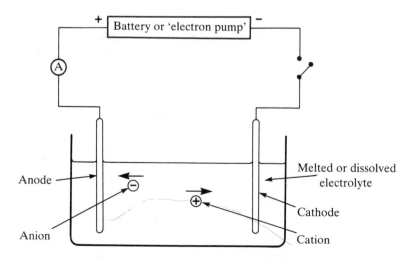

Fig. 13.1   Electrolysis cell

13.2   Selective discharge of ions

Positive metallic ions and hydrogen ions are discharged at the cathode, and negative ions, such as chloride, Cl^-, bromide, Br^-, iodide, I^-, and hydroxide ions, OH^-, are discharged at the anode. For a single substance in the molten state such as lead bromide the situation is very simple; lead ions, Pb^{2+}, collect two electrons at the cathode and metallic lead is deposited:

$$Pb^{2+} + 2e^- \rightarrow Pb$$

At the anode two bromide ions each give up an electron and bromine collects as a gas:

$$2Br^- \rightarrow Br_2 + 2e^-$$

If the electrolyte is dissolved in water the situation is not so straightforward, and the species discharged at the electrode varies according to the conditions. In addition to the ions of electrolyte present there are now ions derived from water $(H^+ + OH^-)$; although these ions will be present only in low concentrations they may be discharged in preference to ions of the electrolyte.

The ease of discharge of each ion is governed by its electrode potential (page 187). If inert (platinum or carbon) electrodes are used the order of discharge of some ions is:

Cations $Ag^+, Cu^{2+}, H^+, Ni^{2+}, Fe^{2+}, Zn^{2+}, Al^{3+}, Mg^{2+}, Na^+, Ca^{2+}, K^+$
Anions $OH^-, I^-, Br^-, Cl^-, SO_4^{2-}, F^-$

This order of discharge of ions in solutions of comparable concentrations is sometimes altered by the conditions of the electrolysis. The main factors involved here are concentration, the nature of the electrodes and overvoltage.

(a) Concentration

If two ions are present, one of which is more easily discharged than the other, then under normal conditions that ion will be discharged first. However, if the ion that requires more energy for discharge is present in very much greater concentration then it will be discharged preferentially. For example, in the electrolysis of a dilute solution of sodium chloride, oxygen will collect at the anode. This results from the discharge of OH^- ions from the aqueous solution:

$$2OH^-(aq) \rightarrow H_2O(l) + \tfrac{1}{2}O_2(g) + 2e^-$$

However, in the electrolysis of a concentrated solution of sodium chloride, the gas collected at the anode is mainly chlorine; the high concentration of chloride ions results in their preferential discharge.

$$2Cl^-(aq) \rightarrow Cl_2(g) + 2e^-$$

(b) Nature of the electrodes and overvoltage

The ease of discharge of the ions is not always the same for every electrode, and change of electrode can alter the order of discharge of the ions. An ion

Table 13.1 Electrolytic processes

Electrolyte	Electrodes	Anode reaction	Cathode reaction	Cell contents	Use
Cold brine	Graphite	$2Cl^-(aq) \rightarrow Cl_2(g) + 2e^-$	$2H^+(aq) + 2e^- \rightarrow H_2(g)$	Increasing concentration of sodium hydroxide with dissolved chlorine	Manufacture of sodium chlorate (I) for domestic bleaches
Hot concentrated brine	Graphite	$2Cl^-(aq) \rightarrow Cl_2(g) + 2e^-$	$2H^+(aq) + 2e^- \rightarrow H_2(g)$	Increasing concentration of sodium hydroxide with dissolved chlorine	Manufacture of sodium chlorate (V) for weedkiller
Brine	Graphite anode mercury cathode	$2Cl^-(aq) \rightarrow Cl_2(g) + 2e^-$	$Na^+ + e^- \rightarrow Na$	Brine continually renewed	Manufacture of chlorine and sodium hydroxide
Molten sodium chloride	Graphite anode steel cathode	$2Cl^-(l) \rightarrow Cl_2(g) + 2e^-$	$Na^+(l) + e^- \rightarrow Na(l)$	Gradually consumed and continually replenished	Manufacture of sodium metal
Aluminium oxide dissolved in molten cryolite	Graphite	Various	$Al^{3+}(l) + 3e^- \rightarrow Al(l)$	Gradually consumed (aluminium oxide continually replenished)	Manufacture of aluminium
Copper sulphate solution	Copper	$Cu(s) \rightarrow Cu^{2+}(aq) + 2e^-$	$Cu^{2+}(aq) + 2e^- \rightarrow Cu(s)$	Copper sulphate concentration remains steady	Purifying copper metal

which cannot be discharged with normal ease at certain electrodes is said to have an *overvoltage* at these electrodes. For example, if platinum electrodes are used in the electrolysis of brine, hydrogen ions are discharged at the cathode, leaving sodium ions in the solution; but if the platinum electrode is replaced by a mercury cathode, then sodium ions are discharged leaving the hydrogen ions in the solution. The hydrogen ions thus require a potential greater than their normal electrode potential for discharge at a mercury surface; hydrogen has an overvoltage with respect to mercury. The nature of overvoltage is not clear; it depends on the ion, the type of electrode and the conditions of the electrolysis.

Another important example is the deposition of zinc from solutions of zinc ions. If a platinum cathode is used then hydrogen collects at the cathode, as would be expected since zinc is higher in the electrochemical series (has a more negative electrode potential) than hydrogen; however, if a zinc cathode is used then zinc is deposited preferentially since hydrogen has an overvoltage at a zinc electrode.

Changing the nature of the anode can also alter the electrolytic process which takes place. The anode collects electrons during electrolysis; when inert electrodes are used these come from the negative ions which are attracted to the positive electrode. In the electrolysis of copper sulphate using platinum electrodes the negative ions attracted to the anode are sulphate ions, SO_4^{2-}, and hydroxide ions, OH^-, from the water. Of these two the OH^- ions are the more easily discharged and they give up their electrons, decomposing to water and oxygen, which is collected. If a copper anode is used then the electrons are provided by the anode itself. The copper releases electrons and dissolves as copper(II) ions, Cu^{2+}, in a reaction that requires less energy than the discharge of OH^- ions.

$$Cu(s) \rightarrow Cu^{2+}(aq) + 2e^-$$

The overall result of electrolysing copper sulphate solution with copper electrodes is that equal amounts of copper are dissolved from the anode and deposited at the cathode; the concentration of the copper sulphate solution remains constant throughout.

Some of the more important electrolytic processes are summarised in Table 13.1.

13.3   Faraday's laws

The quantitative laws relating the amount of electricity passed in electrolysis to the mass or volume of elements discharged at the electrodes were first formulated by Michael Faraday in 1834.

The first law states that the mass of a substance liberated at an electrode during electrolysis is proportional to the quantity of electricity passed.

$$\text{mass (grams)} \propto \text{current (ampères)} \times \text{time (seconds)}$$

or

$$\text{mass} \propto As$$

or

$$\text{mass} = e \times A \times s$$

where e is the *electrochemical equivalent* of the substance. This is defined as the mass of substance liberated by one coulomb of electricity, a quantity that can be measured very accurately.

The second law relates the masses of different substances that are liberated by the same amount of electricity. In modern terms, the passage of one mole of electrons dissolves from the anode or deposits at the cathode the mass of a mole of the substance divided by the number of electrons involved in the electrode reaction. For example, the passage of one mole of electrons deposits one mole of silver atoms (107.88 g) at the cathode, but only one third of a mole of aluminium atoms (9 g) according to the equations:

$$Ag^+(aq) + e^- \rightarrow Ag(s)$$

and

$$Al^{3+}(aq) + 3e^- \rightarrow Al(s)$$

In the discharge of gases account must be taken of the atomicity of the gaseous molecule. For example, oxygen is released at the anode when hydroxide ions are discharged:

$$2OH^-(aq) + 2OH^-(aq) \rightarrow 2H_2O(l) + O_2(g) + 4e^-$$

Here one mole of oxygen molecules is released by the passage of four moles of electrons; or the passage of one mole of electrons (96 500 coulombs) releases a quarter of a mole of oxygen molecules. In terms of the volume of gaseous oxygen molecules released, this is one quarter of a mole or $22.4/4$ dm^3. By similar reasoning it can be seen that half a mole, or 11.2 dm^3 of chlorine gas molecules are released by one mole of electrons in the reaction:

$$2Cl^-(aq) \rightarrow Cl_2(g) + 2e^-$$

The quantity of electricity involved in the passage of a mole of electrons is approximately 96 500 coulombs, and is sometimes called the Faraday constant. This figure is calculated from Faraday's first law and the experimentally determined mass of silver deposited by one coulomb. It is found that 111.8×10^{-5} gram of silver is deposited by one coulomb, therefore a mole of silver atoms (107.88 grams) is deposited by $\dfrac{107.88}{111.8 \times 10^{-5}}$ or approximately 96 500 coulombs.

A further calculation can be made which gives a very reliable estimate of the Avogadro constant, L. The deposition of 107.88 grams of silver requires the passage of one mole of electrons:

$$Ag^+(aq) + e^- \rightarrow Ag(s)$$

The charge on an electron has been accurately measured by Millikan and found to be 1.6×10^{-19} coulombs, giving the total charge on one mole of electrons as $L \times 1.6 \times 10^{-19}$ coulombs; this we have just shown to be 96 500 (96 487) coulombs.

Hence

$$96\,487 = L \times 1.60 \times 10^{-19}$$

or

$$L = \frac{96\,487}{1.60 \times 10^{-19}} \text{ or } 6.03 \times 10^{23}$$

13.4 Conductivity

(a) Electrolytic conductivity

Electrolytic solutions offer resistance to the passage of an electric current in the same way that solid conductors do. The characteristic *resistivity* (rho) of a solution of an electrolyte is defined as the resistance in ohms between the opposite faces of a unit cube of the solution. As for a solid conductor, the resistance of a tube of solution is proportional to the length and inversely proportional to the cross-sectional area, and

$$R = \frac{l}{A} \times \rho$$

where R = resistance, l = length of tube, A = cross-sectional area, and ρ = resistivity. The SI unit for resistivity is the ohm metre (Ω m), but the more convenient ohm centimetre (Ω cm) is generally used.

In studying electrolysis, our interest is in the extent to which a solution *conducts* a current, and the characteristic *conductivity* (kappa) is the usual basis of comparison. Conductivity is the reciprocal of resistivity

$$\kappa = \frac{1}{\rho}$$

and if ρ is in Ω cm units, then κ is in Ω^{-1} cm^{-1} units. The conductivity of an electrolytic solution depends on the nature of the solution and its concentration.

Electrolytic conductivities are measured in a conductivity cell, which is balanced against a variable resistance in a Wheatstone bridge circuit.

The solution must be made up with *conductivity water*—specially purified distilled water having a conductivity of less than 0.5×10^{-6} ohm^{-1} cm^{-1}. The

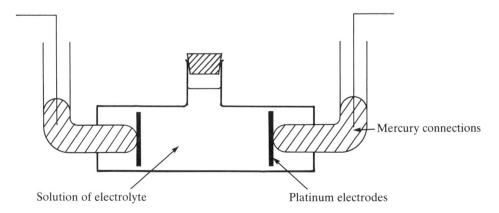

Solution of electrolyte   Platinum electrodes   Mercury connections

Fig. 13.2 A conductivity cell

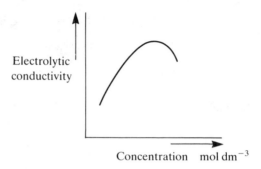

Fig. 13.3 Conductivity of electrolytes

platinum electrodes are coated with platinum black to reduce polarisation and alternating current is used to prevent any 'back e.m.f.'.

The variation of the conductivity of an electrolyte at a constant temperature depends on its concentration; in general it rises to a maximum and then decreases. However, although the same shape of curve is obtained for both strong and weak electrolytes, it must be remembered that the relative conductivities of these two types of electrolytes are so different that the two curves cannot be compared on the same scale.

For strong electrolytes, which are fully ionised, increasing concentration at first results in increasing conductivity, but after a maximum is reached the presence of more ions produces interference. This prevents freedom of movement and results in a reduction of current-carrying capacity.

(b) Molar conductivity

To compare the conductivities of solutions of different concentrations by means of their electrolytic conductivities is not entirely satisfactory, on account of the ionisation and interference effects described above. To allow for these, the idea of *molar conductivity* has been introduced: the conductivity which would be found for a volume, $V \text{ cm}^3$, of solution containing one mole of electrolyte placed between electrodes one centimetre apart. The theoretical volume V required is called the *dilution* of the solution; for a 0.1 M solution V would be $10\,000 \text{ cm}^3$. Molar conductivities are not of course measured directly, they are calculated by multiplying the actual electrolytic conductivity of a solution by the dilution V corresponding to the molar concentration of the electrolyte.

$$\text{Molar conductivity } \lambda = \kappa \times V.$$

Since κ is measured in $\Omega^{-1} \, cm^{-1}$ and V in $cm^3 \, mol^{-1}$ the units of λ are $cm^2 \, \Omega^{-1} \, mol^{-1}$.

The molar conductivity can also be expressed in terms of concentration, which is the reciprocal of dilution:

$$\lambda = \kappa \times V = \kappa \times \frac{1}{\text{concentration/mol cm}^{-3}}$$

The variation of molar conductivity for strong and weak electrolytes can now be studied by plotting molar conductivity against (a) the dilution or (b) the square root of the concentration of the solution:

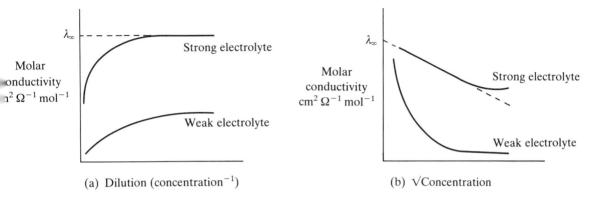

(a) Dilution (concentration^{-1})   (b) $\sqrt{}$Concentration

Fig. 13.4 Variation of molar conductivities for strong and weak electrolytes

The molar conductivity of strong electrolytes steadily increases with increasing dilution, and a maximum value at infinite dilution, λ_∞ can be obtained by extrapolating the curves in (a) or (b) of Fig. 13.4.

For weak electrolytes such an extrapolation is not possible as the conductivity at such low concentrations cannot be accurately measured. However, values for the molar conductivities at infinite dilution can be found indirectly by the application of *Kohlrausch's Law* concerning the independent migration of ions.

(c) Kohlrausch's law

Kohlrausch, working in 1876, measured the molar conductivities of a great number of salts having various combinations of cations and anions.

The molar conductivities of some sodium and potassium salts are given in Table 13.2:

Table 13.2. Molar conductivities of some sodium and potassium salts

Salt	Molar conductivity λ_∞ $cm^2 \, \Omega^{-1} mol^{-1}$ at room temperature
KNO_3	145
$NaNO_3$	121
KCl	150
$NaCl$	126

From these figures it can be seen that there are constant differences for salts in which one ion remains the same:

System	Difference
$\lambda_\infty KCl - \lambda_\infty NaCl$	24
$\lambda_\infty KNO_3 - \lambda_\infty NaNO_3$	24
$\lambda_\infty KCl - \lambda_\infty KNO_3$	5
$\lambda_\infty NaCl - \lambda_\infty NaNO_3$	5

Evidently each ion contributes its own share to the molar conductivity independently of any other ion present, and the molar conductivity of an electrolyte is the sum of the molar conductivities of the ions it forms. For example:

$$\lambda_\infty KCl = \lambda_\infty K^+ + \lambda_\infty Cl^-$$

The molar conductivities of specific ions can be measured, but are not in fact needed to determine the molar conductivity of weak electrolytes, as this can be done by difference. For example, to find that of ethanoic acid, the molar conductivities at infinite dilution of hydrochloric acid, sodium ethanoate and sodium chloride are measured; these are respectively 426, 91 and $126 \, cm^2 \, \Omega^{-1} \, mol^{-1}$.

Now

$$\lambda_\infty HCl = \lambda_\infty H^+ + \lambda_\infty Cl^-$$

and

$$\lambda_\infty NaEt = \lambda_\infty Na^+ + \lambda_\infty Et^-$$

and

$$\lambda_\infty NaCl = \lambda_\infty Na^+ + \lambda_\infty Cl^-$$

also

$$\lambda_\infty HCl + \lambda_\infty NaEt = \lambda_\infty H^+ + \lambda_\infty Cl^- + \lambda_\infty Na^+ + \lambda_\infty Et^-$$

$$\therefore \quad \lambda_\infty H^+ + \lambda_\infty Et^- = \lambda_\infty HCl + \lambda_\infty NaEt - \lambda_\infty NaCl$$

or $\quad 426 + 91 - 126$

or $\quad 391 \, cm^2 \, \Omega^{-1} \, mol^{-1}$

Since the molar conductivity of a weak electrolyte increases as the concentration decreases it can be assumed that at infinite dilution the electrolyte is fully dissociated and therefore the degree of ionisation, α, of a weak electrolyte of concentration $c \, mol \, cm^{-3}$ is given by $\dfrac{\lambda_c}{\lambda_\infty}$. Values for α (degree of ionisation) found by this method are in good agreement with those found by using the depression of freezing point or elevation of boiling point (Chapter 7).

Questions

1. Name the products obtained at the electrodes during the electrolysis of (a) molten sodium chloride (b) concentrated sodium chloride solution (c) dilute sodium chloride solution. In each case give the electrode equations and comment on the liquid or solution remaining in the cell during the electrolysis.

2. What differences would you expect to observe in two silver voltameters connected in series if the electrodes in the first cell were made of platinum while those in the second cell were made of silver?

3. A current of 0.200 A is passed for 20 minutes through two voltameters in series. One voltameter has copper electrodes and contains a solution of copper(II) sulphate, the other has platinum electrodes and contains a dilute solution of sulphuric acid. Calculate to three significant figures:
(*a*) the mass of copper deposited on the cathode of the first voltameter
(*b*) the volume of oxygen (expressed in cm^3 at s.t.p.) liberated at the anode of the second voltameter.
(Take the Faraday constant to be $96\ 500\ C\ mol^{-1}$)   (O)

4. Explain the following observation as fully as you can.
The deposition of 1 mole of silver atoms at the cathode in the electrolysis of aqueous silver nitrate(V) requires $9.65 \times 10^4\ C$ of electricity, but the deposition of 1 mol of gold atoms at the cathode in the electrolysis of an aqueous gold salt requires $2.895 \times 10^5\ C$ of electricity.   (C)

5. (*a*) What do you understand by the term *molar conductivity* as applied to an electrolyte?
(*b*) Show diagrammatically how the molar conductivity changes with increasing dilution for (i) a strong electrolyte (ii) a weak electrolyte. Give one example of each type of electrolyte.
   Account briefly for the changes you have suggested.
(*c*) The molar conductivity of aqueous ethanoic acid of concentration 0.1 mol/l was $4.6\ cm^2\Omega^{-1}mol^{-1}$, and at infinite dilution $352\ cm^2\Omega^{-1}mol^{-1}$.
   Calculate (i) the degree of dissociation of the acid at this concentration, (ii) the pH of the acid solution.   (AEB 1978)

6. (*a*) Describe briefly an experiment you could carry out to illustrate the relationship between the charge on an ion in solution and the quantity of electrical charge required for its discharge.
(*b*) A current of 0.5 A was passed through $200\ cm^3$ of an aqueous solution of silver nitrate of concentration $0.1\ mol\ dm^{-3}$ for one hour. The anode was platinum and the cathode was silver.
(i)   Give the equation for the discharge of the metal ion.
(ii) What mass of silver would be deposited in this time?
(iii) What is the concentration in $mol\ dm^{-3}$ of the solution left with respect to both silver and nitric acid?
(iv) What gas would be discharged at the anode? What changes would occur if the platinum anode was replaced with a silver anode?
(*c*) Starting from a named raw material describe the extraction of a named metal using the process of electrolytic reduction.   (AEB 1980)

7. (*a*) Define the terms (i) *electrolytic conductivity* (*specific conductance*), (ii) *molar conductivity* of an aqueous solution of an electrolyte.
(*b*) The electrolytic conductivity of a saturated aqueous solution of thallium(I) chloride, TlCl, at 25°C is $2.40 \times 10^{-3}\ ohm^{-1}\ cm^{-1}$. The molar conductivities at infinite dilution of thallium(I) hydroxide, sodium hydroxide and sodium chloride are 273, 248 and $126\ ohm^{-1}\ cm^2\ mol^{-1}$ respectively. Estimate (i) the molar conductivity of thallium(I) chloride,

(ii) the solubility of TlCl in water at 25°C in mol dm^{-3}. State any law you assume in your calculation. (O)

8. (a) What factors affect the mass of a given substance liberated during electrolysis?

Why is aluminium extracted by electrolysis of its molten oxide, whereas lead is extracted by reduction of its oxide with carbon?

(b) Explain the meaning of the term *standard electrode potential* and outline its measurement for a metal such as zinc.

Draw a fully labelled diagram of a Daniell cell as you would prepare it to measure its standard e.m.f.

Calculate the e.m.f. of this cell when operating under standard conditions. (C)

Part IV
Chemistry of the Elements

14   Periodicity

14.1   Introduction

The modern form of the Periodic Table with the elements placed in order of increasing atomic number and arranged to show the periodic nature of the electronic structure of the atoms has been briefly described in Chapter 1. This chapter deals with the changes of the physical and chemical properties of the elements as they appear in the horizontal periods of the table.

Each period begins with an element that has a single s electron in a new main energy level and ends with a noble gas whose electronic structure is stable, having all the electronic orbitals of that energy level filled (Chapter 2). The number of elements in the period increases as the number of sub-shells in each energy or quantum level increases.

Period 1, the first short period, contains only two elements, hydrogen ($1s^1$) and helium ($1s^2$), since the first quantum level holds only two s electrons.

Period 2, the second short period; contains eight elements, lithium to neon; a new quantum level, holding p as well as s electrons, is started with lithium ($1s^2 2s^1$) and filled at neon ($1s^2 2s^2 2p^6$).

Period 3, the third short period, also contains eight elements, sodium ($1s^2 2s^2 2p^6 3s^1$) to argon ($1s^2 2s^2 2p^6 3s^2 3p^6$).

Period 4 is the first long period; the third quantum level being expanded by five d orbitals. It contains eighteen elements, starting with potassium ($1s^2 2s^2 2p^6 3s^2 3p^6 4s^1$), incorporating the ten d-block elements, scandium to zinc, and finishing with six p-block elements, gallium to krypton ($1s^2 2s^2 2p^6 3s^2 3p^6 3d^{10} 4s^2 4p^6$).

Period 5 is another long period like Period 4, comprising eighteen elements starting with rubidium and ending with the noble gas xenon.

Period 6, the third long period, is further complicated by the availability of f orbitals in the fourth and fifth quantum levels. It is lengthened by a set of elements which occur between lanthanum, atomic number 57, and hafnium, atomic number 72; these fourteen elements are known as the lanthanoids or f-block elements.

Period 7, the final period, is made up entirely of radioactive elements, and includes the start of a further inset of f-block elements, the actinides; this period is incomplete and contains many man-made elements.

Much can be learned by looking at the relative sizes of atoms and the ions formed from them, as shown in Fig. 14.1. The atomic radii of the atoms of the elements decrease from left to right across each period. The s electrons of the atoms of the first elements of each period are not pulled close to the nucleus, being protected from the positive influence by the screening effect of the filled penultimate shell. As the positive charge on the nucleus of each succeeding atom in the period increases, the outer electrons are pulled closer to the nucleus, the attractive force of the positive charge outweighing the mutual repulsion of the electrons. The atoms therefore become steadily smaller in volume in spite of the increasing numbers of protons, neutrons and electrons present.

The smaller size of the atoms and the increasing attraction of the positive nucleus is reflected in an increase of energy required to remove one electron

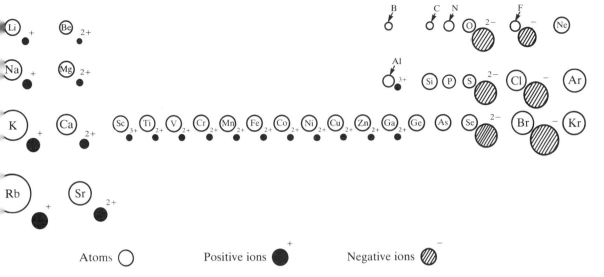

Fig. 14.1 Relative sizes of some atoms and ions

from each succeeding atom. A plot of first ionization energy against atomic number illustrates the periodicity of the electronic structure of the elements. Elements of similar electron configuration appear at the same relative positions on the graphs, with the trend for ionization energy to increase across the table being clearly shown, together with small variations as electrons are removed from new sub-shells; thus the energy required to

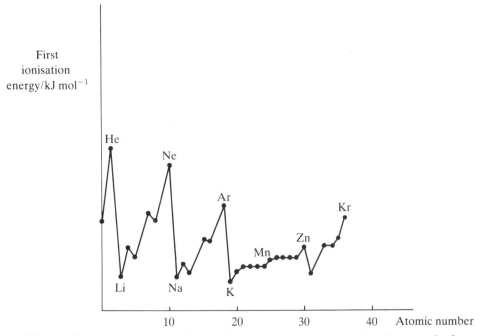

Fig. 14.2 Ionisation energies and atomic numbers for the elements hydrogen to krypton

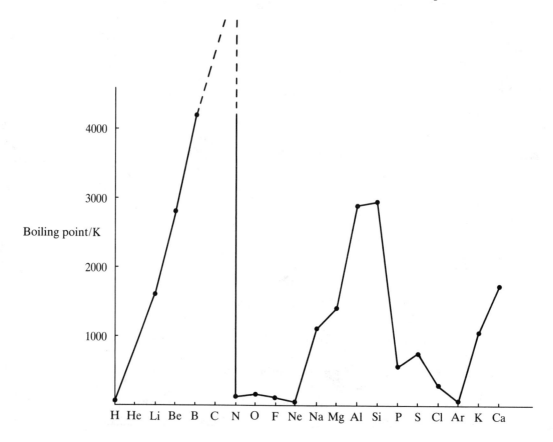

Fig. 14.3   Boiling points of elements H—Ca

remove the 3p electron from aluminium is slightly less than that required to remove one of the paired 3s electrons from magnesium. Again, the very small variation in first ionization energy among the d-block elements emphasises the similar nature of these elements.

The periodicity of the properties of the elements is evident in the similarity of the graphs obtained when any physical property of the elements is plotted against atomic number. In all cases the elements of each group appear at the same relative positions on the graphs. Figure 14.3 shows such a graph, obtained by plotting boiling points against atomic numbers for the first twenty elements.

Figure 14.4 is a modern version of the classic graph obtained by Lothar Meyer in 1869. Meyer plotted the atomic volume of the then known elements against their atomic weight in an attempt to establish a method of classification of the elements. Here *atomic volume* is plotted against atomic number, and more recently discovered elements are included. Atomic volume is found by dividing the relative atomic mass by the density of the solid element. It is thus the volume in cubic centimetres of one mole of the element in the solid state, including the space between the adjacent atoms in the crystal.

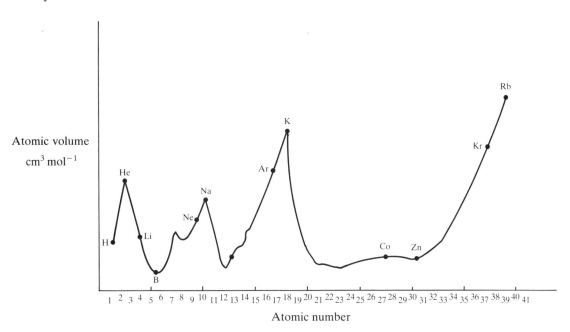

Fig. 14.4 Lothar Meyer graph (modernised version)

14.2 Properties of the elements

Comparing the second and third short periods, it can be seen that the first elements in both periods are metals. Their *s* electrons, distant and shielded from the nucleus, provide a cementing force between the positive ions giving rise to metallic structure (page 61) with mobile electrons for high electrical conductivity, while the ease of removal of these *s* electrons leads to high reactivity and the formation of positive ions in compounds.

The third elements in the second and third periods are more electronegative and less metallic in nature; the electrons in the smaller atoms are drawn closer to the nucleus, reducing the tendency to form positive ions. Boron is in fact non-metallic, while the very small size of the aluminium ion, Al^{3+}, leads to a high degree of covalency in its compounds.

Carbon and silicon, both non-metals, show a further increase in electronegativity. There is now little tendency to form either positive or negative ions by the loss or gain of four electrons, since this would require more energy than could be supplied by the formation of ionic lattice structures. The elements have giant atomic structures with covalent bonding and form covalently bonded compounds.

The succeeding elements in each period become increasingly electronegative as their atomic radii decrease. They are non-metallic low melting point solids or gases of molecular structure, and all form covalent compounds with other elements. The most electronegative elements towards the right of the periods have an increasingly strong attraction for electrons, forming negative ions in combination with metallic elements. These ions, in which electrons are gained to complete the outer octet of electrons to give a noble gas structure, are larger than the parent atoms; the extra electrons enlarge the ion not only by their presence but by exerting a repulsive effect which tends to expand the electron cloud, thus increasing the ionic radius.

Table 14.1 Trends in properties of the elements in Period 3

Element	Na	Mg	Al	Si	P	S	Cl	Ar
Structure	← Giant lattice →				← Molecular →			
Nature	← Metallic →←				Non-metallic →			
Electron structure	2.8.1	2.8.2	2.8.3	2.8.4	2.8.5	2.8.6	2.8.7	2.8.8
Melting Point/K	371	923	932	1683	315	392	172	84
Boiling Point/K	1163	1390	2720	2950	554	718	239	87
Atomic radius/nm	0.157	0.140	0.125	0.117	0.110	0.103	0.099	0.192

Increasing electronegativity ⟶

The elements scandium to zinc in Period 4 comprise a separate set within the period, as can be seen from Fig. 14.2. They all have similar electronegativity, with first ionisation energies increasing only slightly across the set. The atomic radius is fairly constant, showing only a small decrease with increasing positive charge on the nucleus, the screening effect of the extra d electrons being such that the outer s electrons are all pulled towards the nucleus by approximately the same amount. A slight increase of atomic radius is seen in the elements to the right of the group, where the d orbitals become filled. These metals differ from those of Groups I, II and III because of the presence of electrons in the d orbitals in the penultimate shell. Thus the formation of positive ions does not lead to an inert gas structure, and since the d electrons are available for ionic bonding, positive ions of variable charge are possible. The small size of the ions formed, together with the availability of variable numbers of d electrons, leads to the formation of complex anions in which the central metal ions may be present in a number of different oxidation states.

Each period ends with a noble gas of stable electronic configuration, all available energy levels being filled with the maximum number of electrons. The increased shielding effect of these filled orbitals results in an increase in size of the atoms together with a marked lack of reactivity. The elements form very few compounds and, uniquely among elemental gases, exist as atoms rather than as molecules.

14.3 Properties of the hydrides, oxides and chlorides of Periods 2 and 3

(a) Hydrides

The metals of very low electronegativity at the start of each period, with large atoms and low first ionisation energies, form ionic hydrides in which hydrogen accepts an electron to become a negative ion. These ionic hydrides are moderately stable compounds with a salt-like lattice structure, the size of the H^- ion being between those of Br^- and I^-.

Although the heavier members of Group 2 with large atoms also form ionic hydrides, those of beryllium, magnesium and aluminium are covalent, consistent with the smaller size of the atoms and positive ions of these metals, and they are polymeric compounds.

The hydrides of the succeeding non-metallic elements of Periods 1 and 2 are all covalent and of molecular structure. The tetrahedral molecules of the hydrides of carbon and silicon are symmetrical, while those of the hydrides of the following more electronegative elements become increasingly polar.

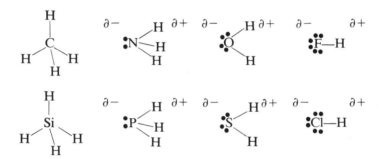

Fig. 14.5 Increasing polarity of hydrides of elements in Periods 2 and 3

The stability of the covalent hydrides increases from left to right across the periods, (although methane is exceptionally stable). As well as forming mononuclear hydrides, the elements in the middle of the periods form a number of polynuclear hydrides such as B_6H_6, Si_2H_6 and P_2H_4; these, with the exception of the numerous hydrides of carbon (page 344), tend to be unstable.

(b) Chlorides

The properties of the chlorides follow a pattern similar to that of the hydrides. The ionic chlorides of Group 1, being crystalline substances of high lattice energy, have high melting points and form neutral solutions. The gradual change to covalency is seen with magnesium chloride which, although mainly ionic, has some covalent character and hydrolyses slightly on solution in water. Beryllium and aluminium both form covalent chlorides with a giant lattice structure, which hydrolyse in water and are soluble in organic solvents. The chlorides of the elements of Groups IV, V and VI are all of covalent molecular structure, those of Period 3 onward being readily hydrolysed by water. They become less stable from left to right as the electronegativities of the elements become more similar; indeed chlorine is not sufficiently electronegative to induce the maximum covalency of sulphur, and a hexachloride is not formed.

(c) Oxides

The same change of bonding from ionic to covalent is seen with the oxides. Those of the metals at the start of the period are high melting point solids with giant ionic structures which, when soluble in water, form alkaline solutions. The oxides of boron, aluminium and silicon are also high melting point solids with giant structures, but, while the former two are amphoteric, silicon dioxide is acidic. The remaining oxides of the non-metals to the right of the table have covalent molecular structures. They are mainly acidic oxides and are discussed more fully in the appropriate chapters.

Table 14.2. Summary of properties of oxides, chlorides and hydrides of Period 3

Formula of oxide	Na_2O	MgO	Al_2O_3	SiO_2	P_2O_3 P_2O_5	SO_2	Cl_2O Cl_2O_7
Structure	←——— Giant lattice ———→ ←——— Molecular ———→						
Nature	←——— Ionic —→ ←——— Covalent ———→						
Formula of chloride	NaCl	$MgCl_2$	$AlCl_3$	$SiCl_4$	PCl_3 PCl_5	S_2Cl_2 SCl_2 SCl_4	Cl-Cl
Structure	←——— Giant lattice —→ ←——— Molecular ———→						
Nature	←— Ionic —→ ←——— Covalent ———→						
Solution in water	pH 7	pH 6	←——— Hydrolyses to acid solution ———→				
Formula of hydride	NaH	MgH_2	AlH_3	SiH_4	PH_3	SH_2	Cl H
Structure	← Giant lattice ———→ ←——— Molecular ———→						
Nature	Ionic ———→ ←——— Covalent ———→						

Questions

1. Plot the molar latent heat of vaporisation of the first twenty elements in the Periodic Table against the atomic number of the elements.
2. Explain why the first ionisation energies of the elements in Period 3 of the Periodic Table increase from sodium to argon.
3. Answer the following, giving *explanations*, and using as examples elements from the period sodium to argon **only**. Write *equations* where appropriate.
   (a) Give **one** example in **each** case of
   (i) an element which forms a basic oxide

(ii) an element which forms an acidic oxide

(iii) an element which forms an amphoteric oxide

(b) Give an example of an element which forms a liquid chloride consisting of simple molecules; the chloride is vigorously hydrolysed by water.

(c) Give an example of an element which forms a solid oxide, insoluble in water, the solid consisting of a giant molecular lattice.   (AEB 1979)

4. The chemistry of the elements and their compounds is often studied by consideration of a period or group of the Periodic Table.

   Indicate the structural differences seen in the elements of (a) the second short period (sodium to argon) and (b) Group IV (carbon to lead)

   Compare the trends in the properties of the chlorides of the elements in each case.   (L)

5. (a) Describe and suggest reasons for the trends across and down the Periodic Table in (i) atomic radius, (ii) atomic volume and (iii) electronegativity.

   (b) Discuss the bonding in (i) sodium chloride and (ii) hydrogen chloride.

   (c) Describe and account for the bonding you would expect to be present in (i) rubidium chloride and (ii) iodine chloride (ICl).   (JMB)

6. The elements E to J have consecutive atomic numbers and their first ionization energies are as follows:

E	F	G	H	I	J
1060	1020	1260	1520	418	590 $kJ\,mol^{-1}$

(a) Which element is likely to be:

(i) a noble gas

(ii) a Group II metal?

(b) Using the letters as symbols, write down ionic formulae for two ionic compounds that are likely to be formed by the union in each case of two of the elements E to J.

(c) Name two physical properties, other than ionization energies, in which elements show a periodic trend.   (O)

15 Hydrogen and water

15.1 Discovery, occurrence and uses of hydrogen

As early as the middle of the seventeenth century a colourless inflammable gas was obtained by Boyle from the reaction of iron with vitriolic acid. In 1766, Cavendish investigated the properties of this 'inflammable air', which was later called hydrogen by Lavoisier, deriving the name from Greek words meaning 'water producer'.

Hydrogen does not occur to any significant extent in the earth's atmosphere. Although it is found in some volcanic gases its very small molecule is not retained in the earth's gravitational field. Hydrogen has been identified in the outer atmosphere of the sun, and it is believed that the tremendous amount of energy we receive from the sun is released by the nuclear fusion of hydrogen atoms to form helium (page 39).

Hydrogen occurs in the combined state in water, organic matter and in petroleum and natural gas, the main sources of the vast quantities of the gas required industrially. It is used in many processes, such as the production of ammonia, methanol and margarine, the reduction of the ores of germanium and tungsten, and the preparation of high-octane motor spirit from petroleum distillates. The very hot flames required for cutting and welding metals, and for the production of synthetic gems, are produced in the oxy-hydrogen flame and the atomic hydrogen torch.

Because of its low density hydrogen is used to fill balloons; it has been used in airships, but on account of the fire risk this use has practically ceased, helium being used instead.

15.2 Properties and reactions of hydrogen

Hydrogen is a colourless and odourless gas, insoluble in water and of very low density. It can be liquefied, with great difficulty, at 20 K, and solidifies at a few degrees lower.

The gas is composed of diatomic molecules in which the orbitals of the $1s$ electrons of each hydrogen atom overlap in a molecular orbital to form a very strong bond; each atom in the molecule thus acquires the structure of a helium atom.

The energy released in the formation of the hydrogen molecule provides the high temperature of the atomic hydrogen welding torch. The gas is first

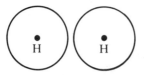

Hydrogen atoms each
with one electron

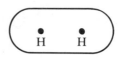

Hydrogen molecule
with s electrons forming σ bond

Fig. 15.1 The hydrogen molecule

atomised by passing it through an electric arc; the atoms are then allowed to recombine at the spot where the heat is required:

$$H(g) + H(g) \rightarrow H_2(g) \qquad \Delta H = -436 \text{ kJ mol}^{-1}$$

As a result of the strength of the bond between the atoms in its molecule hydrogen is comparatively unreactive although it forms compounds with all the elements except the noble gases and, under the right conditions, combines directly with the more electronegative elements.

A mixture of hydrogen and oxygen can be kept indefinitely at room temperature and pressure without reaction, but if a spark or a flame is introduced a violent explosion occurs:

$$H_2(g) + \tfrac{1}{2}O_2(g) \rightarrow H_2O(g) \qquad \Delta H^{\ominus}_{298} = -285 \text{ kJ mol}^{-1}$$

The reaction can be made to take place smoothly by burning a jet of hydrogen in air or oxygen, as in the oxyhydrogen welding torch.

With the halogens, hydrogen reacts explosively with fluorine, burns in chlorine (this reaction is explosive in bright sunlight) and reacts reversibly with bromine and iodine. It also reacts reversibly with nitrogen, in the presence of a catalyst, to form ammonia, and with molten sulphur to form hydrogen sulphide.

Thus the ease of reaction is greatest with highly electronegative elements and diminishes with decreasing electronegativity, as can be seen by comparing the enthalpy of formation of the halogens.

Table 15.1

Molecule	Enthalpy of formation kJ mol^{-1}
HF	−271
HCl	−92
HBr	−36
HI	+27

In all these compounds hydrogen acquires the noble gas structure of helium by sharing electrons in a covalent bond, as it does in its own diatomic molecule; in this respect hydrogen resembles the halogens in behaviour. However, the hydrogen atom has very little attraction for electrons and therefore does not readily form a hydride ion H^-; this is in contrast to the halogens which have considerable electron affinities and readily attract electrons to form ions such as Cl^-. Thus only the most reactive metals with high negative electrode potentials react with hydrogen to form ionic hydrides. For example, calcium hydride is made by heating calcium in hydrogen.

$$Ca(s) + H_2(g) \rightarrow Ca^{2+}(H^-)_2(s) \qquad \Delta H = -189 \text{ kJ mol}^{-1}$$

The hydrogen atom with its single electron can also be compared to the alkali metals, all of which have a single outer electron. However, while alkali metals readily lose their electrons to form positive ions, the formation of an H^+ ion does not take place easily. The first ionisation energies of the alkali metals range from 382 to 525 kJ mol^{-1}, much lower than the 1315 kJ mol^{-1}

required to remove an electron from a gaseous hydrogen atom. The ion, when formed, is also very small, being simply a bare proton; when formed in chemical systems it immediately reacts to form a complex ion in which it gains a pair of electrons in a coordinate bond (page 54). For example, when hydrogen chloride dissolves in water to make an acid solution the following reaction takes place:

$$HCl(g) + H_2O(l) \rightarrow [H_2O \rightarrow H]^+(aq) + Cl^-(aq)$$

The H_3O^+ ion is known as the *oxonium ion* (page 173). A similar reaction leads to the formation of the ammonium ion:

$$H^+(aq) + NH_3(aq) \longrightarrow \left[\begin{array}{c} H \\ | \\ H-N \rightarrow H \\ | \\ H \end{array} \right]^+ (aq)$$

Both in the oxonium ion and in the ammonium ion the hydrogen ion has acquired a helium structure by the gain of a pair of electrons.

Hydrogen is a useful reducing agent since it can so readily remove oxygen from the oxides of metals with positive electrode potentials occurring below it in the electrochemical series. This is particularly valuable in the production of metals such as tungsten and germanium that are required free from carbon.

$$3H_2(g) + WO_3(s) \rightarrow W(s) + 3H_2O(g)$$

Hydrogen gas is used to reduce unsaturated organic compounds (page 341) by *hydrogenation*, which is the addition of hydrogen to a double bond, usually in the presence of a metallic catalyst such as finely divided nickel or platinum.

$$H_2C:CH_2(g) + H_2(g) \rightarrow C_2H_6(g)$$

This reaction is used industrially to harden unsaturated oils for the manufacture of margarine. It was also used in the laboratory to estimate the number of double bonds present in an organic compound, but this is now done by spectroscopic methods.

Many organic compounds are reduced by reactions in which hydrogen is produced; these reactions are carried out *in situ*, that is, in the reaction mixture. For example, nitrobenzene is reduced to aniline in the presence of tin and hydrochloric acid, and esters are reduced to alcohols by the action of sodium on ethanol (page 410). Other reducing agents include sodium amalgam and water, the action of water on an ethereal solution of lithium tetrahydridoaluminate(III), and the action of hydrochloric acid on zinc. The reduction is effected by the electrons released in the reaction, rather than by the hydrogen produced. Thus in the reaction

$$Zn(s) + 2H^+ \rightarrow Zn^{2+}(aq) + H_2(g)$$

the reducing action results from the ionisation of the zinc:

$$Zn(s) \rightarrow Zn^{2+}(aq) + 2e^-$$

15.3   Hydrides

Hydrogen forms covalent hydrides with the elements of Group IV to VII. The stability of these hydrides decreases with the increasing mass of the elements down any group. For example, methane, CH_4, is stable at raised temperatures while plumbane, PbH_4, decomposes at room temperature.

H H:C:H H	H H:N: H	H::O: H:	H:F:
Methane	Ammonia	Water	Hydrogen fluoride
Group IV	Group V	Group VI	Group VII

Fig. 15.2   Covalent hydrides of the head elements of Groups VI–VII

Group I and Group II metals form ionic hydrides which have salt-like structures; most of these are decomposed on heating. The presence of the negative hydrogen ion, H^-, can be shown by fusing the hydrides with alkali halides and electrolysing the melt; hydrogen is evolved at the *anode*. The ionic hydrides react with water to release hydrogen gas; calcium hydride is used as a convenient medium for storing and transporting hydrogen.

$$CaH_2(s) + 2H_2O(l) \rightarrow Ca(OH)_2(s) + 2H_2(g)$$

Boron and aluminium in Group III form complex hydrides such as sodium tetrahydridoborate(III), (sodium borohydride) and lithium tetrahydridoaluminate(III), (lithium aluminium hydride).

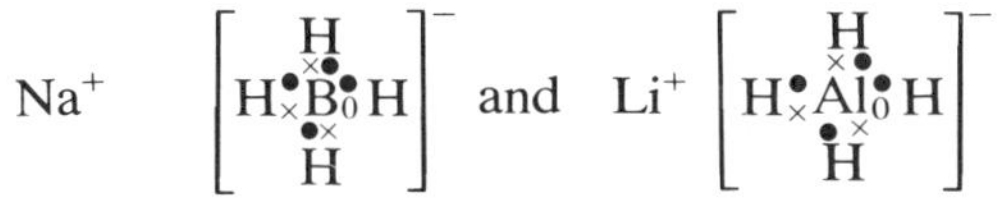

$$Na^+ \left[\begin{array}{c} H \\ H:B:H \\ H \end{array} \right]^- \text{ and } Li^+ \left[\begin{array}{c} H \\ H:Al:H \\ H \end{array} \right]^-$$

These hydrides are prepared in ether; they are used as reducing agents in organic chemistry, like the ionic hydrides of Groups I and II they release hydrogen on reaction with water.

Hydrogen is absorbed or occluded by many of the transition metals (page 319) to form interstitial hydrides of variable composition. The small hydrogen molecule appears to penetrate the metal lattice without disturbing the original crystal structure. The absorption is reversible, the hydrogen being readily recovered in all cases. The occlusion of hydrogen by finely divided palladium is used to obtain very pure samples of the gas.

15.4   The hydrogen bond

Comparing the properties of the simple hydrides of the elements carbon to fluorine, it is noticeable that water is the only liquid hydride in this series. A closer look at Groups V, VI and VII shows that the boiling points of the hydrides of all the head elements are abnormally high (Fig. 15.3).

This results from the formation of hydrogen bonds between the molecules; as has been seen in Chapter 4 there are electrostatic attractions

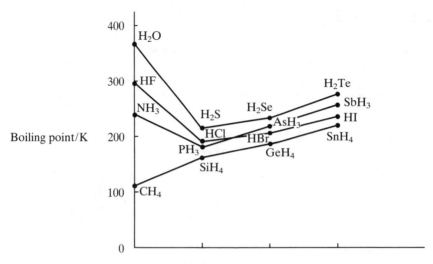

Fig. 15.3 Boiling points of hydrides

between the hydrogen atom in one molecule and an electronegative atom in another molecule. The higher the electronegativity of the second atom the more likely the formation of hydrogen bonds. Thus the tendency increases from nitrogen through oxygen to fluorine, and decreases down any group of elements. The bonds arise where the electronegative atom in the molecule carries a lone pair of electrons; thus there is no such bonding in methane but in the ammonia molecule hydrogen bonds are formed between the nitrogen atom and a hydrogen atom of an adjacent molecule, forming an aggregate of several molecules.

Hydrogen fluoride molecules are extensively linked by hydrogen bonds since the fluorine atom is highly electronegative; the effect is so marked that hydrogen fluoride is liquid just below room temperature, at 19°C.

The extensive hydrogen bonding in water results from the presence of two hydrogen atoms in each molecule together with two lone pairs of electrons on each oxygen atom; thus each molecule can form up to four hydrogen bonds.

Water and other hydrogen bonded liquids have anomalously high boiling points and enthalpies of vaporisation because energy is required to

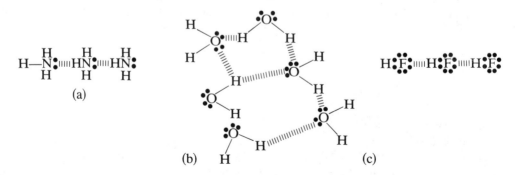

Fig. 15.4 Hydrogen bonding (||||||||) between (a) ammonia molecules (b) water molecules (c) hydrogen fluoride molecules

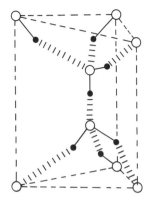

Part of wurtzite structure of ice

Tetrahedral
water molecule

● = Hydrogen atom   ○ = Oxygen atom   ⊣⊣⊣⊣⊣ = Hydrogen bond

Fig. 15.5   Structure of ice

break up the aggregated molecules before they are vaporised as single molecules. The well-known surface tension effects of water result from the inward attraction of the surface molecules by hydrogen bonds to other water molecules, giving rise to the surface 'skin' phenomenon.

The ability of water molecules to form four hydrogen bonds arranged tetrahedrally makes it possible for them to join in a three-dimensional lattice. This occurs in ice, which normally crystallises with a wurtzite structure (page 83) and occasionally a diamond structure.

The three-dimensional structure is more open, and occupies more space, than the aggregated structure of liquid water, so ice is less dense than water and, unusually, the solid floats in the liquid.

15.5   Water

(a) As a solvent

Water is the prime solvent for ionic compounds. Its polar molecule is able to break up the ionic lattice by surrounding both cations and anions and separating the hydrated ions from the crystal. (See also Chapter 8 for energy changes involved.)

Hydrated ions

Fig. 15.6 Water molecules attacking an ionic crystal

The water molecules associated with the hydrated cations may be held simply by loose electrostatic attraction, or, in many cases, particularly with transition metals, by liganding bonds (page 56). The water molecules hydrating anions are frequently hydrogen bonded. Hydrated ions may carry the water molecules with them on crystallisation as water of crystallisation; thus copper(II) sulphate(VI)-5-water, $CuSO_4 \cdot 5H_2O$, is more correctly written $(Cu(H_2O)_4)^{2+}(SO_4 \cdot H_2O)^{2-}$. In other hydrated crystals, such as sodium sulphate(VI)-10-water, $Na_2SO_4 \cdot 10H_2O$, the water molecules form part of the crystal structure without being closely attached to any particular ion.

Many covalent compounds are dissolved by water. These are compounds which can form hydrogen bonds with water, such as the lower molecular mass alcohols and the sugars, both of which contain—OH groups attached to an organic radical. The hydrogen bonds that these groups form with water molecules draw them into solution:

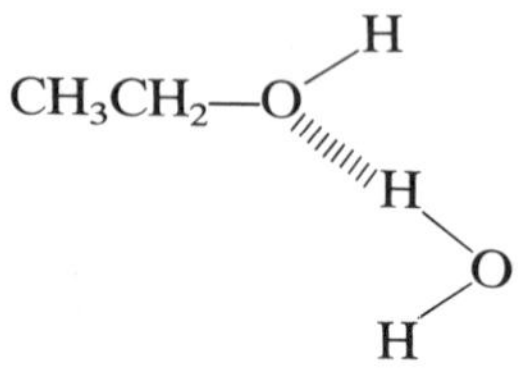

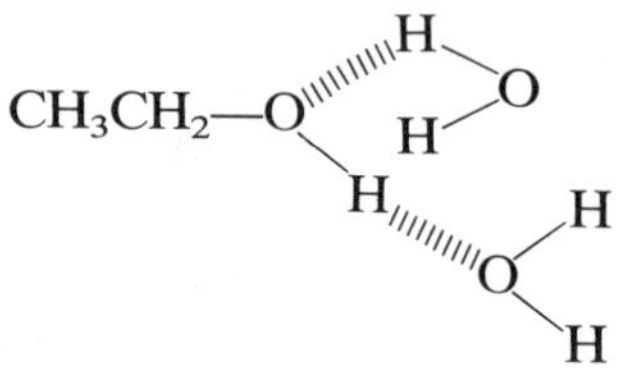

The amines (page 419) of lower molecular mass are similarly drawn into solution by hydrogen bonds formed with the unbonded pair of electrons in the nitrogen atom:

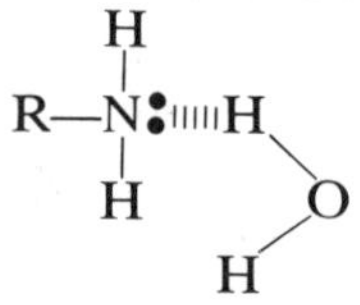

(b) Hard water

Since water is such a universal solvent, natural waters are rarely mineral free. One of the most familiar forms of natural water is *hard water*, which is defined as water which will not readily form a lather with soap. This effect mainly results from the presence of calcium ions and, to a lesser extent, magnesium ions. In some districts iron ions also contribute to the hardness.

Although hard water has many advantages, such as supplying calcium for the healthy development of teeth and bones, it can cause many problems. The calcium ions react with soap to form insoluble calcium stearate. This is familiar as the 'scum' that appears when soap is used in hard water districts. Not only does this increase the quantity of soap necessary for laundering, but is undesirable in itself as it clings to cloth giving a grey dinginess; in districts where the water contains iron this staining may be even more marked. In addition, hard water causes *boiler scale*: deposits of calcium carbonate in boilers and hot water pipes. These incrustations choke water pipes and reduce the efficiency of boilers since they do not transmit heat readily.

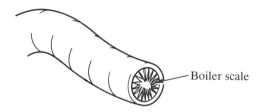

Fig. 15.7 Pipe choked with calcium deposits.

Deposits of calcium carbonate are also responsible for the 'furring' of kettles.

Since hard water is favorable to some operations, such as beer brewing, and unfavorable to others, such as dyeing, tanning and making white pottery, great care is taken over the location of industries which may be affected one way or another.

The presence of calcium hydrogen carbonate causes *temporary hardness*. Rain water containing dissolved carbon dioxide falls on chalk or limestone hills and converts the calcium carbonate into the hydrogen carbonate, which is soluble.

$$CaCO_3(s) + H_2O(l) + CO_2(g) \rightleftharpoons Ca^{2+}(aq) + 2HCO_3^-(aq)$$

This is regarded as temporary because the effect can be removed by boiling the water, the raised temperature causes the calcium hydrogen carbonate to decompose, thus precipitating the calcium as calcium carbonate. To soften water by boiling is not an economic process because the amount of energy required to boil water is very high. A less expensive method is to add slaked lime; the quantity is exactly calculated to remove all the calcium hydrogen carbonate by forming insoluble calcium carbonate.

$$Ca(HCO_3^-)_2(aq) + Ca^{2+}(OH^-)_2(aq) \rightarrow 2Ca^{2+}CO_3^{2-}(s) + 2H_2O(l)$$

Many water works use this method.

Calcium and magnesium ions also dissolve in water in the form of sulphates, causing *permanent hardness*. Water washing over rocks containing gypsum, $CaSO_4 \cdot 2H_2O$, or anhydrite, $CaSO_4$, dissolves some of the minerals; since sulphates are not decomposed at the temperature of boiling water the hardness cannot be removed by boiling. Nor is the calcium precipitated by the addition of slaked lime, so other methods have to be used. These methods have the advantage of softening both temporarily and permanently hard water:

(i) Sodium carbonate (washing soda) is added to precipitate calcium as calcium carbonate.

$$(Na^+)_2(CO_3)^{2-}(s) + Ca^{2+}(aq) \rightarrow 2Na^+(aq) + Ca^{2+}(CO_3^{2-})(s)$$

(ii) Modern water softening powders such as 'Calgon' contain sodium polyphosphates which form complexes with calcium and magnesium ions, preventing them from forming a 'scum' with soap. The complexes have the advantage of being soluble in water so that they are readily rinsed out of materials.

(iii) Much water softening is carried out nowadays by *ion exchange* methods. These involve the use of sodium aluminium silicate clays, such as the zeolites, from which Permutit has been developed. The clays are packed as granules into cylinders, and a slow stream of the hard water is passed through. The calcium ions are retained by the clay, while sodium ions are released. The water emerging contains sodium ions, which do not interfere with the action of soap, while the calcium ions remain behind as calcium aluminium silicate. The clay is regenerated by passing concentrated sodium chloride solution (brine) through it.

(iv) An extention of the ion exchange method can be used to produce pure water by *deionization*. For this the cylinders are packed with specially prepared resins which are capable of extracting both cations and anions from water. In these not only are Ca^{2+} and Mg^{2+} ions removed, but also the SO_4^{2-} and $(HCO_3)^-$ ions. This produces water pure enough for most laboratory purposes, but if absolutely pure water free of all dissolved material is required it must be distilled.

Pure water is very slightly ionised:

$$H_2O(l) \rightleftharpoons H^+(aq) + OH^-(aq)$$

but since the concentration of each ion is only $10^{-7}\,mol\,dm^{-3}$ it is virtually non-conducting.

Everyday uses of distilled or deionised water are for filling steam irons and topping-up car batteries. In some desert districts, such as Kuwait, drinking water has to be produced by distillation, but it is flat and tasteless, lacking the 'tang' of natural water.

(c) Hydrolysis

When a substance dissolves in a solvent, it should be possible to recover the substance unchanged by the appropriate physical means. Many substances, however, are said to 'dissolve' in water when in fact they react with it to form soluble products. The term *hydrolysis* is used to describe some of these reactions, in which a compound divides into two parts, one reacting with the hydrogen of the water and the other with the hydroxyl radical, OH^-. The hydrolysis of salts of strong bases with weak acids, such as sodium ethanoate, and of non-metal chlorides, such as phosphorus trichloride, provide examples:

$$NaEt(s) + H_2O(l) \rightarrow HEt(aq) + Na^+(aq) + OH^-(aq)$$

$$PCl_3(g) + 3H_2O(l) \rightarrow H_3PO_3(aq) + 3H^+(aq) + 3Cl^-(aq)$$

Numerous examples of hydrolysis are described elsewhere in this book.

Questions

1. Describe the bonding in (*a*) sodium hydride, (*b*) hydrogen chloride, (*c*) the hydroxonium ion. In each case show that the hydrogen atom has acquired the electronic structure of a helium atom.

2. Write a brief account of hydrogen as a reducing agent.

3. Hydrogen resembles both the alkali-metals and the halogens in its chemical behaviour. Survey the evidence which leads to this conclusion.

Use the compounds which you have described to illustrate the changes in bond-type of the hydrides of elements in a period of the Periodic Table. (JMB)

4. Explain the terms (a) electrovalent bonding, (b) covalent bonding, (c) hydrogen bonding, illustrating your answer by reference to sodium chloride, tetrachloromethane, and water.

What structural changes take place when sodium chloride is dissolved in water? Discuss the energy changes associated with each of these structural changes. (C)

5. Water is not only one of the most abundant materials in nature, but also one of the most widely used substances in chemistry,

(a) Discuss the role played in chemistry by water.

(b) Relate the chemical uses of water to its properties as a liquid or in the molecular state as appropriate. (L)

16   The *s*-block elements

Table 16.1   Group I and Group II metals: physical data

	Atomic number	Electronic configuration	Ionisation energy $kJ\,mol^{-1}$		Standard electrode potential V	Atomic radius nm (metallic)	Ionic radius nm	m.p. K	b.p K
			First	Second					
Group I									
Li	3	2.1 $--2s^1$	520		−3.03	0.152	0.060	459	1653
Na	11	2.8.1 $---3s^1$	494		−2.71	0.186	0.095	371	1156
K	19	2.8.8.1 $----4s^1$	418		−2.87	0.227	0.133	336	1033
Rb	37	2.8.8.18.8.1 $-----5s^1$	403		−2.93	0.248	0.148	312	952
Cs	55	2.8.18.18.8.1 $------6s^1$	374		−3.02	0.263	0.169	303	943
Group II									
Be	4	2.2 $---2s^2$	899	1757	−1.85	0.112	0.031	1553	3243
Mg	12	2.8.2 $---3s^2$	738	1451	−2.37	0.160	0.061	923	1373
Ca	20	2.8.8.2 $----4s^2$	589	1145	−2.87	0.197	0.097	1124	1513
Sr	38	2.8.18.8.2 $-----5s^2$	549	1064	−2.89	0.215	0.113	1023	1633
Ba	56	2.8.18.18.2 $------6s^2$	502	965	−2.90	0.221	0.135	983	1873

16.1   General characteristics and trends in properties

The metals of Groups I and II of the periodic table are collectively known as the *s*-block elements. The name derives from *s* electrons only in the outermost or bonding orbitals of the atoms of each element.

The single *s* electrons of Group I metals are effectively shielded from the positive nuclei by the penultimate shells of electrons conforming to a noble gas configuration. The resulting poor control of the outer *s* electrons gives rise to atoms of large radius and low density.

A relatively small amount of energy is required to remove the *s* electrons from the atoms to form singly-charged positive ions. The atoms are therefore highly reactive. The reactivity increases from lithium to caesium as the atoms become heavier and the number of completely filled electron shells increases, resulting in larger atomic radii and a progressive decrease in first ionisation energy (Table 16.1).

The comparative ease with which one electron is lost is reflected in the low electronegativity of the elements, decreasing from lithium to caesium.

The ions formed, although smaller than the parent atoms, are still relatively large with very little tendency to regain the valency electron; thus they form ionic compounds of high lattice energy with non-metal atoms and radicals. Only lithium forms a positive ion small enough to exert a distorting effect on the electrons of the negative ion sufficient to weaken the lattice structure. Thus while compounds of the elements sodium to caesium are fully ionic and thermally stable, some lithium compounds show a degree of covalency and are relatively easily decomposed by heat. The metals' low electronegativity is reflected in their strongly basic properties and in those of their oxides and hydroxides. The hydroxides are soluble and fully ionised in aqueous solutions, which are strongly alkaline, the basic properties increasing from lithium to caesium.

The large singly charged positive ions have little attraction for electrons and therefore little polarising effect; thus the ions of sodium to caesium show small tendency to form hydrates in solution and few of their compounds form hydrated crystals. Lithium is exceptional since its much smaller ion, with the positive nucleus screened only by a helium structure of two s electrons, has a much greater polarising power than the other members of the group; it forms hydrated ions of definite composition in which the water molecules are bonded to the ion.

The Group I metals have large negative electrode potentials which increase in negativity from sodium to caesium, in accordance with the decrease of first ionisation energy down the group. Exceptionally, lithium has a value approximately equal to that of caesium; since electrode potentials are measured in aqueous solution the enthalpy of hydration of the lithium ion enhances its negative potential. Thus while lithium is the least reactive of the metals of Group I, its ions in solution are particularly reactive.

Group II metals show both similarities to and differences from the Group I metals, which result from the differences in structure and size of their atoms and ions. Each element has two s electrons available for bonding in the outermost orbit. These are more firmly held by the increased pull of the nuclei, each containing one more proton than the neighbouring Group I metal; thus the atoms have smaller radii and higher first ionisation energies (Table 16.1) while the energy required to lose both electrons to form stable ions is much greater. As a result the metals are less reactive than the corresponding Group I metals. The reactivity again increases down the group from beryllium to barium as the screening effect of the completed electron shells of the larger atoms result in lower first and second ionisation energies.

The metals of Group II are each more electronegative than their neighbouring Group I metals; indeed the ions of the elements at the start of the group are moderately good polarisers; their small size and double positive charge results in a tendency to attract electrons. Thus beryllium and, to a lesser extent, magnesium and calcium exhibit a degree of covalency in their compounds. Similarly the smaller, more highly charged, ions exert a distorting effect on the lattice structure of Group II carbonates and nitrates, rendering them much less stable to heat than those of Group I. The stability of these oxocompounds increases from beryllium to barium as the atomic and ionic radii increase.

The increase in electronegativity of Group II metals is also shown in their reduced basic properties; although still strongly basic they are less so than the Group I metals.

The smaller, more highly charged ions also have a greater attraction for water molecules, and most Group II metals form complex hydrates of definite composition in aqueous solution, the water molecules being carried into the solid compounds as water of crystallisation.

The Group II metals have highly negative electrode potentials, comparable in value to those of Group I, the high enthalpy of hydration compensating for the greater energy required for ionisation. Since all the Group II metal ions are hydrated, the electrode potential becomes progressively more negative from beryllium to barium with no anomalous values such as that for lithium in Group I.

The head elements of any group tend to show a more marked difference in physical properties and chemical reactions from the second elements in the group than is consistent with the general trend. The electronegativity and general properties of the head elements, particularly those of groups to the left of the table, are more consistent with those of the second element in the following group. Thus lithium resembles magnesium in many respects, and beryllium resembles aluminium. The diagonal relationship between lithium and magnesium is discussed at the end of this chapter, and that between beryllium and aluminium is discussed at the end of Chapter 17.

16.2   Occurrence and uses

The elements of Groups I and II are not found free in nature, although most of them are plentiful in the combined state.

The chief source of lithium is from alumino-silicate rocks such as spodumene $(Li^+Al^{3+}(SiO_3{}^{2-})_2)$. Sodium occurs as sodium chloride in the sea and in the rock salt deposits left behind by the drying up of former seas—such deposits are found in Salzburg and Cheshire. There are also deposits of sodium nitrate, Chile saltpetre, in South America. Potassium, as the chloride, is found with magnesium in carnallite, $K^+Cl^-Mg^{2+}(Cl^-)_26H_2O$, and as the nitrate in saltpetre; potassium carbonate (potash) can be extracted from wood ash.

Beryllium, like lithium, is found in alumino-silicate rocks. The ore, beryl, is perhaps best known as the jewels emerald and aquamarine.

Magnesium, besides occurring with potassium in carnallite, is found as magnesite, $MgCO_3$, kieserite, $MgSO_4 \cdot H_2O$, and in sea water as the chloride.

Calcium is widely distributed as calcium carbonate in chalk, limestone and marble, and as calcium sulphate in gypsum and anhydrite.

Strontium and barium are found as sulphates and carbonates together with calcium and magnesium compounds, but in lesser quantities.

The reactivity of the metals of both groups makes them difficult to extract from their ores. The name 'alkaline earths' for the Group II metals derives from the fact that the oxides are so recalcitrant that they were originally taken to be elements. It was not until electrolysis was introduced in the early nineteenth century that isolation of the metals was possible. Sodium was first obtained by Davy (1807) by the electrolysis of molten sodium hydroxide. Nowadays the metals are all extracted by electrolysis of the fused chlorides, using graphite anodes (to resist corrosion by the chlorine) and steel cathodes.

Sodium is the cheapest and most widely used of the alkali metals. Its vapour provides the yellow glow of certain street lamps. Alloyed with potassium it is used as a coolant liquid for nuclear reactors. In the laboratory it is extensively used as a reducing agent for organic compounds; lithium is also used for this purpose as lithium aluminium hydride.

Of the Group II elements, magnesium is the most used in the metallic state. As an alloy with aluminium it provides light and strong construction material for aeroplanes. Burning with a brilliant light, it is used as an illuminant in warfare and in photography. Because of its position high in the electrochemical series, it provides cathodic protection from corrosion for iron pipelines and the hulls of ships (page 324).

Beryllium, having a small atom, is transparent to X-rays and provides a 'window' material for X-ray apparatus. It also forms an alloy with copper which has a remarkably high tensile strength. The remaining elements are little used as metals.

Compounds of elements in both groups are of great importance biologically. Sodium, potassium and lithium salts are all present in body fluids; bones and teeth are composed of calcium compounds, and the presence of magnesium is necessary for the functioning of some hormones. Magnesium is also necessary for the proper functioning of plant enzymes, and it is the central atom of the chlorophyll molecule (page 60). Calcium hydroxide is used to decrease the acidity of soil and various calcium compounds are used as fertilisers.

16.3 Properties and reactions

The Group I elements are soft metals of low densities, low melting points and comparatively low boiling points, having body-centred cubic crystalline structures. The Group II metals are much harder and denser, with higher melting and boiling points than those of Group I. They have smaller atoms bound closely together by the two electrons in their valency orbitals into one or other of the closely-packed systems (page 81), except for barium which has a body-centred structure.

All the metals are good conductors of heat and electricity; they are silvery in colour when pure or freshly cut, but tarnish rapidly on exposure to air, becoming covered with a layer of oxide that hides the metallic lustre. The Group I metals are, in fact, so reactive that they are stored under oil.

With the exception of beryllium and magnesium the elements have emission spectra (page 17) in the visible range, which give rise to the characteristic flame colours seen when small quantities of their compounds are heated in a bunsen flame. For example the following colours can be seen:

Lithium—red
Sodium—golden yellow
Potassium—lilac
Calcium—orange
Strontium—crimson
Barium—apple green

Elements of both groups exhibit only one oxidation state in their compounds. Group I metals form ions with one positive charge by the loss of the single electron from the valency orbital. Group II metals form ions with two positive charges by the loss of the two s electrons in the valency orbital. In each case the resulting ion has the structure of a noble gas. For example:

$$Na \rightarrow Na^+ + e^-$$
$$2.8.1 \quad 2.8$$
$$\text{(neon structure)}$$

and

$$Mg \rightarrow Mg^{2+} + 2e^-$$
$$2.8.2 \quad 2.8$$
$$\text{(neon structure)}$$

As has been seen the ease with which the Group I metals lose one electron is reflected in their low first ionisation energies (Table 16.1) and accounts for their extreme reactivity. The readiness with which the metals combine with electronegative elements by the transference of the valency electron is further enhanced by the high lattice energies of the resulting compounds, making the overall reactions exothermic.

$$2Li(s) + \tfrac{1}{2}O_2(g) \rightarrow Li_2O(s) \qquad \Delta H = -596 \text{ kJ mol}^{-1}$$
$$Na(s) + \tfrac{1}{2}Cl_2(g) \rightarrow NaCl(s) \qquad \Delta H = -411 \text{ kJ mol}^{-1}$$
$$2K(s) + S(s) \rightarrow K_2S(s) \qquad \Delta H = -418 \text{ kJ mol}^{-1}$$
$$Na(s) + \tfrac{1}{2}H_2(g) \rightarrow NaH(s) \qquad \Delta H = -57 \text{ kJ mol}^{-1}$$

Only lithium forms a nitride by direct combination with nitrogen; its small ion is comparable in size to that of the nitride ion, making the compound of sufficiently high lattice energy to overcome the energy required to disrupt the triple bond of the nitrogen molecule.

The reactivity of the metals increases markedly down the group; for example, lithium reacts quietly with water, the reaction with sodium is vigorous, potassium takes fire and both caesium and rubidium react explosively. In all cases the metal releases hydrogen and an alkaline solution is formed.

$$2Na(s) + 2H_2O(l) \rightarrow 2Na^+(aq) + 2OH^-(aq) + H_2(g)$$

Much higher energies are required to form the dispositive ions of the Group II metals, since these are the sum of the first and second ionisation energies (Table 16.1). However Group II elements are reactive metals combining directly and exothermically with electronegative elements to give compounds of high lattice energy. These include direct reactions with nitrogen since the small dipositive ions form nitrides of high lattice energy, making the reactions exothermic.

$$Ca(s) + H_2(g) \rightarrow CaH_2(s) \qquad \Delta H = -189 \text{ kJ mol}^{-1}$$
$$Ba(s) + Cl_2(g) \rightarrow BaCl_2(s) \qquad \Delta H = -891 \text{ kJ mol}^{-1}$$
$$3Mg(s) + N_2(g) \rightarrow Mg_3N_2(s) \qquad \Delta H = -461 \text{ kJ mol}^{-1}$$
$$Mg(s) + \tfrac{1}{2}O_2(g) \rightarrow MgO(s) \qquad \Delta H = -601 \text{ kJ mol}^{-1}$$

Reactivity again increases down the group as the outer electrons are further from the nucleus and more effectively screened from it when the atoms become larger. Magnesium decomposes steam at red-heat but reacts only slowly with cold water; calcium and the succeeding metals in the group attack water with increasing vigour to release hydrogen and form an alkaline solution.

$$Ca(s) + 2H_2O(l) \rightarrow Ca^{2+}(aq) + 2OH^-(aq) + H_2(g)$$

16.4 Selected compounds

(a) Oxides

The normal oxides of both Group I and II are basic oxides containing the O^{2-} ion; for example Na_2O and CaO. Group II oxides are exceptionally stable and have high melting points; magnesium oxide, for example, is used to line high-temperature furnaces.

The heavier metals of Group I also form oxides containing the larger $(O_2)^{2-}$ ions and $(O_2)^-$ ions. Sodium forms a peroxide $(Na^+)_2(O_2)^{2-}$ since its ion is large enough to form a stable lattice with the $(O_2)^{2-}$ ion. Potassium, with an even greater ionic radius, also forms a superoxide $K^+(O_2)^-$. The peroxides and superoxides are strong oxidising agents capable of igniting some organic compounds or even dry filter paper. Peroxides react with water to release hydrogen peroxide, and superoxides to release hydrogen peroxide and oxygen:

$$Na_2O_2(s) + 2H_2O(l) \rightarrow 2Na^+(aq) + 2OH^-(aq) + H_2O_2(aq)$$

$$2KO_2(s) + 2H_2O(l) \rightarrow 2K^+(aq) + 2OH^-(aq) + H_2O_2(aq) + O_2(g)$$

These higher oxides are not formed by lithium or the Group II metals (except barium) since their ions are too small to accommodate the larger electronegative ions in a stable lattice.

The normal oxides of Group I metals dissolve in water giving alkaline solutions containing hydroxide ions.

$$Na_2O(s) + H_2O(l) \rightarrow 2Na^+(aq) + 2OH^-(aq)$$

Group II metal oxides are much less soluble than those of Group I. Beryllium oxide is virtually insoluble, magnesium oxide sparingly so, while calcium and barium oxide both dissolve to give alkaline solutions of their hydroxides.

$$BaO(s) + H_2O(l) \rightarrow Ba^{2+}(aq) + 2OH^-(aq)$$

The uses of the oxides and hydroxides of Groups I and II depend largely on their basic properties. Sodium hydroxide is a white deliquescent solid very soluble in water; concentrated solutions are caustic and can burn the skin severely. In the laboratory its solutions are used to precipitate insoluble metal hydroxides from solutions of their salts, to dissolve aluminium and zinc as aluminates and zincates (page 251) and, in general, to supply hydroxide ions where needed. Industrially it is used to remove sulphur compounds and acids from petroleum distillates, to extract phenols and cresols from coal tar, and in the production of rayon, textiles, paper and soap. Potassium hydroxide is more expensive, it is used when potassium is required in the final product, as in soft soap.

In addition to the use of magnesium oxide as a furnace lining, magnesium hydroxide is sold as a suspension, 'milk of magnesia', used as an antacid.

Calcium oxide, quick lime, is reacted with a calculated quantity of water to produce slaked lime, for which there is considerable demand.

$$CaO(s) + H_2O(l) \rightarrow Ca(OH)_2(s) + heat$$

Slaked lime, as a strong base, is used to recover ammonia in the Solvay process (page 472) by displacing the gas from the ammonium chloride produced in the operation.

$$Ca^{2+}(OH^-)_2(s) + 2NH_4^+Cl^-(aq) \rightarrow Ca^{2+}(aq) + 2Cl^-(aq) + 2H_2O(l) + 2NH_3(g)$$

Since lime is cheap it is used to regulate soil acidity; spread on fields it neutralises the acids set free by decaying organic matter. At the same time it adds valuable calcium salts.

Temporary hardness of water in limestone or chalk districts can be removed by adding calculated quantities of slaked lime.

Building mortar is a paste of sand and slaked lime which sets hard on drying; whitewash is a suspension of slaked lime in water; bleaching powder is made by passing chlorine over slaked lime.

(b) Halides

The metals of both groups react with all the halogens to give a complete range of halides. These are white crystalline compounds, mainly ionic except where the positive metal ion is small enough to exert a polarising effect on the halide ions, thus introducing a degree of covalency. For example, sodium and potassium chloride are both wholly ionic, have high melting points and dissolve to make neutral aqueous solutions. Lithium chloride is partially covalent, has a lower melting point and is hydrolysed by hot water to make an acid solution. This results from the smaller size of the Li^+ ion; the effect is more marked in Group II where the larger positive nucleus leads to ions of smaller radius and greater polarising power. The heavier elements of the group form ionic chlorides; magnesium chloride is partially covalent while the chloride of beryllium, the head element, which has an exceptionally small atomic radius, is completely covalent.

The chlorides of both groups can be made by direct combination of the elements, but they are best prepared by neutralising hydrochloric acid solutions either with solutions of the alkalis or with the metal carbonates, as appropriate, and allowing the resulting solutions to crystallise.

$$HCl(aq) + NaOH(aq) \rightarrow Na^+(aq) + Cl^-(aq) + H_2O(l)$$

$$2HCl(aq) + MgCO_3(s) \rightarrow Mg^{2+}(aq) + 2Cl^-(aq) + H_2O(l) + CO_2(g)$$

Sodium and potassium chloride form ionic crystals with no water of crystallisation, lithium and the Group II metal chlorides are hydrated to varying degrees with the water of crystallisation forming complexes of definite composition with the metal ions. Examples are $LiCl_2 \cdot 2H_2O$, $CaCl_2 \cdot 6H_2O$ and $BaCl_2 \cdot 2H_2O$. The first two deliquesce on exposure to the atmosphere; calcium chloride, which is cheap and readily available, is used to dry gases and in desiccators.

The ionic nature of the halides is also influenced by the size of the halide ion; thus the fluoride of any one of the metals is more ionic than the iodide since the large electron cloud of the iodine atom is more easily distorted than that of the fluorine atom. This tendency is reflected in the melting points of the halides of calcium (Table 16.2):

Table 16.2

Compound	CaF_2	$CaCl_2$	$CaBr_2$	CaI_2
Melting point °C	1360	772	730	575

The high lattice energy of the Group II fluorides renders them insoluble since the energy of hydration of the metal ions is not sufficient to bring the compounds into solution; thus in contrast to the extreme solubility of calcium chloride, calcium fluoride is insoluble.

The uses of the alkali and alkaline earth metal halides are varied. Sodium chloride is one of the most useful chemicals we have. It is essential in our diet and is a valuable food preservative. It provides the raw material for the manufacture of sodium carbonate, sodium hydroxide, chlorine and sodium metal; spread over icy roads it lowers the melting point of the ice, turning it to a salt solution. Strong brine is used to regenerate water-softening resins (page 232).

Potassium chloride is used as a fertiliser and in the manufacture of potassium nitrate; potassium bromide is used medicinally as a sedative and in photography.

The ability of calcium chloride to absorb moisture is used not only to dry the air in desiccators but also to keep cotton thread moist during spinning and to lay dust in coal mines.

Magnesium chloride is required for the electrolytic production of magnesium metal. The anhydrous salt required for this process cannot be made by the direct evaporation of the hydrated salt, for this produces a basic chloride, $Mg(OH)Cl$, by hydrolysis. It is obtained by heating magnesium oxide with carbon in a stream of chlorine.

$$MgO(s) + C(s) + Cl_2(g) \rightarrow MgCl_2(s) + CO(g)$$

(c) Carbonates and hydrogencarbonates

The anhydrous carbonates of both groups are white powdery solids; the thermal stability of the carbonates depends on the size of the metal cation; small ions, especially where doubly charged, exert sufficient polarising power to distort the carbonate ion, making it less stable. Thus in Group I sodium and potassium carbonates are not decomposed by heat, but lithium carbonate decomposes fairly readily.

$$Li_2CO_3(s) \rightarrow Li_2O(s) + CO_2(g)$$

In Group II magnesium carbonate decomposes above 550°C, calcium carbonate at temperatures over 900°C, and barium carbonate only when the temperature exceeds 1360°C. Beryllium carbonate is too unstable to exist at ordinary temperatures.

Sodium and potassium carbonates form various hydrates, the most important of which is washing soda, $Na_2CO_3 \cdot 10H_2O$. The water of crystallisation in sodium carbonate-10-water is not closely bound by the cation in a complex ion of definite composition; the crystals are efflorescent, that is, they steadily lose water of crystallisation on exposure to the atmosphere, leaving a powdery monohydrate $Na_2CO_3 \cdot H_2O$.

Sodium and potassium carbonates are soluble in water; lithium carbonate is much less so, while those of Group II metals are insoluble. Sodium carbonate is used as a primary standard in volumetric analysis to standardise acids, since it can be obtained in a very pure state and can be dried by heating without decomposition.

$$2HCl(aq) + Na_2CO_3(aq) \rightarrow 2NaCl(aq) + CO_2(g) + H_2O(l)$$

Methyl orange is used as the indicator since it is not affected by the weakly acid carbon dioxide produced in the reaction (page 181).

If carbon dioxide is bubbled through solutions of sodium or potassium carbonate it is absorbed with the formation of hydrogencarbonates.

$$Na_2CO_3(aq) + CO_2(g) + H_2O(l) \rightarrow 2NaHCO_3(aq)$$

These can be obtained as solids by evaporation, sodium hydrogencarbonate being better known as the sodium bicarbonate of baking powder. On heating it loses carbon dioxide and water (as steam).

$$2NaHCO_3(s) \rightarrow Na_2CO_3(s) + H_2O(g) + CO_2(g)$$

Hence if introduced into bread or cake mixtures it causes the mixture to 'rise' on cooking. If used alone the residue of sodium carbonate gives an unpleasant soapy taste to the product; to overcome this baking powder also contains a solid acid, usually sodium pyrophosphate or potassium hydrogen tartrate (cream of tartar) which reacts with the carbonate to produce more carbon dioxide (and therefore better rising) and leaves a pleasant tasting residue.

Calcium carbonate, from which the lime described in section (a) is derived, has various crystalline forms and is widespread in nature as chalk, limestone, Iceland spar and marble. The white cliffs of Dover and the marble statues of Greece are both calcium carbonate. Chalk and limestone are quarried on a large scale for use in lime kilns; before Lord Elgin brought the famous Elgin Marbles to the British Museum, the marble of the Parthenon was being burned for lime by the local populace. The conversion of calcium carbonate to calcium oxide is a reversible reaction (page 167):

$$CaCO_3(s) \rightleftharpoons CaO(s) + CO_2(g)$$

but in the limekiln it proceeds steadily to the right as the carbon dioxide is continuously removed.

The numerous small disued limekilns in country districts date from the days of wooden ships and carts, when it was preferable to transport limestone rather than the destructive 'burnt lime'.

Calcium carbonate is also important for making cement and concrete. Cement is made by fusing a mixture of clay and limestone and grinding the clinker formed to powder; a thick paste of the powder with water sets to a hard mass of interlacing calcium aluminium silicate crystals. After setting, cement becomes steadily harder with the passage of years, making it a most valuable building material. Concrete is made by mixing gravel and sand with the cement paste.

Besides all this, calcium carbonate is used in glass making and for the production of sodium carbonate by the Solvay process.

Suspensions of magnesium and calcium carbonate can be brought into solution as hydrogencarbonates, although these are so unstable that the solid salts are not known. It is by this means that the calcium ions that cause temporary hardness of water are brought into solution.

(d) Sulphates and hydrogensulphates

The sulphates of both groups are more stable to heat than the carbonates. Where decomposition does take place at fairly high temperatures sulphur trioxide is evolved leaving a residue of oxide.

$$CaSO_4(s) \xrightarrow{1000°C} CaO(s) + SO_3(g)$$

Group I sulphates are all soluble, but only lithium sulphate, $LiSO_4 \cdot H_2O$, and sodium sulphate, $Na_2SO_4 \cdot 10H_2O$, form hydrates. Sodium sulphate-10-water can be crystallised from solution below 60°C; the crystals are efflorescent. The anhydrous salt is used to dry organic liquids since it takes up so much water and is unreactive.

In Group II the sulphates of both barium and strontium are insoluble and do not form hydrates.

Barium sulphate is used in X-ray examination of internal organs. Although barium salts are normally poisonous the sulphate is so insoluble that it can be swallowed in a porridge by patients before X-ray examination; barium being opaque to X-rays, a picture of the contours of the stomach or bowel is obtained.

Calcium sulphate occurs naturally as gypsum, $CaSO_4 \cdot 2H_2O$, which is partially dehydrated on heating to a hydrate of formula $(CaSO_4)_2H_2O$. This is used as plaster of Paris; mixed to a paste with water it slowly reforms the dihydrate, setting into a firm plaster. It is used both in building and for setting broken limbs.

The most important hydrate of magnesium sulphate is $MgSO_4 \cdot 7H_2O$, which occurs naturally as Epsom salts.

The sulphates increase in solubility from barium to magnesium and beryllium sulphates, both of which are fairly soluble. Calcium sulphate is only sparingly soluble, but sufficiently so to be responsible, together with magnesium sulphate, for the permanent hardness of water (page 231).

Sodium and potassium hydrogensulphates can both be obtained as crystalline monohydrates; sodium hydrogensulphate is used as a solid acid, providing H^+ ions in aqueous solution by hydrolysis:

$$HSO_4^-(aq) \rightleftharpoons H^+(aq) + SO_4^{2-}(aq)$$

(e) Nitrates

The nitrates of both Group I and Group II metals are decomposed on heating, the degree of decomposition depending on the polarising ability of the metal cation. Thus sodium and potassium nitrate(V) are both decomposed to nitrate(III) compounds with the loss of oxygen, while lithium nitrate(V) and the Group II nitrates decompose further losing nitrogen dioxide and leaving a residue of oxide.

$$2NaNO_3(s) \rightarrow 2NaNO_2(s) + O_2(g)$$

$$2LiNO_3(s) \rightarrow Li_2O(s) + 2NO_2(g) + \tfrac{1}{2}O_2(g)$$

$$2Ca(NO_3)_2(s) \rightarrow 2CaO(s) + 4NO_2(g) + O_2(g)$$

Like all nitrates they are very soluble; those of lithium and Group II crystallising as hydrates such as $LiNO_3 \cdot 3H_2O$, $Ca(NO_3)_2 \cdot 4H_2O$, while barium nitrate(V) is unhydrated.

Potassium nitrate, commonly called saltpetre, is a constituent of gunpowder, fuses and fireworks. It is made from the cheaper, but deliquescent, sodium nitrate by mixing the latter with potassium chloride and separating the potassium nitrate by crystallisation (page 110).

Calcium nitrate is a valuable fertiliser, supplying both calcium and nitrogen to the soil; mixed with calcium phosphate(V) it is sold under the name 'Nitrophos'.

16.5 Diagonal relationship between lithium and magnesium

The chemistry of lithium differs in some respects from that of the rest of Group I. Being more akin to that of magnesium, lithium is said to have a diagonal relationship with magnesium. The two metals are about equally electronegative and form ions of approximately the same size. These small ions have great polarising ability; thus many of the compounds of lithium, like those of magnesium, are partially covalent and appreciably soluble in organic solvents, whereas those of the rest of Group I are wholly ionic. Lithium ions also form hydrates of some stability in solution, the molecules of water being carried into the solid state as water of crystallisation.

Lithium salts, like the corresponding magnesium compounds, are deliquescent. The carbonate, hydroxide and nitrate of lithium are all decomposed by heat to leave residues of oxide, as are those of magnesium. Both metals form nitrides and carbides directly, whereas the other Group I metals do not. The metals have similar electrode potentials; that of lithium is abnormally high for Group I metals, as a result of the high enthalpy of hydration of the lithium ion. Thus while lithium, with a higher first ionisation potential than the rest of Group I, is the least reactive element in the group, its ions have a greater reducing power in solution even than caesium ions.

Questions

1. Define an s-block element. Using a scale of $1 \, cm = 0.100 \, nm$, construct circles to represent the sizes of the atoms and ions of the Group I and Group II metals.

2. Make a brief survey of the uses of (*a*) sodium compounds and (*b*) calcium compounds.

3. (*a*) Define the term *first ionisation energy* of an element.

(*b*) The following table gives some information about Group I elements

Element	Atomic radius in nm	Ionic radius in nm	First ionisation energy in kJ mol^{-1}
Lithium	0.152	0.060	519
Sodium	0.186	0.095	494
Potassium	0.231	0.133	418
Rubidium	0.244	0.148	402
Caesium	0.262	0.169	376

Give reasons for

(i)   the difference between the atomic radius and the ionic radius of an element,

(ii) the increase in the atomic radius which occurs as the group is descended,

(iii) the decrease in the first ionisation energy which occurs as the group is descended,

(iv) the high first ionisation energy of lithium.

(*c*) State **two** ways in which the chemical properties of lithium are different from those of the other elements.   (AEB 1978)

4. Outline the main factors responsible for the observed trends and gradations in physical and chemical properties of elements, and their compounds, in the same group of the Periodic Table.

The elements in Group II of the Periodic Table are, in order of increasing atomic number, beryllium, magnesium, calcium, strontium, barium and radium (chemical symbol: Ra). In the light of your knowledge of the chemistry of the first five of these elements, make predictions concerning the following points:

(a) the reaction of radium with water,

(b) the solubility of radium salts in water,

(c) the acid/base behaviour of the oxide and hydroxide of radium,

(d) the type of bonding present in radium compounds and the physical properties of these compounds,

(e) methods that could be used to obtain pure radium from its compounds.   (L)

5. By consideration of the trends in the properties of the Group I elements and their compounds, deduce possible answers to the following questions concerning the element francium (Fr, atomic number 87).

(*a*) Which noble gas would have the same electronic configuration as the francium ion?

(*b*) Give the formula of the compound formed between francium and hydrogen.

(*c*) Write down the equation for the reaction of francium with water.

(*d*) What further reaction would take place if the solution obtained in (*c*) were exposed to the atmosphere?

(*e*) Why would the compound formed between francium and chlorine be soluble in water but insoluble in benzene? (JMB)

6. The properties of the first member of a group of elements in the Periodic Table are not typical of the group as a whole. Discuss this with reference to the chemistry of the elements of Groups I(Li–Cs) and II(Be–Ba).

You should include in your answer specific properties which differentiate lithium and beryllium from other members of their respective groups as well as the reasons for the differentiation. (JMB)

7 Boron and aluminium in Group III

Table 17.1 Physical data

Element	Atomic number	Electron configuration	Ionisation energy $kJ\,mol^{-1}$			Standard electrode potential V	Atomic radius nm (cov.)	Ionic radius nm	m.p. K	b.p. K
			1st	2nd	3rd					
B	5	2.3 $1s^2 2s^2 p^1$	800	2400	3700	—	0.80	—	2300	4200
Al	13	2.8.3 $----3s^2 3p^1$	580	1800	2700	−1.66	0.125	0.045	933	2723
Ga	31	2.8.18.3 $----3d^{10}4s^2 4p^1$	580	2000	3000	−0.52	0.125	0.062	303	2510
In	49	2.8.18.18.3 $----4d^{10}5s^2 5p^1$	560	1800	2700	−0.34	0.150	0.081	429	2320
Tl	81	2.8.18.32.18.3 $----5d^{10}6s^2 6p^1$	590	2000	2900	−0.34 (Tl^+/Tl)	0.155	0.095	577	1740

17.1 General characteristics and trends in properties

The elements of Group III are the first members of the *p*-block elements in any period; each element has a single *p* electron and two *s* electrons available for bonding. The configuration for the electrons in the outermost orbital is ns^2np^1 where *n* is the quantum number. On bonding one *s* electron is promoted to give a configuration ns^1np^2 or sp^2 hybridisation.

Boron atoms are very small, with the result that the sum of the first, second and third ionisation energies is very large, too large to be recovered by the lattice energy released in the formation of ionic compounds. Thus boron is a non-metal and all its compounds are covalent. The remaining members of the group are all metals, with metallic character increasing from aluminium to thallium.

Aluminium, although metallic and less electronegative than boron, exhibits a high degree of covalency in many of its compounds. The sum of its first three ionisation energies is still very large, and the tripositive ion, when formed, is very small. Thus the Al^{3+} ion exerts a strong polarizing effect, attracting the electron charge-cloud of negative ions and inducing covalency in compounds; notably, aluminium chloride is completely covalent.

Boron, having only three valency electrons, forms compounds in which the boron atom is surrounded by only six electrons. As a result, complex compounds are formed in which boron accepts a pair of electrons in a coordinate bond:

$$\begin{array}{ccc} & F \quad H & \\ & | \quad\; | & \\ F{-}B + {:}N{-}H & \longrightarrow & F{-}B{\leftarrow}N{-}H \\ & | \quad\; | & \\ & F \quad H & \end{array}$$

247

Aluminium forms similar complexes but has much less electron-accepting power. (Both aluminium chloride and boron trifluoride are Lewis acids, see page 184).

$$\begin{array}{ccc} & \text{Cl} \quad \text{H} & & \text{Cl} \quad \text{H} \\ & | \quad\;\; | & & | \quad\;\; | \\ \text{Cl}-\text{Al} + \text{:N}-\text{H} & \longrightarrow & \text{Cl}-\text{Al}\leftarrow\text{N}-\text{H} \\ & | \quad\;\; | & & | \quad\;\; | \\ & \text{Cl} \quad \text{H} & & \text{Cl} \quad \text{H} \end{array}$$

The reactivity of the Al^{3+} ion comes from its small size and high charge. It is stabilised in aqueous solution and in some crystalline compounds by hydration. Six molecules of water become liganded to the ion, forming an octahedral complex with considerable enthalpy of hydration.

Such ions are present in crystals of aluminium sulphate-18-water, $(Al^{3+})_2(SO_4^{2-})_3 \cdot 18H_2O$.

The energy evolved on hydration is largely responsible for the relatively high negative electrode potential of aluminium, which is the highest in the group.

The trend in electrode potential in Group III is interrupted at gallium and decreases for indium and thallium. This is due to the lower screening effect of the penultimate shells of electrons which, in these elements, do not have noble gas structures. The penultimate shells of gallium and indium follow a series of d-block elements, and their atoms contain electrons in d orbitals, while the atoms of thallium also contain electrons in f orbitals. All three metals show a tendency to form unipositive ions as a result of the *inert pair effect*. These ions become more stable from gallium to thallium, and Tl^+ ions are more stable than Tl^{3+} ions. The effect arises as a result of the two s electrons in the valency orbital becoming inert, leaving only the p electron to take part in bonding.

Boron, the head element, shows a typical diagonal relationship with silicon. Some of the similarities between the chemistry of boron and of silicon are described in this chapter. Again, exceptionally from the rest of the group, boron has a maximum covalency of four. Like other elements in the first short period it has electrons only in the first and second quantum levels; the remaining members of the group have electrons in the third and subsequent quantum levels and hence vacant d orbitals available for bonding. Thus while aluminium fluoride dissolves in hydrofluoric acid with the formation of a complex ion, $(AlF_6)^{3-}$, boron trifluoride dissolves to form ions with four covalencies, $(BF_4)^-$.

17.2 Boron and compounds of boron

Boron is not a plentiful element and is not found free in the Earth's crust; its chief ores are borax, $Na_2B_4O_7 \cdot 10H_2O$, and colemanite, $Ca_2B_6O_{11} \cdot 5H_2O$;

other borates are found as deposits from hot springs or salt lakes in the Americas, Russia and Turkey. It is extracted by the reduction of its oxide with magnesium, but is difficult to obtain pure.

$$B_2O_3(s) + 3Mg(s) \rightarrow 3MgO(s) + 2B(s)$$

Boron, as a neutron absorber, is used in nuclear reactors. Traces of boron are used for hardening steel, and crystalline boron is used in making transistors.

Crystalline boron is hard and unreactive, but amorphous boron, when heated, reacts with nitrogen, oxygen, sulphur and the halides to form binary compounds.

$$2B(s) + N_2(g) \rightarrow 2BN(s)$$

$$4B(s) + 3O_2(g) \rightarrow 2B_2O_3(s)$$

$$4B(s) + 3S_2(s) \rightarrow 2B_2S_3(s)$$

$$2B(s) + 3Cl_2(g) \rightarrow 2BCl_3(l)$$

The reaction with oxygen is so vigorous that powdered boron is used in pyrotechnical mixtures for flares and propellants and in detonators for explosive charges.

Boron, having only three electrons available for bonding, forms compounds in which it does not gain a complete inert gas structure. These compounds can react with others, such as ammonia, with the formation of a dative covalent, or coordinate, bond:

$$H_3N(g) + BF_3(l) \rightarrow H_3N \rightarrow BF_3(s)$$

The ability to accept electrons makes boron trifluoride a useful catalyst for polymerisation and reforming reactions (page 354).

A series of hydrides of boron can be made, which are unstable and spontaneously inflammable in air; they are similar to the hydrides of silicon in Group IV, with which element boron has a diagonal relationship.

A hydride with boron present as the central atom of an anionic complex is also formed. This is made by the reaction of an alkali metal hydride with B_2H_6 in ether (page 227):

$$2NaH + B_2H_6 \rightarrow 2Na^+(BH_4)^-$$

Sodium tetrahydridoborate(III) releases hydrogen on reaction with water; it is used in organic chemistry as a reducing agent.

The similarity of boron and silicon is shown in the great affinity for oxygen both elements exhibit. Boron oxide forms a giant covalent molecule obtained as a glassy three-dimensional structure, and the borate ion exists in various complex forms. The sodium salt, $Na_2B_4O_7 \cdot 10H_2O$ is commonly known as borax.

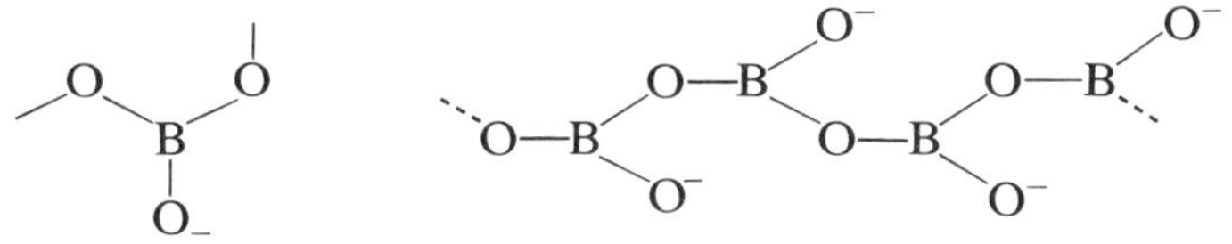

Unit of boron oxide   Borate ion $(B_2O_4)^{2-}$

Fig. 17.1   Boron oxide and borate ion

If a little borax is heated on a loop of platinum wire it loses water and fuses into a colourless glassy bead. Traces of transition metal salts impart characteristic colours to the bead which can be used for identification of the metal ions in the laboratory.

Boron oxide is the anyhdride of trioxoboric(III) acid, H_3BO_3.

$$B_2O_3(s) + 3H_2O(l) \rightarrow 2H_3BO_3(s)$$

Commonly known as boracic acid it is a mild antiseptic used in eye washes. Borax is used for glazing enamel and pottery, as a flux for soldering and in the manufacture of heat-proof glass. In laundering, borax is used to give a good 'finish' to cloth and for producing peroxoborates which act as 'whiteners' in various commercial soap powders.

17.3   Aluminium

Aluminium is the third most plentiful element in the earth's crust. Until 1886 it was one of the most expensive metals, owing to the great difficulty of extracting it from its ore. The oxide is extremely stable and has a very high melting point, making reduction by carbon difficult. The metal only became cheap and universally available after the introduction of electrolysis. The modern method is to electrolyse purified bauxite dissolved in fused cryolite (page 468). The metal is formed by reduction at the cathode:

$$Al^{3+} + 3e^- \rightarrow Al$$

Aluminium is a moderately reactive metal but is resistant to corrosion; on exposure to air it becomes immediately coated with a very thin, tightly bound, layer of oxide which protects it from further attack. This oxide coating can be thickened and made more resistant by making aluminium the anode in an electrolytic bath containing sulphuric or chromic acid. The process, called *anodising*, has the advantage that the layer of oxide will adsorb dyes and so can be coloured.

Aluminium is strong, light and easily moulded, is used extensively in building, in aircraft and ship production, for household utensils and for table tops. In addition it is a good conductor of electricity and, reinforced with steel wire, is used for overhead electric cables.

Aluminium used for building and construction work is often strengthened by alloying with other metals. Duralumin contains small percentages of copper, magnesium and manganese, while magnalium contains magnesium and small amounts of calcium. These alloys, more readily corroded than the metal alone, are often protected by a thin coating of pure aluminium.

Aluminium combines directly with oxygen, sulphur, nitrogen and the halides when heated:

$$4Al(s) + 3O_2(g) \rightarrow 2Al_2^{3+}O_3^{2-}(s)$$

$$4Al(s) + 6S(s) \rightarrow 2Al_2S_3(s)$$

$$2Al(s) + N_2(g) \rightarrow 2AlN(s)$$

$$2Al(s) + 3Cl_2(g) \rightarrow 2AlCl_3(s)$$

Like boron, it has great affinity for oxygen. It is used to reduce the ores of such metals as chromium and manganese that are required in small quantities free from carbon.

A study of Table 17.1 shows that, although still high, the first three ionisation energies of aluminium are lower than those of boron; thus, while boron does not form an ion, aluminium forms a tripositive ion Al^{3+}. The energy required to remove the single p electron is markedly less than that required to remove the pair of s electrons in the formation of this ion, which is *isoelectronic* with those of sodium and magnesium.

$$Na^+ \quad Mg^{2+} \quad Al^{3+}$$
$$2.8 \quad\quad 2.8 \quad\quad 2.8$$

The size of these three ions decrease from sodium to aluminium as the increasing charge on the ion draws the electrons closer to the nucleus; the aluminium ion being small and highly charged exerts considerable polarising power, as described in Section 17.1.

Aluminium, being only weakly metallic, is not readily attacked by dilute acids, although it dissolves in moderately concentrated hydrochloric acid to release hydrogen with the formation of hydrated aluminium ions. With moderately concentrated sulphuric acid sulphur dioxide is formed.

$$2Al(s) + 6HCl(aq) \rightarrow 2Al^{3+}(aq) + 6Cl^-(aq) + 3H_2(g)$$

$$2Al(s) + 6H_2SO_4(aq) \rightarrow 2Al^{3+}(aq) + 3SO_4{}^{2-}(aq) + 6H_2O(l) + 3SO_2(g)$$

Aluminium does not react with either dilute or concentrated nitric acids, which render it inert, possibly by forming a protective layer of oxide. The oxide layer does not protect it from attack by sodium or potassium hydroxide in which it dissolves to form a complex anion with the release of hydrogen.

$$2Al(s) + 2OH^-(aq) + 6H_2O(l) \rightarrow 2Al(OH)_4{}^-(aq) + 3H_2(g)$$

The tendency of aluminium to form complexes is also seen in its hydrides. Aluminium hydride, AlH_3, is a white polymeric solid which cannot be formed by direct combination. Better known is lithium tetrahydridoaluminate(III) (lithium aluminium hydride) in which aluminium is the central atom of an anionic complex. It is made in ether solution by the reaction between lithium hydride and aluminium chloride.

$$4Li^+H^- + AlCl_3 \rightarrow Li^+(AlH_4)^- + 3Li^+Cl^-$$

Like the corresponding boron compound (sodium tetrahydridoborate(III)) it is used as a reducing agent in organic chemistry. It is a strong reducing agent able to reduce carboxylic acids to alcohols (page 406). It is also selective; it will reduce aldehydes to alcohols, but does not attack double bonds. This can be of use in reducing higher unsaturated aldehydes to unsaturated alcohols, since the double bond will remain undisturbed.

The hydrated aluminium(III) ion acts as an acid in solution, reacting with water to form oxonium ions:

$$[Al(H_2O)_6]^{3+}(aq) + H_2O(l) \rightleftharpoons [Al(H_2O)_5(OH)]^{2+}(aq) + H_3O^+(aq)$$

$$[Al(H_2O)_5(OH)]^{2+}(aq) + H_2O(l) \rightleftharpoons [Al(H_2O)_4(OH)_2]^+(aq) + H_3O^+(aq)$$

$$[Al(H_2O)_4(OH)_2]^+(aq) + H_2O(l) \rightleftharpoons Al(H_2O)_3(OH)_3(s) + H_3O^+(aq)$$

The resulting solution is so strongly acidic that aluminium salts of weak acids, such as carbonates, cannot be prepared in solution since they are immediately hydrolysed with the precipitation of aluminium hydroxide.

$$[Al(H_2O)_5(OH)]^{2+}(aq) + CO_3^{2-}(aq) \rightarrow Al(H_2O)_3(OH)_3(s) + H_2O(l) + CO_2(g)$$

17.4 Selected compounds of aluminium

(a) Aluminium oxide

Aluminium oxide occurs naturally as bauxite, the main ore, and in massive crystalline form as the very hard mineral corundum. Clear crystals of corundum provide the gems ruby, sapphire, oriental topaz and oriental emerald, the colours being provided by traces of transition metal ions. Impure corundum, emery, is used as an abrasive.

The hardness, high melting point and great stability of aluminium oxide make it a suitable material for the refractory linings of high temperature furnaces. Specially prepared aluminium oxide is used as an adsorption phase in chromatography, for example, to separate the chlorophylls (page 104). In the laboratory its affinity for water is used in the dehydration of alcohols to alkenes (page 379).

$$C_2H_5OH \xrightarrow[300°C]{Al_2O_3} CH_2{=}CH_2 + H_2O$$

Aluminium oxide is *amphoteric*; it reacts as a base to form salts with acids and also with alkalis to form aluminate(III) ions:

$$(Al^{3+})_2(O^{2-})_3(s) + 6H^+(aq) \rightarrow 2Al^{3+}(aq) + 3H_2O(l)$$

$$(Al^{3+})_2(O^{2-})_3(s) + 2OH^-(aq) + 3H_2O(l) \rightarrow 2[Al(OH)_4]^-(aq)$$
$$\text{aluminate ion}$$

(b) Aluminium hydroxide

Aluminium hydroxide is formed as a white gelatinous precipitate when aqueous ammonia is added to a solution of an aluminium salt:

$$Al^{3+}(aq) + 3OH^-(aq) \rightarrow Al^{3+}(OH^-)_3(s)$$

The gelatinous nature of the precipitate results from hydrogen bonding with water molecules. Ammonia molecules do not form a complex with aluminium(III) ions in solution, hence the hydroxide is not redissolved by addition of excess ammonia. However, like the oxide, aluminium hydroxide is amphoteric, dissolving in strongly basic solutions such as sodium hydroxide, with the formation of the aluminate(III) ion.

$$Al^{3+}(OH^-)_3(s) + OH^-(aq) \rightarrow [Al(OH)_4]^-(aq)$$

The complex ion is hydrated in solution with two water molecules giving it an octahedral structure:

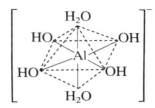

Aluminium hydroxide readily adsorbs dyes; it is used as a mordant in the dyeing industry. The cloth is immersed in a solution which deposits aluminium hydroxide on the threads; when subsequently dipped in the dye the hydroxide absorbs the colour and holds it on the threads of the cloth.

(c) Aluminium halides

The bonding in the halides of aluminium progresses from ionic to covalent as the size of the halide ion increases from fluorine to iodine. Fluorine, being highly electronegative and having a small ion, forms an ionic compound, aluminium fluoride, AlF_3. It is a white anhydrous substance which dissolves in hydrofluoric acid to form a complex acid, H_3AlF_6; the hexafluoroaluminate ion, $[AlF_6]^{3-}$, occurs in cryolite, Na_3AlF_6 (page 468). A similar complex ion, $[AlCl_4]^-$, derived from aluminium chloride, is involved in the Friedel-Crafts reaction in organic syntheses (page 361).

Aluminium chloride forms dimeric covalent molecules, Al_2Cl_6, which do not dissociate even in the vapour phase until a temperature of 400°C is reached; dissociation is complete above 800°C.

$$Al_2Cl_6 \underset{cool}{\overset{heat}{\rightleftharpoons}} 2AlCl_3$$

The dimeric molecules crystallise in a layer lattice held together by van der Waals forces; they are coordinately bonded by the donation of a lone pair of electrons from a chlorine atom of each molecule to the aluminium atom in the other molecule:

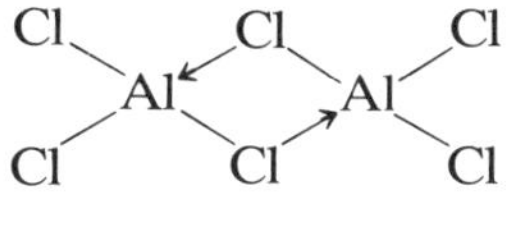

Anhydrous aluminium chloride is made commercially by heating aluminium oxide with carbon in a stream of chlorine:

$$Al_2O_3(s) + 3C(s) + 3Cl_2(g) \rightarrow 2AlCl_3(s) + 3CO(g)$$

or, in the laboratory, by passing chlorine over heated aluminium foil.

The chloride is so readily hydrolysed that moist air must be excluded from the apparatus by the use of the calcium chloride tube shown in the diagram. A container of aluminium chloride should always be opened with great care as a build-up of hydrogen chloride from hydrolysis by water vapour from the air may throw out the contents with some violence.

The reaction of aluminium chloride with water is highly exothermic as the heat of hydration of the small, highly charged, Al^{3+} ion is more than

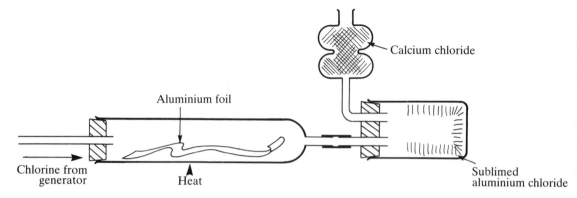

Fig. 17.2 Preparation of aluminium chloride

sufficient to break the covalent Al—Cl bonds. The hydrated aluminium ion reacts further with the water to produce oxonium ions, resulting in a strongly acidic solution.

$$Al_2Cl_6(s) + 12H_2O(l) \rightarrow 2[Al(H_2O)_6]^{3+}(aq) + 6Cl^-(aq)$$

$$[Al(H_2O)_6]^{3+}(aq) + H_2O(l) \rightarrow [Al(H_2O)_5OH]^{2+}(aq) + H_3O^+(aq)$$

Crystals of aluminium(III) chloride-6-water are difficult to obtain since further hydrolysis takes place.

(d) Aluminium sulphate

Aluminium sulphate-18-water forms large transparent crystals in which the molecules of water hydrating the ion are carried into the solid state as water of crystallisation.

Aluminium sulphate is used to treat sewage, which often exists as a colloidal solution difficult to filter or disperse. The small tripositive Al^{3+} ion neutralises the negatively charged 'micelles' or aggregates of molecules present in colloidal sewage, causing them to coagulate and form precipitates. Aluminium sulphate is also effective in clotting blood, being used in 'styptic' pencils to inhibit bleeding.

(e) The alums

Aluminium sulphate forms a double salt with potassium sulphate, potash alum, $K^+Al^{3+}(SO_4{}^{2-})_2 \cdot 12H_2O$. The octahedral crystals are formed when equimolar solutions of the two salts are mixed and allowed to crystallise. A series of alums, all of similar formula, can be made by replacing the potassium ion with other univalent ions of approximately the same ionic radius and/or the aluminium ion with trivalent ions of transitional metal elements of similar ionic radius:

Ammonium alum $(NH_4)^+Al^{3+}(SO_4{}^{2-})_2 \cdot 12H_2O$

Ammonium iron(III) alum $(NH_4{}^+)Fe^{3+}(SO_4{}^{2-})_2 \cdot 12H_2O$

Chrome(III) alum $K^+Cr^{3+}(SO_4{}^{2-})_2 \cdot 12H_2O$

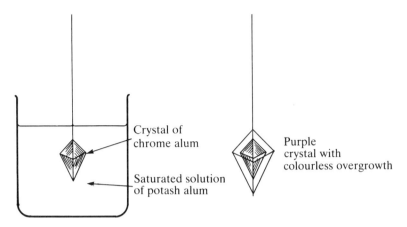

Fig. 17.3 Crystal overgrowths

These alums are *isomorphous* with one another; the crystals are of identical shape with the same crystal angles and proportions. Because the cations comprising them have similar ionic radii they are interchangeable, and the alums form mixed crystals and overgrowths. Thus a purple chrome alum crystal suspended in a saturated solution of potash alum will build up a colourless overgrowth of the same crystal shape.

17.5   The diagonal relationship between beryllium and aluminium

As mentioned in Chapter 16, beryllium in Group II bears a resemblance to aluminium; so strong is this resemblance that when beryllium was first discovered it was thought to have a valency of three.

The beryllium atom is small in comparison with those of the other members of Group II, giving rise to an exceptionally small dipositive ion which has a strong polarising effect comparable to that of the Al^{3+} ion. Beryllium also matches aluminium in the enthalpy of hydration of its ion, electronegativity and electrode potential. The compounds of beryllium are markedly covalent, as are those of aluminium; they are hydrolysed in aqueous solution, giving rise to a small hydrated ion $[Be(H_2O)_4]^{2+}$ which acts as an acid, as does $[Al(H_2O)_6]^{3+}$.

Beryllium is rendered passive by nitric acid in the same way as aluminium; it dissolves in strong alkalis with the release of hydrogen; both its oxide and hydroxide are amphoteric; the beryllate ion $[Be(OH)_4]^{2-}$ is similar to the aluminate ion $[Al(OH)_4]^{-}$. Both metals form complex anions with fluorine, $[BeF_4]^{2-}$ and $[AlF_6]^{3-}$.

Questions

1. Give concise accounts of the chemical principles underlying the extraction of aluminium from bauxite. No details of the manufacturing process need be given unless they serve to illustrate or exemplify points of principle.

The usefulness of a metal is largely dependent on its resistance to corrosion. Briefly explain what is meant by atmospheric corrosion, and outline how for aluminium this form of corrosion is prevented or reduced.   (L)

2. (*a*) Outline the laboratory preparation of anhydrous aluminium chloride starting from aluminium.

(*b*) The relative molecular mass of anhydrous aluminium chloride in benzene solution is 267, but measurements of the density of the vapour are consistent with a relative molecular mass of 133.5. How do you explain these observations?

(*c*) State and explain what happens when anhydrous aluminium chloride is added to water containing a few drops of universal indicator solution and aqueous sodium hydroxide is then added gradually until it is present in excess.

(*d*) Explain why colloidal arsenic(III) sulphide is readily precipitated by a small amount of aqueous aluminium chloride.   (C)

3. State in both cases what anion containing aluminium is formed in (*a*) and (*b*) below and outline the chemistry of the reactions involved.

(*a*) Aqueous sodium hydroxide is added to a solution of an aluminium salt until present in excess.

(*b*) Lithium is heated in a steam of hydrogen and the product added in excess to a suspension of aluminium chloride in ethyoxyethane.

Draw diagrams to illustrate the electronic structures of the anions formed in (*a*) and (*b*).   (C)

4. Write short explanatory notes on the amphoteric nature of oxides.   (L)

5. The hexaaqualuminium(III) ion yields a species having the general formula $Al(H_2O)_x(OH)_y^{charge\ z}$ when treated with a base. State and justify a simple algebraic expression connecting (i) x and y, (ii) y and z.

(JMB)

18 Group IV

18.1 General characteristics and trends in properties

Each of the elements in Group IV is again more electronegative than the corresponding member of Group III, carbon being the most electronegative element in the group. The elements show a steady trend from non-metallic to metallic in character with increasing atomic mass. Carbon and silicon are non-metals, germanium is metalloid, having some metallic and some non-metallic properties, tin and lead are both metals though they each have some non-metallic character, as in the amphoteric nature of their oxides.

Carbon, both as diamond and graphite, has an exceptionally high melting point. Diamond, with its rigid three-dimensional structure, is the hardest natural substance known, and does not conduct electricity. (Graphite is exceptional in this respect.) The structures of diamond and graphite are discussed in more detail in Chapter 5).

The crystalline forms of silicon and germanium are both of the diamond type, but neither of them is as hard nor do they have such high melting points; both are semi-conductors. Tin has three allotropic forms, one of which is again of the diamond type. However, the main crystalline form of tin is metallic, of moderate melting point and a good conductor.

$$
\begin{array}{ccc}
\alpha\text{-tin} & \beta\text{-tin} & \gamma\text{-tin} \\
\text{grey tin} & \text{white tin} & \text{brittle tin} \\
\text{(diamond} & \overset{13.2°C}{\rightleftharpoons} \text{(body-centred} & \overset{161°C}{\rightleftharpoons} \text{(rhombic} \\
\text{structure)} & \text{cubic metallic} & \text{structure)} \\
& \text{structure)} &
\end{array}
$$

Lead has no allotropes; it is a low melting point metal, with a cubic close-packed structure, and a good conductor of electricity.

The elements in Group IV have each two s and two p electrons in the valency orbital, making four electrons available for bonding. They all form tetracovalent compounds with four symmetrically disposed bonds. This results from the promotion of one s electron to the p level to make four equal sp^3 hybridised orbitals (page 51):

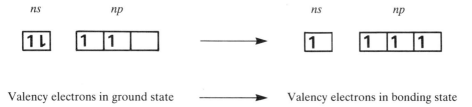

Valency electrons in ground state $\longrightarrow$ Valency electrons in bonding state

None of these elements shows any great tendency to form M^{4+} or M^{4-} ions, since the loss or gain of four electrons involves a very large transfer of energy. Carbon, with a very low atomic radius, is markedly more electronegative than the succeeding members of the group. The high value of the sum of its first four ionisation energies precludes any possibility of the

Table 18.1 Physical data

Element	Atomic number	Electronic configuration	Sum of 1st + 2nd + 3rd + 4th ionisation energies $kJ\,mol^{-1}$	Atomic radius nm (cov.)	M^{2+} ion radius nm	Electronegativity (Pauling scale)	Standard electrode potentials V		m.p. K	b.p. K
							M^{2+}/M	M^{4+}/M^{2+}		
C	6	2.4 $1s^2 2s^2 2p^2$	14 290	0.077	—	2.5	—	—	sub-limes	4120
Si	14	2.8.4 $---3s^2 3p^2$	9990	0.117	—	1.8	—	—	1983	2950
Ge	32	2.8.18.4 $---3d^{10}4s^2 4p^2$	9960	0.122	0.093	1.8	—	—	1210	3100
Sn	50	2.8.18.18.4 $---4d^{10}5s^2 5p^2$	8910	0.141	0.112	1.8	−0.14	+0.15	505	2960
Pb	82	2.8.18.32.18.4 $---5d^{10}6s^2 6p^2$	9420	0.154	0.120	1.8	−0.13	+1.69	600	2024

formation of an ion with four positive charges. Indeed only tin and lead show there is any firm evidence for the existence of 4^+ ions, tin(IV) fluoride and lead(IV) fluoride both having some ionic character.

Germanium, tin and lead also form divalent compounds as a result of the inert pair effect described in Chapter 17. The s electrons in the valency orbital remain paired, leaving the two p electrons to take part in bonding.

Silicon forms only one compound (the very unstable SiO) in which it has an oxidation state of $+2$; germanium(II) and tin(II) compounds are strong reducing agents, readily losing electrons to form germanium(IV) or tin(IV) compounds, while lead(IV) compounds are oxidising agents readily gaining electrons to become lead(II) compounds.

Germanium(II) compounds are almost completely covalent, while there is an increasing degree of ionic character through tin(II) compounds to lead(II) compounds. The relative stabilities of the $+4$ and $+2$ states for tin and lead are reflected in the standard electrode potentials for the M^{4+}/M^{2+} electrode systems.

$$\text{Sn}^{4+}(\text{aq}) + 2e^- \rightleftharpoons \text{Sn}^{2+}(\text{aq}) \qquad E^{\ominus}_{298} = +0.15 \text{ V}$$

$$\text{Pb}^{4+}(\text{aq}) + 2e^- \rightleftharpoons \text{Pb}^{2+}(\text{aq}) \qquad E^{\ominus}_{298} = +1.69 \text{ V}$$

Both tin and lead are moderately reactive metals with electrode potentials of -0.14 V and -0.13 V respectively, although the reactivity of lead is largely suppressed by its tendency to form insoluble coatings of, for example, oxide, chloride or sulphate.

Carbon, as head element, with electrons in s and p orbitals only, cannot expand its covalency beyond four. When bonded to four other atoms its valency orbital is filled. The succeeding elements with electrons in higher quantum levels with available d orbitals can expand their covalencies by forming coordinate bonds. Thus tetrachloromethane is not attacked by water, while silicon tetrachloride is readily hydrolysed:

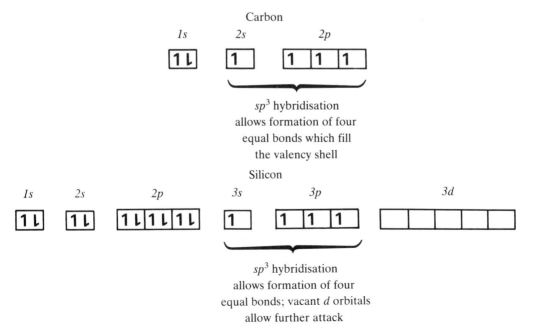

No possible attack by water   Silicon tetrachloride is hydrolysed

18.2   Occurrence and uses

(a) Carbon

Combined carbon is widespread in nature, and the element is found in its two pure forms as diamond and graphite; diamonds are mined chiefly in South Africa and graphite in Mexico, Ceylon and Siberia. The industrial demand for graphite is now so high that it is manufactured, by treating coke at very high temperatures. The structures of diamond and graphite have already been described.

Diamond, although extremely durable, is actually the less stable of the two allotropes, as can be seen from their heats of combustion:

$$C(\text{graphite}) + O_2(g) \rightarrow CO_2(g) \qquad \Delta H = -393.4 \text{ kJ mol}^{-1}$$

$$C(\text{diamond}) + O_2(g) \rightarrow CO_2(g) \qquad \Delta H = -395.5 \text{ kJ mol}^{-1}$$

Being the one of a hardest substances known, diamond is used to cut glass, in rock-drills and as an abrasive for sharpening tools. The high refractive index of diamond makes it the most brilliant of all jewels.

Graphite is, unusually for a non-metal, a good conductor of electricity, as the delocalised electrons in the layer structure are mobile. Since it is also resistant to corrosion it is used as the anode material in electrolytic processes where the anode gas is corrosive, as in the case of chlorine. In nuclear reactors graphite is used as a moderator to control the speed of escaping neutrons (page 39). The layer structure of graphite also makes it a good lubricant, either dry or mixed with oils. The sliding of the layers over one another also accounts for its use in 'lead' pencils since it marks paper.

Charcoal and coke, both impure forms of carbon, were considered to be amorphous, but are now known to be microcrystalline forms of graphite. Wood charcoal, made by heating wood in the absence of air, has a great capacity for absorbing gases and is used in producing a high vacuum and in gas masks. Animal charcoal, from bones, contains calcium phosphate; it will absorb dyes and colouring matters and is used in sugar refining. Carbon black is a form of microcrystalline carbon made by burning oil or natural gas in a limited supply of air; it is used to make printers' ink and to improve the wearing properties of rubber in motor tyres.

Coke, formed from coal in the production of coal gas, is used as fuel, as a reducing agent in blast furnaces, and in the production of water-gas, a mixture of hydrogen and carbon monoxide produced by passing steam over red-hot coke:

$$C(s) + H_2O(g) \rightarrow CO(g) + H_2(g)$$

Carbon occurs in the combined state as coal, which contains 65–85% of carbon together with combined hydrogen, oxygen and minerals, and as petroleum, which is a mixture of liquid hydrocarbons (page 340). Both these 'fossil fuels' are derived from decayed organic matter; the former from tropical forests which existed millions of years ago, and the latter from sea creatures which died and sank to the bed of the oceans, also long ago. The organic residues remained buried under great depths of rocks for many ages until the effects of heat and pressure resulted in the coal and oil we use so freely today.

Combined carbon also occurs as carbonates in mineral rocks such as chalk, limestone and marble (calcium carbonate), dolomite (calcium and magnesium carbonate) and malachite (copper carbonate).

(b) Silicon

Silicon is the second most abundant element in the Earths's crust, being widely distributed as silica and silicates.

The silicates form giant structures built up from ions which have a tetrahedral structure similar to that of silica:

Orthosilicate $(SiO_4)^{4-}$

Pyrosilicate $(Si_2O_7)^{6-}$

Fibrous silicate $(SiO_3)_n^{2n-1}$

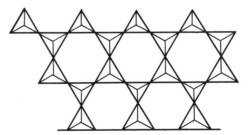

Sheet-like silicate

Fig. 18.1 Examples of silicate ion structures

The more complex silicates form chains or sheets held together by metal ions which give rise to fibrous materials like asbestos or to readily cleaved crystals such as mica or talc. The metal ions may be magnesium, calcium, potassium, sodium or aluminium. An important class of silicates, the zeolites, are clays used in ion exchange columns for producing soft or de-ionised water (page 231).

The element silicon is extracted by reduction of the oxide by magnesium or zinc. It is a high melting point solid, used as a semiconductor in transistors. It forms an alloy with iron, ferrosilicon, which is used in the manufacture of acid-resistant steels.

(c) Germanium

Germanium was not isolated until 1886, although its existence had been predicted by Mendeleyev in 1871 to fill a gap in his Periodic Table of the Elements. Previously many people had questioned and even derided the idea of a periodic classification of the elements, but the discovery of germanium, with properties exactly as forecast, silenced the critics. The modern Periodic Table is a development from Mendeleyev's original classification (page 216).

At first germanium was of academic interest only, but more recently it has become important as a semiconductor in transistors, for which crystalline germanium is required in a very pure form. Its properties are metalloid—intermediate between metallic and non-metallic.

(d) Tin

Tin occurs as cassiterite, SnO_2, in Indonesia, Malaysia, Bolivia, Nigeria and Cornwall, and the metal is extracted by reduction with coke.

Tin melts at the comparatively low temperature of 232°C. Alloys of tin include bronze (tin and copper), solder and pewter (tin and lead) and type metal (tin, antimony and lead). An important use of tin is in plating iron for food containers ('tin' cans); unfortunately if the tin is scratched the can will corrode very quickly since tin is above iron in the electrochemical series, and an electrochemical cell develops in the damaged area with tin as the positive pole, leading to rapid rusting. (page 324).

(e) Lead

The metal occurs in many parts of the world in the combined state as galena, lead sulphide, especially in Australia and Canada. It is extracted by roasting the ore to obtain the oxide and then reducing the oxide with coke:

$$2PbS(s) + 3O_2(g) \rightarrow 2PbO(s) + 2SO_2(g)$$

$$2PbO(s) + C(s) \rightarrow 2Pb(s) + CO_2(g)$$

Being readily extracted, soft and easily worked, and not much subject to corrosion, lead has been used for water pipes and conduits since Roman times, but recently copper and plastic pipes have largely replaced lead for plumbing, partly because they are less liable to burst when frozen and partly because of the slight risk of lead poisoning from lead pipes, especially in soft water districts. Lead is slightly soluble in natural water, but in hard water districts the pipes rapidly acquire a protective coating of lead carbonate which obviates further solution.

Lead has been extensively used as a roofing material, and in alloys with tin as mentioned above. The lead plates in accumulators are hardened by the addition of a little antimony.

Lead tetraethyl has been widely used in petrol to prevent 'knocking', or the violent ignition of petrol in the cylinders of car engines. With the vastly increased number of petrol-driven vehicles, an unacceptable level of lead compounds in the soil is developing and improved grades of petrol containing less tetraethyl lead have been produced. Many of the lead compounds used in paints are also being replaced because of the toxic nature of lead.

18.3 Reactions

Carbon forms many compounds which have a 'back-bone' of self-linked carbon atoms; the linkage results in compounds with a chain-like structure, referred to as *catenated* compounds. The tendency to catenation results from the high energy of the C—C bond making the chains stable; these bonds are stronger than those between atoms of other non-metals. Elements close to carbon in the Periodic Table share the property to a limited extent, but carbon is unique in the length of chains formed and the number and variety of compounds resulting from catenation. Also, the strength of the C—C bond is comparable to the strength of bonds between carbon and other atoms such as hydrogen or oxygen, increasing the stability of the catenated compounds; other non-metal elements form stronger bonds with atoms of other elements than they do between atoms of their own kind thus reducing the strength of the chains.

Table 18.2 Comparison of bond energies

Bond	C—C	Si—Si	C—O	Si—O
Energy $kJ\,mol^{-1}$	346	179	358	368

Carbon also forms a variety of compounds with multiple bonds between adjacent atoms or between carbon atoms and atoms of other elements. These compounds and the catenated carbon compounds are described in the chapters on carbon chemistry.

Carbon as diamond or graphite is unreactive; it is more reactive in the form ot charcoal and other microcrystalline states. All forms of carbon burn, at high enough temperatures, forming carbon dioxide in a plentiful supply of air or carbon monoxide if the supply is limited. Charcoal or, commercially, coke reacts at very high temperature with sulphur making carbon disulphide; the reaction is endothermic and the compound is unstable.

$$C(s) + 2S(s) \rightleftharpoons CS_2(l) \qquad \Delta H_{298}^{\ominus} = +92 \text{ kJ mol}^{-1}$$

Microcrystalline carbon also combines directly with fluorine, silicon and the more reactive metals; the metal compounds contain the dicarbide ion $(C{\equiv}C)^{2-}$

$$C(s) + 2F_2(g) \rightarrow CF_4(g)$$

$$C(s) + Si(s) \rightarrow CSi(s)(\text{carborundum})$$

$$2C(s) + Ca(s) \rightarrow Ca^{2+}(C{\equiv}C)^{2-}(s)$$

The element is not attacked by dilute acids, concentrated hydrochloric acid or alkali hydroxides, but charcoal reduces concentrated oxidising acids.

$$C(s) + 4HNO_3(l) \rightarrow CO_2(g) + 2H_2O(l) + 4NO_2(g)$$

$$C(s) + 2H_2SO_4(l) \rightarrow CO_2(g) + 2H_2O(l) + 2SO_2(g)$$

Carbon is a strong reducing agent; it releases many metals from their oxides and reduces steam to water-gas (page 260).

The remaining elements in Group IV are all more reactive than carbon; their reactions reflect the trend from non-metallic to metallic properties. They all burn at suitable temperatures in oxygen; silicon, germanium and tin yielding oxides in which they have an oxidation state of $+4$ (SiO_2, GeO_2, SnO_2) while lead yields a mixture of lead(II) oxide, PbO, and the complex oxide Pb_3O_4.

They all react directly with chlorine; silicon, germanium and tin forming tetrachloro compounds, while lead forms lead(II) chloride, $PbCl_2$.

Silicon and germanium do not react with dilute acids, but the more metallic tin and lead react to make tin(II) and lead(II) salts.

$$Sn(s) + 2HCl(aq) \rightarrow SnCl_2(aq) + H_2(g)$$

$$Pb(s) + 2HNO_3(aq) \rightarrow Pb(NO_3)_2(aq) + H_2(g)$$

The reaction of lead with both hydrochloric and sulphuric acids is very slow; the metal becomes coated with insoluble precipitates of lead(II) chloride or lead(II) sulphate. Concentrated nitric acid has no effect on silicon; it reacts with germanium and tin forming hydrated oxides and with lead to make divalent salts.

$$3Sn(s) + 4HNO_3(aq) \rightarrow 3SnO_2(s) + 4NO(g) + 2H_2O(l)$$

$$3Pb(s) + 8HNO_3(aq) \rightarrow 3Pb(NO_3)_2(aq) + 2NO(g) + 4H_2O(l)$$

Sodium hydroxide, molten, or in concentrated solution, reacts with all four elements; the vigour of the reaction and the nature of the products vary with the degree of non-metallic character of the elements. Silicon and germanium dissolve with the release of hydrogen and the formation of silicate(IV) and germanate(IV) ions. Tin and lead are more resistant to attack by alkalis, the products of the reactions being stannate(II) and plumbate(II) ions, in which the s electrons of the metal atoms are inert, taking no part in the bonding.

18.4 Selected compounds

(a) Hydrides

Silicon and germanium each form a limited number of hydrides of short chain length and of formulae similar to the very stable catenated hydrides of carbon (the alkanes); they are moderately stable to heat but ignite spontaneously in the air. Tin and lead form only simple hydrides, SnH_4 and PbH_4.

Of the simple hydrides, only methane, CH_4, is stable and unreactive; the thermal stabilities of the remainder decrease steadily from SiH_4 which decomposes above 400°C to plumbane, PH_4, which decomposes at 0°C.

(b) Oxides

The main oxides of Group IV are the monoxides, in which the elements have an oxidation number of +2, and the dioxides, in which the elements are present in the +4 oxidation state. Carbon monoxide differs from the other monoxides in that it is a neutral gas, whereas the remainder of the monoxides are solids with properties varying from acidic to amphoteric; little is known of silicon monoxide whose very existence is not well established.

Carbon monoxide is *isoelectronic* with nitrogen, with a triple bond between the atoms:

$$:C::O^{×}_{×} \quad \text{or} \quad C≡O$$

$$:N::N: \quad \text{or} \quad N≡N$$

It burns to carbon dioxide with the evolution of considerable heat, making it industrially important, especially as a constituent of water gas and as a powerful reducing agent in the reduction of metal oxides in the blast furnace reactions (page 469).

$$Fe_2O_3(s) + 3CO(g) \rightarrow 2Fe(s) + 3CO_2(g)$$

Carbon monoxide combines with transition metals to form carbonyls, by the donation of the pair of electrons on the carbon atom in a coordinate bond.

$$Ni(s) + 4CO(g) \rightarrow Ni(CO)_4(g)$$

Mixed with hydrogen in the correct proportions it is used, with a suitable catalyst, to manufacture methanol (page 375).

$$CO(g) + 2H_2(g) \rightarrow CH_3OH(l)$$

The gas, although odourless, is very poisonous, combining with haemoglobin in such a way as to prevent the uptake of oxygen. It is present in substantial quantities in the exhaust gases from car engines and faulty water heaters and has been the cause of a great many deaths.

Germanium(II) oxide is a black solid, unstable with respect to germanium(IV) oxide, to which it is immediately oxidised in the presence of air; tin(II) oxide is also easily oxidised, while lead(II) oxide is more stable than lead(IV) oxide, being obtained by the thermal decomposition of the dioxide:

$$2PbO_2(s) \rightarrow 2PbO(s) + O_2(g)$$

The monoxides of germanium, tin and lead are amphoteric, giving rise to salts when dissolved in acids and to hydrated ions containing the element in the +2 oxidation state when dissolved in alkalis;

$$PbO(s) + 2HNO_3(aq) \rightarrow Pb(NO_3)_2(aq) + H_2O(l)$$
$$SnO(s) + 2OH^-(aq) + H_2O(l) \rightarrow [Sn(OH)_4]^{2-}(aq)$$

Solutions of germanate(II) ions and stannate(II) ions are powerful reducing agents, while the plumbate(II) ion is not easily oxidised. Hydrated lead(II) oxide is precipitated from solutions of soluble lead salts by the addition of ammonia solution.

$$Pb^{2+}(aq) + 2OH^-(aq) \rightarrow Pb(OH)_2(s)$$

Carbon dioxide, like carbon monoxide, is a gas, while the dioxides of the other Group IV elements are solids. With the exception of lead (IV) oxide, the dioxides are all more stable than the corresponding monoxides; this is consistent with the greater stability generally of the +4 oxidation state in Group IV.

Carbon dioxide, silicon(IV) oxide and germanium(IV) oxide are all predominantly acidic, although germanium(IV) oxide dissolves with difficulty in concentrated acids to give germanium(IV) compounds.

Tin(IV) oxide and lead(IV) oxide are both classed as amphoteric; under certain conditions they can be dissolved in acids to make tin(IV) and lead(IV) compounds, but their main properties are acidic.

Carbon dioxide has a linear molecule with double bonds between the carbon and the two oxygen atoms, $O{=}C{=}O$. The strength of the $C{=}O$ bond accounts for the formation of discrete molecules of carbon dioxide rather than a giant structure with single bonds between carbon and oxygen similar to that found in silicon(IV) oxide (see below).

Carbon dioxide is a dense colourless gas, fairly easily liquefied on compression; if the liquid is cooled by expansion part solidifies. Solid carbon dioxide sublimes at $-78°C$ at atmospheric pressure, making it a useful portable refrigerant which does not require leak-proof containers since no liquid is produced by melting. Cylinders of the compressed gas make safe and effective fire-extinguishers.

Carbon dioxide is a fairly inert gas; it dissolves in water to make a weakly acid solution:

$$CO_2(g) + H_2O(l) \rightleftharpoons H^+(aq) + HCO_3^-(aq)$$

With alkalis, it forms carbonates in solution, but the majority of metal carbonates are insoluble.

$$2NaOH(aq) + CO_2(g) \rightarrow Na_2CO_3(aq) + H_2O(l)$$

The amount of carbon dioxide in the atmosphere remains remarkably constant at about 0.03%; the quantity withdrawn by photosynthesis and the formation of mineral carbonates is balanced by that entering the atmosphere from animal respiration and the burning of fuels.

Silicon(IV) oxide is also wholly acidic, not being attacked by acids (except hydrogen fluoride) but reacting with hot concentrated alkalis to make silicates. With fused sodium hydroxide, sodium silicate is formed.

$$4NaOH + SiO_2 \rightarrow Na_4SiO_4 + 2H_2O$$

Sodium silicate is known as water-glass. Formerly used in solution to preserve eggs, it is the basis for making 'chemical gardens'; the weird growths in these gardens being silicates of metals with coloured ions. If a solution of sodium silicate is acidified, silicon(IV) oxide is precipitated as 'silica gel', which can be filtered off, washed free of acid and dried; it readily absorbs water and is an excellent drying agent for gases and organic liquids.

The Si—O bond is stronger than any other silicon bond except Si—F. This results in a giant molecular structure for silicon(IV) oxide, or silica, which exists in three polymorphic crystalline forms, quartz, tridymite and crystobalite, and in amorphous form as flint. The crystalline forms are built of tetrahedral units in which the silicon atom is surrounded by four oxygen atoms.

$$
\begin{array}{c}
O^- \\
| \\
{}^-O-Si-O- \\
| \\
{}_-O
\end{array}
$$

The different crystalline structures for the forms of silica arise from variations in the way in which the tetrahedra are linked together. These tetrahedra are also the basis of the structure of the many different silicates (Section 18.2.b). Crystalline silica melts at about 1700°C to a transparent viscous liquid which on cooling forms silica glass. This is really a supercooled liquid; the tetrahedral units are retained throughout but, instead of crystallising to a regular structure, remain randomly arranged as in the liquid. Silica glass is resistant to acids and can withstand sudden changes of temperature since it has a low coefficient of expansion.

Ordinary soda glass is made by fusing silica with calcium carbonate and sodium carbonate:

$$(Na^+)_2CO_3^{2-}(s) + SiO_2(s) \rightarrow (Na^+)_2SiO_3^{2-}(s) + CO_2(g)$$

$$Ca^{2+}CO_3^{2-}(s) + SiO_2(s) \rightarrow Ca^{2+}SiO_3^{2-}(s) + CO_2(g)$$

Introducing a proportion of lead or potassium alters the characteristics of the glass—producing, for example, lead crystal glass. The addition of boron produces borosilicate glass.

The ability of silicon to form structures containing chains of alternating silicon and oxygen atoms is utilised in the formation of *silicones*. These most useful and versatile compounds are made by the hydrolysis of the chlorosilanes. Alkyl or aryl chlorides (page 432) are passed over heated silicon in the presence of a copper catalyst.

$$2CH_3Cl + Si \rightarrow (CH_3)_2SiCl_2$$

The chlorosilanes are hydrolysed:

$$(CH_3)_2SiCl_2 + 2H_2O \rightarrow (CH_3)_2Si(OH)_2 + 2HCl$$

The resulting hydroxy compounds polymerise with the elimination of water molecules:

$$
\begin{array}{ccccc}
CH_3 & & CH_3 & & CH_3 \quad CH_3 \\
| & & | & & | \qquad | \\
HO-Si-OH & + & HO-Si-OH & \longrightarrow & HO-Si-O-Si-OH + H_2O \\
| & & | & & | \qquad | \\
CH_3 & & CH_3 & & CH_3 \quad CH_3
\end{array}
$$

Carefully controlled conditions are used to produce chains of the required lengths or cross-linked structures. The products are oils, greases, resins or rubbery compounds which can be used as lubricants, varnishes, paints, polishes or water-proofing materials.

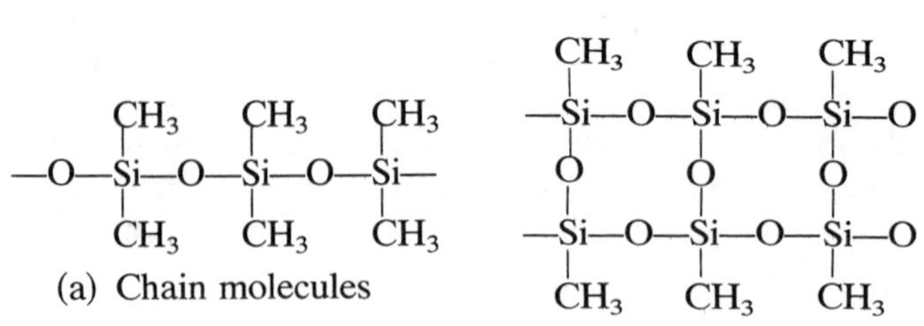

$$\begin{array}{ccc} CH_3 & CH_3 & CH_3 \\ | & | & | \\ -O-Si-O-Si-O-Si- \\ | & | & | \\ CH_3 & CH_3 & CH_3 \end{array}$$

(a) Chain molecules

$$\begin{array}{ccc} CH_3 & CH_3 & CH_3 \\ | & | & | \\ -Si-O-Si-O-Si-O \\ | & | & | \\ O & O & O \\ | & | & | \\ -Si-O-Si-O-Si-O \\ | & | & | \\ CH_3 & CH_3 & CH_3 \end{array}$$

(b) Crosslinked structures

Fig. 18.2 Two dimensional representations of silicone structures

Germanium(IV) oxide and tin(IV) oxide can both be made by reaction with oxygen at high temperatures, while lead(IV) oxide, being less stable than lead(II) oxide, is prepared by the oxidation of PbO or Pb^{2+} salts in alkaline solution. Germanium(IV) oxide reacts with fused sodium hydroxide to form germanate(IV) ions, $[GeO_4]^{4-}$; tin(IV) oxide dissolves in concentrated solutions of alkalis to yield hydrated stannate(IV) ions.

$$SnO_2(s) + 2OH^-(aq) + 2H_2O(l) \rightarrow [Sn(OH)_6]^{2-}(aq)$$

Lead(IV) oxide also yields plumbate(IV) ions on reaction with fused alkali; these ions are formed in their hydrated state as $[Pb(OH)_6]^{2-}$ at the positive plate in a lead accumulator.

The hydrated stannate(IV) and plumbate(IV) ions are octahedral complexes with the metal ion at the centre; the +4 oxidation state of lead is stabilised in these ions.

Lead(IV) oxide has a more amphoteric character than the other dioxides, dissolving in concentrated hydrochloric acid to yield lead (IV) chloride; this rapidly decomposes to lead(II) chloride with the evolution of chlorine. As indicated by this reaction, lead(IV) oxide is a powerful oxidising agent, capable of oxidising Mn^{2+} to MnO_4^- ions, and chloride ions to chlorine. It is thermally unstable, decomposing on heating to dilead(II) lead(IV) oxide (red lead) or on stronger heating to lead(II) oxide:

$$3PbO_2(s) \rightarrow Pb_3O_4(s) + O_2(g)$$

$$2PbO_2(s) \rightarrow 2PbO(s) + O_2(g)$$

Dilead(II)lead(IV) oxide contains lead in two oxidation states, $Pb_2^{II}Pb^{IV}O_4$; heating the oxide in nitric acid yields a mixture of solid lead(IV) oxide suspended in a solution of lead(II) nitrate which can be filtered to obtain PbO_2 as a brown solid.

(c) Halides

Carbon forms numerous compounds with the halogen elements, some of which are described in Chapter 29. All the Group IV elements form tetrahalides with the four main halogen elements (except that lead does not form the tetrabromide or tetraiodide); germanium, tin and lead also form dihalides. The stabilities of the tetrahalides decrease from those of carbon to those of lead, while the stabilities of the dihalides increase from germanium through tin to lead.

Tetrachloromethane is a dense, sweet-smelling colourless liquid, thermally stable and not hydrolysed by water, with which it is immiscible.

Silicon(IV)chloride is stable to heat but is hydrolysed violently on contact with water, with the formation of hydrated silicon(IV)oxide.

$$SiCl_4(l) + 4H_2O(l) \rightarrow SiO_2 \cdot 2H_2O(s) + 4H^+(aq) + 4Cl^-(aq)$$

Germanium(IV) and tin(IV) chlorides behave similarly, while lead(IV) chloride decomposes at room temperature to lead(II) chloride:

$$PbCl_4(l) \rightarrow PbCl_2(s) + Cl_2(g)$$

The elements, other than carbon, also form complex anions involving the halogens. For example, silicon(IV) fluoride is formed by the action of hydrogen fluoride on silicon(IV) oxide. The reaction is carried out by heating calcium fluoride with silicon dioxide and concentrated sulphuric acid.

$$CaF_2(s) + H_2SO_4(l) \rightarrow 2HF(g) + CaSO_4(s)$$

$$4HF(g) + SiO_2(s) \rightarrow SiF_4(g) + 2H_2O(l)$$

The silicon(IV)fluoride can react further with hydrofluoric acid, making hexafluorosilicilic acid containing a hexafluorosilicate(IV) ion.

$$SiF_4 + 2HF \rightarrow 2H^+ + (SiF_6)^{2-}$$

Many other complex ions such as $(GeF_6)^{2-}, (SnF_6)^{2-}, (GeCl_4)^{2-}$ and $(PbCl_4)^{2-}$ are formed.

Germanium(II) and tin(II) chlorides are each less stable than the tetra-chlorides and are strong reducing agents. Tin(II) chloride reduces mercury(II) chloride successively to mercury(I) chloride and to mercury.

$$SnCl_2(aq) + 2HgCl_2(aq) \rightarrow SnCl_4(aq) + (HgCl)_2(s)$$

$$(HgCl)_2(s) + SnCl_2(aq) \rightarrow 2Hg(l) + SnCl_4(aq)$$

Germanium(II) and tin(II) chlorides can only be dissolved in water in the presence of hydrochloric acid since the divalent ions are readily hydrolysed with the precipitation of basic salts such as Sn(OH)Cl. The acid also keeps the tin(IV)chloride in solution.

The dihalides of lead are more ionic in character; they are solids at room temperature and conduct electricity in the fused state. Since they are insoluble they are prepared by precipitation from aqueous solutions containing the appropriate ions.

$$Pb^{2+}(aq) + 2Cl^-(aq) \rightarrow PbCl_2(s)$$

They react with concentrated hydrogen halide acids to form complex acids with lead in the anion. For example, precipitated lead chloride will dissolve in concentrated hydrochloric acid (see above):

$$PbCl_2(s) + 2HCl(aq) \rightarrow 2H^+(aq) + (PbCl_4)^{2-}(aq)$$

(d) Oxosalts of tin and lead

Both tin and lead form oxosalts in the +2 oxidation states; those of tin are rapidly hydrolysed to basic salts. Lead(II) nitrate and lead(II) ethanoate are soluble and provide a source of Pb^{2+} ions in laboratory work, Lead(II) sulphate and lead(II) chromate are insoluble; precipitation of the latter, a bright yellow compound, can be used to identify the presence of lead(II) ions.

18.5   Summary of properties and reactions of Group IV elements

Element	Carbon	Silicon	Germanium	Tin	Lead
Type	Non-metal	Non-metal	Semi-metal	Metal	Metal
Structure	Diamond Graphite	Diamond	Diamond	Diamond Metallic	Metallic
Oxide MO	Stable: reducing agent: neutral gas.	Very unstable.	Unstable reducing agent: mainly acidic:	Readily oxidised: reducing agent: amphoteric.	Stable: amphoteric.
Oxide MO_2	Stable; acidic; gas.	Stable: acidic: solid.	Stable: amphoteric; (mainly acidic).	Stable: amphoteric.	Decomposes on heating: oxidising agent: amphoteric.
Hydride MH_4	Stable	Increasingly unstable —————————————————→			
Chloride MCL_4	Stable unhydro- lysed	Stable rapidly hydrolysed	Stable readily hydrolysed	Moderately stable hydrolysed	Unstable
+4 Oxidation state	Decreasing stability ——————————————————————→				
+2 Oxidation state	— Increasing stability ———————————→				

Questions

1. Compare and contrast the compounds of tin and lead, paying particular attention to their stabilities and their oxidising or reducing properties.
2. Write brief notes on the following:
   (a) Silicates
   (b) Carbides
   (c) Silicones
   (d) Carbonyl compounds
   (e) Glass
3. Survey the chemistry of the Group IV elements (C—Pb) by giving
   (a) a summary of the physical and chemical properties of the elements,
   (b) brief descriptions of preparative routes to the chlorides and oxides.

(c) a discussion of group trends in valencies and bond-types of the chlorides and oxides

(d) a discussion of the special properties of carbon and the ways in which its chemistry differs from the other members of the group. (JMB)

4. (a) *Outline* the preparation of

(i) the dioxides of carbon and silicon,

(ii) the tetrachlorides of carbon and silicon.

(b) How *do* the electronic structures of the elements carbon and silicon (atomic numbers C = 6, Si = 14) and the structures of the dioxides and tetrachlorides help to explain that, at ordinary temperature and pressure,

(i) carbon dioxide is a gas, whilst silicon(IV) oxide (silicon dioxide) is a solid of high melting point,

(ii) carbon tetrachloride does not react with water, whilst silicon tetrachloride reacts vigorously with cold water giving steamy fumes and a gelatinous precipitate? (AEB 1977)

5. 'Whereas the study of the alkali metals reveals the similarities that exist between elements in a particular group of the Periodic Table, the study of the elements C to Pb brings out the differences and dissimilarities between them.' Discuss and justify this statement. (L)

6. Explain why lead(II) chloride is insoluble in dilute hydrochloric acid but will dissolve in concentrated hydrochloric acid.

19 Nitrogen and phosphorus in Group V

Table 19.1 Physical data

Element	Atomic number	Electronic configuration	Atomic radius nm (cov)	m.p. K	b.p. K	Electro-negativity Pauling scale
N	7	2.5 $1s^22s^22p^3$	0.070	63	77	3.0
P	15	2.8.5 $---3s^23p^3$	0.110	317	554	2.1
As	33	2.8.18.5 $---3d^{10}4s^24p^3$	0.118	886 sublimes		2.0
Sb	51	1.8.18.18.5 $-----4d^{10}5s^25p^3$	0.136	903	1910	1.9
Bi	83	2.8.18.32.18.5 $------5d^{10}6s^26p^3$	0.152	545	1832	1.9

19.1 General characteristics and trends in properties of the elements in Group V

The elements of Group V range in character from non-metallic nitrogen and phosphorus to weakly metallic bismuth. Arsenic and antimony are metalloids, having both metallic and non-metallic characteristics. The chemistry of the elements phosphorus to bismuth shows a steady trend, with the oxides progressing from acidic through amphoteric to weakly basic, and the chlorides becoming more salt-like. Nitrogen, the head element, differs sharply from phosphorus and the succeeding elements. It has a relatively high electronegativity, a small atom, a maximum covalency of four, and exists in the free state as a stable diatomic gas. Like carbon and oxygen it readily forms multiple bonds, the nitrogen molecule having a triple bond :N≡N:. It is the great strength of this bond which accounts for the low reactivity of the free element. Phosphorus and the succeeding elements do not readily form such multiple bonds. The single P—P bonds found in phosphorus molecules are not as strong as the nitrogen triple bond, making phosphorus much more reactive.

As members of the third group of the *p*-block elements, each element has two *s* electrons and three *p* electrons in the valency shell. Thus theoretically they could acquire an inert gas structure by the gain of three electrons. However, the amount of energy required to form a negative ion M^{3-} by the gain of three electrons is very large, and it is seldom formed. The nitrogen atom is sufficiently small and electronegative to form nitride ions in combination with the very reactive metals of Groups I and II. Compounds such as

272

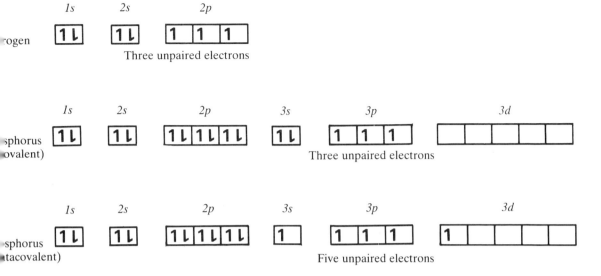

Fig. 19.1 Electron structures of nitrogen and phosphorus

sodium nitride, $(Na^+)_3N^{3-}$, and magnesium nitride, $(Mg^{2+})_3(N^{3-})_2$, have sufficiently high lattice energy to be fairly stable; in these nitrides the small size of the negative ion and large size of the positive ion favour the formation of an ionic compound, in accordance with Fajan's Rule. Some phosphides such as Na_3P can be made, but the larger and less electronegative atom of phosphorus does not readily form a P^{3-} ion, and the phosphides are not wholly ionic.

Nitrogen and phosphorus do not form positive ions, but the larger atoms of arsenic, antimony and bismuth, with electrons in d and f orbitals, all form compounds containing M^{3+} ions as a result of the inert pair effect (Chapter 17). Of these, only Bi^{3+} ions are at all stable.

The main type of bonding in the group is covalent. All the elements form tricovalent compounds involving the p electrons in the valency orbital, such as ammonia, NH_3, and phosphorus trichloride, PCl_3. Phosphorus and the succeeding elements can all expand their covalency to five, since they have available d orbitals to accommodate the extra electrons, while nitrogen has no such orbitals available. (Fig. 19.1).

Thus phosphorus forms two chlorides PCl_3 and PCl_5, while nitrogen forms only the very unstable NCl_3.

However nitrogen can expand its covalency to four by the donation of the paired s electrons in a coordinate bond, as in the formation of ammonia borontrichloride:

$$
\begin{array}{ccc}
\quad H \quad Cl & & H \quad Cl \\
\quad | \quad\ | & & | \quad\ | \\
H\!-\!N\!\overset{\bullet}{\underset{\bullet}{}}\!+\!B\!-\!Cl & \rightarrow & H\!-\!N\!\rightarrow\!B\!-\!Cl \\
\quad | \quad\ | & & | \quad\ | \\
\quad H \quad Cl & & H \quad Cl
\end{array}
$$

Phosphorus does not act as an electron donor so readily.

19.2   Nitrogen

(a) Occurrence, reactions and uses

Nitrogen comprises 78% by volume of the air, and occurs combined as sodium nitrate (caliche) in mineral deposits in a very dry region of Chile. It is essential to life, being a constituent element of proteins. The natural routes by which nitrogen compounds reach the soil, as shown in the nitrogen cycle (Fig. 19.2), are supplemented by the use of vast amounts of artificial fertiliser derived from ammonia made by the Haber process, from coal distillation and from the Chilean nitrates.

Nitrogen is a colourless odourless gas, slightly soluble in water. It is so unreactive that Lavoisier named it 'azote' (the lazy one) and the Germans call it 'stickstoff'. It does not burn, nor will it support combustion in the usual sense, although magnesium will continue to burn in it if previously ignited.

At very high temperatures, such as those provided by an electric arc or a flash of lightning, nitrogen combines directly with oxygen to make nitrogen monoxide:

$$N_2(g) + O_2(g) \rightleftharpoons 2NO(g) \qquad \Delta H_{298}^{\ominus} = +180 \text{ kJ mol}^{-1}$$

The reaction is endothermic and reversible, but some of the nitrogen monoxide combines with oxygen to make nitrogen dioxide which dissolves in rain water and is washed into the soil.

Nitrogen also combines directly with hydrogen, but although the reaction is mildly exothermic it is reversible, since the ammonia formed is not stable at raised temperatures.

$$N_2(g) + 3H_2(g) \rightleftharpoons 2NH_3(g) \qquad \Delta H_{298}^{\ominus} = -92 \text{ kJ mol}^{-1}$$

The reaction is favoured thermodynamically by high pressures and low temperatures (page 165) but the conditions are modified in the manufacture of ammonia (page 475).

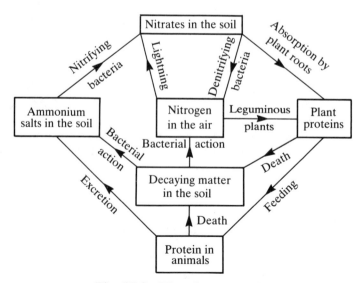

Fig. 19.2   The nitrogen cycle

Nitrogen forms covalent nitrides with a number of elements, such as sulphur, phosphorus, silicon, aluminium and boron. The resulting compounds may be molecular in character or may form solids with giant crystalline lattice structures. Boron nitride occurs in two crystalline forms, an exceptionally hard diamond-like crystal and a substance resembling graphite. In these structures one boron atom and one nitrogen atom together have a total of eight valency electrons, making them isoelectronic with two carbon atoms.

Metals high in the electrochemical series form ionic compounds with nitrogen which contain the N^{3-} ion. These are high melting-point white crystalline solids readily decomposed by water to yield ammonia.

$$Mg_3N_2(s) + 6H_2O(l) \rightarrow 3Mg(OH)_2(s) + 2NH_3(g)$$

$$Li_3N(s) + 3H_2O(l) \rightarrow 3LiOH(aq) + NH_3(g)$$

With transition metals nitrogen forms interstitial nitrides similar to the interstitial hydrides described in Chapter 15. The presence of the nitrogen molecules in the metallic lattice confers hardness and resistance to corrosion on the metal.

The main use of nitrogen is in the manufacture of ammonia and nitric acid. It also provides an inert atmosphere where this is required for welding or other metallurgical operations that require the exclusion of oxygen.

(b) Compounds of nitrogen

Nitrogen forms compounds in which it has oxidation states ranging from -3 to $+5$; these are summarised with examples in Table 19.2

Table 19.2 Oxidation states of nitrogen

Compound or ion		Oxidation number
{ Trioxonitric(V) acid { (Nitric acid)	HNO_3	$+5$
{ Trioxonitrate(V) ion { (Nitrate ion)	NO_3^-	
Nitrogen dioxide	NO_2	
Dinitrogen tetraoxide	N_2O_4	$+4$
{ Dioxonitric(III)acid { (Nitrous acid)	HNO_2	
{ Dioxonitrate(III)ion { (Nitrite ion)	NO_2^-	$+3$
Nitrogen trichloride	NCl_3	
Nitrogen oxide	NO	$+2$
Dinitrogen oxide	N_2O	$+1$
Nitrogen	N_2	0
Hydroxylamine	NH_2OH	-1
Hydrazine	N_2H_4	-2
Ammonia	NH_3	-3
Ammonium ion	NH_4^+	
Nitride ion	N^{3-}	

(i) Compounds of nitrogen with oxidation numbers −3 to −1

These include the ionic nitrides containing N^{3-} ions such as magnesium nitride, Mg_3N_2, and the covalent hydrides of nitrogen. Of the latter, ammonia, in which nitrogen has an oxidation state −3, is the best known.

In the ammonia molecule the three *p* electrons in the valency orbital of nitrogen form covalent bonds with electrons from three hydrogen atoms. The two *s* electrons remain unbonded, giving the molecule an overall tetrahedral shape with the lone pair of electrons occupying one of the corners of the tetrahedron.

Ammonia is a colourless gas with a pungent odour, very soluble in water. It is readily liquefied by compressing and cooling. Liquid ammonia has many of the characteristics of water; for example, there is extensive hydrogen bonding through the lone pair of electrons on the nitrogen atom, which accounts for its relatively high boiling point and molar heat of vaporisation.

$$H\!-\!\overset{\displaystyle H}{\underset{\displaystyle H}{N:}}\cdots H\!-\!\overset{\displaystyle H}{\underset{\displaystyle H}{N:}}\cdots H\!-\!\overset{\displaystyle H}{\underset{\displaystyle H}{N:}}\cdots H\!-\!\overset{\displaystyle H}{\underset{\displaystyle H}{N:}}\cdots H\!-\!\overset{\displaystyle H}{\underset{\displaystyle H}{N:}}$$

These characteristics make ammonia a good refrigerating liquid, but recently it has been replaced by less toxic liquids such as the freons (page 443).

Liquid ammonia is, like water, a polarising solvent, although the degree of ionisation is much less in ammonia:

$$NH_3 + NH_3 \rightleftharpoons NH_4^+ + NH_2^- \qquad K_{NH_3} = 10^{-33}$$

$$H_2O + H_2O \rightleftharpoons H_3O^+ + OH^- \qquad K_W = 10^{-14}$$

A number of reactions take place in liquid ammonia that are analogous to reactions in aqueous solution e.g. acid/base reactions between ammonium salts and metal amides:

$$NH_4^+Cl^-(NH_3) + Na^+NH_2^-(NH_3) \rightarrow Na^+Cl^-(NH_3) + 2NH_3(NH_3)$$

$$H^+Cl^-(aq) + Na^+OH^-(aq) \rightarrow Na^+Cl^-(aq) + H_2O(aq)$$

Liquid ammonia differs from water in that it is not immediately attacked by alkali metals; instead these dissolve in ammonia to give solutions of an intense blue colour which decompose only slowly on standing.

Ammonia gas is stable up to 500°C, above which temperature it decomposes almost completely. Mixed with oxygen and passed over a heated platinum catalyst it is oxidised to nitrogen monoxide.

$$4NH_3(g) + 5O_2(g) \rightarrow 4NO(g) + 6H_2O(l)$$

This reaction is important in the manufacture of nitric acid (page 476). Another important industrial reaction of ammonia is its direct combination with carbon dioxide under pressure to make urea for use as a fertiliser and in the production of urea/formaldehyde plastics.

$$2NH_3(g) + CO_2(g) \rightarrow \begin{matrix} NH_2 \\ NH_2 \end{matrix} C\!=\!O(s) + H_2O(l)$$

Ammonia is a base, forming ammonium ions with protons (H^+ ions) by the donation of the lone pair of electrons from a nitrogen atom in a coordinate bond:

$$H_3N\!:\!(aq) + H^+(aq) + Cl^-(aq) \rightarrow NH_4^+(aq) + Cl^-(aq)$$

It is a sufficiently strong base to gain protons from water although the quantity of ammonium ions formed is small and a solution of ammonia is only weakly basic.

$$NH_3(aq) + H_2O(aq) \rightleftharpoons NH_4^+(aq) + OH^-(aq)$$

The ammonium ion assumes a regular tetrahedral shape in which all the N—H bonds are equal, the positive charge being distributed evenly over the whole ion.

Ammonium salts are widely used; the nitrate, sulphate and phosphate are all valuable fertilisers, ammonium chloride is the electrolyte in dry batteries, ammonium carbonate is the 'sal volatile' of smelling salts, and ammonium nitrate is also used to produce the anaesthetic, dinitrogen oxide

$$NH_4^+NO_3^-(s) \rightarrow N_2O(g) + 2H_2O(l)$$

Under some circumstances the decomposition of ammonium nitrate can be explosive, so the production of dinitrogen oxide has to be carried out with care. The possibility of explosion also makes the storage of ammonium nitrate for use as fertiliser hazardous; for this reason it is frequently mixed with chalk.

Ammonia molecules form complex ions with a number of metal ions in solution, by liganding the ion through a coordinate bond. Examples are $[Cu(NH_3)_4]^{2+}$ (page 331) which is responsible for the deep blue colour that develops when ammonia is added to solutions of copper(II) ions, and $[Ag(NH_3)_2]^+$ (page 56) which forms when silver chloride dissolves in ammonia solution.

Derivatives of ammonia, in which nitrogen has oxidation states of -2 and -1, have one of the hydrogen atoms replaced by a group of atoms. In hydrazine an —NH_2 group replaces a hydrogen atom, giving rise to a bipyramidal molecule:

The compound is a weaker base than ammonia, forming two sets of salts.

$$N_2H_4(aq) + HCl(aq) \rightarrow NH_2NH_3{}^+Cl^-(aq)$$

$$N_2H_4(aq) + 2HCl(aq) \rightarrow (NH_3{-}NH_3)^{2+}(Cl^-)_2(aq)$$

When the hydrogen atom of ammonia is replaced by an OH^- group another weak base, hydroxylamine, NH_2OH, is formed. Both these compounds form condensation products with aldehydes and ketones (page 391). Both are also strong reducing agents. Hydrazine is used as a rocket fuel.

(ii) Compounds of nitrogen with oxidation numbers +1 to +5

Nitrogen forms several gaseous oxides with oxidation states ranging from +1 to +5. Of these the more important are dinitrogen oxide, N_2O, nitrogen monoxide, NO, and nitrogen dioxide, NO_2.

Dinitrogen oxide, commonly known as laughing gas, is an anaesthetic much used in dentistry. The molecule is linear; since it has only a small dipole moment it is thought to be a resonance hybrid of the two forms:

$$\overset{-}{N}{=}\overset{+}{N}{=}O \quad \text{and} \quad \overset{+}{N}{\equiv}\overset{-}{N}{-}O$$

Dinitrogen oxide is stable up to 500°C but above this temperature it is so readily decomposed that it will relight a glowing splint.

$$2N_2O(g) \rightarrow 2N_2(g) + O_2(g)$$

Nitrogen monoxide, NO, is formed in small amounts by the direct combination of nitrogen and oxygen at the very high temperatures of lightning flashes; it subsequently reacts with oxygen to make nitrogen dioxide, which dissolves in rain as nitric(V) acid, and is one of the contributors to nitrates reaching the soil.

$$N_2(g) + O_2(g) \rightarrow 2NO(g)$$

$$2NO(g) + O_2(g) \rightarrow 2NO_2(g)$$

The oxidation of nitrogen monoxide is a rapid reaction. Low concentrations of nitrogen monoxide emitted from car exhausts constitute a pollution problem, both the monoxide and the dioxide being very poisonous.

Nitrogen monoxide is a neutral oxide, insoluble in water; the molecule has an odd electron with a structure that is a resonance hybrid of two forms:

$$:\!\overset{\bullet}{N}\!:\!:\!\overset{\bullet\bullet}{O}\!: \quad \text{and} \quad :\!\overset{\bullet\bullet}{N}\!:\!:\!\overset{\bullet}{O}\!:$$

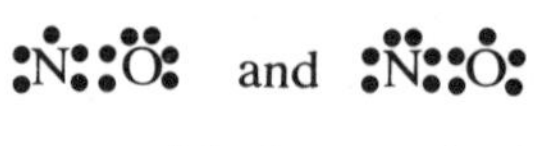

Nitrogen dioxide also has an odd electron in its molecule, which has a non-linear structure with an angle of 134° between the N—O bonds and is a resonance hybrid of two forms:

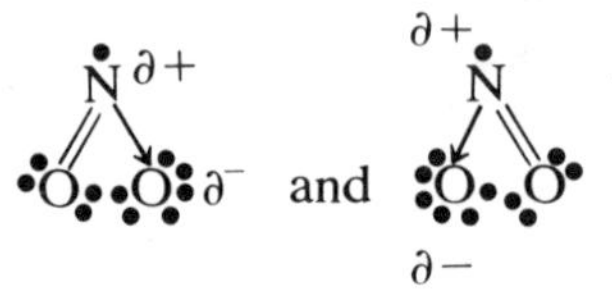

Molecules of NO_2 readily dimerise by forming a weak bond by the pairing of the odd electrons on the nitrogen atoms, the structure again being a resonance hybrid of two forms:

and

Nitrogen dioxide, a dark brown poisonous gas, is in equilibrium with the colourless dimer dinitrogen tetraoxide at moderate temperatures. The proportion of dinitrogen tetraoxide is greater at low temperatures, and when the mixture is heated it becomes progressively darker until a temperature of 150°C is reached, at which point decomposition into nitrogen monoxide and oxygen occurs and the mixture becomes lighter in colour.

$$N_2O_4 \underset{\text{cool}}{\overset{\text{heat}}{\rightleftharpoons}} 2NO_2 \underset{\text{cool}}{\overset{\text{heat}}{\rightleftharpoons}} 2NO + O_2$$

Colourless   Dark brown   Colourless

Nitrogen dioxide is an acidic gas dissolving in water to give a mixture of nitrous and nitric acids:

$$2NO_2(g) + H_2O(l) \rightarrow HNO_2(aq) + HNO_3(aq)$$

Nitrous acid (dioxonitric(III) acid) is known only in solution, being too unstable to be isolated. If required, it is generally prepared *in situ* by adding dilute hydrochloric acid to sodium nitrite; a blue solution containing some nitrous acid is produced, accompanied by brown fumes of oxides of nitrogen resulting from slight decomposition.

$$Na^+(NO_2)^-(aq) + HCl(aq) \rightarrow HNO_2(aq) + Na^+Cl^-(aq)$$

The acid is readily oxidised by passing air through the solution

$$2HNO_2(aq) + O_2(g) \rightarrow 2HNO_3(aq)$$

Sodium nitrite is important in the production of diazonium salts (page 422) used in the dye industry.

Nitric acid (systematically, trioxonitric(V) acid) when pure is a colourless dense oily liquid, which decomposes on heating to nitrogen dioxide and oxygen:

$$4HNO_3(l) \rightarrow 2H_2O(l) + 4NO_2(g) + O_2(g)$$

The concentrated acid is a powerful oxidising agent; dropped on to dry sawdust it will char and then ignite the mass, showing the necessity for careful storage of the acid.

The oxidising properties of concentrated nitric acid enable it to dissolve metals too low in the electrochemical series to be attacked by dilute acids. The nature of the reaction depends on the concentration; with pure concentrated acid nitrogen dioxide is evolved:

$$Cu(s) + 4HNO_3(l) \rightarrow Cu^{2+}(NO_3^-)_2(aq) + 2NO_2(g) + 2H_2O(l)$$

while with 50% acid nitrogen monoxide results

$$3Cu(s) + 8HNO_3(aq) \rightarrow 3Cu^{2+}(NO_3^-)_2(aq) + 2NO(g) + 4H_2O(l)$$

The concentrated acid also oxidises non-metal elements to oxoacids:

$$S(s) + 6HNO_3(l) \rightarrow H_2SO_4(l) + 2H_2O(l) + 6NO_2(g)$$

$$P(s) + 5HNO_3(l) \rightarrow H_3PO_4(l) + H_2O(l) + 5NO_2(g)$$

$$I_2(s) + 10HNO_3(l) \rightarrow 2HIO_3(l) + 4H_2O(l) + 10NO_2(g)$$

Dilute nitric acid is a much milder oxidising agent; it reacts with metals of negative electrode potential to give nitrates, and, in general, oxides of nitrogen rather than hydrogen are formed. Magnesium, with a very negative potential of -2.37 V, releases hydrogen from very dilute solutions:

$$Mg(s) + 2HNO_3(aq) \rightarrow Mg(NO_3)_2(aq) + H_2(g)$$

The dilute acid also oxidises iron(II) ions to iron(III) ions, and hydrogen sulphide to sulphur and water.

$$3Fe^{2+}(aq) + 4H^+(aq) + NO_3^-(aq) \rightarrow 3Fe^{3+}(aq) + 2H_2O(l) + NO(g)$$

$$3H_2S(g) + 2H^+NO_3(aq) \rightarrow 3S(s) + 4H_2O(l) + 2NO(g)$$

Nitric acid is used to nitrate organic compounds; the nitrocompounds are used in the manufacture of dyes and explosives. As an oxidising agent it is used in rocket fuel, in the production of cyclohexanol from cyclohexane, for the manufacture of nylon and in the production of phosphoric and oxalic acids. Vast quantities of nitric acid are also used to make nitrates for fertilisers. The manufacture of the acid is described in Chapter 32.

Measurement of the bond lengths and angles of nitric acid vapour by electron diffraction methods show that the molecule is planar, with bond lengths and angles as shown in Fig. 19.3. From this a completely symmetrical nitrate ion is derived, with a delocalised electronic structure as explained in Chapter 4.

Metallic nitrates are all soluble in water, making them useful sources of metallic ions in the laboratory. They are all decomposed by heat, the degree of decomposition depending on the metal ion present. Nitrates of metals forming ions of low charge and large ionic radius, such as the alkali metals, are reduced only to nitrites with the loss of oxygen, while other metal nitrates yield nitrogen dioxide in addition to oxygen and are reduced to their oxides:

$$2NaNO_3(s) \rightarrow 2NaNO_2(s) + O_2(g)$$

$$2Pb(NO_3)_2(s) \rightarrow 2PbO(s) + 4NO_2(g) + O_2(g)$$

The action of heat on lead nitrate provides a useful laboratory method of preparing nitrogen dioxide, which is separated from the oxygen by passing the gases through a U-tube cooled in ice; the nitrogen dioxide condenses as the liquid dimer, dinitrogen tetraoxide.

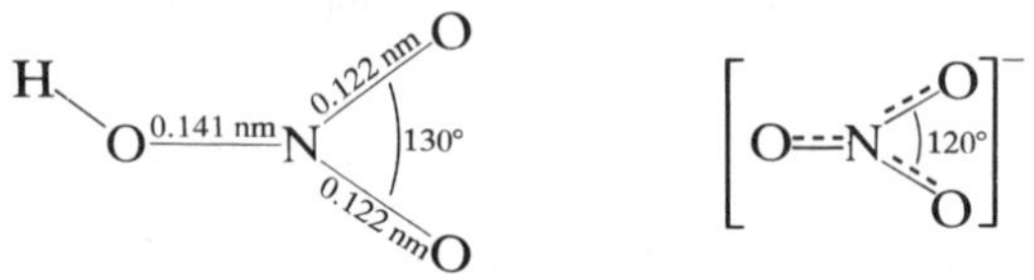

Fig. 19.3 Nitric acid and nitrate ion

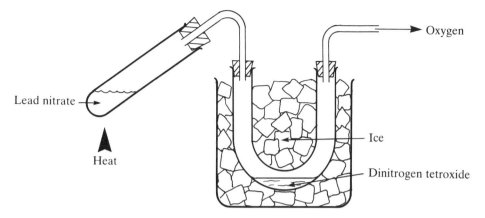

Fig. 19.4 Preparation of dinitrogen tetraoxide

If metal nitrates are heated with concentrated sulphuric acid, the more volatile nitric acid is displaced as colourless fumes which condense to oily droplets.

$$NaNO_3(s) + H_2SO_4(l) \rightarrow HNO_3(l) + NaHSO_4(s)$$

This method was used to manufacture nitric acid from nitrates before the introduction of the Ostwald process (page 476).

19.3   Phosphorus

(a) Occurrence, properties, reactions and uses

Phosphorus, in the combined state, occurs as rock phosphate, $Ca_3(PO_4)_2$, and apatite, $CaF_2 \cdot 3Ca_3(PO_4)_2$. Large deposits of the ores are found in North America, North Africa and the USSR. Calcium phosphate is the principal constituent of bones and teeth, and phosphorus compounds are present in all vegetable and animal tissues. Phosphorus was first isolated from urine in 1669 as a white pasty mass that readily ignited, and also glowed in the dark. Its discoverer, Hennig Brand, was hoping to find gold according to the current theory that most yellow substances contained it; he called this new element phosphorus, from a Greek word meaning light-bearer.

Phosphorus exists in several allotropic forms, the main ones being white and red phosphorus. White phosphorus, which is always formed when phosphorous vapour is condensed, is a yellowish waxy solid composed of tetrahedrally shaped P_4 molecules;

it is spontaneously inflammable in air, and has to be stored under water. White phosphorus is unstable and changes to the red allotrope at all

Table 19.3 Allotropes of phosphorus

	White phosphorus	Red phosphorus
Melting point	44°C	Sublimes at 400°C
Boiling point	280°C	Sublimes at 400°C
Density	1.83 g cm^{-3}	2.2 g cm^{-3}
Solubility in carbon disulphide	Very soluble	Insoluble
Ignition temperature	35°C	260°C
Toxicity	Poisonous	Not poisonous
Phosphorescence	Glows in the dark	—
Reaction with hot sodium hydroxide	Forms phosphine	No reaction

temperatures; this change is slow at room temperature but rapid at temperatures above 260°C in the presence of a trace of iodine as catalyst. The red form of phosphorus is denser than the white and has a macromolecular structure of uncertain constitution. It does not ignite until heated, when it burns with a yellow flame to give a mixture of phosphorus trioxide and phosphorus pentoxide. A comparison of the two allotropes is given above. Both forms of phosphorus react with chlorine forming phosphorus trichloride:

$$2P(s) + 3Cl_2(g) \rightarrow 2PCl_3(l)$$

Red phosphorus is oxidised by nitric(V)acid to phosphoric(V) acid:

$$P(s) + 5HNO_3(l) \rightarrow H_3PO_4(aq) + H_2O(l) + 5NO_2(g)$$

White phosphorus reacts with sodium hydroxide to give phosphine (PH_3):

$$P_4(s) + 3NaOH(aq) + 3H_2O(l) \rightarrow 3NaH_2PO_2(aq) + PH_3(g)$$

Phosphorus is used to make phosphates for fertilisers etc. Small amounts are also used in the manufacture of matches, fireworks, rat poison and phosphor-bronze alloys.

(b) Hydrides of phosphorus

Phosphorus forms both phosphine, PH_3, and diphosphine, P_2H_4, which are analogous to ammonia and hydrazine. Diphosphine, which spontaneously ignites in air, is of little importance. Phosphine is a poisonous gas which is also readily flammable and burns to form phosphorus(V) oxide:

$$4PH_3 + 8O_2 \rightarrow P_4O_{10} + 6H_2O$$

It has the same pyramidal structure as ammonia but phosphorus is not sufficiently electronegative for the lone pair of electrons to form hydrogen bonds. Phosphine therefore has a lower boiling point than ammonia although it has a heavier molecule. Phosphine is only slightly soluble in water, being a very weak base. It forms a phosphonium ion with hydrogen halides, but the salts are not stable as are the ammonium salts:

$$PH_3(g) + HI(aq) \rightarrow PH_4^+I^-(aq)$$

(c) Phosphorus halides

Phosphorus, unlike nitrogen, can expand its valency orbital beyond eight electrons as a result of the availability of empty d orbitals in the atom. Thus, while nitrogen forms only a trichloride utilising the p electrons for bonding, phosphorus forms both a trichloride and a pentachloride.

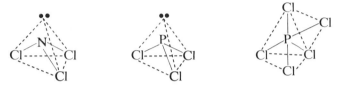

Fig. 19.5   Chlorides of nitrogen and phosphorus

The trichloride is made by passing chlorine over solid white or heated red phosphorus. It distils as a colourless fuming liquid readily hydrolysed to phosphoric acid:

$$P_4(s) + 6CL_2(g) \rightarrow 4PCl_3(l)$$

$$PCl_3(l) + 3H_2O(l) \rightarrow H_3PO_3(aq) + 3HCl(aq)$$

If phosphorus trichloride is treated with excess chlorine, phosphorus pentachloride, PCl_5, is formed as a yellow solid which fumes in moist air and hydrolyses to phosphoric(V) acid, H_3PO_4:

$$PCl_3(l) + Cl_2(g) \rightleftharpoons PCl_5(s)$$

$$PCl_5(s) + 4H_2O(l) \rightarrow H_3PO_4(aq) + 5HCl(aq)$$

Phosphorus pentachloride sublimes at 160°C and dissociates on further heating; in the vapour phase before dissociating the molecule has a bipyramidal shape, while as a solid the structure is partially ionic, $(PCl_4)^+(PCl_6)^-$:

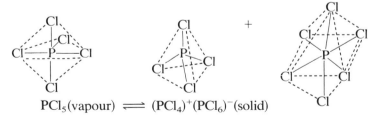

$$PCl_5(\text{vapour}) \rightleftharpoons (PCl_4)^+(PCl_6)^-(\text{solid})$$

Both the trihalide and the pentahalide are used to introduce halogen atoms into organic molecules containing —OH or $>$C$=$O groups:

$$3C_2H_5OH + PCl_3 \rightarrow 3C_2H_5Cl + H_3PO_3$$

$$CH_3COOH + PCl_5 \rightarrow CH_3COCl + POCl_3 + HCl$$

$$(CH_3)_2C{=}O + PCl_5 \rightarrow (CH_3)_2CCl_2 + POCl_3$$

In phosphorus trichloride oxide (phosphorus oxychloride), $POCl_3$, phosphorus is in the +5 oxidation state; the oxygen atom replacing two of the chlorine atoms

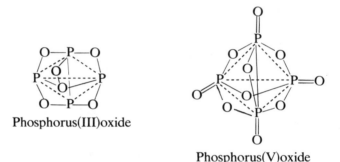

(d) Oxides and oxyacids of phosphorus

Phosphorus forms two oxides, in which it has an oxidation state of +3 to +5. Both oxides exist in the dimerised forms P_4O_6 and P_4O_{10}. Their structures are based on the tetrahedral molecule of white phosphorus (Fig. 19.6):

Fig. 19.6 Structures of phosphorus oxides

Phosphorus(III) oxide, a white solid, is the anhydride of phosphonic acid, to which it is hydrolysed by water:

$$P_4O_6(s) + 6H_2O(l) \rightarrow 4H_3PO_3(l)$$

Phosphorus(V) oxide is a white hygroscopic solid used both as a dehydrating and a drying agent:

$$CH_3CONH_2(s) \xrightarrow{P_4H_{10}} CH_3CN(g) + H_2O(l)$$

It is the anhydride of phosphoric(V) acid. On solution in water it is first converted to polytrioxophosphoric acid known as metaphosphoric acid. This can be obtained as a glassy solid of uncertain structure composed of polymerised units of formula HPO_3. The solution of metaphosphoric acid is converted to phosphoric(V) acid on boiling.

$$P_4O_{10}(s) + 2H_2O(l) \rightarrow 4HPO_3(aq)$$

$$HPO_3(aq) + H_2O(l) \rightarrow H_3PO_4(aq)$$

Pure phosphoric(V) acid is dehydrated on heating to heptaoxodiphosphoric(V) acid (pyrophosphoric acid).

$$2H_3PO_4(l) \rightarrow H_4P_2O_7(s) + H_2O(l)$$

The structures of these acids are based on tetrahedra with central phosphorus atoms (Table 19.4)

Table 19.4   Structures of some oxoacids of phosphorus

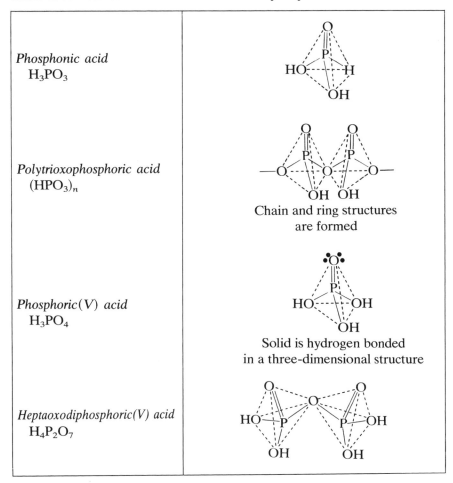

Phosphonic acid
H₃PO₃

Polytrioxophosphoric acid
(HPO₃)ₙ

Chain and ring structures
are formed

Phosphoric(V) acid
H₃PO₄

Solid is hydrogen bonded
in a three-dimensional structure

Heptaoxodiphosphoric(V) acid
H₄P₂O₇

Phosphoric(V) acid is a strong non-volatile acid capable of displacing hydrogen bromide and iodide from metal halides; it is used for this in preference to sulphuric acid which is reduced by the bromide and iodide ions to give a mixed product (page 312).

$$3NaI(s) + H_3PO_4(s) \rightarrow Na_3PO_4(s) + 3HI(g)$$

The acid is tribasic, and is neutralised in three stages to give three salts. Thus on neutralisation with sodium hydroxide the three salts sodium dihydrogenphosphate, disodium hydrogenphosphate, and normal sodium phosphate are formed successively.

$$NaOH(aq) + H_3PO_4(aq) \rightarrow NaH_2PO_4(aq) + H_2O(l)$$

$$NaOH(aq) + NaH_2PO_4(aq) \rightarrow Na_2HPO_4(aq) + H_2O(l)$$

$$NaOH(aq) + Na_2HPO_4(aq) \rightarrow Na_3PO_4(aq) + H_2O(l)$$

(e) Uses of phosphates

Calcium phosphate(V) is widely used for the manufacture of fertilisers. Since it is insoluble it has to be converted to the soluble calcium dihydrogenphosphate by reaction either with sulphuric acid to form 'superphosphate', or with phosphoric acid to give 'triple superphosphate', or with nitric acid to give 'nitrophos'.

$$Ca_3(PO_4)_2 + 2H_2SO_4 \rightarrow Ca(H_2PO_4)_2 + 2CaSO_4$$
$$\text{'superphosphate'} \longrightarrow$$

$$Ca_3(PO_4)_2 + 4H_3PO_4 \rightarrow 3Ca(H_2PO_4)_2$$
$$\text{'triple superphosphate'}$$

$$Ca_3(PO_4)_2 + 4HNO_3 \rightarrow Ca(H_2PO_4)_2 + 2Ca(NO_3)_2$$
$$\text{'nitrophos'} \longrightarrow$$

Phosphates are also important in the food and soft drink industries. They have a pleasant taste and are used in various mineral waters and in baking powders. For the latter, sodium dihydrogen phosphate(V) is used as the solid acid to release carbon dioxide from sodium bicarbonate, the phosphatic residue leaving an agreeable flavour in the cake. Sodium metaphosphate is used in processing cheese. Phosphates are used in the manufacture of water softeners such as Calgon, to rust-proof steel and to fireproof timber and textiles.

Phosphate(V) compounds play an important play in the chemistry of living things, being essential constituents of both DNA (deoxyribonucleic acid) and RNA (ribonucleic acid), which are the molecules that provide and carry the essential information for the synthesis of proteins in the body.

19.4   Uses of arsenic, antimony and bismuth

The chemistry of the last three members of Group V is not considered in detail in this text. They are mostly used in alloys. A small proportion of arsenic is used to harden lead for lead shot; its poisonous oxide is used to destroy pests and as a weed-killer. Antimony forms several useful alloys, being used to harden the lead in accumulator plates and to make type metal, which expands slightly on cooling and so gives a clear impression from the mould. Bismuth makes low melting point alloys such as that used to seal automatic sprinklers, which melts at 70°C and releases water in case of fire; bismuth nitrate is used as a disinfectant and in pharmacy.

Questions

1. Give a brief account of the nitrogen cycle in nature.
2. How does ammonia react with (a) water and (b) sulphuric acid? Compare phosphine and phosphonium salts with ammonia and ammonium salts.
3. This question concerns the elements arsenic (atomic number 33), nitrogen (atomic number 7) and phosphorus (atomic number 15).

(*a*) For what reason are the elements placed in the same group of the Periodic Table?

(*b*) Distinguishing between *s, p* and *d* electrons, give the electronic configuration of the elements.

(*c*) Place the elements in order of decreasing electronegativity (i.e. putting the most electronegative element first) Name one other element which is more electronegative than all three.

(*d*) Give the structures of

(i)   the tetraamminezinc(II) ion;

(ii)   dinitrogen tetraoxide;

(iii) solid phosphorus pentachloride

Outline the preparation of (iii)

(*e*) Explain why

(i)   phosphorus forms a trichloride and a pentachloride but nitrogen forms only a trichloride;

(ii)   ammonia is more basic than phosphine;

(iii) nitrogen is more inert than phosphorus.   (AEB 1976)

4. Find the element antimony Sb (atomic number 51) in the Periodic Table.

(*a*) Which two oxidation states would you expect this element to exhibit in its chlorides?

(*b*) Which of the two chlorides would you expect to have the higher melting point?

(*c*) Explain your answer to (*b*).

(*d*) How does the thermal stability of SbH_3 compare with that of NH_3 and PH_3?

(*e*) How would the acidity of antimony(V) oxide compare with that of phosphorus(V) oxide?   (JMB)

5. Outline the chemical reactions involved in the synthesis of ammonia and of nitric acid starting from nitrogen.

State and explain the variation in the action of water on the trichlorides of the elements of Group V. Why does the nitrogen not form a pentachloride?   (C)

6. It was predicted early in this century that the world faced starvation unless a means of 'fixing' atmospheric nitrogen could be found. What problems do we now face because of the greater availability and use of nitrogen compounds in agriculture?   (L)

7. Outline how one compound for each of four different oxidation states of nitrogen may be obtained, *either* directly *or via* an intermediate compound of different oxidation state, from nitrogen in its pure state.   (L)

20 Oxygen and sulphur in Group VI

Table 20.1 Physical data

Element	Atomic number	Electronic configuration	Atomic radius nm (cov).	Ionic radius nm X^{2-}	m.p. K	b.p. K	Electro-negativity Pauling scale
O	8	2.6 $1s^2 2s^2 2p^4$	0.066	0.140	54	90	3.5
S	16	2.8.6 $---3s^2 3p^4$	0.104	0.184	392	717	2.5
Se	34	2.8.18.6 $----3d^{10} 4s^2 4p^4$	0.114	0.198	490	958	2.4
Te	52	2.8.18.18.6 $----4d^{10} 5s^2 5p^4$	0.132	0.221	723	1663	2.1
Po	84	2.8.18.32.18.6 $------5d^{10} 6s^2 6p^4$	0.152	—	—	—	—

20.1 General characteristics and trends in properties of the Group VI elements

The Group VI elements constitute the fourth group of p block elements. Their small atomic radii and relatively high electronegativity lead to the formation of covalent compounds or compounds in which the elements appear as negative ions. Thus they are mainly non-metallic but, as the electronegativity decreases with increasing atomic radius, some metallic properties develop. Selenium and tellurium are both semi-conductors, tellurium being the better of the two. Although the oxides are predominantly acidic, the existence of four-valent cations, Te^{4+} and Po^{4+}, has been detected in tellurium dioxide and polonium dioxide; both these oxides are partially amphoteric, with polonium dioxide the more basic.

The valency orbitals of the elements each contain four p and two s electrons; thus the formation of four-valent cations must be a result of the inert pair effect (Chapter 17). An inert gas structure results from the formation of X^{2-} ions by the gain of two electrons, as in the metallic oxides, sulphides, selenides and tellurides. The lattice energy of these compounds decreases with increasing size of the negative ion. Thus, while oxygen forms ionic oxides with most metals, sulphur, selenium and tellurium form ionic compounds with only the very reactive metals of Groups I and II. The X^{2-} ions are stable in solid compounds but are unstable in aqueous solution, the O^{2-} ion attracting a proton from a water molecule with the formation of an OH^- ion.

$$O^{2-}(s) + H_2O(l) \rightarrow 2OH^-(aq)$$

The XH^- ion becomes less stable with increasing atomic radius, SeH^- and TeH^- being very unstable. The SH^- ion is formed when the sulphides of

Groups I and II dissolve in water, but is readily hydrolysed with the evolution of hydrogen sulphide gas if the solution is boiled:

$$S^{2-}(s) + H_2O(l) \rightarrow HS^-(aq) + OH^-(aq)$$

$$HS^-(aq) + H_2O(l) \rightarrow OH^-(aq) + H_2S(g)$$

The Group VI elements can also acquire an inert gas structure by the covalent sharing of two electrons in combination with other elements, as in water, hydrogen sulphide and the chlorides of sulphur and tellurium. In all these compounds the lone pairs of electrons on the atoms give a characteristic shape to the structure.

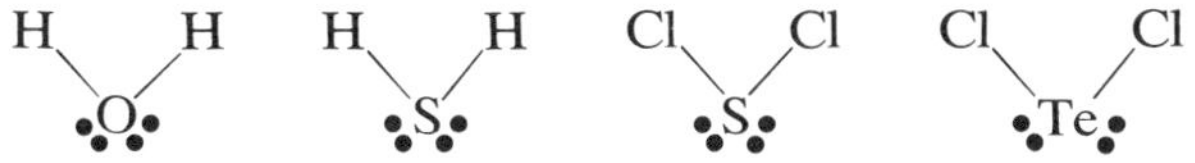

Oxygen and (to a much lesser degree) sulphur form compounds with double bonds, such as carbonyl chloride (phosgene), carbon dioxide and carbon disulphide.

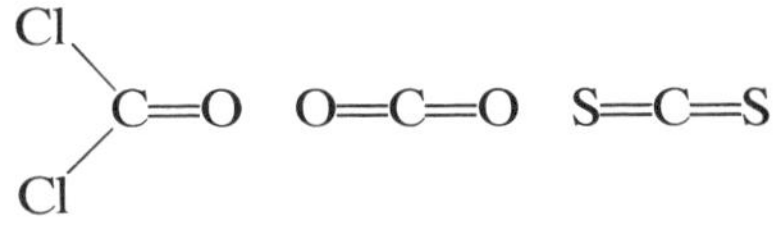

The elements of the group, other than oxygen, can all expand their covalencies since they have available d orbitals to accommodate the extra electrons, forming compounds and complex ions in which they attain oxidation states of $+4$ and $+6$, as in the dioxides, trioxides, halides and oxoacids such as SO_2, SO_3, SeO_3, SF_6, TeF_6, SO_3^{2-} and SO_4^{2-}.

While the oxygen atom cannot expand its normal covalency beyond 2, it may donate a lone pair of electrons in a coordinate bond; thus in the oxonium ion oxygen has a covalency of 3:

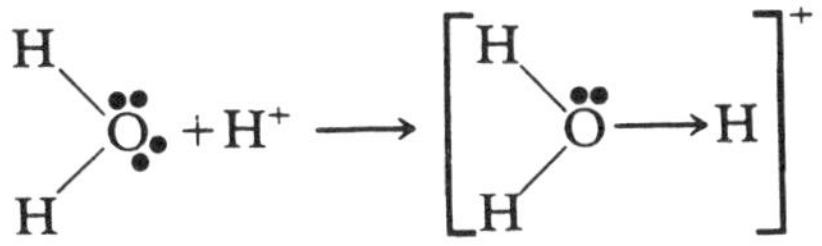

Oxygen, the head element, differs from sulphur and the succeeding elements in more respects than its limited covalency. As a result of its small atomic radius and great electronegativity, the oxygen atom is unique in Group VI in the formation of hydrogen bonds (page 228).

A further difference is the greater tendency of oxygen to form double or triple bonds with itself or other elements, while sulphur tends to form chains of atoms. Thus oxygen molecules O_2 have a multiple bond while sulphur molecules S_8 are linked by single bonds; CO_2 is more stable than CS_2. Linked oxygen atoms occur only as —O—O— groups in the unstable $(O_2)^{2-}$ of peroxides, while sulphur forms long chains of sulphur atoms in plastic sulphur, polysulphides and thionic acids.

In spite of these differences there is sufficient similarity between the two elements for sulphur to form sets of 'thio-' compounds analogous to sets of oxygen compounds. Some examples are:

Cyanates, CNO^-, and thiocyanates, CNS^-.

Alcohols, ROH, and thiols, RSH

Sulphates, SO_4^{2-}, and thiosulphates, $S_2O_3^{2-}$

20.2   Oxygen

Oxygen is the most plentiful element in the earth's outer crust and atmosphere. In the free state it comprises 21% by volume of dry air; in the combined state it forms 89% by mass of water and about 50% of the earth's crust in the form of oxides and silicates. It is an essential component of all living matter, being necessary for both respiration and growth. Its concentration in the air is maintained by the process of photosynthesis, by which green plants take up carbon dioxide and release oxygen.

Oxygen was first isolated by Joseph Priestly in 1774 by the decomposition of mercury oxide, and was shortly after identified as the active part of air by Antoine Lavoisier, who conducted experiments involving careful measurements of mass and volume. This was a great step forward in the study of chemistry, replacing attempts to establish theory by philosophical discussion.

Oxygen is a colourless odourless gas, condensing to a pale blue liquid below 90 K. It is slightly soluble in water, its presence being vital to fish and other marine and freshwater life.

Chemically oxygen is reactive, combining directly with many elements and compounds either in a simple process of oxidation:

$$2Cu(s) + O_2(g) \rightarrow 2CuO(s)$$

$$C_2H_5OH(l) + O_2(g) \rightarrow CH_3COOH(l) + H_2O$$

or in a combustion process, which is oxidation accompanied by the evolution of heat and light:

$$C(s) + O_2(g) \rightarrow CO_2(g) \qquad \Delta H_{298}^{\ominus} = -393 \text{ kJ mol}^{-1}$$

$$C_2H_5OH(l) + 3O_2(g) \rightarrow 2CO_2(g) + 3H_2O(l) \qquad \Delta H_{298}^{\ominus} = -1367 \text{ kJ mol}^{-1}$$

Many substances which oxidise slowly on exposure to air will burn in oxygen when ignited.

A great deal of oxygen is made by the fractional distillation of liquid air (Chapter 32) for use industrially in steel works and for oxyacetylene and oxygen-hydrogen flames for cutting and welding metals. Firemen, divers, mountaineers and astronauts use oxygen in breathing apparatuses, and, as predicted by Priestley, patients suffering from respiratory diseases are sometimes treated with pure oxygen.

The structure of the oxygen molecule is frequently shown as O=O, but the molecule is known to be highly magnetic, a property which arises from the presence of unpaired electrons; a more sophisticated structure in which each atom contains an unpaired electron has been proposed, but this is beyond the scope of this book.

Oxygen has an allotropic form O_3, trioxygen, better known as ozone. Ozone occurs at low concentrations in the upper atmosphere, where it is formed by the action of ultraviolet light on oxygen, and small amounts are formed in electrical discharges and sparks. It is much less stable than oxygen, to which it tends to revert:

$$3O_2 \rightleftharpoons 2O_3 \qquad \Delta H^{\ominus}_{298} = +285 \text{ kJ mol}^{-1}$$

This type of allotropy, where one allotrope is more stable than the other at all temperatures, is known as monotropy (page 93).

The ozone molecule is angular, it exists as a resonance hybrid of two forms:

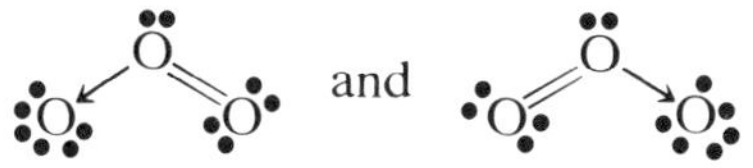

and

Ozone is more reactive than oxygen, and is a powerful oxidising agent and disinfectant. It is used in the ozonolysis of alkenes and has been used to purify air in underground tunnels and to sterilise drinking water.

20.3   Oxides

(a) Normal oxides

These are oxides in which the oxygen atoms are all directly bonded to the element; they are further divided into basic, acidic, amphoteric and neutral oxides.

Basic oxides are ionic oxides of metals in which the metal ions are of low charge and large ionic radius, which react with acids to form salts. They include oxides of Groups I and II metals, such as $(Na^+)_2O^{2-}$ and $Ca^{2+}O^{2-}$, and certain of the oxides of transition elements such as $Cu^{2+}O^{2-}$ and $Fe^{2+}O^{2-}$.

Acidic oxides are covalent in nature, being the oxides of non-metals such as sulphur and of transition metals in which the metal has its highest oxidation number, such as chromium(VI) oxide, CrO_3. These oxides are the anhydrides of the acids they form in solution with water.

$$SO_2(g) + H_2O(l) \rightarrow H_2SO_3(aq)$$

$$CrO_3(s) + H_2O(l) \rightarrow H_2CrO_4(aq)$$

$$SO_3(s) + H_2O(l) \rightarrow H_2SO_4(aq)$$

Amphoteric oxides have some basic and some acidic character; an example is aluminium oxide. The metal ions in such oxides are of high charge and low ionic radius. Other examples are ZnO, Cr_2O_3, SnO and PbO. They react both with acids to form salts and with bases to form complex anions:

$$ZnO(s) + 2H^+(aq) \rightarrow Zn^{2+}(aq) + H_2O(l)$$

$$ZnO(s) + 2OH^-(aq) + H_2O(l) \rightarrow [Zn(OH)_4]^{2-}(aq)$$

Other covalent oxides have no acid/base character; these are neutral oxides such as carbon monoxide, dinitrogen oxide and nitrogen monoxide.

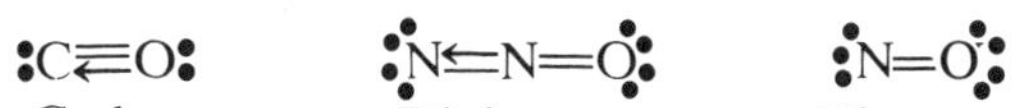

| Carbon monoxide | Dinitrogen oxide | Nitrogen monoxide |

Fig. 20.1   Bonding in some neutral oxides

(b) Higher oxides

These include peroxides and superoxides. They are formed by those metals in Groups I and II, such as potassium and barium, which have sufficiently large ionic radii to accommodate the large peroxide or superoxide ions. The peroxide ion is derived from the hydrogen peroxide molecule, whose structure is such that the two hydrogen atoms do not lie in the same plane.

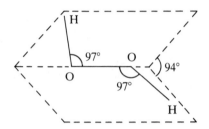

Fig. 20.2   The hydrogen peroxide molecule

The O—O bond is not stable; and H_2O_2 rapidly decomposes to release oxygen. Thus hydrogen peroxide is used as a disinfectant and a bleach, and the metallic peroxides are excellent oxidising agents:

$$2H_2O_2(l) \rightarrow 2H_2O(l) + O_2(g)$$

$$3O_2^{2-}(s) + 2Cr^{3+}(aq) + 4OH^-(aq) \rightarrow 2CrO_4^{2-}(aq) + 2H_2O(l)$$

Barium peroxide is used to prepare hydrogen peroxide by treatment at 0°C with dilute sulphuric acid

$$BaO_2(s) + H_2SO_4(aq) \rightarrow BaSO_4(s) + H_2O_2(aq)$$

An aqueous solution is obtained, after filtration to remove the barium sulphate, which can be concentrated by distillation under reduced pressure.

(c) Suboxides

These are oxides with a low proportion of oxygen in the molecule, involving bonding between atoms of the element as well as between the element and oxygen. An example is tricarbon dioxide, C_3O_2, which exists as a rather unstable gas at room temperature. It has a linear structure:

$$O=C=C=C=O$$

(d) Mixed oxides

Some oxides such as red lead oxide, Pb_3O_4, are thought to exist as two separate oxides combined in stoichiometric proportions. Thus Pb_3O_4 should be written $2PbO·PbO_2$ and named dilead(II)lead(IV) oxide. Similarly magnetite Fe_3O_4 may be regarded as a compound of two oxides of iron $FeO·Fe_2O_3$ and systematically named iron(II) diiron(III) oxide.

20.4 Sulphur

(a) Occurrence, properties and uses

Sulphur occurs in the free state in Louisiana, Texas, Japan and Sicily, the deposits in the United States account for a great part of the world's supply of mineral sulphur. It is extracted from underground deposits by the *Frasch process*. Further quantities are extracted from crude oil and natural gas, both usually containing sulphur compounds. Sulphur is also widespread as the mineral iron pyrites, FeS_2, which is extensively used in the manufacture of sulphuric acid. Many other minerals contain sulphur combined as sulphate, such as gypsum, $CaSO_4 \cdot 2H_2O$, anhydrite, $CaSO_4$, and Epsom salt, $MgSO_4 \cdot 7H_2O$. Sulphur is an important constituent of some proteins, and is released in the decay or certain organic substances as hydrogen sulphide—hence the smell of rotten eggs.

Sulphur is a yellow crystalline solid of low density, insoluble in water. It exists in two main allotropic forms as rhombic and monoclinic crystals.

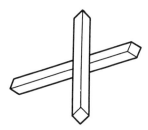

Monoclinic sulphur   Rhombic sulphur

Each of these forms is made up from S_8 molecules in which the sulphur atoms are arranged in a puckered molecule:

Plan of S_8 molecule   Elevation of S_8 molecule

The rhombic crystals are stable at room temperature; they can be transformed into monoclinic crystals by heating to a temperature above 95.6°C—the *transition temperature*. This type of allotropy where one form can be changed into another by changing the physical conditions is known as *enantiotropy* (page 92).

Sulphur also exists in a powdery or amorphous form which has no evident crystalline structure. Amorphous sulphur is precipitated from solutions of sulphur compounds, as in the reaction between hydrochloric acid and sodium thiosulphate:

$$S_2O_3^{2-}(aq) + 2H^+(aq) \rightarrow S(s) + H_2O(l) + SO_2(g)$$

It first appears as a colloidal solution of sulphur, opalescent and milky-looking. As the reaction proceeds more sulphur is precipitated and the mixture becomes more opaque until finally powdery sulphur is deposited as a pale yellow solid. Amorphous sulphur is also obtained as flowers of sulphur, the sublimed solid obtained when sulphur vapour is condensed on a cool surface.

When sulphur is heated slowly to its boiling point out of contact with air it undergoes an interesting series of changes, which can be explained in terms of the changing structure of the molecules as energy is supplied. At 113°C sulphur melts to an amber coloured liquid containing separate S_8 molecules still in ring form. On raising the temperature the liquid darkens and thickens, becoming so viscous that the test tube can be inverted without the sulphur flowing out; the additional energy supplied breaks the ring molecules, forming chains that link together into longer chains of hundreds of atoms which have considerable van der Waals forces between them and become entangled, increasing the viscosity. As the temperature is raised still further the viscous mass becomes liquid again as the additional energy breaks up the chains into smaller units, reducing the viscosity. At 444°C the sulphur boils, yielding a vapour which consists mainly of reformed S_8 molecules. If the nearly boiling liquid sulphur is poured into cold water, so that it is rapidly chilled, a soft rubbery solid, consisting of mixed tangled chains, is formed. This plastic sulphur is not stable, and slowly reverts to rhombic crystals on standing:

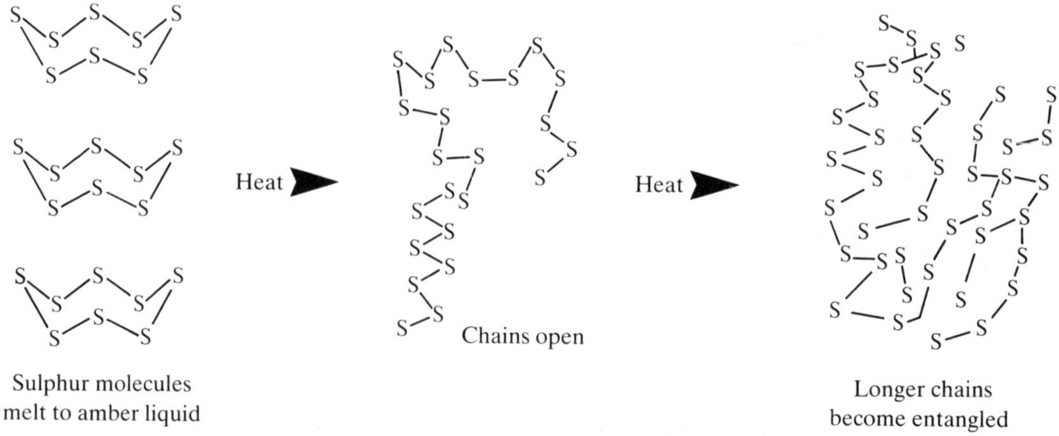

Sulphur molecules melt to amber liquid

Chains open

Longer chains become entangled

Sulphur is a reactive element; when heated it combines directly with with a number of finely divided metals, the reaction being vigorous with metals high in the electrochemical series:

$$Zn(s) + S(s) \rightarrow ZnS(s) \qquad \Delta H^{\ominus}_{298} = -206 \text{ kJ mol}^{-1}$$

Reactive non-metals such as fluorine and chlorine combine directly with heated sulphur:

$$2S(s) + Cl_2(g) \rightarrow S_2Cl_2(l) \text{ (disulphur dichloride)}$$

Hydrogen reacts reversibly to make hydrogen sulphide gas if bubbled through sulphur at its boiling point:

$$H_2(g) + S(s) \rightleftharpoons H_2S(g)$$

At high temperature carbon and sulphur can be made to react endothermically to form the unstable liquid carbon disulphide:

$$C(s) + S_2(s) \rightarrow CS_2(l) \qquad \Delta H^{\ominus}_{298} = +88 \text{ kJ mol}^{-1}$$

In air or oxygen sulphur burns with a bright blue flame producing mainly sulphur dioxide:

$$S(s) + O_2(g) \rightarrow SO_2(g)$$

The burning of sulphur candles was formerly used to disinfect sick rooms where patients with infectious diseases, such as smallpox, had been nursed. This use of burning sulphur as a purifier is mentioned by Homer.

Sulphur is oxidised to sulphuric acid by concentrated nitric acid:

$$S(s) + 6HNO_3(l) \rightarrow H_2SO_4(l) + 6NO_2(g) + 2H_2O(l)$$

If sulphur is boiled with an alkali, a solution of the thiosulphate is finally obtained if the sulphur is present in excess:

$$3S(s) + 6Na^+OH^-(aq) \rightarrow 2(Na^+)_2S^{2-}(aq) + (Na^+)_2(SO_3)^{2-}(aq) + 3H_2O(l)$$

$$(Na^+)_2(SO_3)^{2-}(aq) + S(s) \rightarrow (Na^+)_2(S_2O_3)^{2-}(aq)$$

The sodium sulphide formed in this reaction also forms polysulphide ions in which a chain of sulphur atoms links on to the sulphide ion:

$$(Na^+)_2S^{2-} + xS \rightarrow (Na^+)_2S^{2-}Sx$$

This tendency to catenate has already been illustrated by the chains of sulphur molecules formed in plastic sulphur.

Sulphur is used on a very large scale to produce sulphur dioxide for the manufacture of sulphuric acid. It is also used in the manufacture of matches, fireworks, dyes, drugs and skin ointments. Elemental sulphur is used together with sulphur dichloride to vulcanise rubber. It is also widely used as a fungicide, mainly for dusting grapes and other fruit.

(b) Compounds of sulphur

(i) Oxidation states
The main oxidation states of sulphur and a summary of the compounds containing sulphur in these oxidation states are given in Table 20.2.

(ii) Hydrogen sulphide and sulphides
The hydrogen sulphide molecule, like that of water, is approximately tetrahedral in shape, with the lone pairs of electrons on the sulphur atom occupying two of the tetrahedral sites:

However sulphur is not sufficiently electronegative to allow the formation of hydrogen bonds between the molecules, so, unlike water, hydrogen sulphide in a gas in spite of its higher molecular mass.

Table 20.2 Oxidation states of sulphur

Compound or ion		Oxidation number
Sulphur(VI) oxide (sulphur trioxide)	SO_3	+6
Sulphuric acid (tetraoxosulphuric(VI) acid)	H_2SO_4	
Sulphate ion (tetraoxosulphate(VI) ion)	$SO_4{}^{2-}$	
Sulphur hexafluoride	SF_6	
Sulphur dioxide	SO_2	+4
Sulphurous acid (tetraoxosulphuric(IV) acid)	H_2SO_3	
Sulphite ion (tetraoxosulphate(IV) ion)	$SO_3{}^{2-}$	
Sulphur dichloride oxide	$SOCl_2$	
Thiosulphuric acid	$H_2S_2O_3$	+2
Thiosulphate ion	$S_2O_3{}^{2-}$	
Sulphur dichloride	SCl_2	
Disulphur dichloride	S_2Cl_2	+1
Sulphur	S_8	0
Hydrogen sulphide	H_2S	−2
Sulphide ion	S^{2-}	

The gas is weakly acidic, dissolving in water to dissociate in two stages:

$$H_2S(g) + H_2O(l) \rightleftharpoons H_3O^+(aq) + HS^-(aq)$$

$$HS^-(aq) + H_2O(l) \rightleftharpoons H_3O^+(aq) + S^{2-}(aq)$$

The hydrogen sulphides and sulphides of the very basic Group I metals such as Na_2S are ionic solids; they form transparent colourless crystals which are deliquescent and dissolve in water to give alkaline solutions as a result of hydrolysis.

$$S^{2-}(aq) + H_2O(l) \rightleftharpoons HS^-(aq) + OH^-(aq)$$

Sulphides of other metals have a considerable degree of covalency and are insoluble; many of these sulphides have characteristic colours and their precipitation by hydrogen sulphide can be used to identify metal ions in solution. Those with very low solubility products can be precipitated even from acidic solutions in which the sulphide ion concentration is suppressed by the presence of acid ions.

$$2H_3O^+ + S^{2-} \rightleftharpoons 2H_2O + H_2S$$

Table 20.3 Insoluble sulphides

Precipitated from acid solution		
Sulphide	Colour	Solubility product
Mercury(II) sulphide	red	1.6×10^{-52}
Copper(II) sulphide	black	6.3×10^{-36}
Lead(II) sulphide	black metallic	1.3×10^{-28}
Precipitated from alkaline solution		
Zinc(II)sulphide	white	1.6×10^{-24}
Nickel(II) sulphide	black	2×10^{-26}
Manganese(II) sulphide	buff	1.5×10^{-15}

Those with higher solubility products are precipitated from alkaline solutions in which the sulphide ion concentration is increased. Some examples are given in Table 20.3.

Hydrogen sulphide is a moderately strong reducing agent, being quite readily oxidised to sulphur; for example, iron(III) ions are reduced to iron(II) ions and chlorine to chloride ions.

$$2Fe^{3+}(aq) + S^{2-}(aq) \rightarrow 2Fe^{2+}(aq) + S(s)$$

$$Cl_2(aq) + S^{2-}(aq) \rightarrow 2Cl^-(aq) + S(s)$$

The gas is highly poisonous, the lethal concentration being comparable to that of carbon monoxide or hydrogen cyanide. Anyone overcome by hydrogen sulphide can, however, always be revived by prompt artificial respiration, continued if necessary for several hours.

(iii) Oxides and oxoacids of sulphur

As can be seen in Table 20.2 sulphur forms two main oxides, sulphur dioxide and sulphur(VI) oxide, in which it has oxidation states of +4 and +6 respectively.

Sulphur dioxide is a colourless gas with a pungent irritating odour, soluble in water to form a solution of sulphurous acid (systematically trioxosulphuric(IV) acid) which is so readily decomposed that sulphur dioxide can be recovered from the solution by boiling.

$$SO_2(g) + H_2O \rightleftharpoons H_2SO_3(aq) \rightleftharpoons H^+(aq) + HSO_3^-(aq) \rightleftharpoons 2H^+(aq) + SO_3^{2-}(aq)$$

The burning of sulphur-containing fuels such as coal and heavy oil releases sulphur dioxide into the atmosphere; where concentrations of the gas are high it is a serious air pollutant, causing bronchitis, inhibiting plant life and damaging stonework.

The sulphur dioxide molecule is angular, with identical bonding between the sulphur and oxygen atoms; these are shorter than single bonds and longer than double bonds, indicating a resonance hybrid between two canonical structures (page 477):

Canonical forms   Resonance hybrid

The sulphite ion is tetrahedral, the actual structure again being a resonance hybrid of possible canonical structures such as:

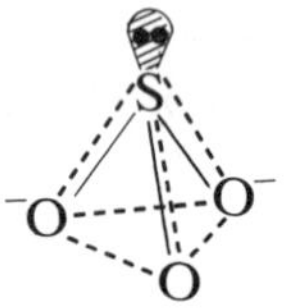

Sulphur dioxide and sulphites are both readily oxidised to sulphur(VI) oxide and sulphates respectively, utilising the lone pair of electrons on the sulphur atom to form six covalent bonds. The oxidation of absolutely pure dry sulphur dioxide is reversible, and takes place only slowly, unless catalysed as in the preparation of sulphur(VI) oxide for the manufacture of sulphuric acid (page 477).

$$2SO_2(g) + O_2(g) \rightleftharpoons 2SO_3(s)$$

Moist sulphur dioxide and solutions of sulphites are reducing agents; the latter even oxidising slowly on standing in air. Although not as strong a reducing agent as hydrogen sulphide, sulphur dioxide in solution reduces iron(III) to iron(II) ions, chromate(VI) to chromium(III) ions, chlorine to chloride ions and manganate(VII) ions to manganese(II) ions.

$$2Fe^{3+}(aq) + SO_3^{2-}(aq) + H_2O(l) \rightarrow 2Fe^{2+}(aq) + SO_4^{2-}(aq) + 2H^+(aq)$$

$$Cr_2O_7^{2-}(aq) + 3SO_3^{2-}(aq) + 8H^+(aq) \rightarrow 2Cr^{3+}(aq) + 3SO_4^{2-}(aq) + 4H_2O(l)$$

$$Cl_2(aq) + SO_3^{2-}(aq) + H_2O(l) \rightarrow 2H^+(aq) + 2Cl^-(aq) + SO_4^{2-}(aq)$$

$$2MnO_4^-(aq) + 5SO_3^{2-}(aq) + 6H^+ \rightarrow 2Mn^{2+}(aq) + 5SO_4^{2-}(aq) + 3H_2O(l)$$

Apart from its major use in the manufacture of sulphuric acid, sulphur dioxide is used to bleach delicate materials such as wool and straw (for hats) which would be damaged by chlorine. Since the bleaching action is a reduction process, such materials slowly become yellowed with age as the pigment is reoxidised by the oxygen of the air.

Sulphur dioxide is also used in the form of sulphite tablets (Campden tablets) as a preservative and anti-oxidant in home jam making and brewing, and as a portable water-sterilising agent.

The sulphur dioxide molecule has a high dipole moment, resulting in strong attractive forces between the molecules of the liquid, which has a boiling point of $-10°C$ and a high molar heat of vapourisation. It was widely used as a commercial refrigerant before safer materials such as the freons (page 443) were developed.

Sulphur(VI) oxide vapour condenses to a crystalline solid, α-SO_3, which slowly changes to long silky needle-like crystals of β-SO_3. In the vapour phase it has a symmetrical planar molecule: both solid forms are polymeric,

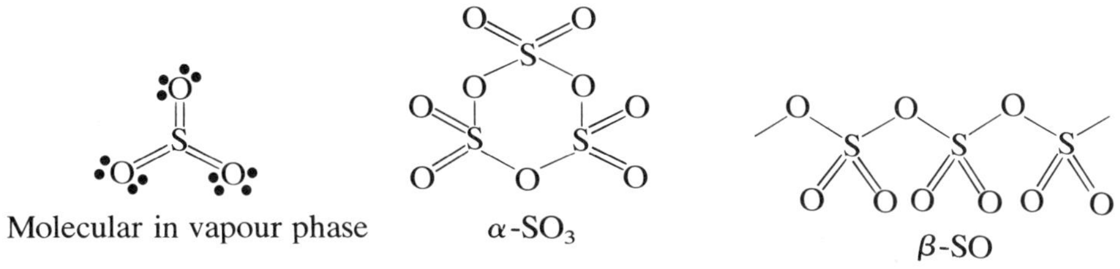

Fig. 20.3   Sulphur(VI) oxide

consisting of tetrahedral units with delocalised bonding in which sulphur is the central atom (Fig. 20.3).

Sulphur(VI) oxide is strongly acidic, reacting violently with water to form sulphuric acid:

$$SO_3(s) + H_2O(l) \rightarrow H_2SO_4(l) \qquad \Delta_{298}^{\ominus}H = -146 \text{ kJ mol}^{-1}$$

Pure sulphuric acid is a dense colourless oily liquid with a relatively high boiling point. The molecules are covalent and tetrahedral in structure with some evidence of hydrogen bonding between them:

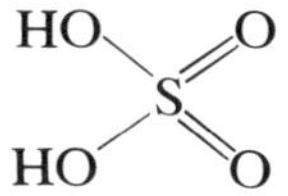

The pure acid has such an affinity for water that it causes severe burns to the skin and destroys cloth or paper by extracting the elements of water from these materials with the evolution of a great deal of heat. The charring of organic molecules is illustrated by the action of the pure acid on sugar:

$$C_{12}H_{22}O_{11}(s) \xrightarrow{\text{H}_2\text{SO}_4} 12C(s) + 11H_2O(l)$$

This destructive effect is utilised in the laboratory preparation of carbon monoxide from methanoic acid, ethandioic acid or salts of these acids:

$$HCOONa(s) + H_2SO_4(l) \rightarrow NaHSO_4(s) + H_2O(l) + CO(g)$$

$$(COONa)_2(s) + H_2SO_4(l) \rightarrow Na_2SO_4(s) + H_2O(l) + CO_2(g) + CO(g)$$

Industrially it is used to convert ethanol to ethoxyethane, and to withdraw the elements of water from a mixture of chlorobenzene and 1,1,1-trichloroethanal in the preparation of DDT (page 439).

The pure acid is a moderately effective oxidising agent. It will oxidise both carbon and sulphur:

$$C(s) + 2H_2SO_4(l) \rightarrow 2H_2O(l) + 2SO_2(g) + CO_2(g)$$

$$S(s) + 2H_2SO_4(l) \rightarrow 2H_2O(l) + 3SO_2(g)$$

and it is used in the laboratory to bring copper and other metals low in the electrochemical series into solution:

$$Cu(s) + 2H_2SO_4(l) \rightarrow CuSO_4(aq) + 2H_2O(l) + SO_2(g)$$

It is reduced by both hydrogen bromide and hydrogen iodide, which are oxidised to halogens in the process; hydrogen iodide may even reduce the sulphur dioxide evolved to hydrogen sulphide:

$$2HBr(g) + H_2SO_4(l) \rightarrow 2H_2O(l) + Br_2(l) + SO_2(g)$$

$$6HI(g) + SO_2(g) \rightarrow 3I_2(s) + 2H_2O(l) + H_2S(g)$$

Since sulphuric acid is non-volatile it displaces the other, more volatile, strong mineral acids from their salts; hydrogen chloride (for hydrochloric acid) and nitric acid being made by the action of the hot concentrated acid on the sodium salts:

$$H_2SO_4(l) + NaCl(s) \rightarrow NaHSO_4(s) + HCl(g)$$

$$H_2SO_4(l) + NaNO_3(s) \rightarrow NaHSO_4(s) + HNO_3(l)$$

The method is not suitable, for making either hydrogen bromide or hydrogen iodide, which, as seen above, are oxidised to the respective halogen.

Concentrated sulphuric acid is also used in the preparation of benzene sulphonic acid (page 361), in addition reactions with alkenes to form alcohols (page 353) and in the manufacture of detergents (page 409).

The acid dissolves and ionises in water in two stages:

$$H_2SO_4(l) + H_2O(l) \rightarrow H_3O^+(aq) + HSO_4^-(aq)$$

$$HSO_4^-(aq) + H_2O(l) \rightarrow H_3O^+(aq) + SO_4^{2-}(aq)$$

The hydrogensulphate ion HSO_4^-, unlike the hydrogensulphite ion, is stable, and under the right conditions hydrogensulphates of the Group I metals can be obtained as crystalline solids; sodium hydrogensulphate constitutes a useful solid acid.

$$NaHSO_4(aq) \rightarrow Na^+(aq) + HSO_4^-(aq) \rightarrow Na^+(aq) + H^+(aq) + SO_4^{2-}(aq)$$

Aqueous sulphuric acid is itself a strong acid, useful in the laboratory as a provider of hydrogen ions for many reactions such as oxidations by manganate(VII) ions. It reacts with metals sufficiently high in the electrochemical series to release hydrogen and with carbonates to yield carbon dioxide, and it is neutralised by metal oxides; in all these reactions sulphates are formed. Most sulphates are soluble in water; exceptions are lead sulphate, barium sulphate and calcium sulphate (the last being very sparingly soluble).

The sulphate ion is tetrahedral, with four equal bonds between the central sulphur atom and the oxygen atoms. The symmetrical structure is a resonance hybrid of structures such as:

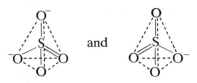

Sulphates are more thermally stable than the corresponding nitrates and carbonates. Sulphates of metals high in the electrochemical series are difficult to decompose by heat, but strong heating of some of the heavier metal sulphates releases a mixture of sulphur dioxide and and sulphur(VI) oxide, leaving a residue of metal oxide. At one time the decomposition of iron(II) sulphate was used to obtain sulphuric acid. The crystals of 'green vitriol' were strongly heated and the gases condensed as sulphuric acid, or 'oil of vitriol'.

$$FeSO_4 \cdot 7H_2O \xrightarrow{heat} FeSO_4(s) + 7H_2O(l)$$

$$2FeSO_4(s) \xrightarrow{heat} Fe_2O_3(s) + SO_2(g) + SO_3(s)$$

$$SO_3(s) + H_2O(l) \rightarrow H_2SO_4(l)$$

Vast quantities of sulphuric acid are now manufactured each year by the Contact process (page 477), since almost every industry uses it in some form or another. The aqueous acid is used for pickling iron and steel (removing the oxide layer before painting, galvanising or tinning), making fertilisers and as the electrolyte in lead accumulators.

(iv) Peroxodisulphuric(VI) acid and potassium peroxodisulphate(VI)

This acid, $H_2S_2O_8$, is made by the electrolysis of ice-cold 50% sulphuric acid with a high current density. It is not sufficiently stable to have been isolated in the pure state; its potassium salt, commonly known as potassium persulphate, is better known.

Both acid and salt are powerful oxidising agents, resulting from the presence in the molecule of an (—O—O—) group similar to that in hydrogen peroxide:

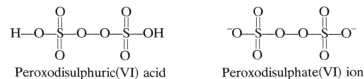

Peroxodisulphuric(VI) acid   Peroxodisulphate(VI) ion

Potassium peroxodisulphate(VI) oxidises iron(II) ions to iron(III) ions, and iodide ions to iodine:

$$2Fe^{2+}(aq) + S_2O_8^{2-}(aq) \rightarrow 2Fe^{3+}(aq) + 2SO_4^{2-}(aq)$$

$$2I^-(aq) + S_2O_8^{2-}(aq) \rightarrow I_2(s) + 2SO_4^{2-}(aq)$$

The essential half-reaction in these reactions is the transfer of two electrons to the peroxodisulphate(VI) ion, giving rise to two sulphate(VI) ions:

$$S_2O_8^{2-} + 2e^- \rightarrow 2SO_4^{2-}$$

(v) Thiosulphuric acid and sodium thiosulphate

The full systematic name for this acid is trioxo(-II) thio(-II) sulphate(VI). The contraction of this to thiosulphate(VI) does not fully reflect the oxidation states of the sulphur atoms so is not wholly correct.

Thiosulphuric acid is unstable,

its sodium salt, $Na_2S_2O_3$, is a mild reducing agent, being used to remove excess chlorine after bleaching fabrics, and in the laboratory to reduce iodine quantitatively to iodide ions.

$$I_2 + 2S_2O_3^{2-} \rightarrow 2I^- + S_4O_6^{2-}$$

Tetrathionate ion

The reaction is rapid, the end-point being readily determined with the use of a starch solution indicator.

Sodium thiosulphate, as 'hypo', is used as a photographic fixing agent; it dissolves excess silver halides, undecomposed by exposure in the camera and subsequent development, by complexing with the silver ion. (The silver from the decomposed halides makes the black or shaded areas of the negative).

(vi) Chlorosulphonic acid and sulphur dichloride oxide
Chlorosulphonic acid, a colourless fuming liquid, is derived from sulphuric acid by the substitution of an —OH group by a chlorine atom:

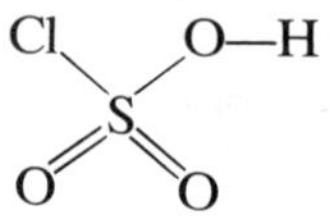

It is used in organic chemistry to introduce a reactive —SO$_2$Cl group into a benzene molecule:

$$C_6H_6 + 2ClOHSO_2 \rightarrow C_6H_5SO_2Cl + HCl + H_2SO_4$$

Sulphur dichloride oxide, SOCl$_2$, is derived, as its systematic name implies, from sulphur dioxide, one of the oxygen atoms being replaced by two chlorine atoms:

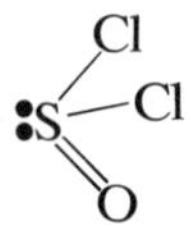

Commonly known as thionyl chloride, it is frequently used to introduce chlorine atoms into organic hydroxy-compounds; it has the advantage over phosphorus pentachloride that the other products of reaction are gaseous (page 460).

$$C_2H_5OH(l) + SOCl_2(l) \rightarrow C_2H_5Cl(l) + SO_2(g) + HCl(g)$$

(vii) Halogen compounds of sulphur
Sulphur forms stable compounds only with the more electronegative halogen elements; the most stable of all is sulphur hexafluoride, in which all six valency electrons of sulphur are involved in bonding (Fig. 20.4) in an octahedral molecule. The heavy inert gas is used as an insulator in high voltage electrical equipment. No equivalent compound of sulphur with chlorine exists, since the sulphur atom is too small to accommodate six chlorine atoms around it.

Sulphur tetrafluoride is less stable than the hexafluoride, being readily hydrolysed by water.

$$SF_4(g) + 2H_2O(l) \rightarrow SO_2(g) + 4HF(g)$$

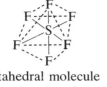

Octahedral molecule   Trigonal bipyramidal   Molecule of
of sulphur hexafluoride   molecule of sulphur   disulphur
                             tetrafluoride   dichloride

Fig. 20.4 Fluorine and chlorine compounds of sulphur

The lone pair of electrons on the sulphur atom make it comparatively reactive, and it is used as a fluorinating agent.

Although sulphur tetrachloride is known, the only stable chloride of sulphur is disulphur dichloride, S_2Cl_2, a red liquid which fumes in moist air as it is hydrolysed.

$$S_2Cl_2(l) + 2H_2O(l) \rightarrow SO_2(g) + 2HCl(g) + H_2S(g)$$

$$SO_2(g) + 2H_2S(g) \rightarrow 2H_2O(l) + 3S(s)$$

It is prepared by passing chlorine over heated sulphur. Its main use is as a solvent for sulphur in the vulcanisation of rubber.

20.5 Selenium, tellurium and polonium

While oxygen and sulphur, the elements of Group VI considered in the previous sections, are both plentiful in nature, selenium, tellurium and polonium are all scarce.

Polonium is a radioactive metal with a half-life of only 138 days. First recognised by Madame Curie, it has been isolated only in minute quantities, twenty-five tonnes of pitchblende yield about one milligram of polonium. Its chemistry has been determined from these minute amounts by the application of sophisticated modern techniques.

Tellurium and selenium, obtained as minor products in the refining of copper, have been used in the same way as sulphur in the vulcanisation of rubber to increase its resistance to oxidation and wear. The elements have been used to increase the corrosion-resistance of lead and magnesium alloys. Selenium is also used in the manufacture of glass, especially red glass, and in alternating current rectifiers and photo-electric cells since its conductivity is greatly increased by the incidence of light.

Questions

1. Write a short account, with examples, of the various types of oxides that occur.

2. (*a*) (i) Explain carefully what you understand by the terms *oxidation* and *reduction*.

(ii) Give, with examples, *two* tests for an oxidising agent and *two* tests for a reducing agent.

(b) The following equations represent two chemical reactions involving oxidation/reduction processes:

(i) $IO_3^-(aq) + 5I^-(aq) + 6H^+(aq) \rightarrow 3I_2(s) + 3H_2O(l)$

(ii) $5As_2O_3(s) + 4MnO_4^-(aq) + 12H^+(aq)$
$$\rightarrow 5As_2O_5(s) + 4Mn^{2+}(aq) + 6H_2O(l)$$

Identifying the oxidising agent(oxidant) and the reducing agent(reductant), and write 'half-equations' showing the donation, or acceptance, of electrons by *each* of them.

Outline how the first reaction is used in titrimetric analysis for the determination of the concentration of iodate ions in aqueous solution.

(AEB 1977)

3. (a) Review the principal oxidation states in which sulphur appears in its compounds and give one example of a suitable compound for each of the oxidation states mentioned.

(b) Outline how one compound for each of four different oxidation states of sulphur may be obtained, *either* directly *or via* an intermediate compound of different oxidation state, from the element sulphur in its pure state.   (L)

4. A compound X of sulphur has the formula SO_xCl_2; 23.8 g of compound X are made to react with water (in excess) and the chloride ion is completely precipitated as 57.4 g of silver chloride.

(a) Calculate the relative molecular mass of compound X and write its molecular formula (i.e. determine x).

(b) In what oxidation state is the sulphur atom in X?

(c) Draw a diagram to illustrate the electronic structure of X. Suggest a shape for the molecule of X.   (C)

5. State, giving reasons, what has been oxidized, what has been reduced and what has not undergone a redox change in the following reactions:

(a) $3Cl_2(g) + 6OH^-(aq) \rightarrow 5Cl^-(aq) + ClO_3^-(aq) + 3H_2O(l)$

(b) $IO_3^-(aq) + 2I^-(aq) + 3Cl^-(aq) + 6H^+(aq) \rightarrow 3ICl(aq) + 3H_2O(L);$

(c) $2Na(s) + H_2(g) \rightarrow 2NaH(s)$

(O)

6. Sulphuric acid is one of the most versatile reagents used in the laboratory. Discuss this statement by referring to reactions which illustrate its behaviour

(a) as an acid,

(b) in the displacement of other acids from their salts,

(c) as a sulphonating agent,

(d) as an oxidising agent,

(e) as a dehydrating agent.   (JMB)

1

The halogens

Table 21.1   Physical data

Element	Atomic number	Electronic configuration	Atomic radius nm	Ionic radius nm X^-	Electro- negativity Pauling scale	Electrode potential V	m. p. K	b. p. K
F	9	2.7 $1s^2 2s^2 2p^5$	0.073	0.136	4.0	+2.85	53	85
Cl	17	2.8.7 $----3s^2 3p^5$	0.099	0.181	3.0	+1.36	172	239
Br	35	2.8.18.7 $--3d^{10}4s^2 4p^5$	0.114	0.195	2.8	+1.06	266	331
I	53	2.8.18.18.7 $----4d^{10}5s^2 5p^5$	0.133	0.216	2.5	+0.54	387	456
At	85	2.8.18.32.18.7 $-----6p^5$	—	—	—	—	—	—

21.1   General characteristics and trends in properties of Group VII elements

The halogens are the last group of elements before the noble gases of Group 0. They are non-metallic, each element being the most electronegative of the period in which it occurs. Fluorine has the highest electronegativity of all elements; those of chlorine, bromine and iodine decrease steadily with increasing atomic mass and atomic radius.

With two *s* electrons and five *p* electrons in the valency shell the elements readily form compounds in which they either gain an electron to form a singly charged negative ion or share an electron in a covalent bond to acquire a noble gas structure. The halide ions are larger than the atoms from which they are derived; the extra electron in the valency orbital causing some mutual repulsion.

As elements, the halogens form diatomic molecules with covalent bonds between the atoms and weak van der Waals forces between the molecules; these forces increase in strength as the size of the atoms and the number of electrons associated with them increase. Thus fluorine and chlorine are gases, fluorine being pale green and chlorine a darker yellowish-green; bromine is a red-brown liquid and iodine a silvery grey solid giving rise to a purple vapour on gentle heating. All four gases or vapours are poisonous, with choking smells and irritant effects on the lungs.

Little is known of astatine, which is a radioactive solid of very short half-life, obtained by the bombardment of bismuth by α-particles.

$$^{209}_{83}\text{Bi} + ^4_2\text{He} \rightarrow ^{211}_{85}\text{At} + 2^1_0 n$$

It is not further considered in this text.

A slight increase in 'metallic' character with decreasing electronegativity of the members of the group is indicated by the presence of I^+ ions as part of some complex ions such as $(IO^+IO_3^-)$ in I_2O_4.

The covalent molecules of the elements are readily soluble in organic solvents, giving rise to solutions of characteristic colours; in tetra-chloromethane solution chlorine is pale green, bromine orange and iodine a deep purple. The elements also all dissolve in water, although iodine is only very sparingly soluble. Fluorine, as a result of its high electronegativity, oxidises water to oxygen.

$$2F_2(g) + 2H_2O(l) \rightarrow 4HF(aq) + O_2(g)$$

Chlorine, bromine and iodine also react with water but to a lesser extent; the molecules disproportionate, giving rise to equilibrium mixtures containing halic(I) acids.

$$X_2(aq) + H_2O(l) \rightleftharpoons H^+(aq) + X^-(aq) + HXO(aq)$$

Where X_2 is iodine the equilibrium is well to the left, little reaction taking place; with chlorine the reaction is more marked and in bright sunlight the steady decomposition of the chloric(I) acid drives the equilibrium wholly to the right with the evolution of oxygen.

$$Cl_2(aq) + H_2O(l) \rightleftharpoons H^+(aq) + Cl^-(aq) + HClO(aq)$$

$$2HClO(aq) \rightarrow 2H^+(aq) + 2Cl^-(aq) + O_2(g)$$

Chlorine, bromine and iodine, having vacant d orbitals, can increase the number of electrons in their valency orbitals beyond eight and form compounds in which they have oxidation numbers ranging from -1 to $+7$.

Fluorine, the head element, has an invariable valency of one; its extreme

Table 21.2 Oxidation states of halogen compounds and ions

Oxidation number	Fluorine	Chlorine	Bromine	Iodine
+7		$HClO_4, ClO_4^-$ Cl_2O_7	$HBrO_4, BrO_4^-$	HIO_4, IO_4^- IF_7
+5		$HClO_3$ ClO_3^-	$HBrO_3$ BrO_3^- BrF_5	HIO_3 IO_3^- I_2O_5 IF_5
+3		$HClO_2$ ClF_3	BrF_3	ICl_3
+1		$HClO$ ClO^- ClF	BrF $BrCl$	ICl IBr
0	F_2	Cl_2	Br_2	I_2
−1	HF HF^- $*RF$	HCl Cl^- RCl	HBr Br^- RBr	HI I^- RI

* R = hydrocarbon radical

electronegativity gives rise to the formation of compounds in which the other elements attain their highest valency states. Thus sulphur forms a hexafluoride and iodine a heptafluoride, whilst oxygen in dioxygen difluoride attains an unusual oxidation state of +2.

The halogens are all very reactive elements, with iodine, bromine and chlorine showing a steady trend in increasing reactivity, while fluorine is markedly the most reactive. This high reactivity of fluorine can be explained as resulting from the relatively low bond energy of the molecule, leading to ease of dissociation into atoms, reinforced by the high energy of the bonds formed with other elements. Comparing the bond energies of the four halogens with the energy of the covalent bonds each makes with carbon, it can be seen that the formation of tetrafluoromethane is much more exothermic than that of the other tetrahalogenomethanes.

Table 21.3

Halogen	F—F	Cl—Cl	Br—Br	I—I
Bond energy $kJ\,mol^{-1}$	158	242	193	151
Halogen/carbon bonds	C—F	C—Cl	C—Br	C—I
Bond energy $kJ\,mol^{-1}$	485	339	209	218

Similarly, in the formation of metal compounds, the low dissociation energy of the fluorine molecule is reinforced by the high energy of the ionic lattice formed, resulting in extremely exothermic reactions.

The oxidising powers of the halogens are reflected in their positive electrode potentials. Fluorine is the strongest oxidising agent, with the greatest ability to accept electrons, while iodine is weakest.

$$I_2(s) + 2e^- \rightarrow 2I^-(aq) \qquad E = +0.54\,V$$

$$F_2(g) + 2e^- \rightarrow 2F^-(aq) \qquad E = +2.85\,V$$

Thus one halogen will displace another halogen lower in the group from its salts in solution:

$$Cl_2(aq) + 2Br^-(aq) \rightarrow 2Cl^-(aq) + Br_2(aq)$$

$$Br_2(aq) + 2I^-(aq) \rightarrow 2Br^-(aq) + I_2(aq)$$

21.2 Occurrence, extraction and uses

None of the halogens occurs free in nature; all are found in the combined state as salts. The name halogen is derived from two Greek words *hals* (sea-salt) and *gennao* (I produce).

The main source of chlorine is sodium chloride from the sea or from rock-salt residues of ancient seas in such places as Stassfurt and Cheshire. Deposits of salt are also found on the shores of salt lakes, like the Dead Sea and the Great Salt Lake of Utah. Chlorine is extracted commercially by the electrolysis of concentrated brine (page 474) and in the laboratory by the oxidation of concentrated hydrochloric acid by potassium manganate(VII) or manganese(IV) oxide.

$$4HCl(aq) + MnO_2(s) \rightarrow MnCl_2(aq) + 2H_2O(l) + Cl_2(g)$$

Bromine also comes from the sea, although bromides are present in far lower concentrations than chlorides. It is extracted by treating acidified sea-water with chlorine, displacing bromine as the free element. Bromine is then blown out as vapour by passing air through the mixture, and finally extracted from the air as a liquid.

Concentrations of iodine salts in the sea are very low, but it is absorbed by certain seaweeds, notably laminaria, in sufficient quantities to make its commercial extraction sometimes worthwhile. Far greater amounts are however obtained from sodium iodate, which occurs as an impurity in Chile saltpetre. Solutions containing the iodate ion are reduced by hydrogen-sulphite ions to free iodine.

$$2IO_3^-(aq) + 5HSO_3^-(aq) \rightarrow I_2(s) + 3HSO_4^-(aq) + 2SO_4^{2-}(aq) + H_2O(l)$$

The concentration of HSO_3^- introduced has to be controlled, since excess would further reduce the iodine to iodide ions.

Fluorine is not found in sea-salts but is plentiful in the United States and elsewhere as the mineral fluorspar (CaF_2); it also occurs in Greenland as cryolite Na_3AlF_6, which is used in the electrolytic production of aluminium (page 468). Production of fluorine is difficult because of its highly reactive nature; electrolysis of fused potassium hydrogenfluoride (page 311) is used.

Increasing quantities of fluorine are required at the present time for the manufacture of fluorinated hydrocarbons which are used as refrigerating fluids, for pressurising aerosols, manufacturing non-stick plastics and making insecticides (see also Chapter 29). Uranium hexafluoride is used to separate out the radioactive isotope of uranium.

Chlorine is used in great quantities for the sterilisation of both drinking water and swimming baths; it is used as a bleaching agent, in the manufacture of dyes, plastics, weedkillers, antiseptics and insecticides. Both chlorine and bromine are used in the manufacture of fire extinguisher fluids.

Bromine is used as silver bromide in photography, as a constituent of many medical drugs and antiseptics, and in the manufacture of dyes. Large quantities have been used in making ethylene dibromide, a component of the 'lead fluid' additive employed to improve the octane number of petrol; the purpose is to avoid the deposit of lead oxides in the engines by forming lead(II) bromide which is discharged in the exhaust.

Iodine is used in medicine both as a constituent of antiseptic ointments and in radioactive tracers (page 40); it finds further use in the manufacture of dyes and in the production of films for colour photography.

Halogen compounds are found in living systems. They have varying effects on human metabolism; traces of fluorides in the diet appear to help the healthy development of teeth; chloride ions are the principal anions in body fluids and blood plasma; bromides are present in small amounts which, if increased, have a sedating effect; iodine compounds are essential to the efficient working of the thyroid. This last effect is so marked that in districts where the natural supply of iodine is low, iodised salt is sold to help maintain the required amount for good health.

21.3   Halogen compounds

(a) The halides

The halogen elements form numerous halides with both metals and non-metals, in which they have an oxidation number -1. The structures of these halides vary from ionic to covalent across the Periodic Table. The strongly metallic elements of Group I and II form ionic halides, while the weakly metallic and non-metal elements form partially or wholly covalent halides (Table 21.4).

Table 21.4   Structure of chlorides of Period 3, sodium to argon

Chloride	Na^+Cl^-	$Mg^{2+}(Cl^-)_2$	$AlCl_3$	$SiCl_4$	PCl_3	SCl_2
Bonding	Ionic	Ionic with some covalency	Mainly covalent some ionic character	Covalent	Covalent	Covalent

The nature of the bonding in metallic halides also varies with the halogen in the compound; fluorine, being the most electronegative element with the greatest attraction for electrons, forms the highest number of ionic compounds, while chlorine, bromine and iodine show an increasing tendency to form metal halides with some covalent character as the size of the electron cloud increases and becomes more easily distorted. Thus metals with ions of small radius and relatively higher charges, such as Be^{2+}, Mg^{2+}, Al^{3+} and Fe^{3+}, form halides, other then fluorides, with varying degrees of covalency. For example, calcium fluoride is ionic, magnesium chloride is partially covalent and aluminium bromide is almost wholly covalent. Similarly silver fluoride is ionic and soluble in water, while silver chloride, bromide and iodide are all partially covalent and become more insoluble as the degree of covalency increases.

The precipitation of silver halides by the addition of silver nitrate solution is used to identify chlorides, bromides and iodides in solution.

$$Ag^+(aq) + Cl^-(aq) \rightarrow AgCl(s)$$

Silver chloride is precipitated as a white flocculent solid, the silver bromide precipitate is cream coloured and silver iodide is yellow. Addition of ammonia solution will redissolve silver chloride by removing silver ions as complex $[Ag(NH_3)_2]^+$ ions. Silver bromide, having a lower solubility product, is only partially dissolved, whilst the solubility product of silver iodide is so low that it remains insoluble even in ammonia solution.

The halides of the more reactive metals are best prepared by the neutralisation of the appropriate hydrohalic acid with a solution of the metal hydroxide, followed by crystallisation of the salt from solution.

$$NaOH(aq) + HCl(aq) \rightarrow NaCl(aq) + H_2O(l)$$

The halides of the alkali metals, other than lithium, are deposited without water of crystallisation.

Hydrated halides of other metals can be crystallised from solutions made by dissolving excess of the metal oxide or carbonate in dilute solutions of the appropriate hydrohalic acid; the solutions are filtered, concentrated and allowed to cool with the deposition of the hydrated salts.

$$CuCO_3(s) + 2HCl(aq) \rightarrow Cu^{2+}(aq) + 2Cl^-(aq) + H_2O(l) + CO_2(g)$$

$$Cu^{2+}(aq) + 2Cl^-(aq) \rightarrow CuCl_2 \cdot 2H_2O(s)$$

$$FeO(s) + 2HCl(aq) \rightarrow Fe^{2+}(aq) + 2Cl^-(aq) + H_2O(l)$$

$$Fe^{2+}(aq) + 2Cl^-(aq) \rightarrow FeCl_2 \cdot 6H_2O(s)$$

Many of the hydrated metal halides are deliquescent, making it necessary to effect the final crystallisation in a desiccator.

Insoluble metal halides can be obtained by mixing a solution of a soluble salt of the metal required with the appropriate hydrohalic acid or sodium halide solution.

$$AgNO_3(aq) + NaBr(aq) \rightarrow AgBr(s) + NaNO_3(aq)$$

$$Pb(CH_3COO)_2 + 2HCl(aq) \rightarrow PbCl_2(s) + 2CH_3COOH(aq)$$

Where soluble halides are required in the anhydrous state they are conveniently prepared by passing the halogen vapour over the heated metal:

$$2Al(s) + 3Cl_2(g) \rightarrow 2AlCl_3(s)$$

The reactions with fluorine and chlorine are highly exothermic. The vigour of reaction decreases with bromine and iodine vapour. With fluorine and chlorine, metals forming more than one type of ion yield the halide with the metal in its higher oxidation state:

$$2Fe(s) + 3Cl_2(g) \rightarrow 2FeCl_3(s)$$

If the anhydrous salt of the metal in a lower oxidation state is required the reaction between the hydrogen halide gas and the metal is used.

$$Fe(s) + 2HCl(g) \rightarrow FeCl_2(s) + H_2(g)$$

Halide ions occur in many complex anions, being liganded (page 56) to the central atom or ion; examples include $[CuCl_4]^{2-}$, $[PbCl_6]^{4-}$, and $[SnF_6]^{2-}$. Many similar complexes containing bromide and iodide ions as ligands occur, but they are generally of lower stability than the chloride complexes. Fluoride ions form complexes with beryllium and boron ions, $(BeF_4)^{2-}$ and $(BF_4)^-$, but similar complexes are not formed with the larger halide ions.

An interesting property of the fluoride ion, not shown by the other halide ions, is the ability to make a hydrogen bond with a molecule of hydrogen fluoride, giving rise to an acid ion $(F\text{---}H \cdots F)^-$. Thus if hydrogen fluoride is passed over sodium or potassium fluorides, acid salts are formed.

$$K^+F^-(s) + HF(g) \rightarrow K^+(F\text{---}H \cdots F)^-(s)$$

The acid corresponding to these salts, HF_2, is a much stronger acid than the normal hydrofluoric acid, HF.

Halides of non-metal elements are generally prepared by passing the halogen vapour over the heated element. If excess of the more electronegative halogens is used, compounds containing the non-metal element in its higher oxidation state result. Thus chlorine passed over heated phosphorus yields first the liquid trichloride, which is further oxidised to phosphorus pentachloride in the presence of excess chlorine.

$$P_4(s) + 6Cl_2(g) \rightarrow 4PCl_3(l) \xrightarrow{4Cl_2(g)} 4PCl_5(s)$$

Both fluorine and bromine also form pentavalent halides, but iodine, being the least electronegative, forms only the triiodide. Similarly fluorine forms a hexafluoride with sulphur while the most oxidised chloride of sulphur is sulphur tetrachloride.

The important metal and non-metal halides are described in more detail in the chapters on the elements concerned. The halogen derivatives of organic compounds are discussed in the relevant chapters on carbon chemistry.

(b) Hydrogen halides

All four of the halogens will react directly with hydrogen to form hydrogen halides, the reaction with fluorine being explosive even at low temperatures.

$$H_2(g) + Cl_2(g) \rightarrow 2HCl(g)$$

The reaction with chlorine is explosive in bright sunlight, but a jet of hydrogen will burn smoothly in chlorine gas. The reaction between bromine and hydrogen requires a catalyst at 250°C, while that between iodine and hydrogen requires a catalyst at higher temperatures still, and is reversible. The direct reaction is used industrially in the manufacture of hydrogen chloride for hydrochloric acid, but in the laboratory indirect methods are used. Hydrogen fluoride and hydrogen chloride are obtained from ionic metallic halides by the action of hot concentrated sulphuric acid, which displaces the more volatile acids.

$$CaF_2(s) + H_2SO_4(l) \rightarrow CaSO_4(s) + 2HF(g)$$

$$NaCl(s) + H_2SO_4(l) \rightarrow NaHSO_4(s) + HCl(g)$$

Although sulphuric acid displaces hydrogen bromide and hydrogen iodide from salts this method cannot be used to prepare them, the hydrogen halides released in the reaction being immediately oxidised by the sulphuric acid to free halogens. In the case of hydrogen iodide, which is the stronger reducing agent, the sulphuric acid may even be reduced to hydrogen sulphide in a series of reactions.

$$H_2SO_4(l) + NaI(s) \rightarrow NaHSO_4(s) + HI(g)$$

$$H_2SO_4(l) + 2HI(g) \rightarrow I_2(s) + SO_2(g) + 2H_2O(l)$$

$$SO_2(g) + 4HI(g) \rightarrow 2I_2(s) + S(s) + 2H_2O(l)$$

$$S(s) + 2HI(g) \rightarrow I_2(s) + H_2S(g)$$

Hydrogen bromide and hydrogen iodide are therefore prepared in the laboratory either by the action of phosphoric(V) acid, which is less readily reduced, on the sodium salts:

$$H_3PO_4(l) + 3NaBr(s) \rightarrow 3HBr(g) + Na_3PO_4(s)$$

or by the hydrolysis of the trihalides. Thus hydrogen bromide is made by adding bromine drop by drop to a mixture of red phosphorus and water:

$$2P(s) + 3Br_2(l) \rightarrow 2PBr_3(l)$$

$$PBr_3(l) + 3H_2O(l) \rightarrow 3HBr(g) + H_3PO_3(aq)$$

Since iodine is a solid the method is modified for the preparation of hydrogen iodide, water being added drop by drop to a mixture of iodine and red phosphorus. The mixture of iodine and phosphorus is less liable to sudden reaction than a mixture of bromine and phosphorus.

Hydrogen chloride, hydrogen bromide and hydrogen iodide are all colourless gases of very similar properties which show regular modifications with increasing atomic number of the halogen. Hydrogen fluoride, however, is a colourless fuming liquid boiling at 19.5°C; the very electronegative fluorine atom induces hydrogen bonding between the molecules, making it less volatile:

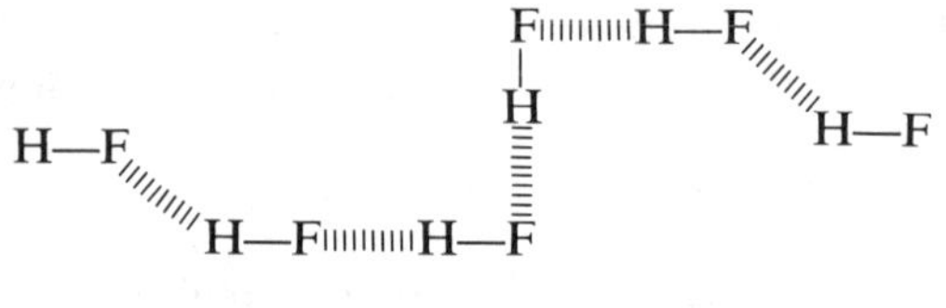

Hydrogen fluoride differs in many other respects from the rest of the hydrogen halides, as can be seen from the data in Table 21.5.

Hydrogen fluoride and hydrogen chloride are both stable to heat, whereas hydrogen bromide and, to a greater extent, hydrogen iodide dissociate at raised temperatures as a result of the lower bond strengths and greater bond lengths in these molecules:

$$2HI(g) \rightleftharpoons H_2(g) + I_2(g)$$

The variations in bond strengths are also reflected in the strengths of the acids produced when the hydrogen halides dissolve in water; the energy required to break the H—X bond is supplied by the heat of formation of the oxonium ion:

$$HCl(g) + H_2O(l) \rightarrow H_3O^+(aq) + Cl^-(aq)$$

This is sufficient to overcome the bond strengths of the hydrogen chloride bromide and iodide, which are all strong acids showing a slight increase in

Table 21.5 Physical properties of hydrogen halides

	HF	HCl	HBr	HI
Bond energy $kJ\,mol^{-1}$	554	431	336	299
Bond length nm	0.086	0.128	0.142	0.160
Dipole moment Debyes	1.9	1.03	0.78	0.38
Acid dissociation constant	6.7×10^{-4}	very large	very large	very large
m.p. °C	−83.1	−115	−87	−50
b.p. °C	19.4	−85	−67	−36

strength HCl<HBr<HI. Hydrogen fluoride, however, with a markedly higher bond strength and internal hydrogen bonding, is a weak acid only partially dissociated into ions in solution:

$$HF(g) + H_2O(l) \rightleftharpoons H_3O^+(aq) + F^-(aq)$$

Although only weakly acidic in aqueous solution, anhydrous hydrogen fluoride is a most dangerous liquid, causing severe burns to the flesh and even attacking glass and silica:

$$SiO_2(s) + 4HF(g) \rightarrow SiF_4(g) + 2H_2O(l)$$

The acid will react further with silicon tetrafluoride, forming hexafluorsilicic acid, H_2SiF_6, containing the complex ion $(SiF_6)^{2-}$.

The ability of hydrogen fluoride to attack glass is utilised in etching processes. The article to be marked is coated with wax, through which the design is cut, and is exposed to the vapour of hydrogen fluoride; finally the removal of the wax with hot water leaves the design etched into the glass.

The lower bond strengths in hydrogen bromide and hydrogen iodide also account for the greater ease with which they are oxidised, as observed in the account of their preparation above. A concentrated solution of hydriodic acid in fact acts as a reducing agent, being used, with red phosphorus as a catalyst, to reduce ethers to iodoalkanes.

$$R{-}O{-}R'(g) + 2HI(aq) \xrightarrow[\text{phosphorus}]{\text{red}} RI(l) + R'I(l) + H_2O(l)$$

In contrast, concentrated hydrochloric acid is only attacked by strong oxidising agents such as potassium manganate(VII) or manganese(IV) oxide.

$$2MnO_4^-(s) + 10HCl(aq) + 6H^+(aq) \rightarrow 2Mn^{2+}(aq) + 8H_2O(l) + 5Cl_2(g)$$

$$MnO_2(s) + 4HCl(aq) \rightarrow Mn^{2+}(aq) + 2Cl^-(aq) + 2H_2O(l) + Cl_2(g)$$

Aqueous solutions of all the hydrohalic acids exhibit the normal properties of dilute acids; they react with carbonates to release carbon dioxide and with reactive metals to release hydrogen, they turn blue litmus red and conduct an electric current. They all form constant-boiling mixtures with water (page 99), that of hydrogen chloride containing 22.4% of hydrogen chloride and boiling at 110°C, providing a method of obtaining solutions of approximately known molarity to be finally standardised against a primary standard such as sodium carbonate.

(c) Halogen oxides

Compounds of fluorine with oxygen are more correctly termed fluorides than oxides since fluorine is the more electronegative element. For the same reason they are not the anhydrides of corresponding acids. The best known is oxygen difluoride which is made by passing fluorine into dilute alkali:

$$2F_2(g) + 2OH^-(aq) \rightarrow F_2O(g) + H_2O(l) + 2F^-(aq)$$

The molecule has an angular structure similar to that of dichlorine oxide, which is one of a number of oxides of chlorine, all of which are explosively unstable, having positive heats of formation. The structures of the oxides of chlorine are given in Table 21.6.

Chlorine dioxide is paramagnetic, having an odd electron in its structure; it has some use in industry as an oxidising agent.

These oxides are all acid anhydrides, dissolving in water to give rise to various acids:

$$Cl_2O(g) + H_2O(l) \rightarrow 2HOCl(aq)$$

$$2ClO_2(g) + H_2O(l) \rightarrow HClO(aq) + HClO_4(aq)$$

$$Cl_2O_6(l) + H_2O(l) \rightarrow HClO_3(aq) + HClO_4(aq)$$

$$Cl_2O_7(g) + H_2O(l) \rightarrow 2HClO_4(aq)$$

Table 21.6 Structure of some oxides of chlorine

Oxide	Structure
Dichlorine oxide (gas)	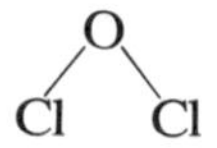
Chlorine dioxide (gas)	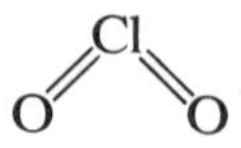
Chlorine hexoxide (liquid)	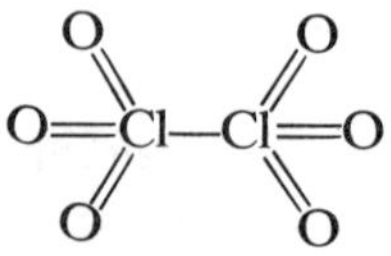
Chlorine heptoxide (gas)	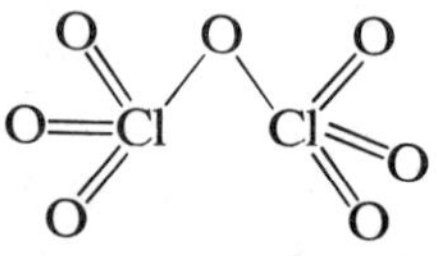

The oxides of bromine are also acid anhydrides and very unstable, while iodine forms a pentoxide derived from iodic(V) acid (HIO_3) which is stable up to 300°C. The white solid, I_2O_5, is a powerful oxidising agent, used in the estimation of small concentrations of carbon monoxide in the air.

$$I_2O_5(aq) + 5CO(g) \rightarrow I_2(s) + 5CO_2(g)$$

The liberated iodine is titrated against standard sodium thiosulphate solution (page 301).

(d) Oxoacids of chlorine, bromine and iodine and their salts

Iodic(V) acid, HIO_3, and the chloric acids are the more important of the halic acids. The chloric acids whose structures are given in Table 21.7 increase in acid strength as the proportion of oxygen in the molecule increases.

The halic(I) acids HClO, HBrO and HIO are formed when the halogens react with water; although only weak acids and too unstable to be isolated, they are oxidizing agents in solution:

$$2I^-(aq) + H^+(aq) + HOCl(aq) \rightarrow I_2(aq) + H_2O(l) + Cl^-(aq)$$

$$2Fe^{2+}(aq) + H^+(aq) + HOCl(aq) \rightarrow 2Fe^{3+}(aq) + H_2O(l) + Cl^-(aq)$$

Chlorine passed into dilute aqueous solutions of sodium hydroxide forms a mixture of the sodium salt of chloric(I) acid together with sodium chloride:

$$Cl_2(g) + 2NaOH(aq) \rightarrow Na^+ClO^-(aq) + Na^+Cl^-(aq) + H_2O(l)$$

Such solutions are the main components of domestic bleaches (Domestos, Parazone etc.)

The chlorate(I) ion is also formed when chlorine is passed over moist calcium hydroxide, producing a complex compound known as bleaching

Table 21.7 Structures of the chloric acids and chlorate ions

Acid		Ion
Oxochloric(I) acid (Chloric(I) acid)	H—O—Cl	 $\left(\overset{\cdot\cdot}{\underset{\cdot\cdot}{Cl}}{-}O \right)^-$
Dioxochloric(III) acid (Chloric(III) acid)	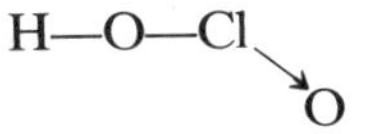 H—O—Cl, with →O	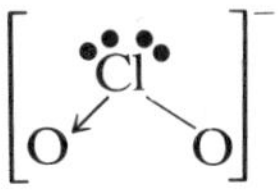 $\left[\overset{Cl}{O \quad O} \right]^-$
Trioxochloric(V)acid (Chloric(V) acid)	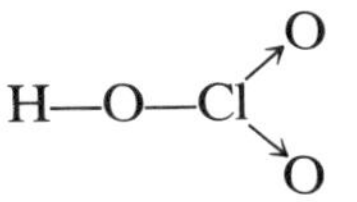 H—O—Cl, with O and O	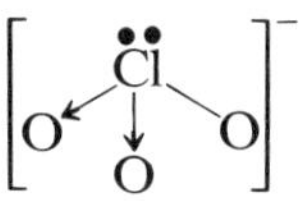 $\left[\overset{Cl}{O \quad \downarrow \quad O} \, O \right]^-$
Tetraoxochloric(VII) acid (Chloric(VII) acid)	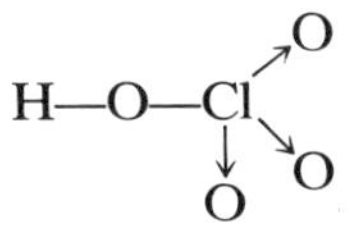 H—O—Cl, with O, O and O	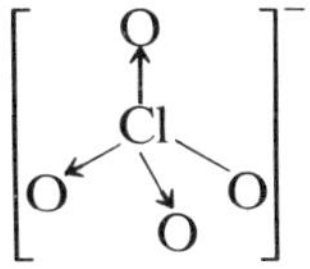 $\left[\overset{O}{\underset{O}{O \; Cl \; O}} \right]^-$

powder, from which chlorine is released by the addition of acid; this provides a useful method of storing chlorine. Bleaching powder is also used in the manufacture of trichloromethane.

Chlorate(I) ions from both the metal salts and the acid disproportionate on heating to a mixture of chloride and chlorate(V) ions:

$$3(OCl)^-(aq) \rightarrow 2Cl^-(aq) + (ClO_3)^-(aq)$$

Thus when chlorine is passed into hot concentrated solutions of sodium hydroxide a mixture of sodium chloride and sodium chlorate(V) is obtained.

$$3Cl_2(aq) + 6Na^+OH^-(aq) \rightarrow 5Na^+Cl^- + Na^+ClO_3^-(aq) + 3H_2O(l)$$

The sodium chlorate(V), used as a weed-killer, is separated by fractional crystallisation. Potassium chlorate(V), is used as the oxidising agent in matches and in some fireworks.

Both sodium chlorate(I) and sodium chlorate(V) are manufactured by the electrolysis of brine. The chlorine released at the anode dissolves in the alkaline solution resulting from the electrolysis (page 206); conditions are varied to produce the desired product.

Bromate(I) ions and iodate(I) ions disproportionate even in cold solutions. Thus iodine mixed with cold potassium hydroxide solution yields potassium iodate(V):

$$I_2(s) + 2K^+OH^-(aq) \rightarrow K^+I^-(aq) + K^+IO^-(aq) + H_2O(l)$$

$$3K^+IO^-(aq) \rightarrow 2K^+I^-(aq) + K^+IO_3^-(aq)$$

Potassium iodate(V) can be isolated in a very pure state for use as a primary standard in volumetric analysis; it releases iodine quantitatively from excess of acidified potassium iodide solution.

$$IO_3^-(aq) + 5I^-(aq) + 6H^+(aq) \rightarrow 3I_2(aq) + 3H_2O(l)$$

The concentration of sodium thiosulphate solutions can thus be determined by titration (page 301) against the iodine released by a known mass of the iodate.

All three halate(V) salts are decomposed by heat to yield oxygen; the ease of decomposition increases with increasing atomic mass of the halogen:

$$2KClO_3(s) \rightarrow 2KCl(s) + 3O_2(g)$$

If potassium chlorate(V) is carefully heated and kept just below the decomposition temperature the chlorate(V) ion disproportionates to yield a mixture of potassium chloride and potassium chlorate(VII).

$$4KClO_3(s) \rightarrow KCl(s) + 3KClO_4(s)$$

21.4   Interhalogen compounds

Covalent compounds containing atoms of two different halogens can be made by reacting the halogens together in the required proportions under suitable conditions. All the possible compounds with diatomic molecules have thus been prepared except one between iodine and fluorine.

The interhalogen compounds are polarised as a result of the different electronegativities of the various halogen atoms. Thus while an iodine molecule or a chlorine molecule is symmetrical, that of iodine monochloride has a dipole moment:

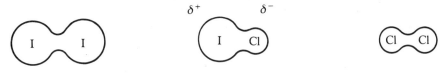

| Iodine molecule | Iodine monochloride molecule | Chlorine molecule |

The interhalogen compounds in Table 21.8 have all been prepared. They are volatile compounds of great reactivity; bromine trifluoride, BrF_3, one of the most reactive substances known, is a vigorous fluorinating agent. Iodine heptafluoride, IF_7, is less reactive than the other interhalogen compounds and is useful as a mild fluorinating agent; it is the only interhalogen compound in which the maximum covalency of one of the halogens is exerted.

Table 21.8 Interhalogens

ClF (gas)	BrF (gas)	
	BrCl (gas)	ICl (liquid)
ClF_3(gas)	BrF_3 (liquid)	ICl_3 (solid)
	BrF_5 (liquid)	IF_5 (liquid)
		IF_7 (gas)

21.5 Exceptional properties of fluorine and its compounds

It has been seen in this chapter that fluorine, typically as the head element of a group, shows some marked differences in nature from the rest of the elements in the group. These differences are summarised as:

(a) It has only one oxidation state of -1.

(b) It forms covalent compounds in which the other elements exert their maximum covalency (PF_5, SF_6, IF_7).

(c) The fluorides of non-metal elements are not as readily hydrolysed as the other non-metal halides, nor are fluorinated organic compounds hydrolysed by water or alkalis.

(d) As a result of its high electronegativity, fluorine forms ionic compounds with metals where other halides are predominantly covalent ($Al^{3+}(F^-)_3$, Ag^+F^-).

(e) Silver fluoride is soluble and calcium fluoride is insoluble, in contrast to the behaviour of the other halides of these metals.

(f) Fluorine combines directly with carbon and with nitrogen, whereas the other halogens do not; furthermore, nitrogen trifluoride, unlike the other nitrogen trihalides, is a stable exothermic compound.

(g) Hydrogen fluoride has an abnormally high boiling point compared to the other hydrogen halides, as a result of the hydrogen bonding induced by the extreme electronegativity of fluorine; similarly it has a much higher bond strength and forms only a weak acid in solution. Hydrogen fluoride forms acid salts of the type KHF_2^- as a result of hydrogen bonding.

Questions

1. Give the oxidation states of the halogen atoms in each of the following: $NaClO$ $KClO_4$ IF_7 $NaBrO_3$ I_2O_5 Cl_2O_7 KBr.
2. What is the effect of passing chlorine into (a) cold dilute sodium hydroxide solution, (b) hot concentrated sodium hydroxide solution? Give equations for each reaction.
3. Describe how you would test for the presence of chloride, bromide and iodide ions in solution.
4. (a) Illustrate the trends in the chemical properties of the halogens F, Cl, Br and I by considering how these elements react with (i) metals, (ii) hydrogen, (iii) aqueous sodium hudroxide.
   (b) Discuss the bonding in hydrogen fluoride and explain how this accounts for one physical property and one chemical property of this hydride which may be described as 'abnormal'.   (O)
5. State and explain what you would observe in the following experiments. Write balanced equations for the reactions that occur.
   (a) Aqueous chlorine is shaken with aqueous potassium iodide in the presence of tetrachloromethane (carbon tetrachloride) and the mixture is then allowed to stand.
   (b) Solid potassium iodide is heated with concentrated sulphuric(VI) acid.
   (c) Aqueous potassium chloride and aqueous silver nitrate are mixed and aqueous ammonia is then added.
   (d) Potassium chlorate(V) is slowly heated to approximately 450°C.   (C)
6. Comment on the following statements.
   (a) The hydrides of chlorine, bromine and iodine dissolved in water are strong acids, and each forms only one potassium salt; hydrogen fluoride forms a weaker acid, and forms both potassium fluoride and potassium hydrogen difluoride (KHF_2).
   (b) When chlorine is dissolved in cold sodium hydroxide solution and acidified silver nitrate solution added, only one half of the chlorine which has dissolved is precipitated as silver chloride. When the sodium hydroxide solution is hot, up to five-sixths of the chlorine can be thus precipitated.
   (c) Sulphuric acid is a weaker acid than hydrochloric acid, but hydrogen chloride is evolved when concentrated sulphuric acid is added to sodium chloride.
   (d) Bromine reacts rapidly with hex-1-ene (C_6H_{12}) in the dark, but light is necessary for its reaction with the isomer cyclohexane.   (L)

2

The *d*-block elements

Table 22.1 Physical data

Element	Sc	Ti	V	Cr	Mn	Fe	Co	Ni	Cu	Zn
Atomic number	21	22	23	24	25	26	27	28	29	30
Atomic radius (*nm*).	0.161	0.145	0.132	0.137	0.137	0.124	0.125	0.125	0.128	0.133
Ionic radius (*nm*) M^{2+}		0.090	0.088	0.080	0.088	0.076	0.074	0.072	0.069	0.074
Melting point K	1673	1950	2190	2176	1517	1812	1768	1728	1356	693
Density g/cm^3	3.10	4.43	6.07	7.19	7.21	7.87	8.79	8.90	8.92	7.13

22.1 General characteristics and trends in properties of the *d*-block elements

The *d*-block elements are those in the central block of the fourth, fifth and sixth periods between Group II and Group III. Up to argon (atomic number 18) the electrons of the successive elements have filled the 1*s*, 2*s*, 2*p*, 3*s* and 3*p* orbitals, making three short periods. The third quantum level has *d* orbitals available, but electrons do not enter these until the 4*s* shell is filled; hence in the fourth period potassium and calcium are followed by ten elements, scandium to zinc, whose electrons successively fill the five 3*d* orbitals. These ten elements all have very similar physical and chemical properties, modified by the number and arrangement of *d* electrons.

Similar *d*-block elements occur in the fifth and sixth periods, where the 4*d* and 5*d* orbitals are filled. The main part of this chapter will be concerned with the elements of the fourth period (the first long period).

The *d*-block elements are dense lustrous metals, with characteristic metallic properties, being malleable, ductile and good conductors of heat and electricity; they are not strongly basic, some having positive electrode potentials. Those that have unfilled or partially filled *d* orbitals or form ions in which the *d* orbitals are only partially filled have some special characteristics and are defined as *transition* metals. In the first long period these include the elements scandium to copper but excludes zinc which has all the *d* orbitals completed both in the metal atoms and in zinc(II) ions.

The presence of unpaired *d* electrons in the transition elements or their ions gives rise to *paramagnetism*, or magnetic permeability greater than in a vacuum. (Where paramagnetism is very strongly developed, as in iron, cobalt and nickel and their alloys, it is termed ferromagnetism.) The unpaired *d* electrons also have the ability to absorb light of definite wavelengths, with the result that hydrated ions of the transition metals are coloured, the colour being that of the reflected unabsorbed wavelengths and varying from metal to metal according to the number and location of unpaired d electrons. This effect is enhanced or modified by the presence of ligands other than water molecules surrounding the ions; thus $[Cu(H_2O)_4]^{2+}$ is blue while $[Cu(NH_3)_4]^{2+}$ is a more intense and deeper blue.

The transition metals form compounds in which they exhibit different oxidation states; the unpaired d electrons require only small amounts of energy to promote them to act as bonding electrons in addition to the $4s$ electrons in the outermost orbits, thus enabling the metals to form compounds in which some or all of their d electrons are utilized in bonding.

22.2   Electron configuration and oxidation states

The electron configurations and oxidation states of the elements scandium to zinc are given in Table 22.2, the principal oxidation states being in bold face. The convention of showing the first eighteen electrons $1s^2 2s^2 2p^6 3s^2 3p^6$ as 'argon core' or Ar is adopted in this table.

It is seen in Table 22.2 that the d orbitals are filled according to Hund's rule; that is, each successive orbital is occupied by a single electron before any orbital is filled by a second electron. The most stable arrangements of d electrons are those where either all the d orbitals are empty, or all are occupied by single electrons or all contain paired electrons. Thus in the metallic atoms of chromium and copper the $4s$ shells contain only one electron, the other being promoted to a d orbital.

Table 22.2

Element	Electron configuration after argon 3d electrons	4s electrons	Oxidation states
Scandium	Ar [_ _ _ _ ↑]	[↑↓]	**3**
Titanium	Ar [_ _ _ ↑ ↑]	[↑↓]	2 3 **4**
Vanadium	Ar [_ _ ↑ ↑ ↑]	[↑↓]	1 2 3 4 **5**
Chromium	Ar [↑ ↑ ↑ ↑ ↑]	[↑]	1 2 **3** 4 5 **6**
Manganese	Ar [↑ ↑ ↑ ↑ ↑]	[↑↓]	1 **2** 3 4 5 6 **7**
Iron	Ar [↑ ↑ ↑ ↑ ↑↓]	[↑↓]	1 **2 3** 4 5 6
Cobalt	Ar [↑ ↑ ↑ ↑↓ ↑↓]	[↑↓]	1 **2 3** 4 5
Nickel	Ar [↑ ↑ ↑↓ ↑↓ ↑↓]	[↑↓]	1 **2** 3 4
Copper	Ar [↑↓ ↑↓ ↑↓ ↑↓ ↑↓]	[↑]	**1 2** 3
Zinc	Ar [↑↓ ↑↓ ↑↓ ↑↓ ↑↓]	[↑↓]	**2**

Similarly the most stable oxidation states of manganese are +7 when all the *d* electrons are used in bonding and +2 when only the 4*s* electrons are used, leaving all five *d* orbitals singly filled. Iron(II) ions, in which only the 4*s* electrons are lost, are readily oxidised to iron(III) ions by the promotion of the sixth *d* electron to a bonding electron, leaving the five *d* orbitals singly filled.

Although copper(II) salts are more familiar in the laboratory, the *covalent* copper(I) compounds are more stable than the corresponding copper(II) compounds. Strong heating converts copper(II) oxides, sulphides and chlorides to those of copper(I):

$$4CuO(s) \rightarrow 2Cu_2O(s) + O_2(s)$$

$$2CuCl_2(s) \rightarrow 2CuCl(s) + Cl_2(g)$$

while copper(II) bromide decomposes to copper(I) bromide on quite gentle heating. Copper(II) iodide has not been isolated; the addition of potassium iodide to copper(II) sulphate solution resulting in the precipitation of copper(I) iodide and the release of iodine.

$$2Cu^{2+}(aq) + 4I^-(aq) \rightarrow 2CuI(s) + I_2(aq)$$

22.3   Physical properties and uses of the transition metals

The atomic radius of a transition element of the first long period, in general, decreases with increasing atomic number; the extra *d* electrons present in the atoms have little screening effect, resulting in the increased positive charge of the atoms of each successive element pulling the 4*s* electrons closer to the nucleus.

Since the size of the atoms decreases as their mass increases, there is a considerable increase in density of the elements from scandium to copper, which like cobalt and nickel, is of high density. The ready availability of the outer electrons leads to high electrical conductivity and strong metallic bonding (page 61), the latter accounting for the high melting point, tensile strength, malleability and ductility of the elements. These properties vary and determine the special uses made of each metal.

Titanium, vanadium and chromium are less malleable and ductile than iron, cobalt and nickel, while manganese is fairly brittle. The harder and more brittle metals are widely used for hardening steels, each conferring particular properties on the steel. Thus titanium, of low density and high melting point, is used for the manufacture of light, strong steels, very resistant to corrosion, for spacecraft, supersonic aircraft and nuclear reactors. Vanadium and chromium steels are exceptionally hard, and are used for high-speed tools and ball-bearings. Chromium, with a pleasing appearance, is used to plate iron. All are used in the production of the many types of stainless steel.

Iron, in conjunction with various proportions of carbon, is of course the basic component of steel, the essential material for all forms of heavy engineering from reinforced concrete work to shipbuilding and, with the modifications outlined in the previous paragraph, for the production of most of the equipment of modern civilisation. Wrought iron is still used for ornamental work.

As mentioned above, iron, cobalt and nickel are all ferromagnetic; magnets with special characteristics of strength and permanency are made by alloying iron with the requisite proportions of cobalt, chromium and tungsten.

Nickel is alloyed with iron and chromium to make the nichrome wire used in the elements of electric fires, with copper and iron to make Monel metal for turbine blades and propellors, and with copper and zinc to make nickel silver for coinage.

Copper as the metal is both useful and ornamental. Its low resistance and great ductility make it the most suitable material we have for electrical wiring, for which it must be of great purity. The green domes of buildings such as the Regency palaces in Brighton are of copper, which becomes coated with a layer of basic sulphate or carbonate. Brass is an alloy of zinc and copper, and bronze is tin and copper. A wide range of engineering components and domestic utensils and ornaments are made from these alloys.

22.4   General chemistry and selected compounds

(a) General chemistry

Of the first series of transition metals, scandium to copper, the most reactive is manganese, followed by iron and chromium.

Titanium and vanadium are very hard and resistant to chemical attack at ordinary temperatures, but combine with oxygen, nitrogen and chlorine at raised temperatures and react with hot concentrated nitric acid to form titanium(IV) oxide, TiO_2, and vanadium(V) oxide, V_2O_5. Titanium(IV) oxide is an inert substance of brilliant whiteness, used as a pigment; vanadium(V) oxide is a valuable catalyst.

Chromium is resistant to chemical attack and is not tarnished in air, while manganese is reactive, forming trimanganese tetraoxide,* Mn_3O_4, on heating in oxygen or air, and reacting not only with dilute acids but also with warm water, yielding hydrogen and compounds containing Mn(II) ions. Chromium reacts only slowly with cold dilute acids, although it is attacked by hot acids and will decompose steam when heated. Both metals are oxidised to sulphates by hot concentrated sulphuric acid. Concentrated nitric acid attacks manganese but renders chromium passive.

Iron, while less reactive than manganese, is still a fairly reactive metal, combining with oxygen to form triiron tetraoxide, Fe_3O_4, and reacting with many other non-metal elements, such as sulphur or chlorine, with some vigour when heated.

$$Fe(s) + S(s) \rightarrow FeS(s)$$

$$2Fe(s) + 3Cl_2(g) \rightarrow 2FeCl_3(s)$$

Iron is oxidised by hot concentrated sulphuric acid but, like chromium, is

*Systematically manganese(II) dimanganese(III) oxide.

rendered passive by nitric acid. It reacts with water only at red heat to yield Fe_3O_4 in a reversible reaction.

$$3Fe(s) + 4H_2O \rightleftharpoons Fe_3O_4(s) + 4H_2(g)$$

With dilute non-oxidising acids the metal dissolves to form hydrogen and iron(II) ions, but the latter are oxidised rapidly to iron(III) ions in solution, especially at the surface where the concentration of dissolved oxygen is higher.

$$Fe(s) + 2HCl(aq) \rightarrow FeCl_2(aq) + H_2(g)$$

$$Fe^{2+}(aq) \rightarrow Fe^{3+}(aq) + e^-(aq)$$

Pure iron does not corrode in dry air, but its rusting in damp conditions is a phenomenon only too well known, costing the country many millions of pounds every year both in damage and in preventive measures.

The process is electrolytic (Fig. 22.1) and is accelerated if the moisture from the air is electrically conducting, as in places near the sea where the air is full of salt spray or in winter when icy roads are treated with salt. The presence of carbon dioxide or other acidic compounds is also contributory. It has been found that iron rusts when the moisture on its surface contains differing concentrations of oxygen; electrons are transferred from the iron at points where the oxygen content is low to points at which the oxygen content is higher. The reactions are complex, but can be represented by the half reactions:

$$Fe(s) \rightarrow Fe^{2+} + 2e^- \quad \text{(anodic area)}$$

$$O_2(g) + 2H_2O(l) + 4e^- \rightarrow 4OH^- \quad \text{(cathodic area)}$$

Thus iron is oxidised to Fe^{2+} ions at points of *lower* oxygen content; diffusion of Fe^{2+} ions and OH^- ions results in the formation of iron(II) hydroxide, $Fe(OH)_2$. Oxidation to iron(III) oxide follows, the rust appearing in the hydrated form as $Fe_2O_3 \cdot xH_2O$. Thus a piece of iron half immersed in water rusts most rapidly at points a little *below* the surface where less dissolved oxygen is present: similarly a partially painted piece of iron will start to rust under the paint, where the oxygen content is less.

The protection of iron by galvanising, covering with a layer of zinc metal,

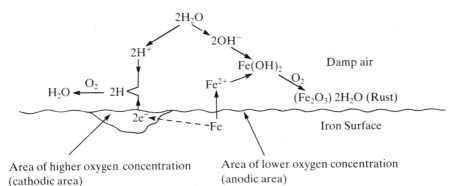

Area of higher oxygen concentration (cathodic area)

Area of lower oxygen concentration (anodic area)

Fig. 22.1 Rusting of iron

is particularly effective because zinc has a more negative potential than iron:

$$Zn^{2+}(aq) + 2e^- \rightleftharpoons Zn(s) \qquad E^\ominus = -0.76 \text{ V}$$

$$Fe^{2+}(aq) + 2e^- \rightleftharpoons Fe(s) \qquad E^\ominus = -0.44 \text{ V}$$

Thus if the zinc is scratched so that the surface of the iron is exposed the zinc will corrode rather than the iron since the reaction taking place will be:

$$Zn(s) + Fe^{2+}(aq) \rightleftharpoons Zn^{2+}(aq) + Fe(s)$$

This is the principle on which the use of sacrificial anodes is based: strips of metal such as magnesium or aluminium, of more negative electrode potential than iron, are affixed to the hulls of ships or to buried pipelines, and become corroded away in place of the iron. On the other hand, if the iron is protected by a layer of tin, as for canned goods (zinc being too poisonous to use), a scratch in the tin exposing the iron to the air actually promotes corrosion since iron has a more negative electrode potential than tin.

Cobalt, nickel and copper are progressively less reactive than iron, as might be expected from their electrode potentials (Table 22.3). All three metals resist corrosion by air and are not attacked by water.

Table 22.3   Electrode potentials for the metals Cr–Cu

Electrode system	Cr^{2+}, Cr	Mn^{2+}, Mn	Fe^{2+}, Fe	Co^{2+}, Co	Ni^{2+}, Ni	Cu^{2+}, Cu
$E^\ominus V$	−0.91	−1.91	−0.44	−0.28	−0.25	+0.34

Copper, having a positive electrode potential, is not attacked by dilute acids, while cobalt and nickel react only slowly. Cobalt and nickel, like iron, are both rendered passive by concentrated nitric acid, but copper is oxidised to copper nitrate, the nature of the reaction depending on the concentration of the acid. Moderately concentrated nitric acid (50% by volume) forms copper(II) nitrate with the evolution of nitrogen monoxide, while the concentrated acid yields nitrogen dioxide.

$$3Cu(s) + 8HNO_3(aq) \rightarrow 3Cu(NO_3)_2(aq) + 4H_2O(l) + 2NO(g)$$

$$Cu(s) + 4HNO_3(l) \rightarrow Cu(NO_3)_2 + 2H_2O(l) + 2NO_2(g)$$

Hot concentrated sulphuric acid is similarly reduced by copper with the evolution of sulphur dioxide and the formation of copper(II) sulphate.

$$Cu(s) + 2H_2SO_4(l) \rightarrow CuSO_4(aq) + 2H_2O(l) + SO_2(g)$$

The transition metals in general are not attacked by alkalis either in the fused condition or in aqueous solution; nickel crucibles are used for laboratory work involving the use of alkalis.

Transition metals form interstitial compounds with non-metals, particularly where the non-metal atom is small; hydrides, carbides and nitrides occur in which the small non-metal atoms fit into the spaces in the metallic lattice without any great distortion of the crystal form. The compounds are not of any definite composition; hydrogen can be recovered by heating the hydride under reduced pressure.

The carbides and the nitrides of the metals are very hard, of high melting point and resistant to corrosion, while still retaining their electrical conductivity.

(b) Selected compounds

The acid/base character of the oxides of the first series of the transition elements changes from left to right across the series. The more stable oxides of the metals to the left of the series are the acidic oxides of higher oxidation number such as vanadium(V) and manganese(VII) oxide; lower oxides, which are either amphoteric or basic, are less stable.

Towards the right of the series basic oxides predominate, with cobalt, nickel and copper forming only basic oxides.

Table 22.4 Oxides of the first series of transition metals

	TiO	VO	CrO	MnO	FeO	CoO	NiO	Cu_2O CuO
Sc_2O_3	Ti_2O_3 TiO_2	V_2O_3 VO_2 V_2O_5	Cr_2O_3 CrO_3	Mn_2O_3 MnO_2 MnO_3 Mn_2O_7	Fe_2O_3	Co_2O_3	Ni_2O_3	

The most important oxide of vanadium is vanadium(V) oxide, V_2O_5, containing vanadium in its only stable oxidation state of $+5$. The oxide is acidic, dissolving in alkalis to form the vanadate(V) ion, VO_3^-.

A solution of ammonium vanadate(V) can be reduced progressively to illustrate the other oxidation states of vanadium. The solution is acidified and heated in the presence of granulated zinc; as the reduction proceeds the colour of the solution changes from yellow through green to blue, then to green and finally violet as the oxidation state of vanadium changes from $+5$ through $+4$ and $+3$ to $+2$ (see Table 22.5).

Table 22.5

Ion	VO_3^-	VO^{2+}	V^{3+}	V^{2+}
Oxidation state	$+5$	$+4$	$+3$	$+2$
Colour	yellow	blue	green	violet

If the solution is filtered from the zinc and left to stand, the colour changes are reversed as the unstable vanadium(II) ions are progressively oxidised back to the vanadium(V) state by the action of dissolved oxygen.

Vanadium(V) oxide is an important catalyst, notably for the oxidation of sulphur dioxide in the contact process for the manufacture of sulphuric acid (Chapter 32).

Whereas the highest oxide of vanadium is the most stable, this is not so with chromium. Chromium(VI) oxide is a powerful oxidising agent which loses oxygen on heating to form the more stable chromium(III) oxide.

$$4CrO_3(s) \rightarrow 2Cr_2O_3(s) + 3O_2(g)$$

It is a wholly acidic oxide, red in colour, very soluble in water to give a strongly acidic solution with powerful oxidising properties known as 'chromic acid'. The acid, H_2CrO_4, has not been isolated, but sodium and potassium salts containing the chromate(VI) ion, CrO_4^-, are familiar in the laboratory as oxidising agents. Solutions of potassium and sodium chromate(VI) are yellow; on acidification the colour changes to orange as a result of the formation of the dichromate(VI) ion.

$$2CrO_4^{2-}(aq) + 2H^+(aq) \rightleftharpoons Cr_2O_7^{2-}(aq) + H_2O(l)$$

The colour change is reversed on the addition of alkali, with the reformation of CrO_4^{2-} ions.

Alkali metal chromates are crystallised from alkaline solutions, while the dichromates can be obtained from acid solutions. The use of sodium dichromate is usually favoured for organic oxidations (page 451) since it is more soluble; during oxidation the dichromate(VI) ion is reduced to the green chromium(III) ion:

$$Cr_2O_7^{2-}(aq) + 14H^+(aq) + 6e^- \rightleftharpoons 2Cr^{3+}(aq) + 7H_2O(l)$$

The standard electrode potential for the half reaction is +1.33 V.

Solutions of both potassium and sodium chromate(VI) are used in analysis to identify metals which form insoluble chromates. The chromates are precipitated from solutions containing the metal ions; barium chromate(VI) is pale yellow, lead chromate(VI) is bright yellow and silver chromate(VI) is red.

$$CrO_4^{2-}(aq) + 2Ag^+(aq) \rightarrow Ag_2CrO_4(s)$$

Chromium(III) oxide is amphoteric, dissolving in acids to form stable salts and in alkalis to form chromate(III) ions.

$$Cr_2O_3(s) + 6HCl(aq) \rightarrow 2CrCl_3(aq) + 3H_2O(l)$$

$$Cr_2O_3(s) + 6OH^-(aq) + 3H_2O(l) \rightarrow 2[Cr(OH)_6]^{3-}$$

The Cr^{3+} ion is hydrated in solution, and chromium(III) salts crystallise as hydrates, which have a number of isomeric forms (page 332). Chromium(III) sulphate forms alums with alkali metal sulphates (page 254).

Manganese, with main oxidation states of +2 and +7, forms stable salts containing Mn^{2+} ions. These, even when hydrated, are not highly coloured, the $[Mn(H_2O)_6]^{2+}$ ion being pale pink. They can be made by dissolving manganese(II) oxide, which is wholly basic, in the appropriate acid.

$$MnO(s) + H_2SO_4(aq) \rightarrow MnSO_4(aq)$$

The most important oxide of manganese is the naturally occurring dioxide (manganese(IV) dioxide) which is found as pyrolusite. The black solid is a powerful oxidising agent, releasing chlorine from hydrochloric acid:

$$4HCl(aq) + MnO_2(s) \rightarrow MnCl_2(aq) + Cl_2(g) + 2H_2O(l)$$

It also acts as a catalyst, promoting the rapid decomposition of both potassium chlorate and hydrogen peroxide:

$$2H_2O_2(aq) \xrightarrow{\text{MnO}_2} 2H_2O(l) + O_2(g)$$

Fused with potassium hydroxide and an oxidising agent, such as potassium chlorate, manganese(IV) oxide is converted to potassium manganate(VI), a green solid.

$$3MnO_2(s) + 6OH^-(s) + ClO_3^-(s) \rightarrow 3MnO_4^{2-} + 3H_2O(l) + Cl^-(aq)$$

In aqueous solution the manganate(VI) ion is only stable in strongly alkaline conditions; if the melt is crushed and boiled in water, through which carbon dioxide is bubbled to remove the OH^- ions, the manganate(VI) ions disproportionate to manganate(VII) ions and manganese(IV) oxide (page 201):

$$3MnO_4^{2-}(aq) + 2H_2O(l) \rightleftharpoons 2MnO_4^-(aq) + MnO_2(s) + 4OH^-(aq)$$

Crystalline potassium manganate(VII) (potassium permanganate) can be obtained by filtering the mixture to remove the unwanted manganese(IV) oxide, and evaporating the purple solution.

Potassium manganate(VII) is an important oxidising agent, being reduced to Mn^{2+} ions by the gain of five electrons.

$$MnO_4^-(aq) + 8H^+(aq) + 5e^- \rightarrow Mn^{2+}(aq) + 4H_2O(l) \qquad E^{\ominus} = +1.51\ V$$

It oxidised alkenes to dihydric alcohols (page 353) and concentrated hydrochloric acid to chlorine:

$$2KMnO_4(s) + 16HCl(aq) \rightarrow 2KCl(aq) + 2MnCl_2(aq) + 8H_2O(l) + 5Cl_2(g)$$

In volumetric analysis it is used to estimate iron salts, oxalates and oxalic acid, its intense colour allowing it to be used as its own indicator. It cannot easily be obtained pure enough to be used as a primary standard, and its solution must be standardised before use in analysis.

$$MnO_4^-(aq) + 5Fe^{2+}(aq) + 8H^+(aq) \rightarrow Mn^{2+}(aq) + 5Fe^{3+}(aq) + 4H_2O(l)$$

$$2MnO_4^-(aq) + 5(COO)_2{}^{2-}(aq) + 16H^+(aq)$$
$$\rightarrow 2Mn^{2+}(aq) + 10CO_2(g) + 8H_2O(l)$$

The latter reaction takes place at about 60°C; it is slow at first but as it proceeds it is catalysed by the presence of Mn^{2+} ions (autocatalysis). The solutions must be acidified with sulphuric acid, since the manganate(VII) ions oxidise hydrochloric acid.

The acid from which MnO_4^- ions are derived has not been isolated, and the oxide Mn_2O_7 is dangerously unstable. The trend towards greater stability of the basic oxides continues through iron, cobalt and nickel to copper; the latter three form no acidic oxides or oxoanions corresponding to those of chromium and manganese. Potassium and barium salts containing the ferrate(VI) ion, FeO_4^{2-}, can be prepared. The ion is dark red in colour and a powerful oxidising agent; neither the acid nor the oxide corresponding to the ion are known. Iron(III) oxide and iron(II) oxide are virtually wholly basic, as is the main oxide of cobalt, CoO, while nickel(II) oxide is completely basic.

The chief oxides of iron are Fe_2O_3, occurring as haematite, and Fe_3O_4, occurring as magnetite. The latter, known as triirontetraoxide, or, systematically, as iron(II)diiron(III) oxide, is a mixed oxide containing FeO and Fe_2O_3 in equal proportions: its magnetic properties were known in very early times by the Chinese, who recognised a stone which would give an attraction to a needle. The needle could then be used, when suspended, to guide ships to the south.

Haematite, iron(III) oxide, is the main ore from which iron is obtained in the blast furnace. The pure compound is used to polish jewellery (jewellers' rouge); it is a rust-coloured solid with an ionic structure, $(Fe^{3+})_2(O^{2-})_3$, analogous to that of aluminium oxide. In the laboratory it can be obtained as the residue from heating iron(II) sulphate.

$$2FeSO_4(s) \rightarrow Fe_2O_3(s) + SO_2(g) + SO_3(s)$$

As with other transition metals the Fe^{3+} ion is small and highly charged, and thus tends to be readily hydrated. Crystalline salts of iron(III) compounds all contain water of crystallisation. Iron(III) sulphate-9-water, $Fe_2(SO_4)_3 \cdot 9H_2O$, is prepared by oxidising iron(II) sulphate with hydrogen peroxide in acid solution:

$$2FeSO_4(aq) + H_2SO_4(aq) + H_2O_2(aq) \rightarrow Fe_2(SO_4)_3(aq) + 2H_2O(l)$$

Iron(III) chloride crystallises from solution as $FeCl_3 \cdot 6H_2O$. The hydrated ion in solution, $[Fe(H_2O)_6]^{3+}$, is violet coloured; this violet colour can also be seen in iron(III) alums such as $KFe(SO_4)_2 \cdot 12H_2O$ (page 254). Iron(III) chloride solutions develop a yellow colour as a result of the hydrolysis of the hydrated ion:

$$[Fe(H_2O)_6]^{3+}(aq) + H_2O(l) \rightleftharpoons [Fe(H_2O)_5OH]^{2+}(aq) + H_3O^+(aq)$$

$$[Fe(H_2O)_5OH]^{2+}(aq) + H_2O(l) \rightleftharpoons [Fe(H_2O)_4(OH)_2]^+(aq) + H_3O^+(aq)$$

$$[Fe(H_2O)_4(OH)_2]^+(aq) + H_2O(l) \rightleftharpoons [Fe(H_2O)_3(OH)_3](s) + H_3O^+(aq)$$

The hydrolysis leads eventually to the precipitation of iron(III) hydroxide unless the solution is made strongly acid to keep the equilibrium to the left.

Anhydrous iron(III) chloride is made by passing dry chlorine over iron filings (page 310); the black iridescent crystals are deliquescent and rapidly turn yellow on exposure to moist air with the formation of the hydrated ion. Anhydrous iron(III) chloride, like aluminium chloride, is covalent in structure, forming dimerised molecules, Fe_2Cl_6, in the vapour state.

By contrast anhydrous iron(II) chloride is mainly ionic. It is prepared by passing dry hydrogen chloride over heated iron filings;

$$2HCl(g) + Fe(s) \rightarrow FeCl_2(s) + H_2(g)$$

Iron(II) salts are formed in solution by the action of dilute acids on iron, the lower oxidation state resulting from the reducing conditions.

$$Fe(s) + H_2SO_4(aq) \rightarrow FeSO_4(aq) + H_2(g)$$

Fe^{2+} ions are hydrated in the same way as Fe^{3+} ions; $[Fe(H_2O)_6]^{2+}$, with six liganding water molecules, is green in colour. The iron(II) ion is normally so readily oxidised to iron(III) ions that stored samples of the salts rapidly become contaminated; however a double sulphate $FeSO_4(NH_4)_2SO_4 \cdot 6H_2O$ can be obtained by crystallisation from solutions containing equimolar concentrations of iron(II) sulphate and ammonium sulphate. This double sulphate is, unlike iron(II) sulphate-7-water, neither efflorescent nor oxidised on exposure to the atmosphere, and, as it retains the normal reactions of iron(II) sulphate, it is used as a primary standard in the titrimetric estimation of potassium manganate(VII) solutions.

$$5Fe^{2+}(aq) + MnO_4^-(aq) + 8H^+(aq) \rightarrow 5Fe^{3+}(aq) + Mn^{2+}(aq) + 4H_2O(l)$$

Cobalt and nickel, consistent with their main oxidation states, both form salts containing tripositive and dipositive ions. The more stable sets of salts are those of cobalt(II) and nickel(II), where the ions are formed by the loss of the two 4*s* electrons. However a number of cobalt(III) compounds can be made in which the Co^{3+} ions are stabilised by complexing ligands. These complexes are similar to those of chromium(III) ions (page 332).

Copper, the last of the first row of *d*-block elements to be classed as a transition metal, is the only one to form compounds in which it has an oxidation state of +1. Although covalent compounds of copper(I) are more stable than those of copper(II), the Cu^+ ion disproportionates in aqueous solution, forming metallic copper and copper(II) ions.

$$Cu^{2+}(aq) + e^- \rightleftharpoons Cu^+(aq) \qquad E^\ominus = +0.15 \text{ V}$$

$$Cu^+(aq) + e^- \rightleftharpoons Cu(s) \qquad E^\ominus = +0.52 \text{ V}$$

$$2Cu^+(aq) \rightleftharpoons Cu(s) + Cu^{2+}(aq) \qquad E^\ominus = +0.37 \text{ V}$$

Thus if copper(I) oxide is dissolved in sulphuric acid, copper is precipitated and a solution of copper(II) sulphate results.

$$Cu_2SO_4(aq) \rightarrow CuSO_4(aq) + Cu(s)$$

Copper(I) oxide, Cu_2O, is formed as a red solid when Cu^{2+} ions in Fehling's or Benedict's solutions are reduced by aldehydes or sugars. (page 389).

Copper(I) halides can be prepared as solids which remain stable if kept dry and out of contact with air. Copper(I) iodide is precipitated when potassium iodide solution is added to copper(II) sulphate solution (see Section 22.2 above). Copper(I) chloride is made by boiling a mixture of copper turnings and copper(II) chloride in concentrated hydrochloric acid. When the mixture is poured into excess cold water the chloride is precipitated as a white solid.

$$Cu(s) + CuCl_2(aq) \rightarrow 2CuCl(s)$$

Copper(II) oxide, CuO, the familiar black oxide of copper, is obtained by heating copper(II) carbonate, which occurs naturally as malachite.

Copper, having a positive electrode potential (+0.34 V), is not attacked by dilute acids; its salts are best prepared by dissolving the oxide or carbonate in acid:

$$CuO(s) + H_2SO_4(aq) \rightarrow CuSO_4(aq) + H_2O(l)$$

$$CuCO_3(s) + 2HCl(aq) \rightarrow CuCl_2(aq) + H_2O(l) + CO_2(g)$$

The characteristic colour of copper(II) salts in solution stems from the blue hydrated copper(II) ion $[Cu(H_2O)_6]^{2+}$; the addition of aqueous sodium hydroxide to these solutions results in the formation of copper(II) hydroxide as a gelatinous blue precipitate.

Table 22.7

Hydroxide	Colour	Further reactions
Chromium(III) hydroxide $Cr(OH)_3$	Pale green	Dissolves in excess NaOH to form deep green chromite ion $[Cr(OH)_6]^{3-}$
Manganese(II) hydroxide $Mn(OH)_2$	White	Rapidly darkens on oxidation to $Mn_2O_3(s)$
Iron(II) hydroxide $Fe(OH)_2$	Green (gelatinous)	Darkens on oxidation
Iron(III) hydroxide $Fe(OH)_3$	Red (gelatinous)	
Cobalt(II) hydroxide $Co(OH)_2$	Violet	Soluble in ammonia solution forming orange/brown complex ion $Co(NH_3)_6^{2+}$
Nickel(II) hydroxide $Ni(OH)_2$	Green	Soluble in ammonia solution forming blue complex ion $[Ni(NH_3)_6]^{2+}$
Copper(II) hydroxide $Cu(OH)_2$	Blue	Soluble in ammonia solution forming deep blue complex ion $[Cu(NH_3)_4]^{2+}$

Hydroxides of characteristic colour precipitated from aqueous solutions of transition metal salts can be used to identify the metal ions present (Table 22.7).

22.5   Complex ions and coordination compounds

The small, highly charged, ions of transition metals have a strong attraction for lone pairs of electrons which is enhanced by the presence of vacant d orbitals; as a result they form many complex ions in which the ligands may be water molecules, ammonia molecules, chloride ions, etc., as described in Chapter 4. Thus the ions are hydrated in aqueous solution by water molecules which are carried on crystallisation into the solid lattice as water of crystallisation. The water molecules liganding the ion can be displaced by other liganding species if the latter form a more stable complex, the stability of any particular ion being described in terms of its stability constant K (page 58). Thus the formation of tetraaquacopper(II) ions is represented by the equation:

$$Cu^{2+}(aq) + 4H_2O(l) \rightleftharpoons [Cu(H_2O)_4]^{2+}(aq)$$

the stability constant for this reaction being:

$$K = \frac{\{[Cu(H_2O)_4]^{2+}\}}{[Cu^{2+}][H_2O]^4}$$

Addition of ammonia drop by drop to a solution of copper(II) ions causes a change of colour from pale blue to deep royal blue as the liganding water molecules are successively replaced by ammonia molecules to give finally tetraamminecopper(II) ions, $[Cu(NH_3)_4]^{2+}$:

$$[Cu(H_2O)_4]^{2+}(aq) + NH_3(aq) \rightleftharpoons [Cu(H_2O)_3NH_3]^{2+} + H_2O(l)$$

$$[Cu(H_2O)_3NH_3]^{2+}(aq) + NH_3(aq) \rightleftharpoons [Cu(H_2O)_2(NH_3)_2]^{2+} + H_2O(l)$$

$$[Cu(H_2O)_2(NH_3)_2]^{2+}(aq) + NH_3(aq) \rightleftharpoons [Cu(H_2O)(NH_3)_3]^{2+} + H_2O(l)$$

$$[Cu(H_2O)(NH_3)_3]^{2+}(aq) + NH_3(aq) \rightleftharpoons [Cu(NH_3)_4]^{2+} + H_2O(l)$$

A stability constant for each of these reactions can be written; the complete displacement of the water molecules by ammonia molecules is accounted for by the higher stability constant for the formation of tetraamminecopper(II) ion:

$$K = \frac{\{[Cu(NH_3)_4]^{2+}\}}{[Cu^{2+}][NH_3]^4}$$

Similarly the addition of concentrated hydrochloric acid to an equal quantity of copper(II) sulphate solution causes a change of colour from blue through green to yellow as the water molecules are successively displaced from $[Cu(H_2O)_4]^{2+}$ by the chloride ions with the formation of $[Cu(Cl)_4]^{2-}$; the stability constant of the tetrachlorocuprate(II) ion being greater than that of the tetraaquacopper(II) ion.

In general the ligands forming the complex of higher stability constant will replace those forming the complex of lower stability constant. This rule may be reversed where water molecules are the ligands; if the yellow solution containing $[Cu(Cl)_4]^{2-}$ ions is greatly diluted, the overwhelming concentration of water molecules now present results in the displacement of the chloride ions and the reappearance of the blue $[Cu(H_2O)_4]^{2+}$ ions.

As a consequence of the great stability of some complex ions, the reactions of the central ion of the complex are modified; iron(III) ions in solution normally give a blood-red colour on addition of ammonium thiocyanate solution as a result of the formation of $[Fe(CNS)]^{2+}$ ions.

$$Fe^{3+}(aq) + (CNS)^-(aq) \rightarrow [Fe(CNS)]^{2+}(aq)$$

However, solutions of potassium hexacyanoferrate(III), $K_3Fe(CN)_6$, give no such colouration since the iron(III) ion is protected by the liganding CN^- ions in the complex $[Fe(CN)_6]^{3-}$; the addition of sodium fluoride solution to solutions of iron(III) ions has a similar effect, a very stable complex being formed with the fluoride ions as ligands, $[FeF_4]^-$.

Hydrated chromium(III) chloride crystallises in three different forms, each of which reacts differently with silver nitrate(V) solution. Solutions of the grey-blue crystals $[Cr(H_2O)_6]^{3+}(Cl^-)_3$ react with silver nitrate(V) with the precipitation of all three chloride ions as silver chloride; the light green $[Cr(H_2O)_5Cl]^{2+}(Cl^-)_2 \cdot H_2O$ reacts to precipitate two-thirds of the chloride ions, while the dark green $[Cr(H_2O)_4Cl_2]^+(Cl^-)_2 \cdot 2H_2O$ yields only a third of its chloride ions as silver chloride.

As was seen in Chapter 4, the geometric shape of the complex ion varies with the nature and number of ligands present; while both tetraaminecop-per(II) and tetraaquacopper(II) ions have a square planar configuration,* the presence of four ligating species usually gives rise to a tetrahedral shape, as in tetrachlorocobaltate(II)ions:

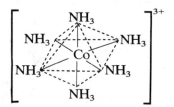

The presence of six ligating species results in an octahedral shape for the ion, as in hexaamminecobalt(III) ions:

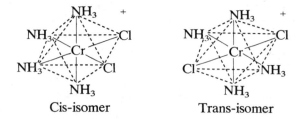

Where such an ion contains more than one type of ligand, geometrical isomerism (Chapter 24) arises. The complex ion $[Cr(NH_3)_4Cl_2]^+$ has a cis-form containing two adjacent chloride ions and a trans-form containing two chloride ions diagonally opposed to one another:

Cis-isomer   Trans-isomer

Both these isomers have a plane of symmetry, hence they are not optical isomers.

Optical isomerism does however arise when the ligating species is bidentate. The chelated ion (page 56) $Cr(en)_3^{3+}$, where (en) represents an ethane-1,2-diamine molecule is an example of this.

---

* These ions are stabilised in aqueous solution by the presence of two additional liganded water molecules resulting in an octahedral ion:

Fig. 22.2 Mirror images of chelated Cr^{3+} ions

There are a number of large planar complexes in which the ligands are organic molecules of some size. Both iron(II) and nickel(II) occur as the central ions of such complexes. In haem (page 60) the iron(II) ion is surrounded by pyrrole rings:

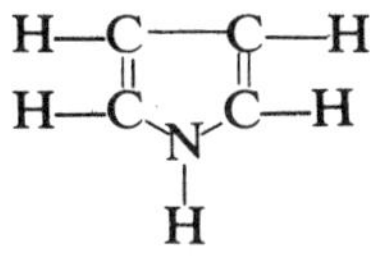

The precipitation of a red solid by the addition of butanedione dioxime to solutions of nickel salts made alkaline with ammonia constitutes a sensitive test for the presence of nickel(II) ions. The precipitated solid occurs as flat plate-like crystals resulting from the formation of nickel butanedione dioxime (dimethylglyoxime):

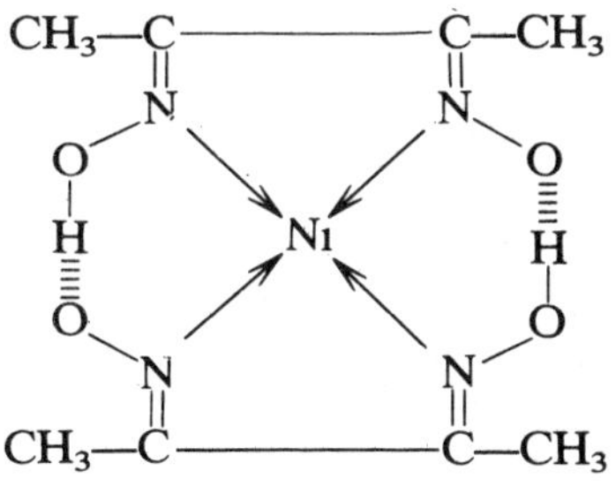

The brilliant colour of the dye Monastral Blue results from the formation of a coordinated copper compound.

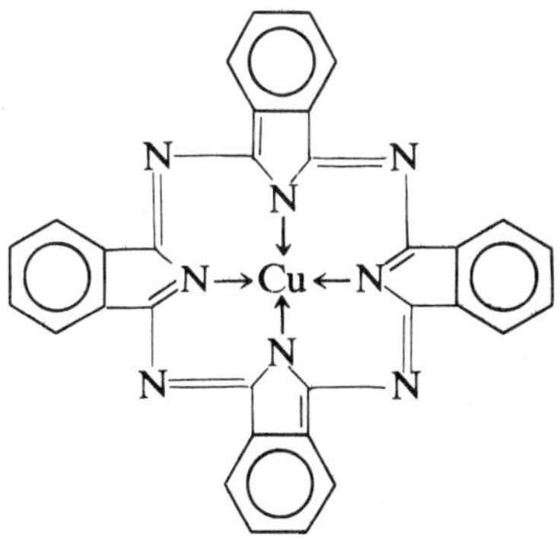

Coordinate compounds of transition metals with carbon monoxide contain the metals in the elemental state with zero oxidation number. Pentacarbonyl iron(0) and tetracarbonyl nickel(0) are both toxic liquids. The coordinate bonds to the metal atoms are made by the donation of the lone pairs of electrons on the carbon atoms of the carbon monoxide.

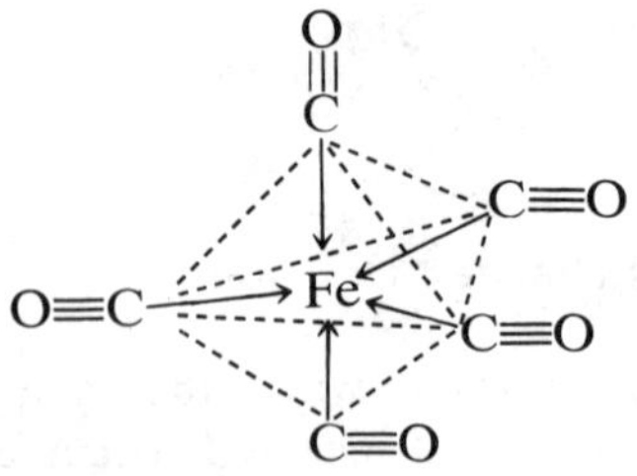

Trigonal bipyramidal
pentacarbonyl iron(O)

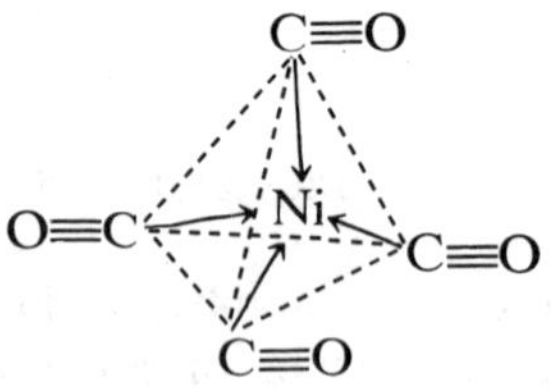

Tetrahedral
tetracarbonyl nickel(O)

Fig. 22.3

Tetracarbonyl nickel(0) is used to manufacture very pure nickel; the carbonyl is readily decomposed by heat to yield the metal and carbon monoxide.

22.6 Catalytic activity

Transition metals and their compounds are well-known for their **catalytic activity**; finely divided iron promotes the reaction between hydrogen and nitrogen in the Haber process for ammonia (Chapter 32), nickel catalyses the hardening of fats and oils for margarine by hydrogen, and platinum, dispersed on asbestos fibre, catalyses many gas reactions.

The catalytic activity of the metals depends on the adsorption of the reacting gases on the metal surface with the formation of bonds between the metal and the adsorbed molecules. This weakens the internal bonding of the molecules, reducing the energy required to disrupt them and enabling them to react more readily. Such catalysts are used in the finely divided state to gain the greatest possible surface area; they are easily 'poisoned' by the adsorption of impurities, necessitating careful purification of the reactant gases.

The catalytic action of transition metal ions frequently depends on the variable valency of the metals, resulting in the formation and decomposition of intermediate compounds between the ions and the reactants. Many of the reactions catalysed are redox reactions; thus the oxidation of iodides by potassium peroxodisulphate takes place only slowly:

$$2I^-(aq) + S_2O_8^{2-}(aq) \rightarrow I_2(aq) + 2SO_4^{2-}(aq) \quad \text{(slow)}$$

The addition of iron(III) ions to the solution results in a much faster reaction; the catalytic effect could be the result of the reaction taking place in two fast stages:

(i) The oxidation of iodide ions to iodine by the Fe^{3+} ions which are thereby reduced to Fe^{2+} ions:

$$2I^-(aq) + 2Fe^{3+}(aq) \rightarrow I_2(s) + 2Fe^{2+}(aq) \quad \text{(fast)}$$

(ii) The Fe^{2+} ions now available are rapidly oxidised back to Fe^{3+} ions with the reduction of the peroxodisulphate ions:

$$2Fe^{2+}(aq) + S_2O_8^{2-}(aq) \rightarrow 2Fe^{3+}(aq) + 2SO_4^{2-}(aq) \quad \text{(fast)}$$

The overall result is the production of iodine molecules and sulphate ions as in the single uncatalysed reaction, but much more rapidly.

22.7 Zinc, cadmium and mercury

These metals occupy positions at the end of the three rows of *d*-block elements, and may be considered as a separate group. They are not classified as transition elements since they have completely filled *d* orbitals both in the metallic state and as ions.

Zn   2.8.18.2

Cd   2.8.18.18.2

Hg   2.8.18.32.18.2

All form doubly charged positive ions, as do the Group II metals, to which they bear some resemblance; differences arise because, while zinc, cadmium and mercury each have eighteen electrons in the penultimate shell, the Group II metals have only eight electrons.

The zinc group metals form smaller ions than those of corresponding members of Group II, as a result of the lower screening power of the shells containing eighteen electrons. This smaller size of potential ions results in a greater tendency for the compounds of zinc, cadmium and mercury to be covalent; the oxides, halides and sulphides are all almost wholly covalent. Where ions are formed, as in the nitrates and sulphates, they enter into a great many complexes, usually with coordination numbers of 2 and 4; $[Zn(NH_3)_2]^{2+}$ is linear while $[CdI_4]^{2-}$ is tetrahedral.

Zinc and cadmium, whose chemistry is very similar, usually occur together as sulphides or carbonates in ores such as zinc blende. They have negative electrode potentials and are attacked by dilute non-oxidising acids, while mercury has a positive electrode potential, is resistant to attack by these acids, and has a rather different chemistry.

Zinc and cadmium are white lustrous metals which, although reactive, are resistant to corrosion; zinc is protected by a layer of carbonate and cadmium by a layer of oxide. Both metals are used as protective coatings for iron; zinc being the covering for galvanised iron, while cadmium is used particularly to protect radio and other electrical components. Both metals form a number of valuable alloys; zinc and copper together make brass while cadmium added to copper makes it tougher and stronger. Pure zinc is used as the combined cell containers and negative electrodes of dry batteries; cadmium/nickel plates are used in alkaline storage batteries.

Zinc and cadmium dissolve in dilute non-oxidising acids to yield salts.

$$Zn(s) + 2HCl(aq) \rightarrow ZnCl_2(aq) + H_2(g)$$

They reduce both concentrated and dilute nitric acid to oxides of nitrogen.

The addition of aqueous sodium hydroxide to solutions of zinc and cadmium salts results in the precipitation of hydroxides.

$$Zn^{2+}(aq) + 2OH^-(aq) \rightarrow Zn(OH)_2(s)$$

$$Cd^{2+}(aq) + 2OH^-(aq) \rightarrow Cd(OH)_2(s)$$

Zinc hydroxide is amphoteric and redissolves in excess sodium hydroxide, but cadmium hydroxide is wholly basic and does not redissolve in excess alkali.

$$Zn(OH)_2(s) + 2OH^-(aq) \rightarrow [Zn(OH)_4]^{2-}(aq)$$

Both hydroxides dissolve in ammonia solution with the formation of ammino complexes.

$$Cd(OH)_2(s) + 4NH_3(aq) \rightarrow [Cd(NH_3)_4]^{2+}(aq) + 2OH^-(aq)$$

Zinc oxide is also amphoteric, and zinc metal, like aluminium, dissolves in sodium hydroxide solution liberating hydrogen and forming a zincate ion.

$$Zn(s) + 2OH^-(aq) + 2H_2O(l) \rightarrow [Zn(OH)_4]^{2-}(aq) + H_2(g)$$

Zinc chloride is a deliquescent solid used as a flux in soldering and as a timber preservative; zinc sulphate is used in the electrolytic galvanising of iron; cadmium sulphate is used in the construction of the Weston cell which provides a reference standard for e.m.f. determinations (page 193).

Mercury, being the only liquid metal, has special uses on this account; it is used in barometers, thermometers, and electrical switches. It forms compounds known as amalgams with other metals; sodium and zinc amalgams are used as reducing agents in the laboratory, tin amalgam is used as a dental filling. Many organo-mercury compounds are used as agricultural fungicides and as timber preservatives.

Mercury forms two sets of compounds, with oxidation states of +2 and +1 respectively.

Dimercury(I) compounds contain the ion $(Hg-Hg)^{2+}$, in which there is a metal-to-metal bond. Mercury(I) chloride, which has a linear structure, is used in the construction of the calomel electrode, which is more convenient as a standard against which to measure electrode potential than the hydrogen electrode. (page 193)

Mercury(II) compounds include the red sulphide occurring as the ore, cinnabar. Mercury(II) chloride, which is almost wholly covalent, is soluble in hot water but only sparingly so in cold water. It is readily reduced to mercury by tin(II) chloride (page 269).

$$SnCl_2(aq) + 2HgCl_2(aq) \rightarrow Hg_2Cl_2(s) + SnCl_4(aq)$$

$$SnCl_2(aq) + Hg_2Cl_2(s) \rightarrow 2Hg(l) + SnCl_4(aq)$$

Questions

1. Make a table representing the arrangement of electrons in the $3d$ and $4s$ orbitals of the following metals and their ions in the $+1, +2$ and $+3$ oxidation state (where appropriate): chromium, manganese, iron, cobalt, nickel and copper.
2. Explain why zinc is not classed as a transition metal although it is a d-block element. Compare the chemistry of zinc and magnesium to illustrate why zinc may be classed as a Group IIB metal.
3. The elements scandium to zinc are generally listed in a separate 'block' of the Periodic Table.

(*a*) Review the major characteristics common to this block of elements which are not generally shown by elements in other blocks of the Periodic Table. Explain how the chemical properties of these elements or their compounds can be understood in terms of electronic structure.

(*b*) Illustrate these characteristics by reference to ONE member of the series scandium to zinc, and discuss critically the extent to which this member exhibits all the characteristics outlined in your answer to (*a*).

(L)

4. (*a*) What is the electronic structure of the manganese atom?

(*b*) How many unpaired electrons are there in the manganese atom?

(*c*) What is the highest valency state exhibited by manganese?

(*d*) What is the most stable oxidation state exhibited by manganese in aqueous solution?

(*e*) State two properties (other than variable valency) characteristic of transition elements which are exhibited by manganese. (O)

5. Interpret the following observations, writing balanced chemical equations where possible and indicating the composition of any complex ions that may be formed.

(i) Copper(II) hydroxide dissolves in aqueous sulphuric acid to give a pale blue solution but gives a yellow solution when dissolved in concentrated hydrochloric acid.

(ii) An aqueous solution of copper(II) sulphate reacts with an excess of potassium iodide to give a white precipitate and a red-brown coloured solution. The red-brown coloured solution can be decolourised with sodium thiosulphate(VI).

(iii) When an aqueous solution of ammonia is gradually added to a pale blue aqueous solution of copper(II) sulphate, a pale blue precipitate, A, forms which redissolves in an excess of aqueous ammonia to give a deep blue solution, B. Reduction of B with hydrazine sulphate gives a colourless solution, C. When C is acidified with dilute sulphuric acid, a pale blue solution (as in A) and a dark brown precipitate, D, are formed. D does not dissolve in dilute non-oxidising acids. (C)

6. (*a*) When manganese(IV) oxide is fused with potassium hydroxide in the presence of potassium nitrate and the cooled residue extracted with dilute sulphuric acid, a purple solution A is obtained. A becomes a green solution B when treated with concentrated potassium hydroxide solution but this green solution becomes a purple solution C when chlorine is bubbled through it.

(i) What are the formulae of, and oxidation states of the manganese in, the species responsible for the colours in solutions A and B?

(ii) Write equations for the conversions of solution A into solution B and of solution B into solution C.

(*b*) When manganese(IV) oxide is heated with concentrated hydrochloric acid, a gas is evolved. Identify this gas and write an equation for the reaction. (JMB)

7. (*a*) Analysis of a complex salt $Co(NH_3)_x Cl_y$ yielded the following results by mass:

$$Co = 23.55\% \qquad NH_3 = 33.93\% \qquad Cl = 42.51\%$$

Calculate the values for *x* and *y*.

(b) Outline how the percentage values quoted above for ammonia and chlorine might have been determined in the laboratory.

(c) Cobalt is a *transition metal*. What is meant by this term?
   Explain why transition elements

(i) form stable complex ions very easily,

(ii) produce coloured ions.

(d) Name two catalysts containing transition elements and give for each one an example of its industrial use.   (AEB 1978)

Part V
Carbon Chemistry

23 Hydrocarbons

23.1 Introduction

Carbon forms a vast number of covalent compounds ranging from the gas methane, with one carbon atom in the molecule, to the complex proteins with more than a million atoms in the molecule. Since these compounds were first obtained from plants or animals, they were thought to be produced by some 'vital force', and were termed 'organic' in contrast to such compounds as metallic salts which were 'inorganic'. In 1829 Wöhler synthesised urea in the laboratory from 'inorganic' chemicals, showing that there was no 'vital' difference between 'organic' and 'inorganic' compounds after all. However, the term 'organic chemistry' was retained as a matter of convenience to denote the chemistry of the covalent carbon compounds.

To understand something of the nature of carbon chemistry it is best to start with simple molecules composed of carbon and hydrogen only. Some of these *hydrocarbons* are obtained in quantity by the fractional distillation of crude oil (page 479), ranging from gases through light petrols to heavy oils and waxes, leaving the tarry bitumens in the residue. Hydrocarbons from crude oil are used not only as fuels and lubricants, but also as the basic materials of the petrochemical industry which provides a great range of products including plastics, nylon, insecticides, varnishes, paints and drugs. The simplest hydrocarbon of all is methane, which occurs widely as natural gas, as firedamp in coal mines and as marsh gas bubbling up from swamps and bogs. The formula of methane is CH_4, and its molecule has the characteristic tetrahedral shape given by the symmetrical distribution of the covalent bonds between carbon and hydrogen (page 51).

Fig. 23.1   Methane molecule

It is the first member of the homologous series called the *alkanes*. A *homologous* series is a set of compounds all having similar structures and similar chemical properties, the molecular composition varying from member to member by one atom of carbon and two atoms of hydrogen. The physical properties show a regular change reflecting the regular increase in molecular mass.

Many other homologous series are derived from the alkanes by the substitution of atoms of other elements or groups of atoms known as functional groups (Chapter 25), which give the series their characteristic properties.

Compounds derived from the alkanes are known as *aliphatic* compounds, they include alcohols, acids, nitrogen compounds and many others.

Another set of compounds, the *aromatic* compounds, are derived from the *arenes*, which are hydrocarbons based on benzene, C_6H_6, which has a ring structure.

Some of the reactions and properties of these compounds are described in this and further chapters on carbon chemistry; to help understand them a short summary of types of reaction and reaction mechanisms is given below.

23.2   Types of reaction

Organic compounds undergo reactions of four main types.

(a) Addition reactions

Addition reactions are those in which two or more molecules combine together to give a single product. They are typical of unsaturated compounds containing a double bond. Alkenes react with hydrogen by addition to form alkanes:

$$CH_2{=}CH_2 + H_2 \rightarrow CH_3CH_3$$

Aldehydes and ketones form addition compounds with hydrogen cyanide:

$$(CH_3)_2C{=}O + HCN \longrightarrow (CH_3)_2C\overset{\displaystyle OH}{\underset{\displaystyle CN}{\big<}}$$

Many polymerisation reactions are addition reactions.

(b) Substitution reactions

When an atom or a functional group in a molecule is replaced by a different atom or functional group the reaction is said to be a substitution reaction. 1-Bromobutane reacts with sodium hydroxide solution to yield butan-1-ol by the substitution of the bromine atom by a hydroxyl group.

$$C_4H_9Br(l) + OH^-(aq) \rightarrow C_4H_9OH(aq) + Br^-(aq)$$

Aromatic compounds undergo many substitution reactions in which one hydrogen atom in the benzene ring is replaced by a different atom or group of atoms; the nitration of benzene results in the replacement of a hydrogen atom by a nitro group:

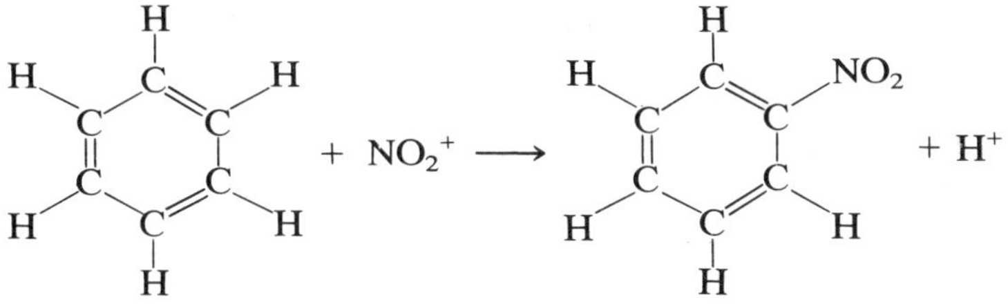

benzene   nitrobenzene

(c) Elimination reactions

Elimination reactions are those in which a small molecule is removed from a larger one leaving an unsaturated compound; they are the reverse of addition reactions. For example, the alkyl alcohols can be dehydrated by passing them over heated aluminium oxide; the products are alkenes and the eliminated molecule is water.

$$CH_3CH_2CH_2OH \xrightarrow{Al_2O_3} CH_3CH=CH_2 + H_2O$$

In some cases the elimination reaction is the second stage of an overall reaction that starts with an addition reaction. For example, the so-called *condensation* reactions of aldehydes and ketones follow this pattern.

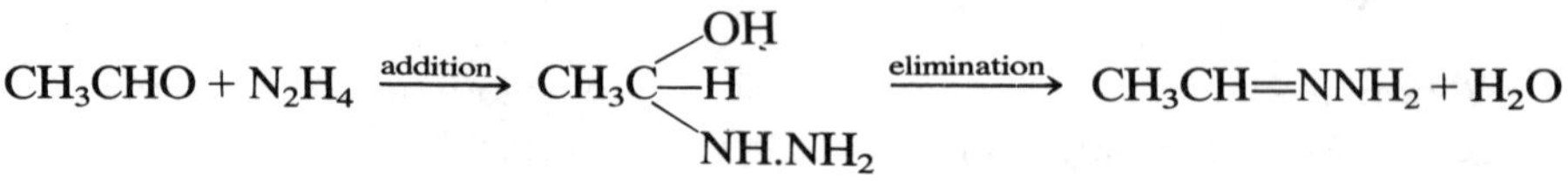

$$CH_3CHO + N_2H_4 \xrightarrow{addition} CH_3C\overset{OH}{\underset{NH.NH_2}{-}}H \xrightarrow{elimination} CH_3CH=NNH_2 + H_2O$$

Condensation polymers (page 413) are also formed by this mechanism.

(d) Rearrangement reactions

These may involve a rearrangement of the carbon skeleton or the migration of a functional group or an aklyl or aryl radical to a new position in the molecule. An example is the conversion of ammonium cyanate to urea.

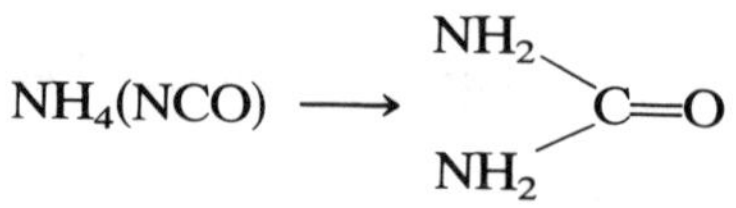

$$NH_4(NCO) \longrightarrow \overset{NH_2}{\underset{NH_2}{>}}C=O$$

The manufacture of phenol and propanone from (1-methylethyl) benzene involves the formation of an intermediate compound which undergoes rearrangement to yield the required products:

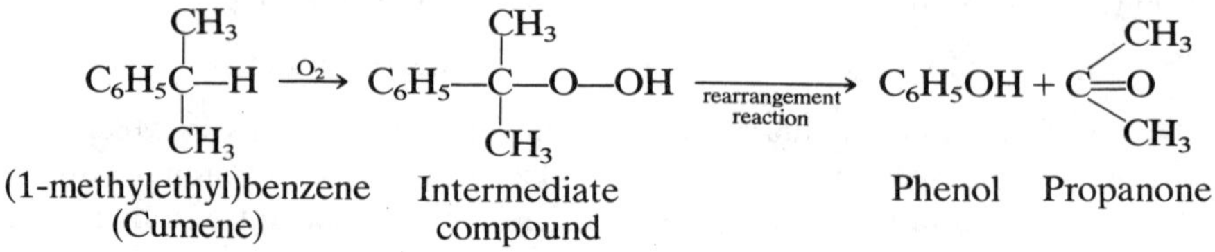

$$C_6H_5\overset{CH_3}{\underset{CH_3}{C}}H \xrightarrow{O_2} C_6H_5\overset{CH_3}{\underset{CH_3}{C}}-O-OH \xrightarrow[reaction]{rearrangement} C_6H_5OH + \overset{CH_3}{\underset{CH_3}{C}}=O$$

(1-methylethyl)benzene   Intermediate   Phenol   Propanone
(Cumene)   compound

23.3   Terminology for reaction mechanisms

Organic reactions involve the breaking of covalent bonds followed by the making of new bonds. The reaction may take place in a single step or in a series of two or more steps; the way in which it takes place is called the reaction mechanism. A study of the rate of reaction is often used to determine the nature of the steps (Chapter 9). The terms used in describing mechanisms are given below.

(a) Homolytic fission

The homolytic fission of a covalent bond between two atoms in a molecule is a symmetrical fission in which each atom retains one electron from the bond, forming free radicals. For example, the bond between the two carbon atoms in ethane may be broken to yield two methyl radicals each containing an unpaired electron.

$$H_3C - CH_3 \rightarrow CH_3\cdot + CH_3\cdot$$

Other examples are:

$$Cl_2 \rightarrow Cl\cdot + Cl\cdot$$

$$CH_3Cl \rightarrow CH_3\cdot + Cl\cdot$$

In each case the $\cdot$ represents an unpaired electron.

(b) Heterolytic fission

A covalent bond such as that between two carbon atoms or a carbon atom and another atom X can be broken *heterolytically*; this is an unsymmetrical splitting in which both electrons of the bond are retained by one of the atoms. The fission may occur in two ways; the carbon atom may retain both electrons, thus forming a negatively charged *carbanion*,

$$-\overset{|}{\underset{|}{C}}-X \rightarrow -\overset{|}{\underset{|}{C}}:^- + X^+$$

or the atom X may retain both electrons, leaving a positively charged *carbonium* ion.

$$-\overset{|}{\underset{|}{C}}-X \rightarrow -\overset{|}{\underset{|}{C}}{}^+ + :X^-$$

(c) Homolytic reagents

Homolytic reagents have a single unpaired electron which can be used to form a covalent bond; they include radicals such as $CH_3\cdot$, $C_6H_5\cdot$, $H\cdot$ and $Cl\cdot$, formed as a result of homolytic fission.

(d) Nucleophilic reagents

These are substances which can donate a pair of electrons to electron-deficient atoms such as the carbon atoms in carbonium ions. They include neutral molecules with an unshared pair of electrons like NH_3, H_2O and alcohols; other nucleophilic reagents are negatively charged ions such as OH^-, CN^-, Cl^- and carbanions.

(e) Electrophilic reagents

As their name suggests, these are reagents which can accept a pair of electrons to form a covalent bond; they include molecules such as SO_3, BF_3 and $AlCl_3$, and molecules like Cl_2 and Br_2 that are easily polarised. Positive ions such as H_3O^+, $NO_2{}^+$ and carbonium ions are also electrophilic reagents.

(f) The inductive effect

The electrons forming a covalent bond between unlike atoms are not shared equally; there is an attraction of the electron pair towards the more electronegative atom. Thus, in an alkyl halide, the electrons are attracted away from the carbon towards the halide atom, giving rise to a dipole moment (page 47).

The effect, which is permanent, is represented either by an arrowhead on the bond or by denoting a fractional charge on each of the atoms:

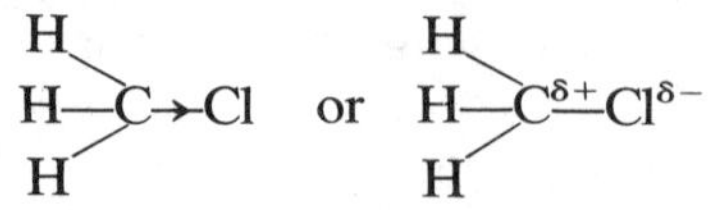

$$
\begin{array}{cc}
\text{H} & \text{H} \\
\text{H---C} \rightarrow \text{Cl} & \text{or} \quad \text{H---C}^{\delta+}\text{Cl}^{\delta-} \\
\text{H} & \text{H}
\end{array}
$$

Atoms and groups which withdraw electrons from the carbon atom to which they are attached are said to have a $-I$ effect; they include Cl, Br, I, OH, CN, C=O, COOH, C_6H_5 and NO_2 groups. Those groups that are electron-donating ($+I$ effect) are fewer; they include CH_3, C_2H_5, C_3H_7 etc.

(g) Curved arrows

These are used to denote the movement of a pair of electrons during a reaction. The tail of the arrow is placed where the electrons come from and the head at the place where they go to. For example the hydrolysis of methyl bromide:

$$OH^- + CH_3Br \rightarrow CH_3OH + Br$$

is represented as:

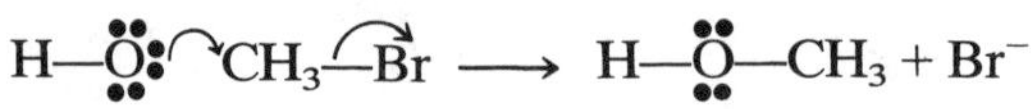

$$\text{H---Ö:} \frown \text{CH}_3 \frown \text{Br} \longrightarrow \text{H---Ö---CH}_3 + \text{Br}^-$$

23.4   The alkanes

This homologous series comprises compounds all of general formula C_nH_{2n+2}; each member differs from the previous one by the addition of a —CH_2— group, and the carbon atoms are linked together as shown in Table 23.1.

The series continues through hexane, heptane, octane, nonane ... up to long chain compounds such as $C_{22}H_{46}$ which are waxy solids. These are the so-called straight-chain hydrocarbons—in reality the tetrahedral distribution of the carbon bonds gives rise to a zig-zag chain:

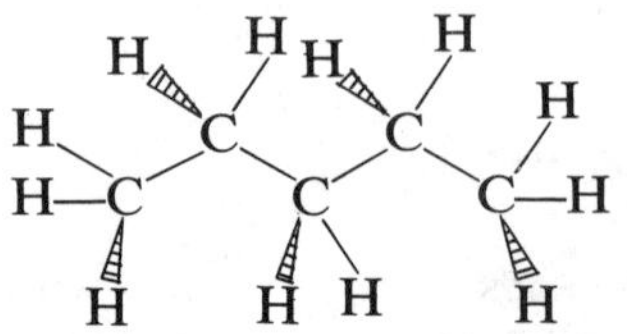

The names of the alkanes are related to the number of carbon atoms present. Thus the prefix eth- indicates the presence of two carbon atoms and

Table 23.1

Alkane	Molecular formula	Structural formula	b.p. °C	State
Methane	CH_4	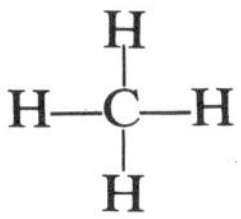	−162	
Ethane	C_2H_6	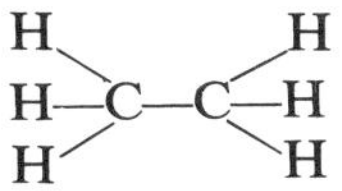	−89	Gases
Propane	C_3H_8	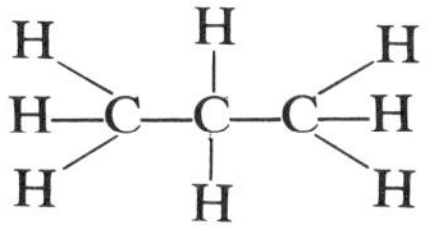	−45	
Butane	C_4H_{10}	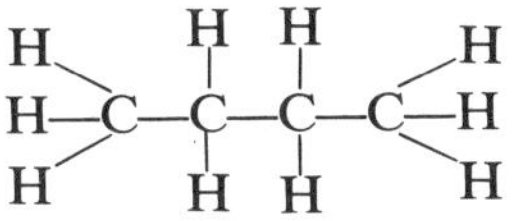	0	
Pentane	C_5H_{12}	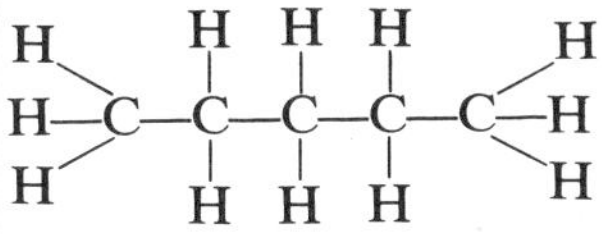	36	Liquid

the prefix pent- the presence of five carbon atoms. The *alkyl* groups derived from the alkanes, $-CH_3$, $-C_2H_5$, $-C_3H_7$ ... are called methyl, ethyl, propyl ... and so on as the number of carbon atoms in the chain increases.

When there are four or more carbon atoms in the chain the possibility of branched chain structures arises. These compounds, which have the same molecular formula but different structural formulae, are called *structural isomers* (Chapter 24). Pentane, C_5H_{12}, has three isomeric structures:

(a) Normal pentane

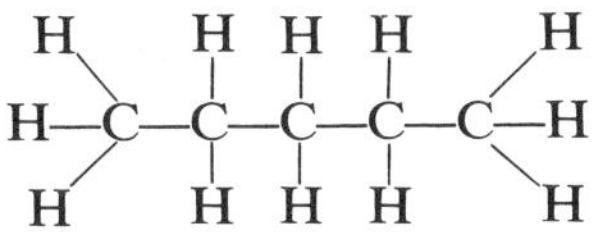

(b) 2-methylbutane

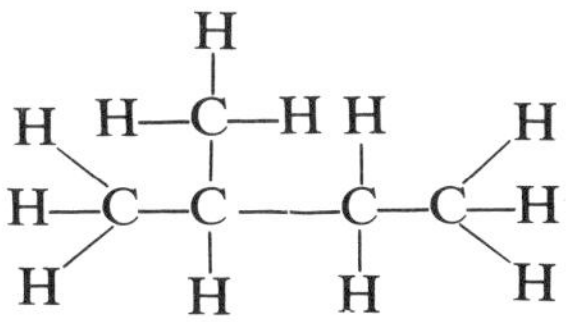

(c) 2,2-dimethylpropane

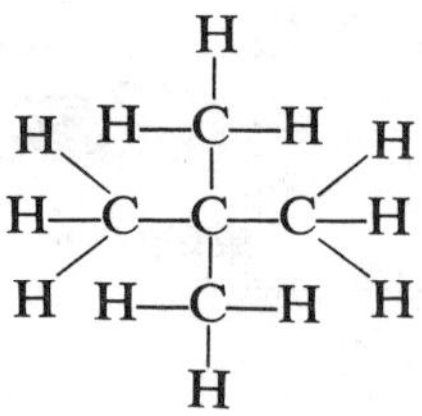

$$\begin{array}{c} \text{H} \\ | \\ \text{H} \quad \text{H--C--H} \quad \text{H} \\ \diagdown \quad | \quad \diagup \\ \text{H--C--C--C--H} \\ \diagup \quad | \quad \diagdown \\ \text{H} \quad \text{H--C--H} \quad \text{H} \\ | \\ \text{H} \end{array}$$

Clearly the naming of these compounds needs some explanation. All these isomers contain five carbon atoms and can therefore be called pentanes. To distinguish them a systematic method of naming is used:

(i) The longest consecutive chain of carbon atoms is chosen. Thus (b) becomes a butane ($4 \times$ C) and (c) becomes a propane ($3 \times$ C). The carbon atoms in these chains are given numbers 1, 2, 3, 4 for (b) and 1, 2, 3 for (c).

(ii) The structure is examined for the presence of 'side-groups'—groups replacing a hydrogen atom attached to one of the carbon atoms in the main chain. In this case (b) has a methyl group attached to carbon atom '2' and (c) has two methyl groups attached to carbon atom '2'. Then, following the system of naming, (b) becomes 2-methylbutane because the methyl group is attached to the second carbon atom of a butane chain; the molecular structure would be identical if the methyl group were attached to carbon '3', but the lower number is always chosen for the naming. It makes no difference whether the side-group is drawn above or below the chain. (c) becomes 2,2-dimethylpropane. The prefix di- indicates the presence of two methyl groups and the 2,2- shows that both are attached to the '2' carbon atom of the propane molecule.

A few more examples, including different side-groups, will help to make the system clear, but models of the various structures should be studied if it is to be fully understood (Table 23.2). Other examples will be explained as they arise.

All the alkanes, whether solid, liquid or gas, are colourless and odourless compounds which are completely immiscible with, and less dense than, water. As each carbon atom in the molecule is bonded by a shared pair of electrons to four other atoms, the molecules are said to be *saturated*; these saturated compounds are relatively unreactive.

A further series of saturated hydrocarbons, the cycloalkanes, have ring structures in which the two end carbon atoms of an alkane molecule are joined, with the elimination of two hydrogen atoms. Their fully saturated nature can be seen from the structural diagrams, (Fig. 23.2).

Two of the more important cycloalkanes are cyclopropane, an anaesthetic, and cyclohexane, which is the raw material for the manufacture of some types of nylon. Above cyclohexane the rings are unstable.

Unlike benzene, the molecules of cyclohexane are not planar; the tetrahedral distribution of the carbon bonds leads to molecules which are either boat-shaped or chair-shaped:

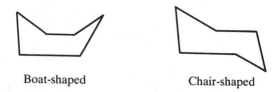

Boat-shaped   Chair-shaped

Table 23.2 Isomeric structures

Structure	Name
	2-methylbutane
	2,3-dimethylbutane
	2-chloropropane
	1-bromo-2-chloropropane

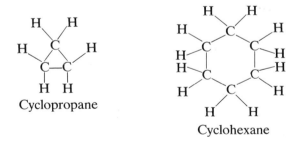

Cyclopropane   Cyclohexane

Fig. 23.2 Cycloalkanes

23.5 Reactions of alkanes

(a) Burning

Complete combustion of alkanes yields carbon dioxide and water; in a plentiful supply of air the lower members burn with a blue flame.

$$CH_4 + 2O_2 \rightarrow CO_2 + 2H_2O \quad \Delta H^{\ominus}_{298} = -890 \text{ kJ mol}^{-1}$$

$$C_9H_{20} + 14O_2 \rightarrow 9CO_2 + 10H_2O \quad \Delta H^{\ominus}_{298} = -6125 \text{ kJ mol}^{-1}$$

CH_4 is natural gas and C_9H_{20} is a molecule typical of paraffin oil. The large quantities of heat generated in these reactions make the hydrocarbons excellent fuels; however, the carbon dioxide and water produced account for the stuffiness of unventilated rooms warmed by oil heaters.

(b) Halogenation

Since no more atoms can be added to the saturated hydrocarbons, *halogenation* takes place by a substitution reaction in which a hydrogen atom is replaced by a halogen atom. These reactions take place only slowly unless activated by sunlight (ultraviolet light); under such conditions alkanes react with chlorine, bromine and, less readily, with iodine.

Chlorine reacts with methane to form a series of compounds in which successive hydrogen atoms are replaced by chlorine atoms:

$$CH_4 + Cl_2 \rightarrow CH_3Cl + HCl$$
(Chloromethane)

$$CH_3Cl + Cl_2 \rightarrow CH_2Cl_2 + HCl$$
(Dichloromethane)

$$CH_2Cl_2 + Cl_2 \rightarrow CHCl_3 + HCl$$
(Trichloromethane)

$$CHCl_3 + Cl_2 \rightarrow CCl_4 + HCl$$
(Tetrachloromethane)

The reaction is explosive in bright sunlight, and a mixture of products is obtained which are not easy to separate, so this is not a good method of preparing the individual compounds.

The reaction takes place by a free radical mechanism leading to a chain reaction. The ultraviolet light activates the chlorine molecule, causing homolytic fission:

$$Cl_2 \xrightarrow{\text{u.v. light}} Cl\cdot + Cl\cdot$$

The chlorine free radical is very reactive and attacks a methane molecule, producing a methyl radical which in turn attacks another chlorine molecule, giving rise to a further $Cl\cdot$ free radical:

$$Cl\cdot + CH_4 \rightarrow CH_3\cdot + HCl$$

$$CH_3\cdot + Cl_2 \rightarrow Cl\cdot + CH_3Cl$$

The regenerated $Cl\cdot$ free radical can now attack another methane molecule and a chain of reactions has been started which will continue until a chain-breaking reaction occurs, usually the reaction together of two free radicals:

$$Cl\cdot + Cl\cdot \rightarrow Cl_2 \quad \text{or} \quad Cl\cdot + CH_3\cdot \rightarrow CH_3Cl$$

and
$$\cdot CH_3 + \cdot CH_3 \rightarrow C_2H_6$$

Under the influence of ultraviolet light, new chains of reaction are, of course, continually started.

(c) Cracking reactions

At high temperatures, in the region of 500°C, alkanes will 'crack' to form new molecules of lower molecular mass; these new molecules frequently have a lower hydrogen:carbon ratio than the alkanes, and are said to be *unsaturated*. Hexane can be cracked to a mixture of butane and ethene, C_2H_4, and butane may break up further to yield ethene and hydrogen:

$$C_6H_{14} \rightarrow C_4H_{10} + C_2H_4$$

$$C_4H_{10} \rightarrow 2C_2H_4 + H_2$$

The cracking process was originally developed to produce petrol from the heavier fractions of crude oil, but the reactions are also a valuable source of hydrogen and ethene, an important starting material for many products such as ethanol and plastics. To obtain the required products catalysts may be used, and steam is sometimes added; the reactions are further examples of chain mechanisms, such fragments as $\cdot CH_3$, $\cdot C_2H_5$ and $\cdot CH_2 \cdot$ being made and reformed into various molecules.

23.6 The alkenes and alkynes

These are the 'unsaturated' hydrocarbons whose molecules do not contain sufficient hydrogen atoms to complete the four bonds each carbon atom is capable of making:

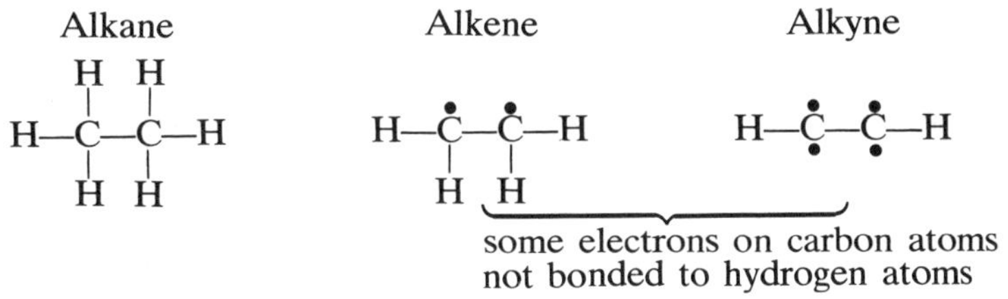

Alkane   Alkene   Alkyne

some electrons on carbon atoms
not bonded to hydrogen atoms

Study of these compounds shows that the electrons not utilised in bonding hydrogen atoms form 'double' and 'triple' bonds (page 45) between the carbon atoms, composed of four and six shared electrons respectively. Thus:

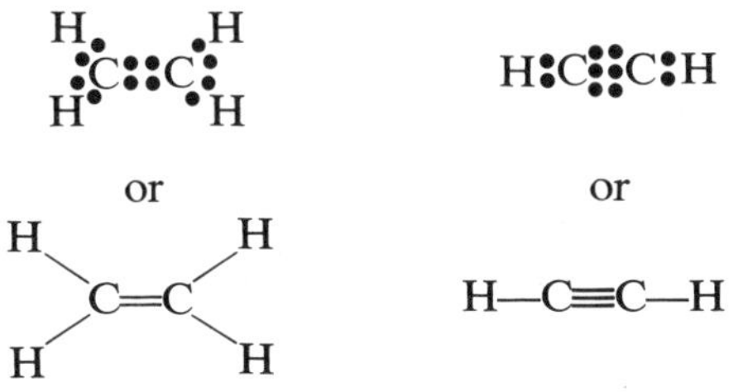

Ethene   Ethyne

The high density of electrons between the carbon atoms draws the carbon nuclei together, thus shortening and strengthening the bond and giving shape to the molecule. The ethene molecule is planar with an angle of approximately 120° between the C—H bonds, while the ethyne molecule is linear (page 52).

Table 23.3

Bond	Bond length, nm	Bond strength, kJ mol^{-1}
C—C	0.154	346
C=C	0.135	611
C≡C	0.121	835

The mutual repulsion of the electrons and the strain of the distortion from the normal tetrahedral shape partly offset the strengthening of the bonds so that the double bond is not twice as strong as the single bond, nor is the triple bond three times as strong.

Table 23.4

	Molecular formula	Structural formula
Ethene	C_2H_4	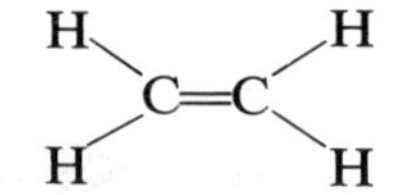
Propene	C_3H_6	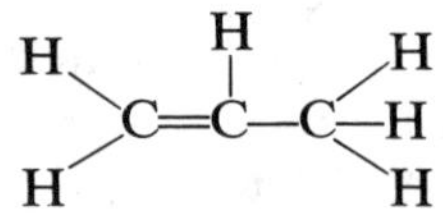
Butene	C_4H_8	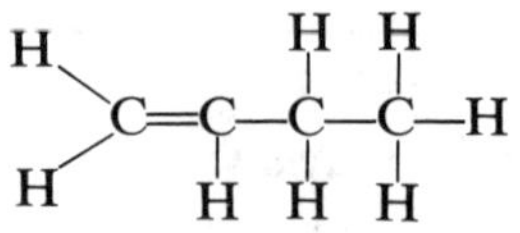
etc		

There are very many of these unsaturated compounds. One homologous series follows ethene, the general formula of the molecules being C_nH_{2n}.

Obviously, with a molecule containing four or more carbon atoms, isomers can arise from the position of the double bond as well as from the different arrangement of the carbon atoms. The naming of these isomers follows the same convention as described above, the position of the double bond being denoted by the number of the carbon atom preceding it.

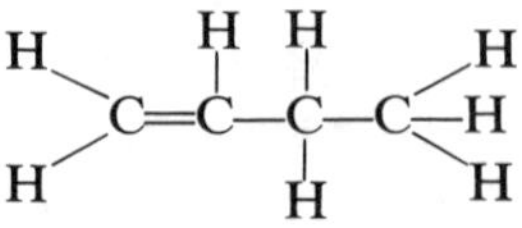

But-1-ene

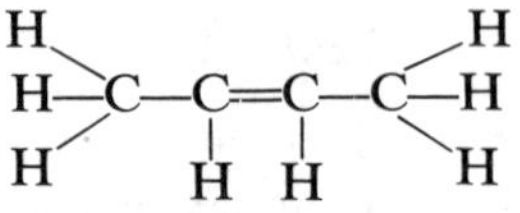

But-2-ene

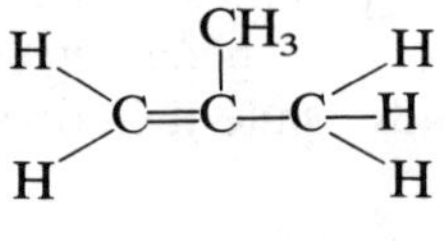

2-methylpropene

In addition, there is no free rotation of atoms linked by a double bond, so but-2-ene exhibits yet another type of isomerism, *stereoisomerism*, (Chapter 24), depending on the arrangement of the two methyl groups on the carbon atoms linked by the double bond. There are similar homologous series of triple bonded compounds, but in this text only ethyne (the acetylene of the oxy-acetylene torch) will be considered. Also the main reactions of the alkenes will be described with respect to ethene as a typical molecule.

These unsaturated compounds are more reactive than the alkanes. Since there is not a full complement of hydrogen atoms in the molecule, their reactions are mainly addition reactions in which diatomic molecules or other groups add to the compound across the double or triple bonds.

23.7   Reactions of alkenes and alkynes

(a) Burning

Unless there is a plentiful supply of air, unsaturated compounds tend to burn with a smoky yellow flame. The products of complete combustion are carbon dioxide and water:

$$C_2H_4 + 3O_2 \rightarrow 2CO_2 + 2H_2O$$
$$C_2H_2 + 2\tfrac{1}{2}O_2 \rightarrow 2CO_2 + H_2O$$

(b) Hydrogenation

Hydrogen is readily added to a hydrocarbon having a double bond, the product being the corresponding alkane:

$$C_2H_4 + H_2 \rightarrow C_2H_6$$
$$\text{Ethene} \qquad \text{Ethane}$$

The reaction is carried out under moderate pressure in the presence of a nickel catalyst.

(c) Halogenation

The halogens react with alkenes to form dihalides; the reaction is rapid with chlorine, less so with bromine and very slow with iodine.

Ethene, bubbled through bromine held under a layer of water, forms a colourless oily liquid, 1,2-dibromoethane.

$$C_2H_4 + Br_2 \rightarrow CH_2BrCH_2Br$$

This compound is named as a disubstituted ethane; the 1,2-notation indicates that the bromine atoms are on different carbon atoms.

The mechanism is the electrophilic addition of bromine to ethene. The electrons of the double bond induce polarisation of the bromine molecule:

$$Br_2 \rightarrow Br^{\delta+}\!-\!Br^{\delta-}$$

The positively charged bromine atom acts as an electrophilic reagent and a carbon–bromine bond is formed using one of the electron pairs of the carbon–carbon double bond; the negative bromide ion is released leaving a carbonium ion which then reacts with the Br^- ion to form the final product:

$$H_2C=CH_2 + \overset{\delta+}{Br}-Br^{\delta-} \longrightarrow \overset{+}{C}-C-H + Br^- \longrightarrow CH_2BrCH_2Br$$

The formation of pure 1,2-dibromoethane occurs only if an inert solution is used. If bromine water is used the OH^- ions are in competition with the Br^- ions produced during the reaction, and some 2-bromoethanol is formed as well as 1,2-dibromoethane:

$$CH_2^+-CH_2Br + OH^- \rightarrow CH_2BrCH_2OH$$

If sodium chloride solution is used the presence of the Cl^- results in the formation of mainly 1-chloro-2-bromoethane since the Cl^- ion attacks the carbonium ion more readily than the Br^- ion:

$$CH_2^+-CH_2Br + Cl^- \rightarrow CH_2ClCH_2Br$$

The appearance of these products confirm the mechanism of the reaction.

Ethyne also is readily halogenated; the triple bond in the molecule leads to the addition of two halogen molecules:

$$HC\equiv CH + 2Br_2 \rightarrow CHBr_2CHBr_2$$
$$1,1,2,2\text{-tetrabromoethane}$$

(d) Reactions with hydrogen halides

The hydrogen halides saturate alkenes to form monohalogen derivatives; in contrast to its reaction with the halogens, ethene reacts readily with hydrogen iodide and hydrogen bromide, but only slowly with hydrogen chloride:

$$C_2H_4 + HBr \rightarrow CH_3CH_2Br$$
$$\text{Bromoethane}$$

Hydrogen bromide is already a polar molecule; electrophilic attack results in the formation of a bond between the carbon and the positively charged hydrogen atom, with the formation of a carbonium ion:

$$H_2C=CH_2 + H-Br \rightarrow H_2C^+-CH_3 + Br^-$$

The released bromine ion reacts with the carbonium ion to make 1-bromoethane:

$$H_2C^+-CH_3 + Br^- \rightarrow CH_2BrCH_3$$

The addition of hydrogen halides to propene is of interest because the halogen atom adds to the middle carbon, that is, the more highly substituted carbon atom, while the hydrogen atom joins the less highly substituted carbon atom.* This results from the inductive effect of the methyl group, which is electron-donating, causing the carbon of the $-CH_2$ group to become

*I.e. that carrying the greater number of hydrogen atoms.

slightly negatively charged:

$$CH_3 \overset{\delta+}{\rightarrow} \overset{\delta-}{\underset{\underset{H}{|}}{C}} = CH_2$$

The electrophilic attack of the positive hydrogen of the hydrogen bromide molecule is thus directed towards the less substituted carbon atom:

$$CH_3 \overset{\delta+}{\rightarrow} CH \overset{\delta-}{=} CH_2 + H-Br \rightarrow CH_3 \overset{+}{C}H-CH_3 + Br^-$$
$$\rightarrow CH_3CHBrCH_3$$
$$\text{2-bromopropane}$$

This effect was found empirically by Markownikov in 1875 and is summarised in his rule which may be generalised as: 'In the addition of unsymmetrical adducts to unsymmetrical alkenes, halogen, or the more negative group, becomes attached to the more highly substituted of the unsaturated carbon atoms.'

Alkynes react with two molecules of hydrogen halides to give fully saturated compounds, Markownikov's rule again being followed.

$$HC{\equiv}CH + 2HBr \rightarrow CH_2{=}CHBr + HBr \rightarrow CH_3CHBr_2$$
$$\text{1,1-dibromoethane}$$

The rate of reaction of the second molecule of hydrogen bromide is very slow. This is of importance in the reaction of ethyne with hydrogen chloride, which gives a good yield of chloroethene (vinyl chloride) for the manufacture of PVC.

$$CH{\equiv}CH + HCl \rightarrow CH_2{=}CHCl$$

(e) Reaction with sulphuric acid

Ethene is absorbed by fuming sulphuric acid to form ethyl hydrogen sulphate which can then be hydrolysed to ethanol.

$$CH_2{=}CH_2 + H.HSO_4 \rightarrow CH_3CH_2HSO_4 \xrightarrow{H_2O} C_2H_5OH + H_2SO_4$$

The reaction is of great importance, being used to manufacture ethanol from ethene.

(f) Oxidation by potassium manganate(VII)

Compounds with carbon–carbon double bonds are readily oxidised by potassium manganate(VII) in either alkaline or acid conditions. The function of the manganate(VII) ion is to supply oxygen, which it does most effectively in acid conditions. The equation given below is not 'balanced' with respect to MnO_4^- ions, but, to indicate the source of oxygen, the conditions are given over the arrow. The oxidation product is immediately hydrolysed by the aqueous solution to give a dihydric alcohol or *diol*. Ethene yields ethane-1,2-diol, which is the ethylene glycol used as an antifreeze additive in car radiators.

$$CH_2{=}CH_2 + \bar{O} + H_2O \xrightarrow[OH^-]{KMnO_4} CH_2OHCH_2OH$$

(g) Ozonolysis

Ozone, bubbled into a solution of ethene in an inert solvent, forms an unstable product which is readily hydrolysed with the complete rupture of the double bond:

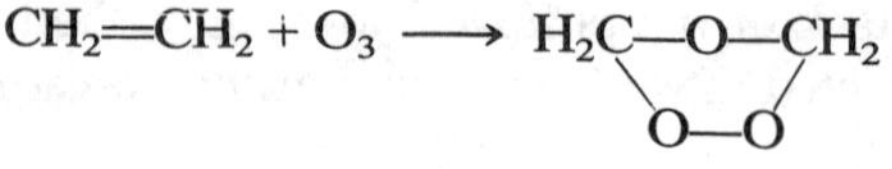

$$CH_2{=}CH_2 + O_3 \longrightarrow \underset{\underset{\displaystyle O{-}O}{\big|\qquad\big|}}{H_2C{-}O{-}CH_2}$$

This reaction is used to determine the position of a double bond in large molecules such as $R'HC{-}CHR''$, in which R' and R'' represent different alkyl or aryl groups. The hydrolysis of the ozonide forms two aldehydes (Chapter 26):

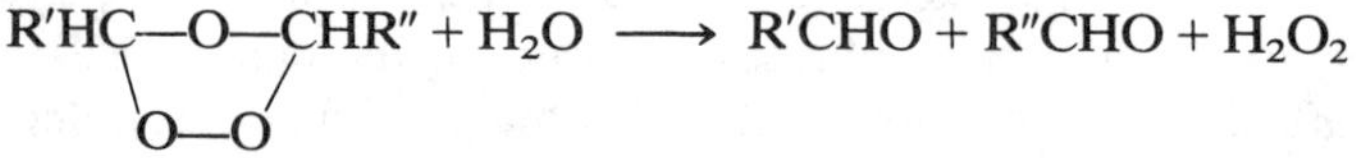

$$\underset{\underset{\displaystyle O{-}O}{\big|\quad\big|}}{R'HC{-}O{-}CHR''} + H_2O \longrightarrow R'CHO + R''CHO + H_2O_2$$

Aldehydes are easily identified, hence the nature of R' and R'' can be found and the position of the double bond in the original molecule located.

(h) Polymerisation reactions

Unsaturated hydrocarbon molecules form addition polymers. An *addition polymer* is a big molecule made up from identical simple molecules, the monomers, which join to make a continuous structure. The polymerisation of alkenes is a chain reaction which proceeds through a free radical mechanism. The polymerisation of ethene, for example, is started by a catalyst or chain initiator which may be an ion or a free radical. One such radical is the phenyl radical, $C_6H_5\cdot$, derived from the breakdown of dibenzoyl peroxide, $(C_6H_5COO)_2O_2$:

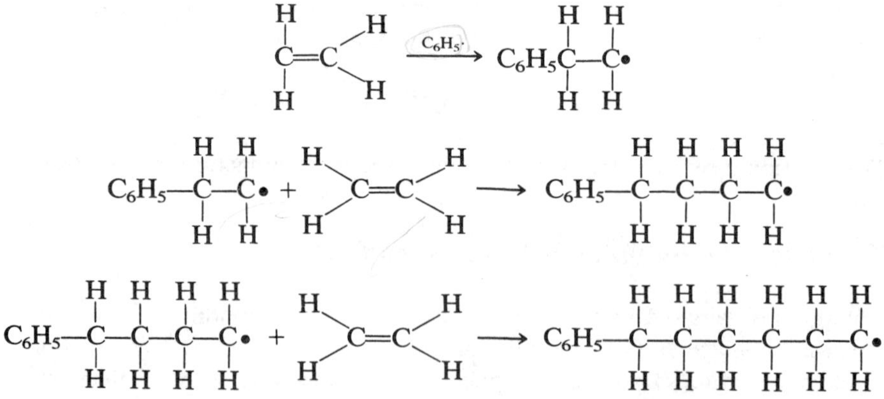

The reaction continues until the chain is terminated either by two chains linking together or by a second free radical reacting with the active end of the polymer chain. The length of the chain can be regulated by controlling the temperature and pressure conditions and the amount of the initiator added.

A catalyst composed of aluminium triethyl and titanium tetrachloride developed by Ziegler in 1953 is used in the manufacture of polyethene,

Groups of chain molecules

Fig. 23.3   Crystalline polymer

polypropylene and other addition polymer plastics. It has the advantage of promoting long chain molecules at normal temperatures and pressures. The linear long chain molecules form partially crystalline polymers with strong van der Waals forces between the molecules.

These polymers are termed thermoplastic since they can be softened by heat and moulded and remoulded in a reversible process. Adjustments to conditions of pressure and to the composition of the catalyst during the manufacture of the plastic can be made. These modify or increase the length of the polymer chain, to give more crystallinity, higher density and higher softening point to the product if a tougher more resilient plastic is required.

A great variety of addition products can be made by modifying the basic ethene molecule, providing plastics of different types and properties.

It is not difficult to see that the nature of the side group can modify the properties of the plastic produced, and that the positions of these groups can also modify the properties if they are large enough to hinder free rotation about the carbon atom. Alteration of the conditions of manufacture can ensure that the groups are added either at random (atactically) or all on the same side of the chain (isotactically):

$$-\overset{|}{\underset{|}{C}}-\overset{|}{\underset{|}{C}}-\overset{|}{\underset{|}{C}}-\overset{|}{\underset{|}{C}}-\overset{|}{\underset{|}{C}}-\overset{|}{\underset{|}{C}}-\overset{|}{\underset{|}{C}}-\overset{|}{\underset{|}{C}}-$$

or on alternately opposite sides of the chain (syndiotactically):

$$-\overset{|}{\underset{|}{C}}-\overset{|}{\underset{|}{C}}-\overset{|}{\underset{|}{C}}-\overset{|}{\underset{|}{C}}-\overset{|}{\underset{|}{C}}-\overset{|}{\underset{|}{C}}-\overset{|}{\underset{|}{C}}-\overset{|}{\underset{|}{C}}-\overset{|}{\underset{|}{C}}-\overset{|}{\underset{|}{C}}-$$

Compounds with a triple bond also undergo polymerisation reactions; the most important example of the polymerisation of ethyne is the formation of benzene when the gas is passed through a hot tube:

$$3C_2H_2 \rightarrow C_6H_6$$

Ethyne   Benzene

Table 23.5 Plastics

Monomer	Plastic	Uses
Chloroethene CH_2=$CHCl$ (vinyl chloride)	Polyvinylchloride (PVC) 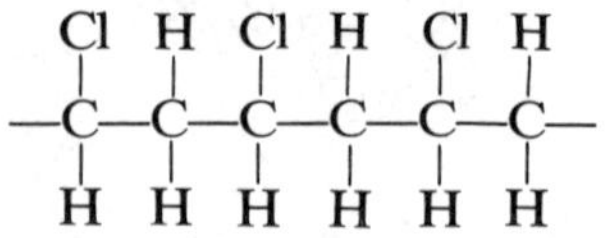 Cl H Cl H Cl H —C—C—C—C—C—C— H H H H H H	Upholstery Records Insulating materials
Tetrafluoroethene F_2C=CF_2	PTFE 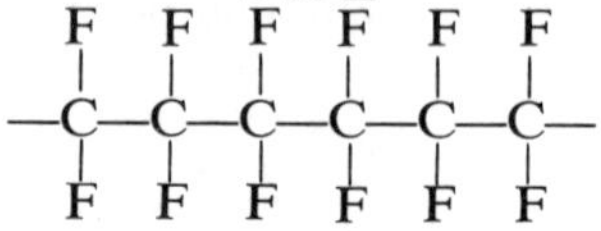 F F F F F F —C—C—C—C—C—C— F F F F F F	Teflon for non-stick pans Artificial joints Insulating materials
Propene H_2C=$CHCH_3$	Polypropylene 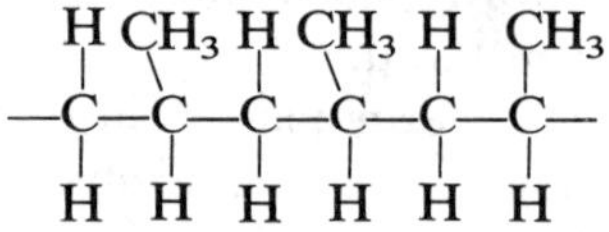 H CH_3 H CH_3 H CH_3 —C—C—C—C—C—C— H H H H H H	Household articles Insulators
Phenylethene H_2C=CHC_6H_5 (styrene)	Polystyrene 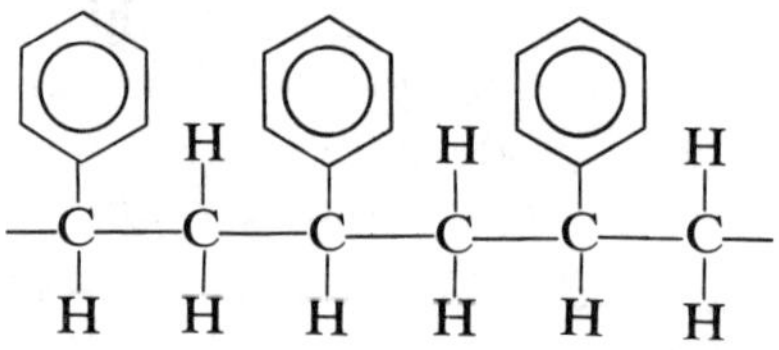 	Refrigerator parts Insulating materials Radio parts Toys Decorative goods
Methyl-2-methylpropenoate H_2C=CCH_3COOCH_3 (methyl methacrylate)	Perspex 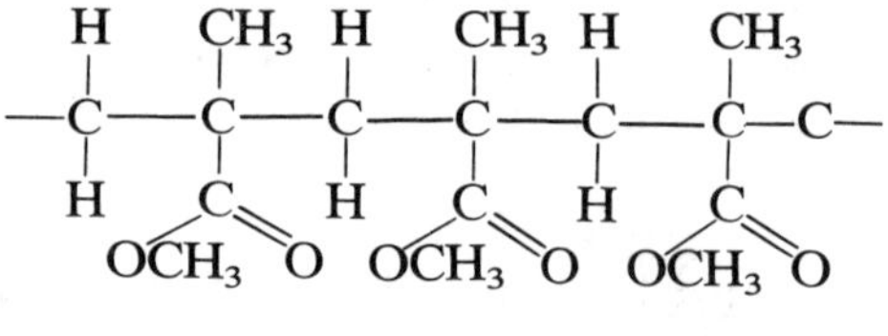 H CH_3 H CH_3 H CH_3 —C—C—C—C—C—C—C— H C H C H C OCH_3 O OCH_3 O OCH_3 O	Substitute for glass Decorative goods
Propenenitrile H_2C=$CHCN$ (Acrylonitrile)	Polyacrylonitrile 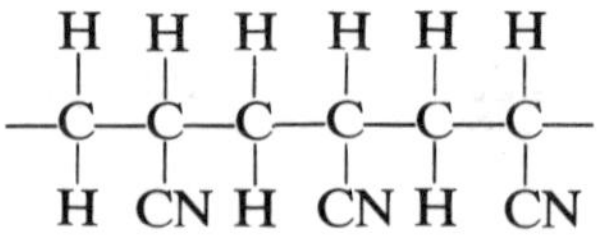 H H H H H H —C—C—C—C—C—C— H CN H CN H CN	Courtelle Acrilan

23.8 The arenes

These compounds, originally extracted from aromatic natural oils, have ring structures of a highly unsaturated yet stable nature. The simplest member of the arenes is benzene, C_6H_6, and many attempts have been made to find a structure to fit the molecular formula. An early suggestion was:

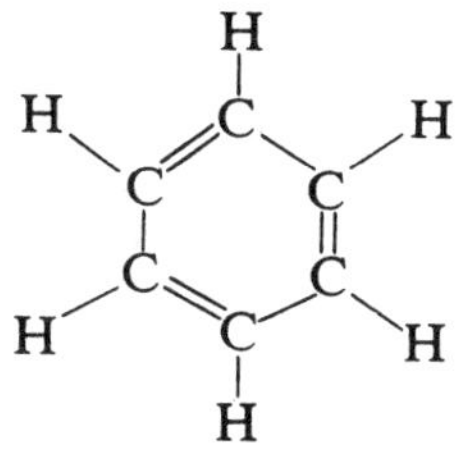

$$
\begin{array}{c}
\text{H} \\
\text{H}-\overset{\text{C}}{\underset{\text{C}}{}}\text{--}\overset{\text{C}}{}\text{--H} \\
\text{H}-\text{C} \qquad \text{C-H} \\
\text{H}
\end{array}
$$

with the three double bonds accounting for the unsaturation. This fits the valency requirements of the carbon and hydrogen atoms, but does not reflect the properties of benzene, which fails to decolourise potassium manganate(VII) solution or to react rapidly with bromine water as do the alkenes and alkynes.

In 1872 Kekulé suggested that the molecule existed as two rapidly interchanging structures in which the double bonds alternated between different positions:

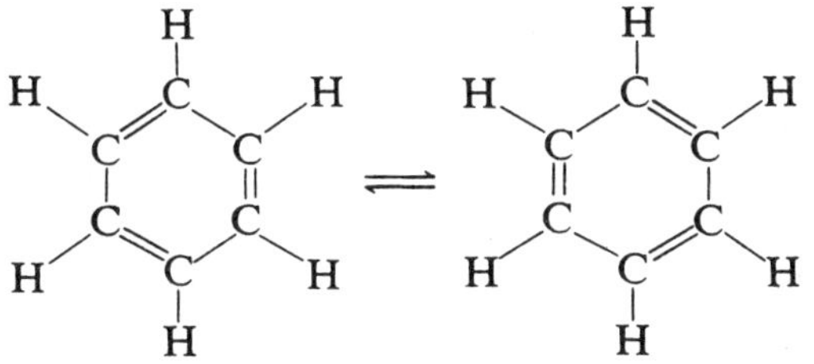

$$
\rightleftharpoons
$$

Nowadays a resonance structure (page 60) is postulated, in which the electrons are delocalised and do not form double bonds between specific carbon atoms, as in the two *canonical forms* of the Kekulé structure. Canonical forms differ only in the arrangement of the electrons; the atomic nuclei remain in the same relative positions. This structure is represented as:

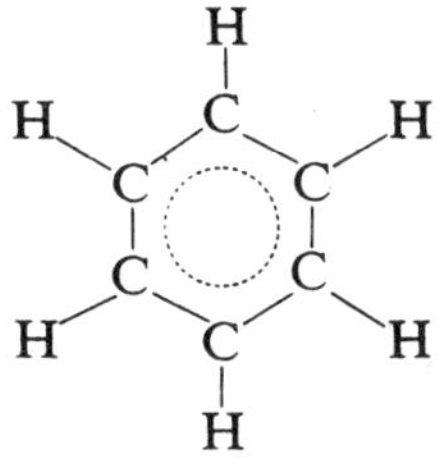

or more simply as ⬡

and the following evidence supports this model:

(i) Chemical properties: benzene burns with the smoky flame characteristic of unsaturated compounds, but does not give other reactions typical of unsaturated compounds.

(ii) X-ray diffraction measurements show that the C—C bonds are all the same length (0.140 nm) which is shorter than that of single C—C bonds (0.154 nm) and longer than that of double C=C bonds (0.135 nm).

(iii) The heat of hydrogenation of benzene is 208 kJ mol^{-1}, not $3 \times 120 \text{ kJ mol}^{-1}$ as might be expected by analogy with the hydrogenation of cyclohexene to cyclohexane, where one double bond reacts with hydrogen:

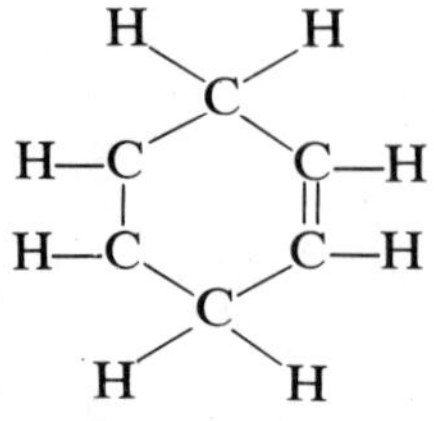

$$+ \; H_2 \longrightarrow \quad \begin{array}{c} C_6H_{12} \\ \text{Cyclohexane} \end{array} \quad \Delta H^{\ominus}_{298} = -120 \; \text{kJ mol}^{-1}$$

Cyclohexene

$$+ \; 3H_2 \longrightarrow \quad \begin{array}{c} C_6H_{12} \\ \text{Cyclohexene} \end{array} \quad \Delta H^{\ominus}_{298} = -208 \; \text{kJ mol}^{-1}$$

Benzene

(iv) The experimentally determined heat of formation of benzene is $-5514 \; \text{kJ mol}^{-1}$, which is greater than the value of $-5310 \; \text{kJ mol}^{-1}$ which can be calculated by supposing that the molecule contains six C—H bonds, three C—C bonds and three C=C bonds. The extra energy accounts for the observed stability of the benzene molecule and is called the *resonance energy*.

Recent work suggests that a variety of structures may exist in liquid benzene, as, for example:

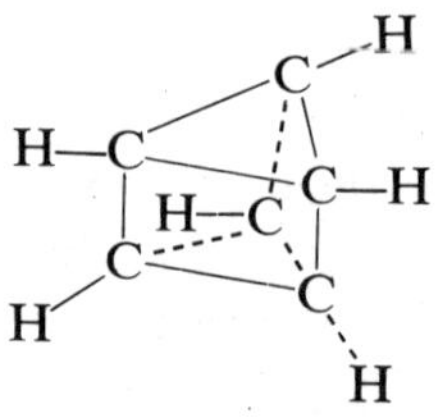

A number of hydrocarbon derivatives of benzene occur in which one hydrogen atom is replaced by alkyl groups. Some examples best illustrate the naming of these; where the traditional name is well known this is given in brackets:

Methylbenzene
C_7H_8 (toluene)

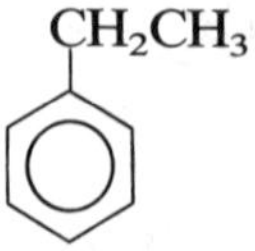

Ethylbenzene
C_8H_{10}

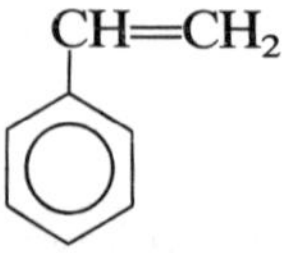

Phenylethene
C_8H_8 (styrene)

The removal of a hydrogen atom from the benzene ring leaves a *phenyl* group; thus methylbenzene could be named phenylmethane.

When two hydrogen atoms in the benzene molecule are replaced the naming has to be more precise, as positional isomerism occurs (Chapter 24). The carbon atoms are numbered starting from one which has a substituted group attached; thus the three isomers of dimethylbenzene are:

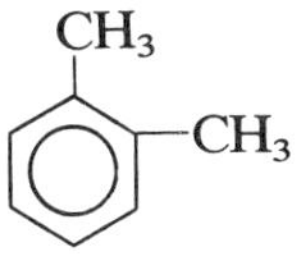

1,2-dimethylbenzene   1,3-dimethylbenzene   1,4-dimethylbenzene
(*ortho*-xylene)   (*meta*-xylene)   CH₃
                                                          (*para*-xylene)

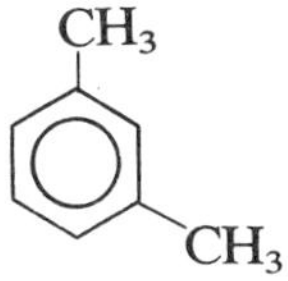

Since the molecule is symmetrical, methyl groups in the 1,5- and 1,6-positions would give products identical to 1,3-dimethylbenzene and 1,2-dimethylbenzene. The older name of xylene is still used and the positions indicated by *ortho*-, *meta*- and *para*- prefixes as shown.

Benzene, toluene and xylene are all sweet-smelling liquids; benzene is very poisonous and should not be handled in the open laboratory.

23.9   Reactions of arenes

(a) Burning

Arenes burn with a characteristic smoky flame in air; complete combustion in oxygen yields carbon dioxide and water.

$$2C_6H_6 + 15O_2 \rightarrow 12CO_2 + 6H_2O$$

(b) Hydrogenation

Benzene can be hydrogenated to cyclohexane by passing its vapour mixed with hydrogen over a heated nickel catalyst. This is a valuable source of cyclohexane for the nylon industry, as benzene is readily available from both the coal distillation and petroleum industries.

(c) Halogenation

The chlorination of benzene, if carried out in the presence of iron filings, is a substitution reaction with a hydrogen atom being replaced by a chlorine atom:

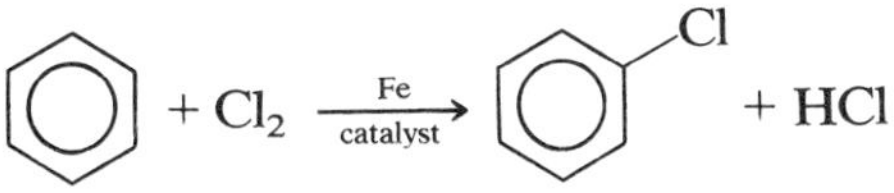

The iron catalyst forms iron(III) chloride which promotes the polarisation of the chlorine molecule to give a Cl^+ ion. The benzene molecule assumes one of the canonical forms of the Kekulé structure and undergoes electrophilic attack by the Cl^+ ion:

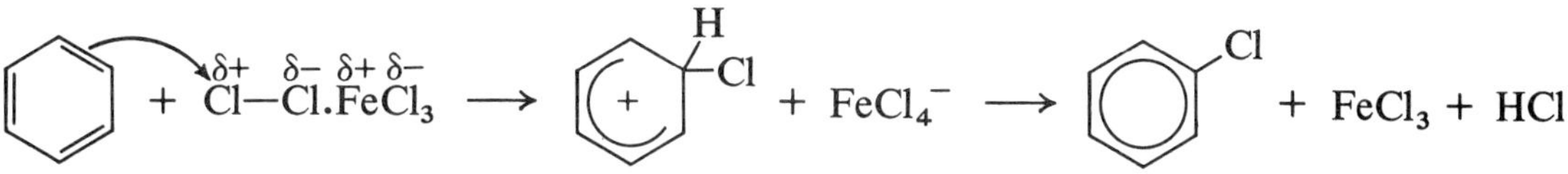

This reaction will also take place with bromine, and, less readily, with iodine. Methylbenzene (toluene) undergoes halogenation with greater ease than benzene, the substitution of the hydrogen atom taking place preferentially at the 2, 4 or 6 positions in the benzene ring. This results from the +I inductive effect of the methyl group, which is electron-donating, facilitating the attack of the Cl^+ ion by increasing the electron density of the benzene ring. The effect on the canonical forms of methylbenzene may be represented thus:

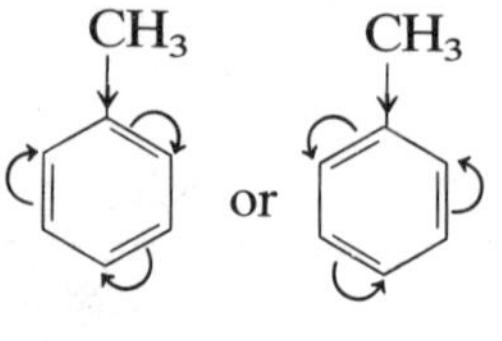

The curved arrow indicates a tendency of the electrons to become located at the 2,4,6 carbon atoms rather than at the 3 or 5 carbon atoms. Thus two isomers are formed:

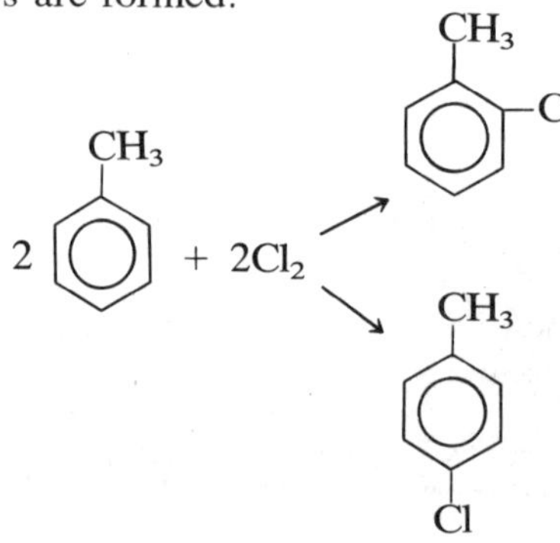

Chloro-2-methylbenzene

+ HCl

Chloro-4-methylbenzene

+ HCl

If the conditions are changed, and chlorine is passed into boiling methylbenzene, without a halogen carrier but irradiated with bright sunlight, then the chlorine attacks the methyl side-chain by a free radical mechanism similar to that described for the reaction of chlorine with methane:

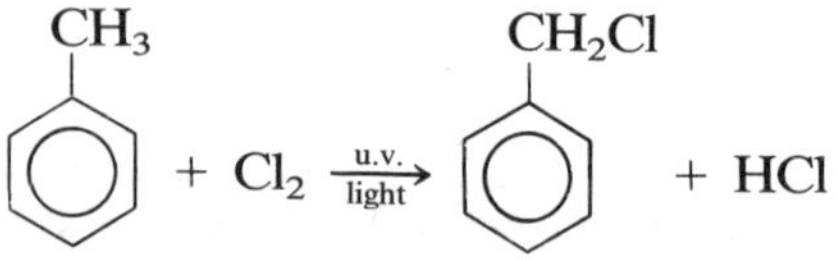

If the chlorination of benzene is carried out as above, in bright sunlight and without a halogen carrier, then an addition reaction occurs yielding hexachlorocyclohexane, which is used as an insecticide under the name of gammexane:

$$C_6H_6 + 3Cl_2 \xrightarrow[\text{light}]{\text{u.v.}} C_6H_6Cl_6$$

(d) Nitration

Benzene does not react directly with nitric acid, but nitration can be effected by using a mixture of concentrated sulphuric acid and nitric acid, kept at a temperature below 55°C (Chapter 30). The acids react to produce the positive nitronium ion NO_2^+:

$$2H_2SO_4 + HNO_3 \rightarrow NO_2^+ + 2HSO_4^- + H_3O^+$$

Electrophilic substitution follows to produce a yellow oil with a sweetish almond-like smell, nitrobenzene:

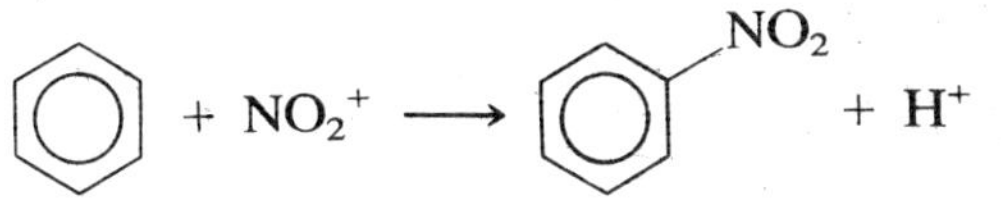

If fuming nitric acid is used, and the nitrating mixture kept under reflux for some time, then further substitution occurs, giving first 1,3-dinitrobenzene and finally 1,3,5-trinitrobenzene, both of which are solids of an explosive nature.

Methylbenzene (toluene) is more easily nitrated than benzene; the presence of the methyl group again facilitating substitution at the 2 and 4 carbon atoms of the benzene ring:

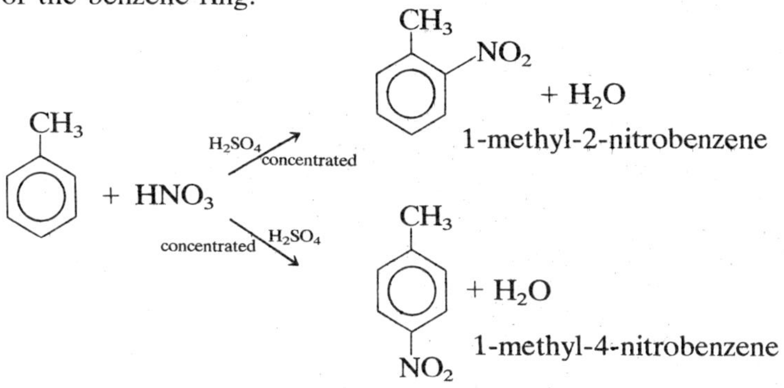

In this case further nitration occurs much more readily, giving 1-methyl-2,4,6-trinitrobenzene, which is better known as the high explosive trinitrotoluene (TNT).

(e) Sulphonation

Benzene reacts with concentrated sulphuric acid if the two liquids are agitated and heated together under reflux for a period of about twenty hours. The product is benzene sulphonic acid, which is an important starting material for other substituted benzene compounds as the sulphonate group is readily replaced:

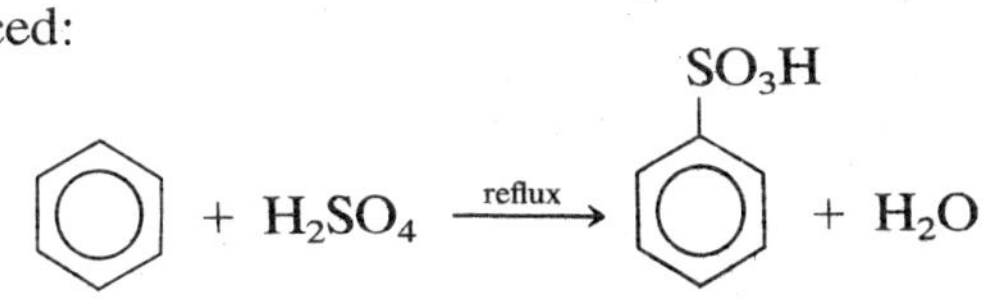

(f) Alkylation

The addition of an alkyl side chain to the benzene molecule is a reaction used to obtain arenes other than benzene, since they are not always readily available from natural or industrial sources. One way of accomplishing this is by the Friedel–Crafts reaction (page 253). The benzene is treated with a halogenoalkane in the presence of aluminium chloride which acts as a catalyst. If methyl iodide is used the product is methylbenzene:

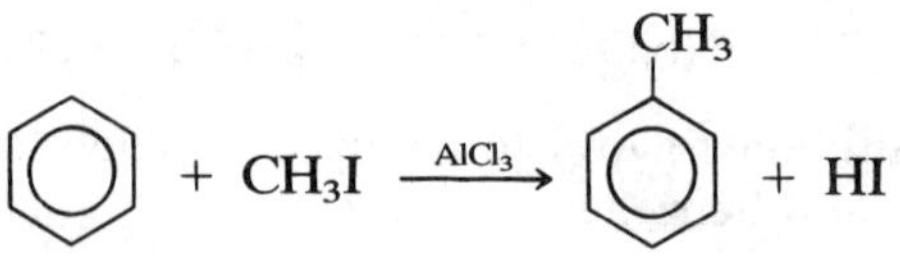

Similarly ethyl iodide will give ethylbenzene:

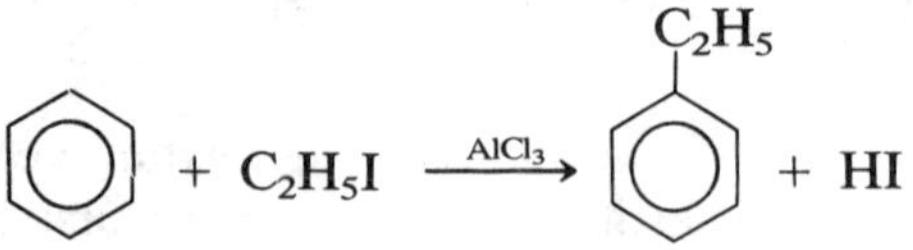

The reaction can also be used for the further alkylation of arenes which already have a side chain molecule. Methylbenzene can thus be alkylated to dimethylbenzene.

(g) Oxidation

The benzene ring is not readily oxidised, and, as we have seen, benzene does not react with acidified potassium manganate(VII). However, the alkyl side-chains of alkylated benzenes are oxidised in mild conditions to aldehydes, and by powerful oxidising agents to carboxylic acids. Methylbenzene vapour, mixed with air and passed over manganese dioxide at high temperatures, is oxidised to benzaldehyde:

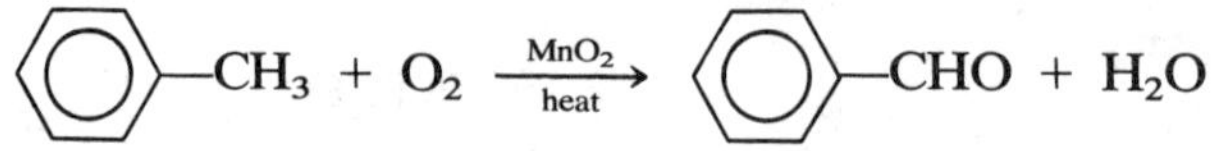

benzaldehyde

Methylbenzene, boiled under reflux with acidified potassium manganate(VII) solution or another powerful oxidising agent such as chromic acid, is oxidised to benzoic acid:

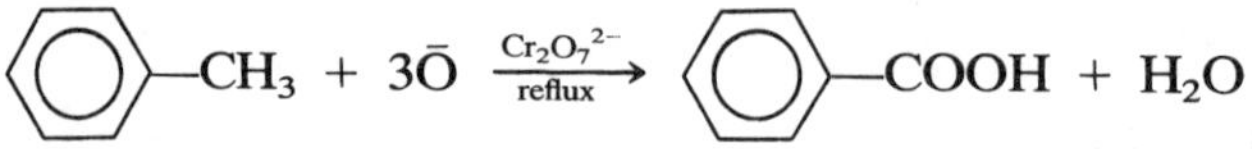

(h) Uses of arenes

Benzene is used in the manufacture of aniline, phenol, DDT, dyes, gammexane, nylon and detergents. It is a good solvent for oils, fats and resins, although recently its poisonous nature has brought restrictions on its use. The substituted arenes, methylbenzene and dimethylbenzene, are also good solvents with the advantage of being less toxic, and methylbenzene (toluene) is used in the manufacture of benzaldehyde, dyes, TNT and drugs.

Questions

1. Define
   (i) an homologous series
   (ii) structural isomerism
2. Suggest a chemical test to distinguish between
   (a) an alkane and an alkene
   (b) an alkene and an alkyne
   (c) an alkene and an arene

3. Distinguish between 'addition' and 'substitution' reactions giving one example of each.

4. What is meant by
   (a) homolytic fission
   (b) a free radical
   (c) hetereolytic fission?

5. Comment on the fact that if chlorine is reacted with methyl benzene in the presence of iron filings the product is a mixture of 1-methyl-2-chlorobenzene and 1-methyl-4-chlorobenzene, whereas if chlorine is passed into boiling methyl benzene in bright sunlight the product is chloromethylbenzene.

6. A mixture of $100 \, cm^3$ of ethene (ethylene) and ethane at a pressure of 99.6 kPa (750 mm Hg) and 27°C was treated with bromine, under conditions which favoured its reaction with alkene, but not with alkane. The results of such an experiment showed that 0.285 g of bromine has been used up.
   (a) State *two* conditions necessary for the above reaction to occur.
   (b) Give the name and structural formula of the product of this reaction.
   (c) Give the name and structure of an isomer of the product referred to in (b), and outline *one* method for its preparation.
   (d) Calculate the percentage composition by volume of the gaseous mixture.
   (e) Write the equations, state the necessary conditions and name the intermediate compounds formed when ethene is converted into (i) butane, (ii) ethanal(acetaldehyde), (iii) a named dicarboxylic acid.
                                                                          (AEB 1978)

7. Summarize the reactions, if any, of simple alkanes, alkenes, alkynes and of benzene with the following reagents:
   (a) bromine,
   (b) potassium permanganate, and
   (c) sulphuric acid.
   Clear structural formulae of reaction products should be given and approximate experimental conditions stated.
       For each of the types of hydrocarbon, select one reaction and briefly discuss its mechanism.   (L)

8. (a) Write an equation for the electrophilic reaction of hydrogen bromide with propene.
   (b) Give the mechanism for the reaction.
   (c) Give the structural formula for an isomer of the product of this reaction.
   (d) Explain briefly why this isomer is not a major product in the reaction of hydrogen bromide with propene under these conditions.
   (e) Write a structure for the product that you would predict to be formed in the reaction of propene with nitrosyl chloride, $NOCl$ (which reacts as NO^+Cl^-).   (JMB)

9. When ethene (ethylene) is bubbled through an aqueous solution containing bromine and potassium chloride, it is found that CH_2BrCH_2Br and CH_2BrCH_2Cl are formed but no CH_2ClCH_2Cl. How may this be explained?   (O)

24 Isomerism

24.1 Introduction

When two or more compounds, exhibiting different chemical or physical properties, have the same molecular formula, they are termed isomers. Each isomer is a different compound, but it is often convenient to speak of the isomeric forms of the compound corresponding to the molecular formula. Isomerism is of particular importance in organic chemistry, and several examples of organic compounds existing in two or more isomeric forms have already been mentioned.

This chapter is concerned with the various types of isomerism, the main classes being structural isomerism and stereoisomerism.

In structural isomerism, the atoms of each isomer are linked together in a different structural form. In stereoisomerism, the atoms of each isomer are linked in the same form; the difference lies in the spatial arrangement of the links. The two main classes can be further divided with respect to the ways in which the isomers are defined.

24.2 Structural isomerism

There are several types of structural isomerism.

(a) Chain isomerism (nuclear isomerism)

The difference in structure arises from the arrangement of the carbon atoms in the skeletal chain. For example, there are three possible arrangements of the carbon chain for the molecule C_5H_{12}:

Pentane ($CH_3CH_2CH_2CH_2CH_3$)   2-methylbutane (($CH_3)_2CHCH_2CH_3$)

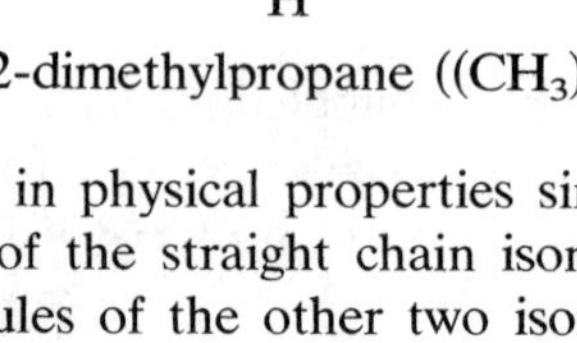

2,2-dimethylpropane (($CH_3)_4C$)

These three isomers differ in physical properties since the van der Waals forces between molecules of the straight chain isomer are much stronger than those between molecules of the other two isomers. Thus the boiling point of 2,2-dimethylpropane is much lower than that of normal pentane.

(b) Position isomerism

There are isomeric compounds in which a functional group occurs in different positions in the molecules. For example, the hydroxyl group in the alcohol C_4H_9OH can occupy two different positions in the straight chain molecule:

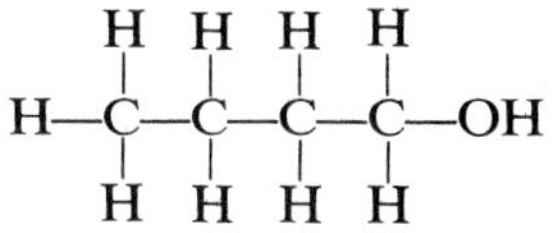

$$
\begin{array}{cccc}
\text{H} & \text{H} & \text{H} & \text{H} \\
| & | & | & | \\
\text{H—C—C—C—C—OH} \\
| & | & | & | \\
\text{H} & \text{H} & \text{H} & \text{H}
\end{array}
\qquad
\begin{array}{cccc}
\text{H} & \text{H} & \text{H} & \text{H} \\
| & | & | & | \\
\text{H—C—C—C—C—H} \\
| & | & | & | \\
\text{H} & \text{OH} & \text{H} & \text{H}
\end{array}
$$

Butan-1-ol ($CH_3CH_2CH_2OH$)   Butan-2-ol ($CH_3CH(OH)CH_2CH_3$)
(primary alcohol)   (secondary alcohol)

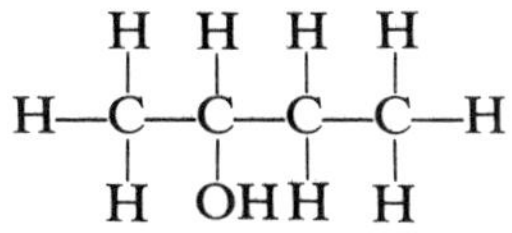

A third isomer is formed when there is a change in the skeletal chain:

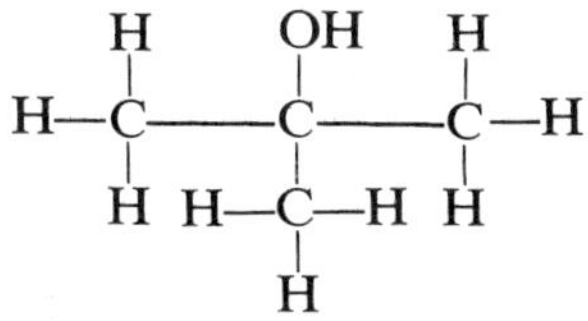

$$
\begin{array}{ccc}
\text{H} & \text{OH} & \text{H} \\
| & | & | \\
\text{H—C} & \text{—C—} & \text{C—H} \\
| & | & | \\
\text{H} & \text{H—C—H} & \text{H} \\
& | & \\
& \text{H} &
\end{array}
$$

2-methylpropan-2-ol (($CH_3)_3COH$)
(tertiary alcohol)

The physical and chemical properties of these isomers differ. The melting points and boiling points are affected both by the shapes of the molecules, as with the isomeric alkanes, and by differences in hydrogen bonding (page 228). The chemical differences shown by primary, secondary and tertiary alcohols are described on page (379).

Isomerism resulting from the varying position of the functional group also arises in aromatic compounds. Thus there are three possible positions for the —OH group with respect to the —CH_3 group in the cresols:

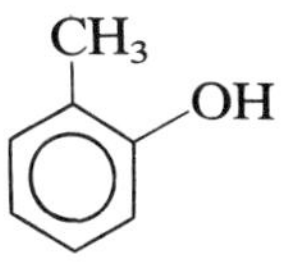

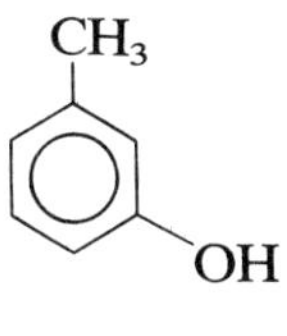

2-methylphenol   3-methylphenol   4-methylphenol
(o-cresol)   (m-cresol)   (p-cresol)

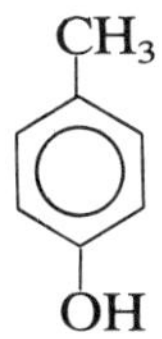

These cresols show differences in physical properties, which are more marked in crystalline structure and therefore melting point than in boiling point. In the case of the phthalic acids, $C_6H_4(COOH)_2$, chemical differences are more evident; for example, only benzene-1,2-dicarboxylic acid forms an acid anhydride, this being 'sterically' impossible for the other two isomers—

the —COOH groups are too distant from one another:

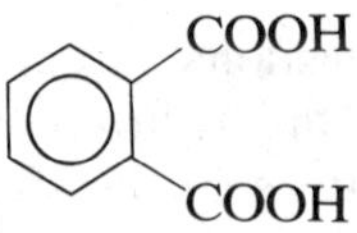

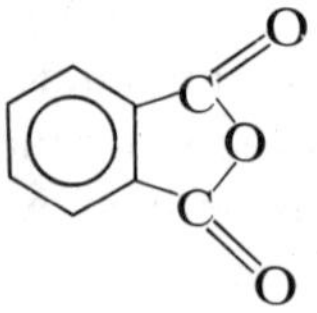

Benzene-1,2-dicarboxylic acid   Benzene-1,2-dicarboxylic anhydride

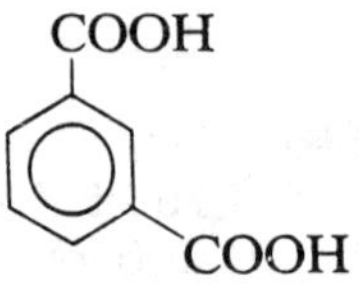

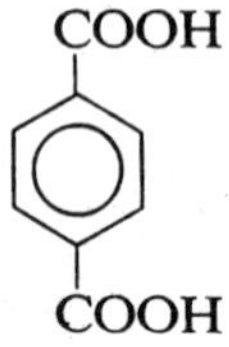

Benzene-1,3-dicarboxylic acid   Benzene-1,4-dicarboxylic acid

(c) Metamerism

Some compounds have isomeric forms in which a functional group joins different combinations of alkyl or aryl radicals. Thus the isomeric ethers of molecular formula $C_4H_{10}O$ are methoxypropane and ethoxyethane:

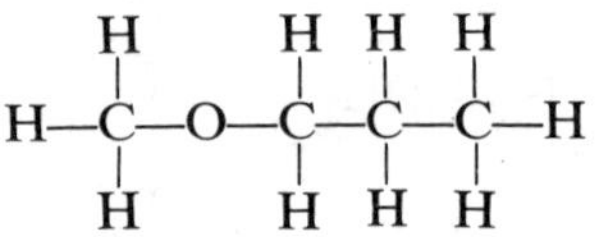

Methoxypropane
(methylpropyl ether)
($CH_3OCH_2CH_2CH_3$)

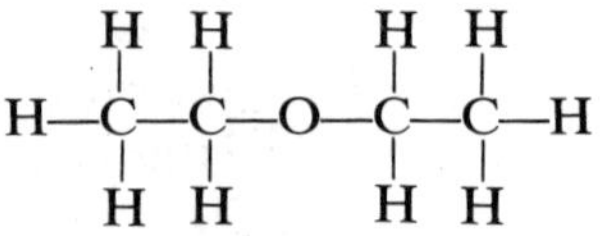

Ethoxyethane
(diethyl ether)
(($C_2H_5)_2O$)

Similar isomers arise with esters of carboxylic acids. For example, phenyl ethanoate is isomeric with methyl benzoate:

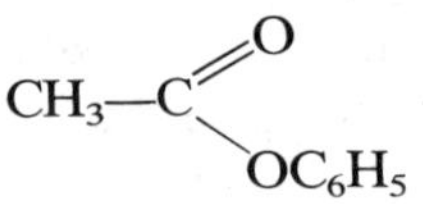

Phenyl ethanoate
($CH_3CO_2C_6H_5$)

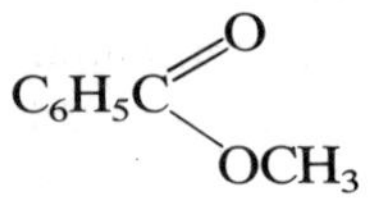

Methyl benzoate
($C_6H_5CO_2CH_3$)

(d) Functional group isomerism

In these isomers the presence of different functional groups, with the molecular formulae remaining identical, indicates that the isomers belong to different homologous series and therefore have distinctly different chemical properties. For example, each of the ethers is isomeric with an alcohol:

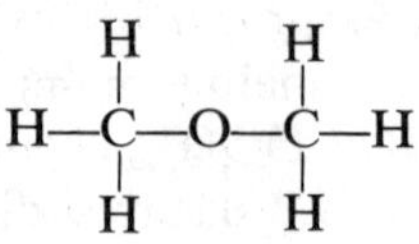

Methoxymethane (($CH_3)_2O$) 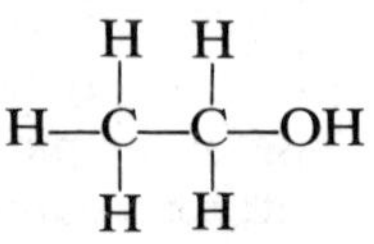   Ethanol (CH_3CH_2OH)

Similarly every aldehyde, except methanal and ethanal, is isomeric with a ketone:

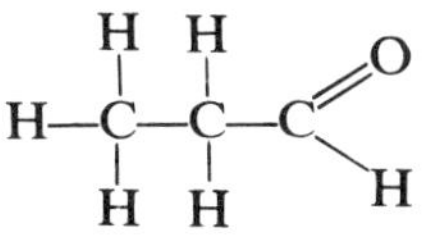

Propanal (CH_3CH_2CHO)

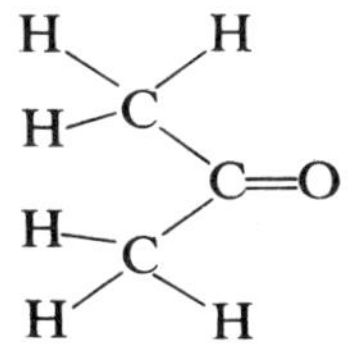

Propanone (($CH_3)_2C{=}O$)

(e) Tautomerism

A special case of functional group isomerism arises when two isomers are in dynamic equilibrium, as in ethyl 3-oxobutanoate (ethyl acetoacetate) which exists in a keto- and an enol- form. This dynamic equilibrium is known as tautomerism:

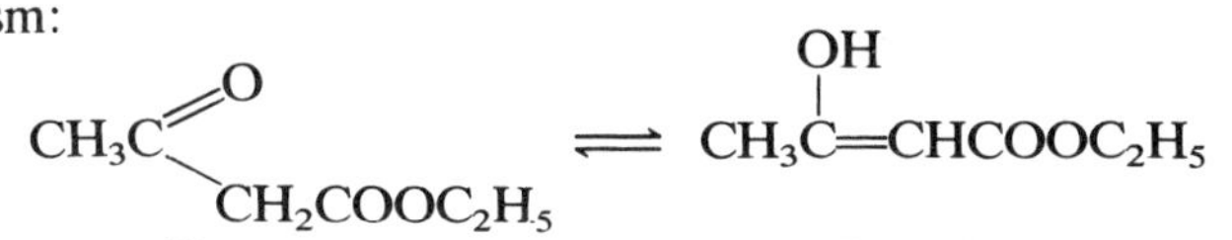

Keto-form

Enol-form

Thus the compound forms a yellow derivative with 2,4-dinitrophenyl-hydrazine (characteristic of the carbonyl group), decolourises bromine water (characteristic of the double bond, —C=C—) and also gives a violet colouration with iron(III) chloride (characteristic of the enol group —C(OH)=C—, page 382). Pure samples of each compound have been isolated from the mixture, which contains about 93% of the keto- form at room temperature.

24.3 Stereoisomerism

Certain compounds have identical molecular and structural formulae but nevertheless exhibit isomerism. This is termed stereoisomerism since it results from the spatial configuration of the atoms or groups. It may be geometrical or optical.

(a) Geometrical isomerism

In molecules where there can be no free rotation about one of the covalent bonds, geometrical isomers occur. For example, compounds such as but-2-ene contain a double bond which does not allow free rotation of the carbon atoms it joins, hence there are two isomers of identical structure. In one, the *cis*-isomer, both methyl groups are on the same side of the bond while in the *trans*-isomer they are diagonally opposed:

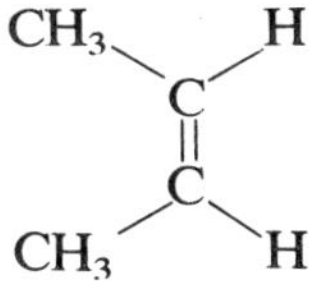

Cis-but-2-ene
m.p. −139°C

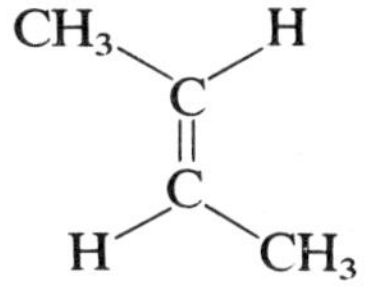

Trans-but-2-ene
m.p. −106°C

The two compounds have similar boiling points and similar chemical properties.

If the methyl groups are replaced by carboxylic acid groups, giving 1,4-butenedioic acid, then the two isomers differ in chemical and in physical properties. Thus the *cis*-isomer can form an anhydride while this is sterically impossible in the *trans*-isomer:

Cis-1,4-butenedioic acid
(maleic acid)

Cis-1,4-butenedioic anhydride
(maleic anhydride)

Trans-1,4-butenedioic acid
(fumaric acid)

(b) Optical isomerism

Light is propagated by electro-magnetic waves, the waves normally oscillating transversely in all directions at right angles to that in which the light is travelling. Thus the transverse section of a beam of light should show oscillations in all possible directions, although the beam is usually represented as a single ray.

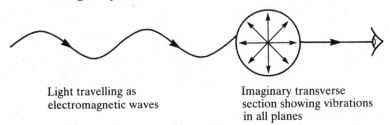

Light travelling as
electromagnetic waves

Imaginary transverse
section showing vibrations
in all planes

Certain crystals can be used, as in Polaroid, to suppress the wave oscillations of light in all planes except one. Calcite, as Iceland Spar, can be made into a Nicol prism which is particularly effective; the light emerging from the prism, vibrating in one plane only, is said to be *plane polarised*.

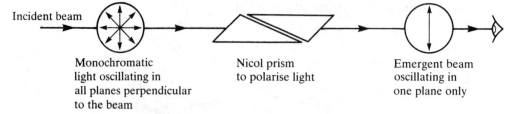

Incident beam

Monochromatic
light oscillating in
all planes perpendicular
to the beam

Nicol prism
to polarise light

Emergent beam
oscillating in
one plane only

Fig. 24.1 Polarisation of light

If a beam of plane polarised light is passed through certain crystalline substances, the plane of polarisation is twisted or rotated. Some of these substances, such as sodium iodate or potassium thiocyanate, lose this property when melted or dissolved in water, indicating that it depends on the arrangement of atoms or ions in the crystal. However, many organic compounds are capable of rotating the plane of polarisation both when pure and in solution, indicating that the property of optical activity arises from the arrangement of the atoms in the individual molecules.

To measure the degree of rotation a polarimeter is used. This has a tube to hold a solution of the optically active compound after the Nicol prism (Fig. 24.1), between this and the eyepiece there is an analysing prism identical to the first. Light emerging from the optically active solution will not pass through this second prism until it is rotated into the new plane of oscillation, and the field in the eyepiece remains dark. If the analysing prism has to be rotated to the right to allow light to be seen again then the compound is termed dextro-rotatory; if it has to be rotated to the left then the compound is laevo-rotatory. A scale on the eyepiece measures the angle of rotation; in practice, since it is easier to detect maximum darkness than maximum brightness, the angle is measured between extinctions with and without the tube.

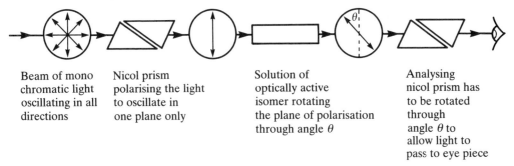

Beam of mono	Nicol prism	Solution of	Analysing
chromatic light	polarising the light	optically active	nicol prism has
oscillating in all	to oscillate in	isomer rotating	to be rotated
directions	one plane only	the plane of polarisation	through
		through angle θ	angle θ to
			allow light to
			pass to eye piece

Fig. 24.2   Polarimeter

The angle of rotation can be shown to be proportional to the length of the tube and the concentration of the solution for light of a given wavelength at a given temperature. The monochromatic light is usually supplied by a sodium lamp, which gives yellow light of a single wave-length. Thus for a solution of an optically active compound the specific rotation α, (defined as the angle of rotation of plane polarised light of wave-length of the sodium D line produced by a column of liquid ten centimeters long containing one gram of the compound per cubic centimeter at 20°C), is given by $[\alpha]_D^{20} = xv/lm$ where x is the observed rotation in degrees, v is the volume of liquid containing m grams of compound and l is the length in decimetres of the column of liquid.

In most cases of optical activity in organic compounds the property is dependent on the tetrahedral distribution of bonds about carbon atoms in a molecule. When all four groups attached to a central carbon atom are different this gives rise to the existence of isomeric molecules that are mirror images of one another, as in butan-2-ol, the sugars and the amino-acids.

An example of the effect of four different groups attached to a central carbon atom is shown in Fig. 24.3, the optical isomers of 2-aminopropanoic acid or alanine.

$$
\begin{array}{ccc}
CH_3 & & CH_3 \\
| & & | \\
C & & C \\
H_2N \diagup \diagdown COOH & HOOC \diagup \diagdown NH_2 \\
H & & H
\end{array}
$$

Fig. 24.3   Optical isomers of 2-aminopropanoic acid

This is better understood if models are made of the two structures; it will be found that one cannot be exactly superimposed on the other. The central carbon atom is said to be asymmetric, and is frequently marked with an asterisk in molecular formulae— $CH_3C^*HNH_2COOH$

Such isomers have identical physical and chemical properties, except that:

(i) One isomer rotates the plane of polarised light to the right (dextro-rotatory) while the other rotates it to the left (laevo-rotatory). These were formerly known as d and l isomers, and are now denoted as (+) and (−) isomers.

(ii) Crystalline derivatives of the isomers have asymmetric structures which are mirror images of one another, known as enantiomorphs (a term also applied to the isomers themselves).

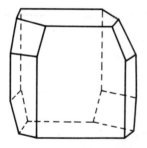 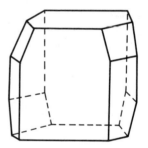

Fig. 24.4   Enantiomorphs of sodium ammonium 2,3-dihydroxybutane-dioate (sodium ammonium tartrate)

(iii) Bacteria and moulds behave differently towards optically active isomers. For example, *penicillium glaucum* will consume (+) 2-hydroxypropanoic acid ((+) lactic acid) whilst it scarcely attacks the (−) enantiomorph.

(iv) Optically active isomers frequently have different physiological effects; (−) nicotine is more toxic than (+) nicotine, and (−) adrenalin is more active than (+) adrenalin in contracting the blood capillaries.

When a compound containing an asymmetric carbon atom is produced by living matter, it is often found to be almost wholly one isomer. For example, natural glucose is dextrorotatory, hence its name dextrose, while the 2-hydroxypropionic acid (lactic acid) occurring in muscles is the (−) isomer. The souring of milk, on the other hand, produces equal quantities of the (+) and (−) forms of lactic acid, giving rise to an optically inactive solution. Such a solution contains a *racemic* mixture, denoted by (±), and the compounds are known as racemates or racemic compounds. Equal quantities of the (+) and (−) forms are always found if the compounds are synthesised in the laboratory.

Several techniques are available for separating optical isomers from a racemic mixture:

(i) Hand-picking of crystalline derivates

Pasteur resolved (±) 2,3-dihydroxybutanedioic acid (tartaric acid) by preparing crystals of the sodium ammonium salt (Fig. 24.4) and laboriously picking out the left-hand and right-hand mirror images. This very tedious and difficult process has been used in a few cases only.

(ii) Bacterial action

A solution of the racemate is prepared and a suitable bacterium or mould chosen to grow in it; one isomer is selectively consumed, leaving a solution containing the other. This method is of limited usefulness since the organisms will grow only in very dilute solutions and easily become poisoned.

(iii) Formation of chemical derivatives

This is the most effective method; it involves the reaction of the racemic mixture with an optically active compound to give products that are no longer optical isomers and have different physical properties. For example, a mixture of (+) acid and (−) acid may be separated by reaction with an optically active base to form two esters:

$$((+) \text{ acid} - (-) \text{ base}) \text{ ester} + H_2O$$

$$(\pm) \text{ acid} + (-) \text{ base}$$

$$((-) \text{ acid} - (-) \text{ base}) \text{ ester} + H_2O$$

The esters can now be separated by fractional distillation or crystallisation, and then hydrolysed to regain the separate (+) acid and (−) acid. Suitable bases include quinine and strychnine. Similarly optically active bases can be separated by reaction with one of the 2,3-dihydroxybutanedioic acid isomers.

(iv) Chromatography

Racemic mixtures of suitable boiling points, such as butan-2-ol, can be separated by vapour phase chromatography, using an optically active stationary phase.

The examples quoted above have been of simple cases of optically active compounds, but many compounds such as amino acids and sugars contain more than one asymmetric carbon atom in the molecule, giving rise to a number of isomers which are optically active to different extents. For example, 2,3-dihydroxybutanedioic acid has two asymmetric carbon atoms:

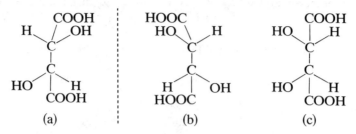

Fig. 24.5 Optical isomers of 2,3-dihydroxybutanedioic acid

This gives rise to a total of four isomers. (a) and (b) are the (+) and (−) forms of the acid, which, in equal proportions, form the (±) racemic isomer. The fourth isomer (c) is optically inactive by internal compensation, since the upper half is a mirror image of the lower half. Formerly mesotartaric acid, systematically it is (*meso*) 2,3-dihydroxybutanedioic acid.

Questions

1. Give the names of the compounds of the following formulae.
(i) $CH_3CH_2CHClCH_3$
(ii) $CH_2ClCH_2CH_2CH_3$
(iii) $CHCl_2CH_2CH_2CH_3$
(iv) $CH_2ClCH_2CH_2CH_2Cl$
One of these compounds exhibits a type of isomerism not shown by the others.
(i) Which compound is this?
(ii) What is the type of isomerism called?
(iii) What special feature of the molecule of this compound makes this isomerism possible? (L)
2. Explain briefly why compound A exhibits optical isomerism but compound B does not.

$$CH_3-\overset{\overset{\displaystyle H}{|}}{\underset{\underset{\displaystyle OH}{|}}{C}}-COOH \qquad CH_3-\overset{\overset{\displaystyle H}{|}}{\underset{\underset{\displaystyle COOH}{|}}{C}}-COOH$$

A   B

3. An organic compound *A* of molecular formula $C_6H_{12}O_2$ was found to rotate the plane of polarisation of polarised light. When it was refluxed with aqueous sodium hydroxide, it was converted into the compounds *B* and *C*. When this mixture was distilled, the distillate contained *C* which was optically active and was oxidised by chromic acid to *D* with molecular formula C_4H_8O. *D* gave a positive response to the triiodomethane (iodoform) reaction, and formed a crystalline derivative, *E*, with an aqueous solution of sodium hydrogensulphite, but it did not reduce an ammoniacal solution of silver nitrate.

The compound B was isolated in the anhydrous state by evaporating the residual liquid to crystallisation, and then heating the crystals gently to drive off the water of crystallisation. When this product was mixed with ethanol and concentrated sulphuric acid and then distilled, a sweet smelling liquid F, with a relative molecular mass of 88, collected in the receiver.

Deduce the structural formulae of the substances A, B, C, D, E and F, giving your reasons. Name three of the substances and write equations for any three of the reactions described above.

Give the structural formula of an isomer of A, which belongs to the same class as A and, like A, exists in two optically active forms.   (O)

4. (a) Starting from ethanal, suggest a reaction scheme for the preparation of 2-aminopropanoic acid (alanine) $CH_3CH(NH_2)COOH$ stating the necessary reagents and conditions.

   (b) Give the structure of the products obtained by reacting 2-aminopropanoic acid with
      (i) aqueous sodium hydroxide solution,
      (ii) dilute hydrochloric acid,
      (iii) ethanoyl chloride (acetyl chloride),
      (iv) an acidified aqueous solution of sodium nitrite.

   (c) 2-aminopropanoic acid is said to exhibit *optical isomerism*. What is meant by this statement? Explain why this term can be applied to this amino acid. Give the name and structure of an alkane which also exhibits optical isomerism.   (AEB 1979)

5. Draw structural diagrams of the isomers of the following molecules, stating the type of isomerism exhibited: 2-bromo-2-chloroethanoic acid; 1,4-butenedioic acid; chloropropane; pentane (C_5H_{12}); two esters of formula $C_4H_8O_2$; an ether and an alcohol of formula $C_6H_{14}O$.

25   Alcohols, phenols and ethers

25.1 Functional groups

A great variety of compounds can be derived from the alkanes and arenes by the substitution of a functional group for one of the hydrogen atoms. A functional group is an atom or group of atoms which gives distinctive properties to the molecules, forming a homologous series (or class of compounds), all of similar chemical nature and modified only by the length of the alkyl chain or the nature of the aryl group to which the functional group is attached. Those derived from alkanes are called aliphatic compounds, and those from arenes are called aromatic compounds. A list of functional groups with examples of the homologous series they give rise to is given below.

Table 25.1

Class of compound	Functional group	Example	Formula
Alcohols	—OH	Ethanol Phenylmethanol Phenol	CH_3CH_2OH $C_6H_5CH_2OH$ C_6H_5OH
Aldehydes	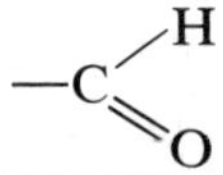 $-C\begin{smallmatrix}H\\ \\O\end{smallmatrix}$	Ethanal Benzaldehyde	CH_3CHO C_6H_5CHO
Ketones	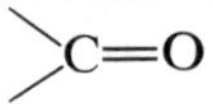 $>C=O$	Propanone Phenylethanone	$(CH_3)_2C=O$ $C_6H_5C=OCH_3$
Ethers	—O—	Ethyoxyethane	$(C_2H_5)_2O$
Carboxylic acids	—COOH	Ethanoic acid Benzoic acid	CH_3COOH C_6H_5COOH
Nitro compounds	—NO$_2$	Nitroethane Nitrobenzene	$C_2H_5NO_2$ $C_6H_5NO_2$
Amines	—NH$_2$	Ethylamine Phenylamine (aniline)	$C_2H_5NH_2$ $C_6H_5NH_2$
Nitriles (or cyanides)	—CN	Ethanenitrile* (or methylcyanide)	CH_3CN
Sulphonic acid	—SO$_3$H	Benzene sulphonic acid	$C_6H_5SO_3H$
Halides	—X	Iodoethane Chlorobenzene	C_2H_5I C_6H_5Cl

* The alternative naming here depends on whether the carbon atom of the cyanide group is counted as part of the carbon chain or not. In this text the term 'nitrile' will be used generally.

The chemistry of the compounds containing these functional groups will be introduced in the appropriate places in the succeeding chapters; this chapter will be concerned with the characteristics of alcohols, phenols and ethers.

25.2 Alcohols

(a) Monohydric alkyl alcohols: structure and properties

This class of compounds, with general formula $C_nH_{2n+1}OH$, derive their special properties from the chemistry of the hydroxyl group. The best known is ethanol, more familar as ethyl alcohol, the prime ingredient of wines, spirits and beer. It has been made by fermenting fruit or grain rich in sugar or starch (which can be converted to sugar) ever since man, in prehistoric times, stumbled on the reaction:

$$C_6H_{12}O_6 \rightarrow 2C_2H_5OH + 2CO_2$$

Ethanol is also familiar as methylated spirit, which is mainly ethanol with about 10% methanol, a small amount of dye and an unpleasant chemical (pyridine) added to make it unfit for drinking. Thus it can be sold without tax for burning in spirit lamps and stoves and as a solvent for polishes and varnishes. Ethanol is used for making ethanoic acid and trichloroethanal. It has been used as a fuel in internal combustion engines, and this use may be extended as petrol becomes scarce.

The formulae of a number of alcohols are given below, with their names; the general suffix -ol is used together with a prefix deriving from the alkyl radical to which the hydroxyl group is attached. The naming follows the system described for the alkanes (page 346) with the formal numbering of the carbon atoms. It will be noticed that there are now two kinds of isomerism; the structural or nuclear isomerism of the alkyl radical and the positional isomerism resulting from the position of the —OH group. Thus there are four isomers of the alcohol containing four carbon atoms—see Table 25.2. These isomers are better understood if models are made.

The position of the hydroxyl group in the molecule modifies the properties of the alcohols; they are differentiated by calling those with a —CH$_2$OH group primary alcohols, those with a $\underset{R}{\overset{R}{\diagdown}}$CHOH grouping secondary alcohols and those with a R—$\overset{\overset{R}{\diagdown}}{\underset{R}{\diagup}}$C—OH grouping tertiary alcohols.

The lower molecular weight monohydric alcohols are colourless liquids with distinctive smells; they are all poisonous except ethanol, which can be tolerated in limited quantities. Methanol in any quantity causes paralysis, blindness and death.

Methanol and ethanol are completely miscible with water; the solubilities of the heavier alcohols decrease as the carbon chain grows and the 'oily' nature of the molecules increases and overcomes the effect of the —OH group which forms hydrogen bonds with water (page 230). The boiling

Table 25.2

Name	Formula	Structure	Boiling point °C
Methanol	CH_3OH	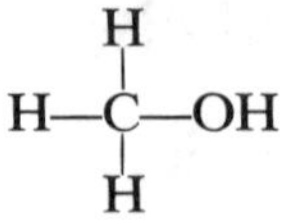 $$H-\underset{\underset{\displaystyle H}{\mid}}{\overset{\overset{\displaystyle H}{\mid}}{C}}-OH$$	64
Ethanol	C_2H_5OH	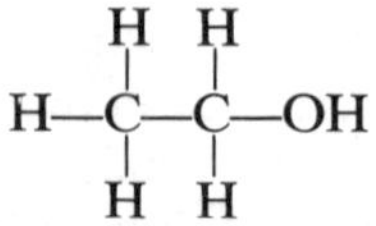 $$H-\underset{\underset{\displaystyle H}{\mid}}{\overset{\overset{\displaystyle H}{\mid}}{C}}-\underset{\underset{\displaystyle H}{\mid}}{\overset{\overset{\displaystyle H}{\mid}}{C}}-OH$$	78
Propan-1-ol	C_3H_7OH	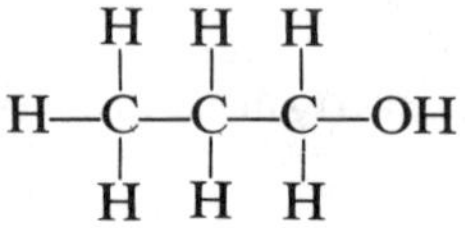 $$H-\overset{H}{\underset{H}{C}}-\overset{H}{\underset{H}{C}}-\overset{H}{\underset{H}{C}}-OH$$	97
Propan-2-ol	$CH_3CHOHCH_3$	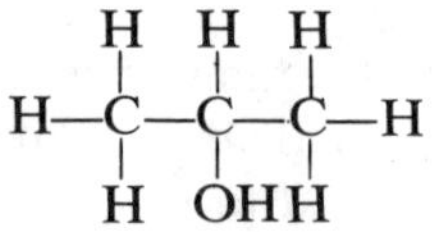 $$H-\overset{H}{\underset{H}{C}}-\overset{H}{\underset{OH}{C}}-\overset{H}{\underset{H}{C}}-H$$	82
Butan-1-ol	C_4H_9OH	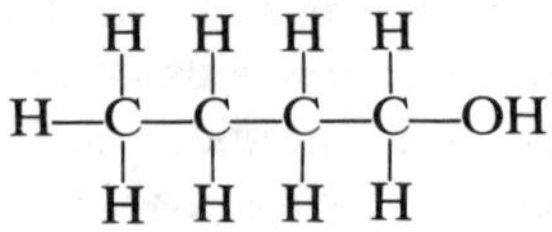 $$H-\overset{H}{\underset{H}{C}}-\overset{H}{\underset{H}{C}}-\overset{H}{\underset{H}{C}}-\overset{H}{\underset{H}{C}}-OH$$	117
Butan-2-ol	$CH_3CHOHCH_2CH_3$	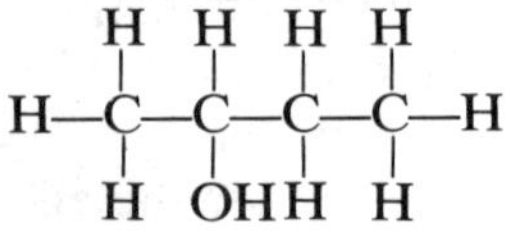 $$H-\overset{H}{\underset{H}{C}}-\overset{H}{\underset{OH}{C}}-\overset{H}{\underset{H}{C}}-\overset{H}{\underset{H}{C}}-H$$	100
2-methyl-propan-1-ol	$(CH_3)_2CHCH_2OH$	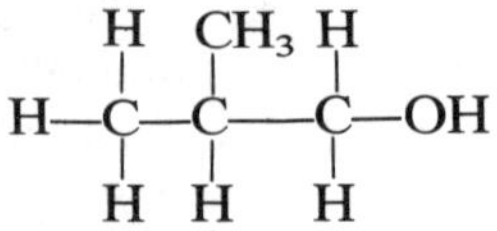 $$H-\overset{H}{\underset{H}{C}}-\overset{CH_3}{\underset{H}{C}}-\overset{H}{\underset{H}{C}}-OH$$	108
2-methyl-propan-2-ol	$(CH_3)_3COH$	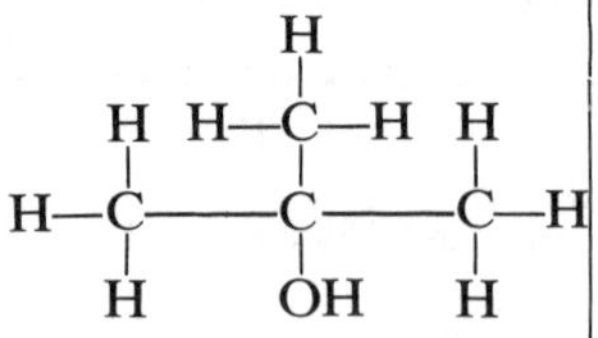 $$H-\overset{H}{\underset{H}{C}}-\overset{\overset{\displaystyle H-C-H}{}}{\underset{OH}{C}}-\overset{H}{\underset{H}{C}}-H$$	83

points of the alcohols are all higher than those of the alkanes of corresponding molecular mass; this results from hydrogen bonding between the
molecules of the alcohols (page 78).

(b) Reactions of the alcohols

Most of the reactions given in this section are illustrated by equations
relating to ethanol, as a typical alcohol in wide general use.

(i) *Oxidation*

The alcohols are readily flammable, and on complete combustion burn to carbon dioxide and water; ethanol, as methylated spirits, burns with a clear blue flame:

$$C_2H_5OH(l) + 3O_2(g) \rightarrow 2CO_2(g) + 3H_2O(l)$$

The products of controlled chemical oxidation depend on the conditions of the reaction. Warmed with potassium dichromate and dilute sulphuric acid, a primary alcohol yields an aldehyde:

$$C_2H_5OH(l) + \bar{O} \xrightarrow[H_2SO_4]{Cr_2O_7^{2+}} CH_3CHO(l) + H_2O(l)$$

Ethanol   Ethanal

If the mixture is refluxed with excess of the potassium dichromate, or with an acid solution of potassium manganate(VII), the oxidation is carried a stage further and a carboxylic acid is formed:

$$CH_3CHO(l) + \bar{O} \xrightarrow[H_2SO_4]{MnO_4^-} CH_3COOH(l)$$

Oxidation reactions of secondary and tertiary alcohols differ, and are described in (c) later.

Alcohols can be oxidised to aldehydes by passing the vapour over finely divided copper at 300°C. A dehydrogenation reaction results:

$$C_2H_5OH(l) \xrightarrow[300°C]{Cu} CH_3CHO(l) + H_2(g)$$

Bacterial oxidation of aqueous ethanol exposed to air is responsible for the souring of wines, turning them to vinegar:

$$C_2H_5OH(l) + O_2(g) \rightarrow CH_3COOH(l) + H_2O(g)$$

Ethanoic acid

This reaction is used to produce non-wine (malt) vinegars by the oxidation of ethanol, produced by brewing, to ethanoic acid.

(ii) *Reaction with phosphorus halides*

The addition of phosphorus pentachloride to alcohols gives rise to clouds of hydrogen chloride and the formation of a chloroalkane:

$$C_2H_5OH + PCl_5 \rightarrow C_2H_5Cl + POCl_3 + HCl$$

This reaction is mainly used to identify the presence of an —OH group in a molecule. If it is required to make chloro-, bromo- or iodo-alkanes, the use of the phosphorus trihalide is preferred. The hydroxyl group is replaced by a halogen atom with the formation of phosphoric acid:

$$3C_2H_5OH + PI_3 \rightarrow 3C_2H_5I + H_3PO_3$$

The bromo- or iodo-alkanes can also conveniently be made by adding the halogen to a mixture of the alcohol and red phosphorus.

(iii) *Reaction with hydrogen halides*

Concentrated hydrogen iodide reacts readily with alcohols to form alkyl iodides. The ease of reaction diminishes with hydrogen bromide and yields are low with hydrogen chloride. Water is the second product, so the presence of a dehydrating agent is helpful; thus bromoethane can be made by refluxing ethanol with concentrated sulphuric acid and potassium bromide, the sulphuric acid both generating the hydrogen bromide and taking up the water (page 449).

$$H_2SO_4(l) + KBr(s) \rightarrow KHSO_4(s) + HBr(g)$$

$$HBr(g) + C_2H_5OH(l) \rightarrow C_2H_5Br(l) + H_2O(l)$$

(iv) Reaction with thionyl chloride
Refluxing the alcohol with sulphur dichloride oxide in the presence of a base, such as pyridine, gives good yields of alkyl chlorides. The base removes hydrogen chloride, thus allowing the reaction to go to completion.

$$C_2H_5OH(l) + SOCl_2(l) \rightarrow C_2H_5Cl(l) + SO_2(g) + HCl(g)$$

(v) The iodoform reaction
If a few drops of ethanol are gently warmed with an alkaline solution of iodine in potassium iodide, the colour of the iodine is discharged and, on cooling, yellow crystals of triiodomethane (iodoform) will slowly form. They have a distinctive shape, viewed under a microscope, and a characteristic odour. Only molecules containing a $CH_3\!-\!C\!\!\begin{smallmatrix} \diagup H \\ \diagdown OH \end{smallmatrix}$ or a $CH_3\!-\!\overset{|}{C}\!=\!O$ group respond to this test; thus neither methanol nor propan-1-ol will form triiodomethane, while ethanol and propan-2-ol will do so. A similar reaction with chlorine in alkaline solution yields trichloromethane (chloroform) (page 442).

(vi) Reaction with chlorine
Chlorine, bubbled into ethanol, first causes oxidation to ethanal; this is followed by substitution in the methyl group. The final product is trichloroethanal (chloral), Cl_3CCHO, a pungent-smelling liquid used as a sleep-inducing drug and in the manufacture of DDT (dichlorodiphenyltrichloroethane).

(vii) Reaction with carboxylic acids
Alcohols react with carboxylic acids to form esters (page 408). Ethanol warmed with pure ethanoic acid in the presence of a small quantity of concentrated sulphuric acid yields ethyl ethanoate:

$$C_2H_5OH + CH_3COOH \xrightarrow{\;H_2SO_4\;} CH_3COOC_2H_5 + H_2O$$

(viii) Reaction with concentrated sulphuric acid
This most important reaction yields ethers if the alcohol is present in excess or alkenes if there is an excess of sulphuric acid. Ethanol, refluxed with concentrated sulphuric acid, first forms ethyl hydrogen sulphate:

$$C_2H_5OH + H.HSO_4 \rightarrow C_2H_5\!-\!HSO_4 + H_2O$$

Further reaction with excess ethanol takes place to form ethoxyethane (diethylether):

$$C_2H_5\begin{array}{c} \diagup OH \\ \diagdown OSO_2 \end{array} + C_2H_5OH \rightarrow C_2H_5\!-\!O\!-\!C_2H_5 + H_2SO_4$$

In general:

$$2ROH \rightarrow R\!-\!O\!-\!R + H_2O$$

When the sulphuric acid is present in excess then the ethyl hydrogen sulphate is dehydrated to ethene:

$$C_2H_5\begin{array}{c} \diagup OH \\ \diagdown OSO_2 \end{array} \xrightarrow{\text{H}_2\text{SO}_4} CH_2\!=\!CH_2 + H_2SO_4$$

or, in general, an elimination reaction takes place:

$$RCH_2CH_2OH \rightarrow RCH\!=\!CH_2 + H_2O$$

(ix) Dehydration to ethene
Ethanol can also be dehydrated by passing its vapour over phosphorus(V) oxide or aluminium oxide:

$$C_2H_5OH \xrightarrow{\text{P}_2\text{O}_5} CH_2\!=\!CH_2 + H_2O$$

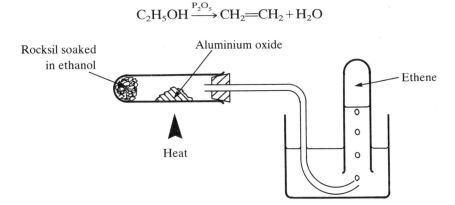

Fig. 25.1   Dehydration of ethanol

(c) Primary, secondary and tertiary alcohols

Primary, secondary and tertiary alcohols can be distinguished by the products they yield on oxidation. Primary alcohols give rise to aldehydes under mild oxidation conditions, such as the use of alkaline potassium manganate(VII).

$$CH_3CH_2CH_2OH + \overset{-}{O} \xrightarrow[\text{OH}^-]{\text{MnO}_4^-} CH_3CH_2CHO + H_2O$$
$$\text{Propan-1-ol} \qquad\qquad\qquad \text{Propanal}$$

Secondary alcohols are more resistant to oxidation and give rise to ketones:

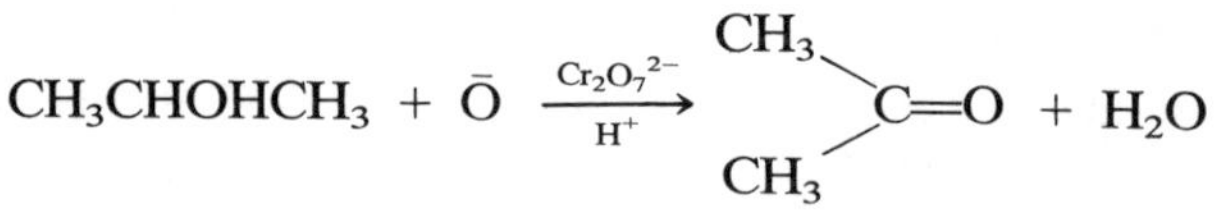

$$CH_3CHOHCH_3 + \bar{O} \xrightarrow[H^+]{Cr_2O_7^{2-}} \begin{array}{c} CH_3 \\ \diagdown \\ \diagup \\ CH_3 \end{array} C{=}O + H_2O$$

Tertiary alcohols are difficult to oxidise, but prolonged reaction with a strong oxidising agent can break up the molecule to smaller units of ketones, aldehydes or even carbon dioxide:

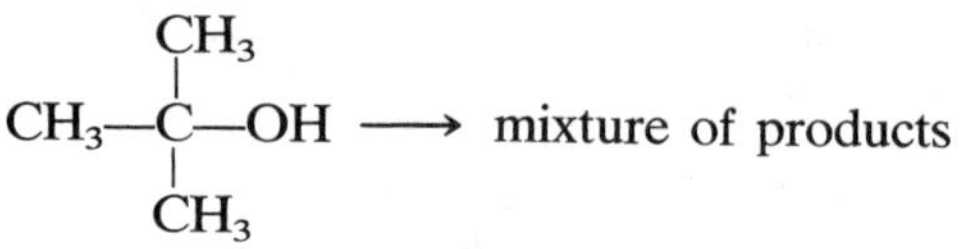

$$\begin{array}{c} CH_3 \\ | \\ CH_3{-}C{-}OH \\ | \\ CH_3 \end{array} \longrightarrow \text{ mixture of products}$$

A secondary alcohol of particular interest is butan-2-ol, whose molecule contains an asymmetric carbon atom (C^*), that is, a carbon atom which is bonded to four different groups, $CH_3CH_2C^*H(OH)CH_3$. These four groups can be arranged in space to give two molecules (stereoisomerism) that are mirror images of one another—enantiomorphs.

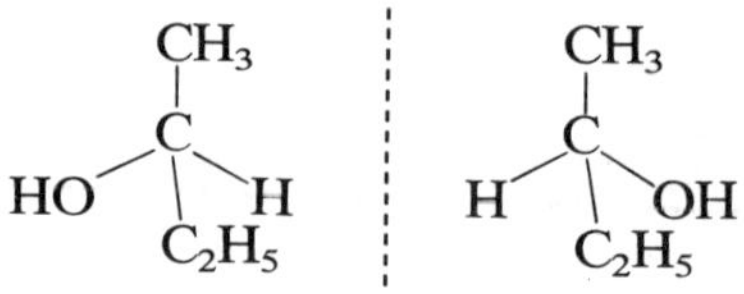

$$\begin{array}{cc} CH_3 & CH_3 \\ | & | \\ C & C \\ HO \diagup | \diagdown H & H \diagup | \diagdown OH \\ C_2H_5 & C_2H_5 \end{array}$$

Fig. 25.2   Optical isomers of butan-2-ol

The separate isomers have the property of rotating the plane of polarisation of plane polarised light in opposite directions, and are said to be optically active (page 368).

(d) Polyhydric alcohols

Alcohols with more than one hydroxyl group in the molecule occur, and of these the more important are ethane-1,2-diol and propane-1,2,3-triol. The former, known as ethylene glycol, is used as anti-freeze in car radiators. It is made from ethene (page 353) in the laboratory, or industrially by direct hydration of epoxyethane:

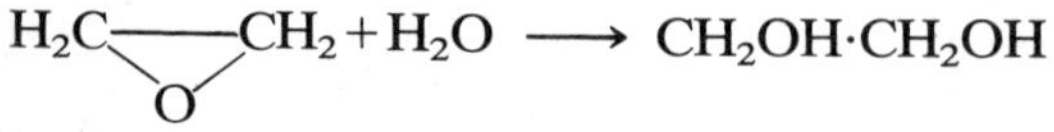

$$H_2C\underset{O}{\diagdown\diagup}CH_2 + H_2O \longrightarrow CH_2OH{\cdot}CH_2OH$$

Oxidation of ethan-1,2-diol gives rise to mixed intermediate products and finally results in the formation of ethanedioic acid (oxalic acid):

$$\begin{array}{c} COOH \\ | \\ CHO \end{array} \longrightarrow \begin{array}{c} COOH \\ | \\ COOH \end{array}$$

Ethanedioic acid

$$\begin{array}{c} CH_2OH \\ | \\ CH_2OH \end{array} \longrightarrow \begin{array}{c} CHO \\ | \\ CH_2OH \end{array} \longrightarrow \begin{array}{c} CHO \\ | \\ CHO \end{array}$$

$$\begin{array}{c} COOH \\ | \\ CH_2OH \end{array}$$

Hydroxyethanoic acid (glycollic acid)

Most of the intermediate products are so readily oxidised that they are difficult to isolate, but reasonable yields of hydroxyethanoic acid can be obtained. An important use of ethane-1,2-diol is in the manufacture of terylene (page 412).

Propane-1,2,3-triol is commonly known as glycerol or glycerine, a colour-less, viscous, sweet-tasting liquid. Vegetable oils and fats are esters of propane-1,2,3-triol, which is obtained by the hydrolysis of these oils in the manufacture of soap (page 409). The hydroxyl groups undergo the normal reactions of such groups; thus, triesters are formed:

$$\begin{array}{c} CH_2OH \\ | \\ CHOH \\ | \\ CH_2OH \end{array} + 3CH_3COOH \longrightarrow \begin{array}{c} CH_2O.OCCH_3 \\ | \\ CHO.OCCH_3 \\ | \\ CH_2O.OCCH_3 \end{array} + 3H_2O$$

Propane-1,2,3-triol   Propane-1,2,3-triethanoate

Propane-1,2,3-triol, as glycerol, is used in the manufacture of explosives, using nitric acid to form trinitroglycerin, and in the manufacture of resins by polymerisation with aromatic dicarboxylic acids. It is also a constituent of many cosmetics and toilet preparations.

(e) Aromatic alcohols

In these compounds the alcohol group is on a side-chain, and they can just as properly be regarded as alkyl alcohols with a substituted phenyl group. The simplest is phenylmethanol (benzyl alcohol), C_6H_5—CH_2OH, which is similar to ethanol in its reactions. It is less soluble in water than ethanol because of the size of the phenyl group, and differs from ethanol in that the benzene nucleus can undergo substitution reactions giving rise to ortho- and para-derivatives. On oxidation by potassium chromate(VI) in aqueous sul-phuric acid it readily yields benzaldehyde; prolonged treatment under reflux results in the formation of benzoic acid:

$$\underset{\text{Phenylmethanol}}{\begin{array}{c} CH_2OH \\ \bigcirc \end{array}} \xrightarrow[H_2O]{H^+/Cr_2O_7{}^{2-}} \underset{\text{Benzaldehyde}}{\begin{array}{c} CHO \\ \bigcirc \end{array}} \xrightarrow[H_2O/reflux]{H^+/Cr_2O_7{}^{2-}} \underset{\text{Benzoic acid}}{\begin{array}{c} COOH \\ \bigcirc \end{array}}$$

25.3 Phenols

(a) Structure and properties

When the hydroxyl group is attached directly to the benzene nucleus the resulting compound is phenol. Many of the reactions of phenol are similar to those of ethanol, but in some respects the properties differ. Phenol forms a very weakly acid solution when dissolved in water:

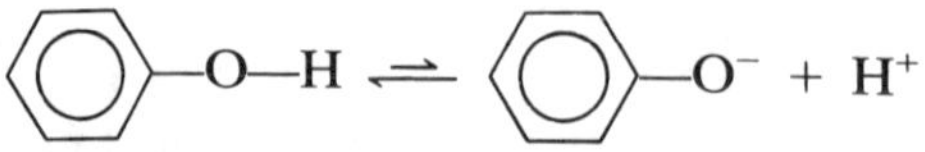

$$\text{C}_6\text{H}_5\text{-O-H} \rightleftharpoons \text{C}_6\text{H}_5\text{-O}^- + \text{H}^+$$

Electrons from the oxygen atom of the hydroxyl group are drawn into the resonance of the benzene ring, thus loosening the bond between the oxygen and the hydrogen atom, making ionisation of the hydrogen easier. Ethanol has no such electron-withdrawing group, so no ionisation of the hydrogen atom takes place and its solution in water is neutral.

Phenol is a pinkish low melting point solid, which has strongly antiseptic properties. It was used, as carbolic acid, in dilute (3%) solution by Lister in 1867, and dramatically reduced the number of deaths from infection following surgical operations and childbirth. Phenol itself is corrosive, and the solid can cause unpleasant burns on the skin. Modern antiseptics such as Dettol or TCP are less corrosive derivatives of phenol. It is an immensely useful compound, being used in the manufacture of methyl 2-hydroxybenzoate (oil of wintergreen) 2-ethanoyloxybenzoic acid (aspirin), azo dyes, cyclohexanol for nylon, and weedkillers such as 2,4-dichloro-phenoxyethanoic acid.

Phenols can be detected by the intense colours that are developed when a few drops of neutral iron(III) chloride solution are added to a concentrated solution of the compound. This coloration is given by all compounds containing a $\overset{\text{OH}}{\underset{\diagup\;\;\diagdown_{\text{C}}}{\text{C}}}$ grouping.

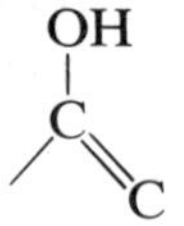

Phenols with more than one hydroxyl group on the benzene ring are well known, the most familiar being pyrogallol or benzene-1,2,3-triol which is used in the laboratory to absorb oxygen, and as a photographic developer:

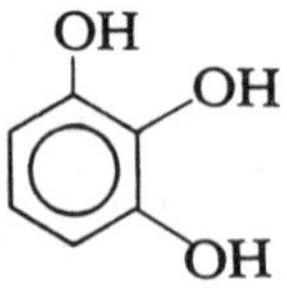

(b) Reactions of phenol

(i) Reaction with alkali
The weakly acidic nature of phenol means that, unlike ethanol, it will react with sodium hydroxide; sodium phenate being formed:

$$\text{C}_6\text{H}_5\text{-OH} + \text{NaOH} \longrightarrow \text{C}_6\text{H}_5\text{-O}^-\text{Na}^+ + \text{H}_2\text{O}$$

(ii) Reaction with sodium

Sodium phenate is also formed by the action of sodium on melted phenol, a reaction similar to that of ethanol with sodium.

$$2 \left\langle\bigcirc\right\rangle\!\!-\!OH + 2Na \longrightarrow 2 \left\langle\bigcirc\right\rangle\!\!-\!\bar{O}\overset{+}{N}a + H_2$$

(iii) Reaction with phosphorus pentachloride

Phenol produces hydrogen chloride with phosphorus pentachloride, indicating the presence of the —OH group but the reaction produces only small yields of chlorobenzene and is not used in the preparation of this chemical.

(iv) Hydrogen halides

In contrast to the alcohols, phenols do not react with hydrogen halides.

(v) Ester formation

Phenols do not react readily with carboxylic acids, but if acyl chlorides are used then esters can be formed. Phenol is *acylated* to phenylethanoate by the use of ethanoyl chloride:

$$\left\langle\bigcirc\right\rangle\!\!-\!OH + CH_3COCl \longrightarrow \left\langle\bigcirc\right\rangle\!\!-\!O.OCCH_3 + HCl$$
$$\text{phenylethanoate}$$

and *benzoylated* to phenyl benzoate by means of benzoyl chloride:

$$\left\langle\bigcirc\right\rangle\!\!-\!OH + \left\langle\bigcirc\right\rangle\!\!-\!COCl \longrightarrow \left\langle\bigcirc\right\rangle\!\!-\!O.OC\!\!-\!\left\langle\bigcirc\right\rangle + HCl$$
$$\text{phenylbenzoate}$$

The benzoylation is best carried out in alkaline solution to remove the hydrogen chloride and enable the reaction to go to completion.

(vi) Reduction

Reduction of phenol is effected by passing the vapour over hot zinc, benzene being formed:

$$C_6H_5OH \xrightarrow[\text{Zinc}]{\text{heat}} C_6H_6 + ZnO$$

(vii) Halogenation

The halogens do not affect the hydroxyl group but replace hydrogen atoms in the benzene ring with halogen atoms. If phenol is treated with excess bromine water a white precipitate of 2,4,6-tribromophenol is formed immediately.

The presence of the hydroxyl group makes the addition of halogens easier as did the methyl group of methylbenzene (page 360). The drawing of one of the lone pairs of electrons into the resonance structure activates the carbon atoms in the 2,4,6-positions of the benzene ring, making electrophilic attack by bromine easier. The possible canonical forms of phenol at the moment of reaction may be regarded as follows:

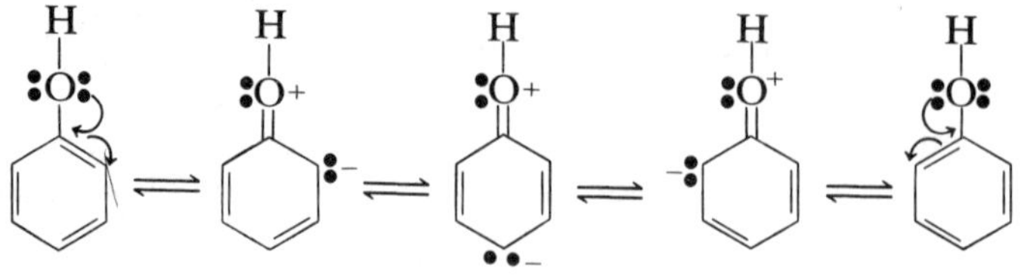

(viii) Nitration

This also takes place much more readily than the nitration of benzene. Dilute nitric acid yields a mixture of 2- and 4-nitrophenols.

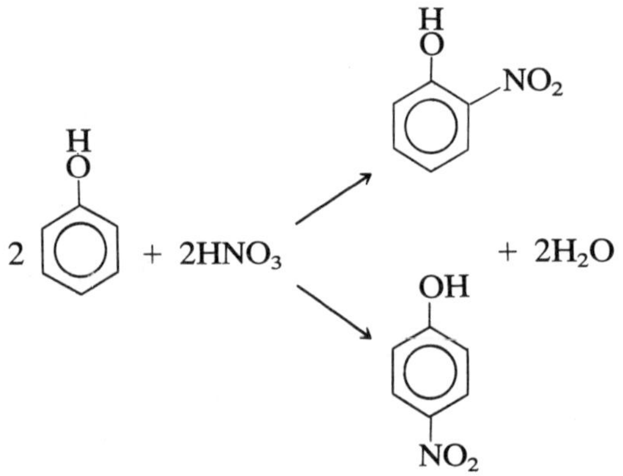

2-nitrophenol exhibits the phenomenon of internal hydrogen bonding (*chelation*) which is not possible with the 4-derivative. The chelated 2-nitrophenol is more volatile than the 4-nitrophenol, and also less soluble, as hydrogen bonding with water is hindered:

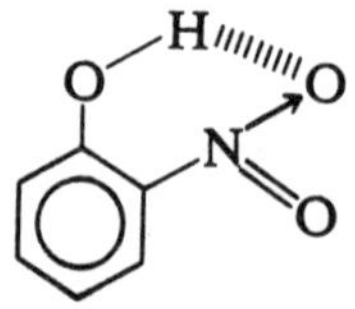

More stringent conditions lead to further nitration, and phenol warmed with a mixture of concentrated nitric and sulphuric acids forms 2,4,6-trinitrophenol, a yellow solid known as 'picric acid'. Picric acid is used in explosives and should always be stored damp. It is used in the laboratory to identify amines and to detect protein fibres in textiles.

(ix) Sulphonation

Phenol is readily sulphonated by reaction with concentrated sulphuric acid at 100°C, a mixture of 2-hydroxy and 4-hydroxy benzene sulphonic acids being formed.

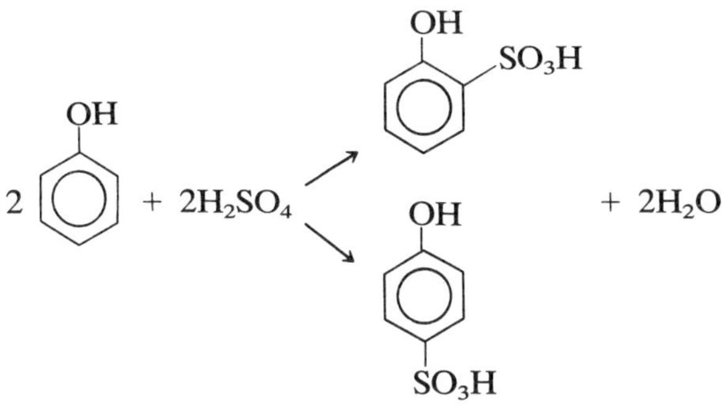

As in the bromination and nitration reactions the attack on the benzene nucleus is electrophilic. The concentrated sulphuric acid dissociates to yield sulphur(VI) oxide; the sulphur atom in this molecule can accept a pair of electrons. Thus the mechanism for the formation of the 2-hydroxy derivative can be represented as:

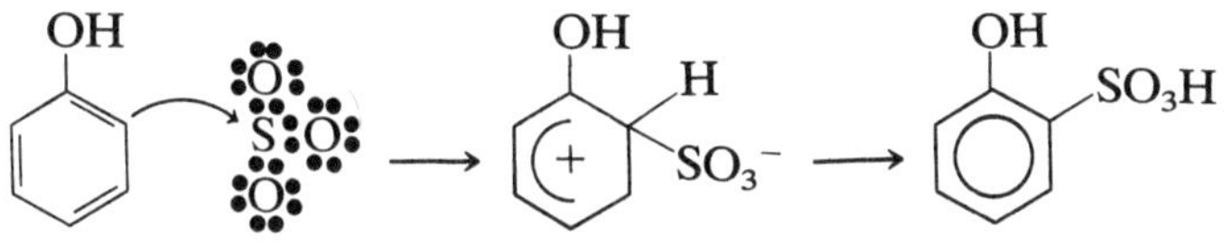

(x) *Hydrogenation*
This is a most important reaction since it is one of the methods of producing cyclohexanol for the manufacture of nylon (page 428). The vapour of phenol together with hydrogen is passed over a nickel catalyst:

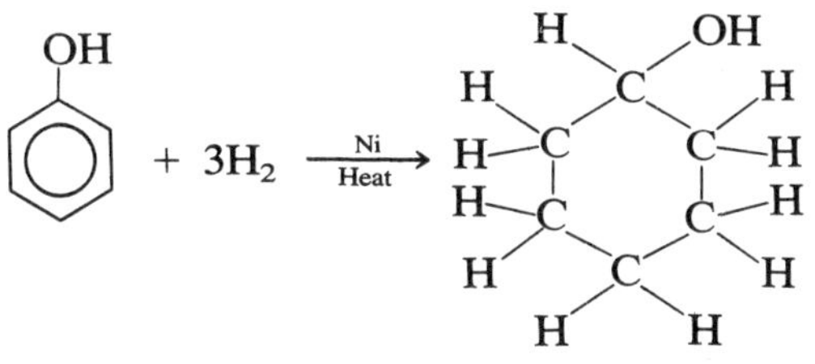

Cyclohexanol

25.4   Ethers

A number of ethers are known; the most important is ethoxyethane or diethyl ether, commonly known as ether and still used as an anaesthetic. It is a colourless volatile liquid with a characteristic odour, only partly miscible with water. Apart from its use as an anaesthetic, it is an excellent solvent, with the advantage that its low boiling point allows it to be distilled away from the solute very easily. Although ether is comparatively inert chemically, its vapour is not only highly flammable but, mixed with air, dangerously explosive. The vapour is heavy and does not disperse quickly, so that explosions have been caused at a considerable distance from the liquid by the drifting of a pocket of ether vapour across the floor or along a drainage channel.

The most important reaction of ether is its reduction to iodoethane when heated with concentrated hydriodic acid:

$$C_2H_5-O-C_2H_5 + 2HI \rightarrow 2C_2H_5I + H_2O$$

Ethoxyethane

This reaction is used to identify unknown ethers, as an ether with two different alkyl groups attached to the oxygen will yield two different iodoalkanes, which can be identified by fractional distillation.

$$CH_3-O-C_2H_5 + 2HI \rightarrow \quad CH_3I \quad + \quad C_2H_5I \quad + H_2O$$

Methoxyethane   Iodomethane Iodoethane

Ethers exhibit a type of isomerism known as *metamerism*, which depends on the presence of two different alkyl groups. Ethyoxyethane is diethyl ether, $(C_2H_5)_2O$; it is isomeric with methoxypropane, $CH_3-O-C_3H_7$. In addition, ethyoxyethane is also structurally isomeric with butanol, C_4H_9OH—in fact, every ether has an isomeric alcohol. To summarise:

$$C_4H_9OH \text{ Alcohol (structural isomer)}$$

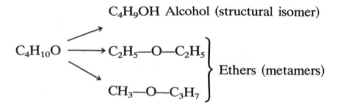

Questions

1. Describe the behaviour of ethanol with concentrated sulphuric acid under different conditions, giving the structural formulae of the products that are formed in each case.
2. What is meant by primary, secondary and tertiary alcohols? How may they be distinguished from one another in the laboratory?
3. Compare and contrast the properties of phenol and ethanol.
4. Suggest a reaction scheme for converting ethanol to ethane-1,2-diol in two stages.
5. Account for the ease of bromination of phenol.
6. A substance, X, has the following composition by mass.
   carbon 52.2%   hydrogen 13.0%   oxygen 34.8%
   (a) What is the empirical formula of X?
   (b) When completely vaporised, 0.023 g of X is found to occupy a volume of 11.2 cm³ (corrected to s.t.p.). What is the relative molecular mass of X?
   (c) What is the molecular formula of X?
   (d) Write down two different formulae which are consistent with your answer to (c).
   (e) X reacts with iodine in aqueous, alkaline solution to produce a yellow precipitate of triiodomethane (*iodoform*). Write the structural formula of substance X and show by means of equations how the triiodomethane is formed.   (JMB)

7. (a) Give the structures and the names of the alcohols which have the molecular formula C_3H_8O.

   (b) Describe how you would distinguish between these alcohols by chemical tests.   (O)

8. Three organic compounds containing a hydroxyl group are A Ethanol (CH_3CH_2OH) B Propan-2-ol ($CH_3CHOHCH_3$) C Phenol (C_6H_5OH).

   (a) What is the principal product (name and formula) when each of these compounds reacts with a mixture of sodium dichromate and sulphuric acid? (If there is no simple reaction, say so.)

   (b) What are the organic products, if any, when each of these compounds reacts with iodine and potassium hydroxide?

   (c) One of the compounds reacts with aqueous bromine to form a precipitate. Which compound is this? What is the formula of the precipitate formed?

   (d) For each of the compounds A, B and C above, a compound exists in which the hydroxyl group has been replaced by a bromine atom. Select *One* of these bromine compounds and describe briefly how you would prepare it in the laboratory from the appropriate compound A, B or C.   (L)

9. A monohydric aliphatic alcohol and an ether are isomeric, having a relative molecular mass of 74.

   (a) Suggest *one* possible structure for the alcohol and *one* for the ether.

   The ether is almost insoluble in water but dissolves in concentrated aqueous acid and is decomposed by hot, concentrated, aqueous hydrogen iodide.

   (b) Suggest a reason for the solubility of the ether in acid.

   (c) Write an equation for the decomposition of the ether by hydrogen iodide.

   The latent heat of evaporation of the ether is $+26 \text{ kJ mol}^{-1}$ and of the alcohol $+44 \text{ kJ mol}^{-1}$.

   (d) Suggest an explanation for this difference in terms of the inter-molecular bonding present in the liquids.   (C)

26 Aldehydes, ketones and carbohydrates

26.1 Aldehydes and ketones

(a) General properties

Aldehydes and ketones both have the characteristic functional group $>C=O$, the carbonyl group, and consequently they have similar chemical reactions, and are conveniently studied together. Differences in the properties arise because aldehydes have a hydrogen atom attached to the carbon atom of the carbonyl group, whilst ketones have two alkyl or aryl groups.

Aldehyde   Ketone

Aldehydes are identified in nomenclature by the ending -al and the ketones by -one, the full names following the normal systematic rules. Some examples are given below, with traditional names where these are commonly used.

Table 26.1 Aldehydes and ketones

Aldehydes			Ketones		
Name	*Formula*	*b.p. °C*	*Name*	*Formula*	*b.p. °C*
Methanal (formaldehyde)	$H—CHO$	-21			
Ethanal (acetaldehyde)	CH_3CHO	21			
Propanal	CH_3CH_2CHO	48	Propanone (acetone)	$CH_3{-}C{=}O{-}CH_3$	56
Butan-1-al	$CH_3CH_2CH_2CHO$	75	Butan-2-one (methylethyl ketone)	$CH_3{-}C{=}O{-}C_2H_5$	80
2-methylpropanal	$CH_3{-}CHCHO{-}CH_3$	64	Pentan-3-one (diethylketone)	$C_2H_5{-}C{=}O{-}C_2H_5$	101.5
Benzaldehyde	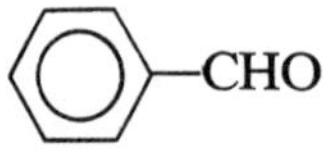 ⬡—CHO	178	Phenylethanone (acetophenone)	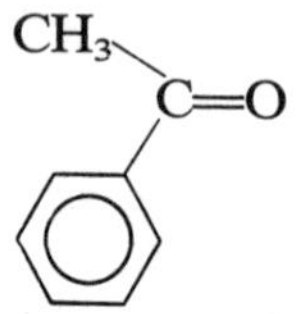 $CH_3{-}C{=}O{-}$⬡	20 (m.p.)

Aldehydes and ketones of low molecular mass are colourless flammable liquids of low boiling points, with characteristic pungent odours. Methanal, ethanal and propanone are completely miscible with water; the solubility is reduced as the extension of the carbon chain increases the oily character. They burn to give a mixture of carbon dioxide and water.

$$2CH_3CHO + 5O_2 \rightarrow 4CO_2 + 4H_2O$$

Aldehydes are so readily oxidised that they reduce copper(II) ions in Fehling's or Benedict's solutions to copper(I) oxide and the silver(I) ions in Tollen's reagent to silver. These reactions provide tests to distinguish aldehydes from ketones; ketones, not being readily oxidised, are not reducing agents and do not respond to the tests.

The aromatic aldehyde benzaldehyde, although very readily oxidised, responds only slowly to these tests; it is much less soluble than ethanal and does not mix readily with the aqueous reagents.

Both ethanal and benzaldehyde can be oxidised to the corresponding acids by passing the heated vapour mixed with air over a copper or nickel catalyst:

$$2CH_3CHO + O_2 \xrightarrow[\text{heat}]{\text{Cu}} 2CH_3COOH$$

Benzaldehyde will in fact oxidise at room temperature:

$$2C_6H_5CHO + O_2 \rightarrow 2C_6H_5COOH$$
Benzoic acid

Ketones, however, can only be oxidised by employing drastic conditions such as refluxing with a concentrated solution of sodium dichromate and sulphuric acid. The products are mixtures of carboxylic acids resulting from the total disruption of the ketone.

(b) Addition reactions of aldehydes and ketones

The unsaturated carbonyl group $>C=O$ undergoes addition reactions, but, in contrast to the alkanes, it is attacked by nucleophilic reagents rather than by electrophilic reagents (page 343).

The carbon–oxygen bond is polarised as a result of the greater electronegativity of the oxygen atom (inductive effect) leaving the carbon atom with a fractional positive charge:

$$>C^{\delta+}{=\!\!=}O^{\delta-}$$

Hence the carbon atom of the carbonyl group can be attacked by the donation of a lone pair of electrons from a nucleophilic reagent. For example, the first step in the addition of HCN to ethanal is:

The intermediate ion with a negative charge reacts with a proton in a second step to form the final addition product, a cyanohydrin:

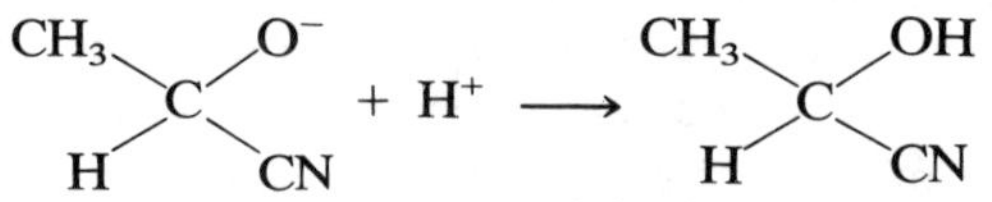

2-hydroxypropanenitrile

This reaction cannot take place except in the presence of a strong base which will react with the weakly acidic HCN to provide the cyanide ions:

$$HCN + OH^- \rightarrow CN^- + H_2O$$

The addition of HCN to aldehydes and ketones is important, providing a method of synthesising hydroxy-substituted acids. Thus the conversion of ethanal to 2-hydroxypropanenitrile can be followed by hydrolysis of the $C{\equiv}N$ group yielding 2-hydroxypropanoic acid. Another important aspect is that an extra carbon atom is added to the alkyl chain.

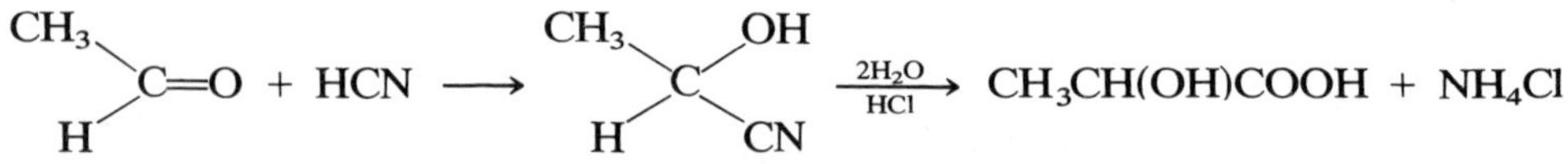

Similar mechanisms allow the addition of sodium hydrogen sulphite to form crystalline derivatives; these are used in the purification of aldehydes and ketones. The nucleophilic attack takes place through the lone pair of electrons on the sulphur atom:

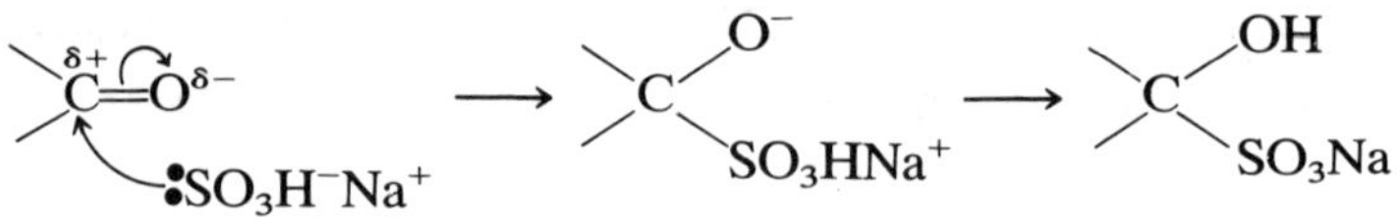

Thus for propanone:

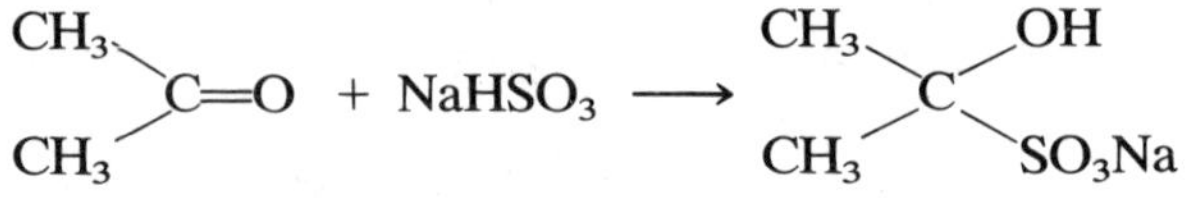

The ammonia molecule reacts similarly with the carbonyl group, but the products are too unstable to be isolated; they rapidly form resinous polymers:

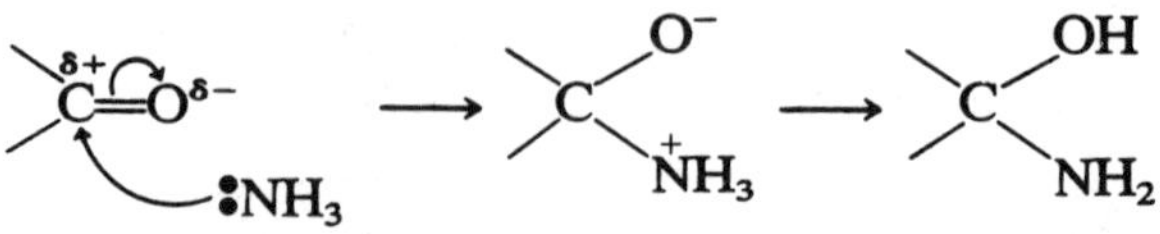

The reductions of aldehydes and ketones to alcohols by lithium tetra-hydridoaluminate(III) (lithium aluminium hydride), $LiAlH_4$, are addition reactions following nucleophilic attack. The AlH_4^- ion is regarded as a hydride ion carrier (:H^-); hence the reaction, which requires the presence of dilute acid, takes place in the following steps:

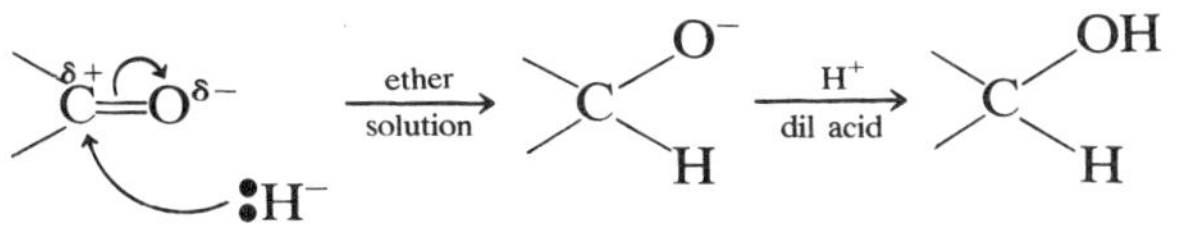

Aldehydes are reduced to primary alcohols while ketones are reduced to secondary alcohols:

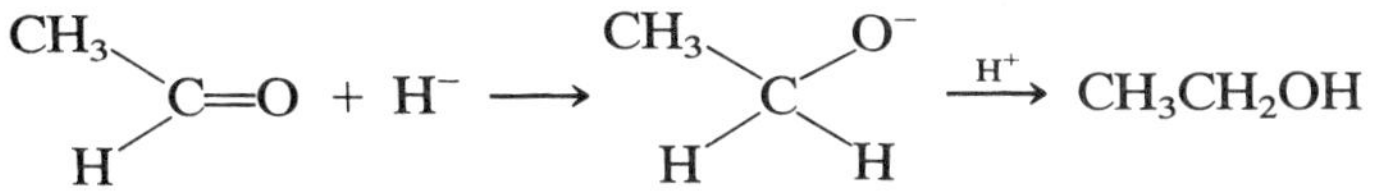

Ethanal

Ethanol

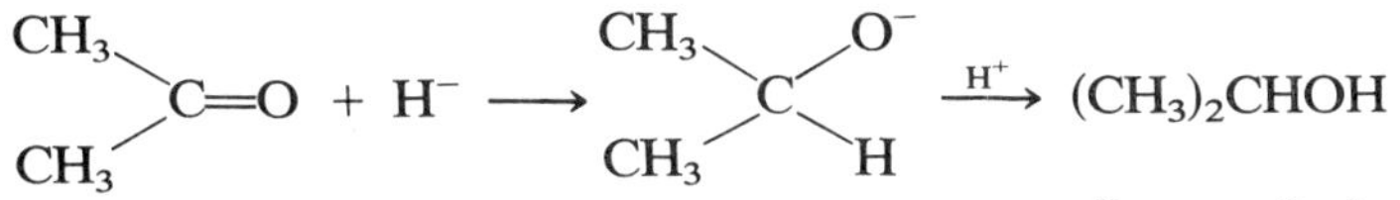

Propanone

Propan-2-ol

More drastic reduction of ketones, for example using zinc and hydrochloric acid, results in the production of hydrocarbons:

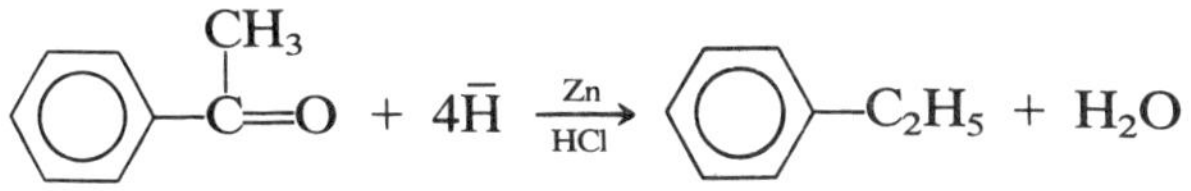

Phenylethanone

Ethylbenzene

This reaction is known as Clemmensen's reaction, after its originator.

(c) Condensation reactions of aldehydes and ketones

While ammonia reacts with the carbonyl group by nucleophilic attack to make addition compounds which later polymerise, derivatives of ammonia such as hydroxylamine, NH_2OH, hydrazine, NH_2NH_2, and substituted hydrazines make stable compounds with aldehydes and ketones by the so-called condensation reactions. These start with nucleophilic attack on the carbonyl group by the electron-donating reagents which results in addition reactions followed by the elimination of small molecules, usually water (page 342).

In general, the reaction between the carbonyl group and NH_2X, where $X = -OH$, $-NH_2$, C_6H_5NH-, or $2,4-(NO_2)_2C_6H_3NH-$, can be represented:

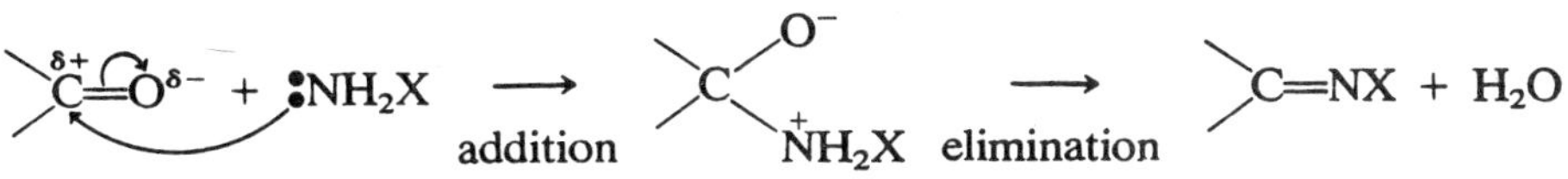

addition   elimination

Thus with hydroxylamine, oximes are formed:

Ethanal oxime

With hydrazine, hydrazones result:

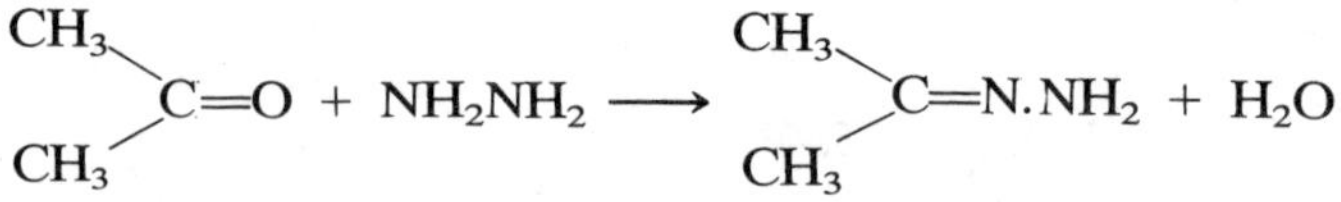

propanone hydrazone

The most important of all these reactions are those with 2,4-dinitro-phenylhydrazine, resulting in the formation of bright yellow crystalline derivatives of sharp melting points which can be used to identify the aldehydes and ketones, Some typical derivatives and their melting points are given in Table 26.2.

Table 26.2 Carbonyl derivatives

Carbonyl compound	2,4-Dinitrophenylhydrazone derivative
CH₃CHO Ethanal	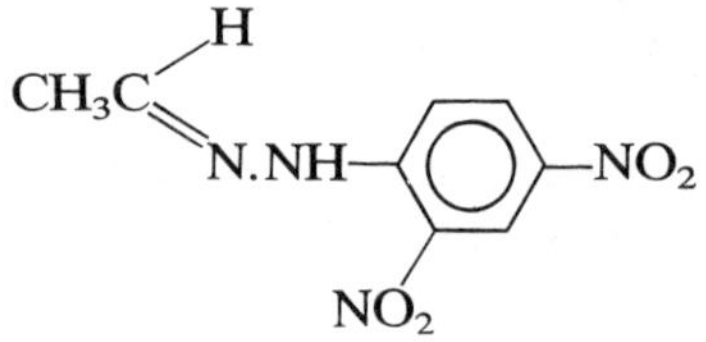
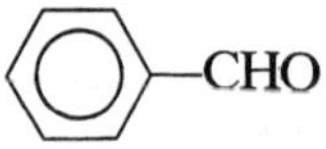 Benzaldehyde	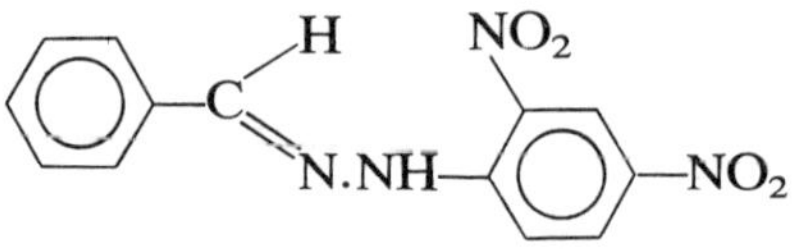
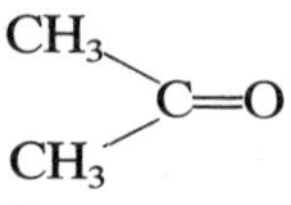 Propanone	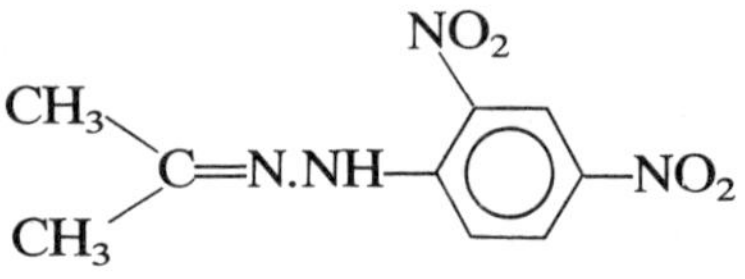

(d) Reactions involving halogens

Phosphorus pentachloride reacts with the carbonyl group to replace the oxygen atom with two chlorine atoms; no evolution of hydrogen chloride occurs as with alcohols or other compounds containing hydroxyl groups.

$$CH_3CHO + PCl_5 \rightarrow CH_3CHCl_2 + POCl_3$$
1,1-dichloroethane

$$(CH_3)_2C{=}O + PCl_5 \rightarrow (CH_3)_2CCl_2 + POCl_3$$
2,2-dichloropropane

Elemental halogens attack the alkyl or aryl radicals rather than the carbonyl group. Substitution reactions in which a hydrogen atom is replaced by a halogen atom take place, the products varying with the conditions.

Chlorine bubbled through ethanal or propanone yields fully substituted products:

$$CH_3CHO + 3Cl_2 \rightarrow CCl_3CHO + 3HCl$$
Trichloroethanal
(chloral)

$$(CH_3)_2C{=}O + 6Cl_2 \rightarrow (CCl_3)_2C{=}O + 6HCl$$

In the presence of a halogen carrier (page 359) the substitution in the benzene ring of benzaldehyde takes place in the meta-position, forming 2-chlorobenzaldehyde, since the $-C{<}^H_O$ group is meta-directing as is the nitro-group (page 417).

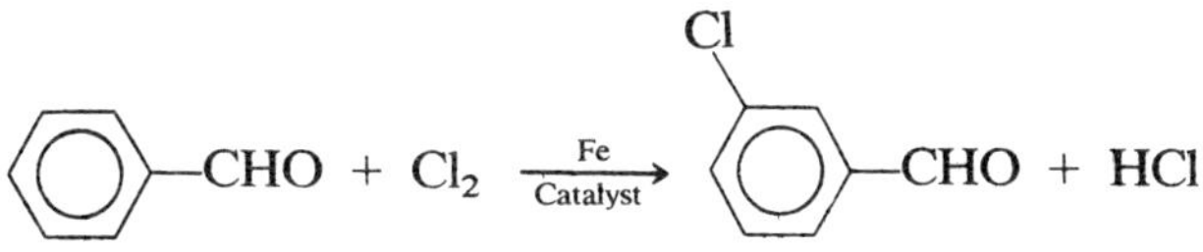

If no catalyst is present the hydrogen atom of the carbonyl group is replaced and benzoyl chloride is formed.

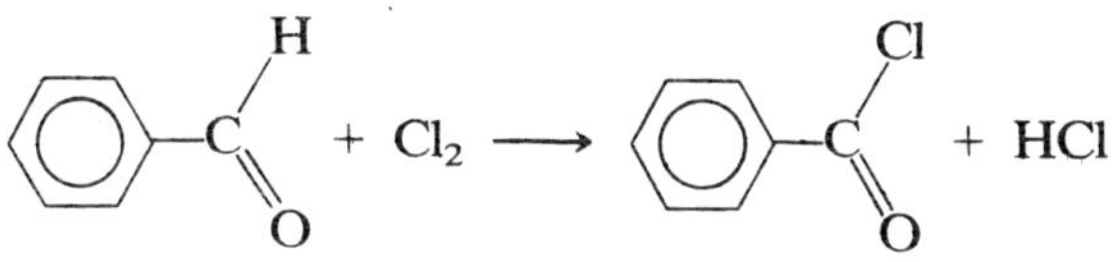

The reactions of propanone with bromine and iodine in dilute acid solution take place slowly and have been extensively used to study rates or reaction (page 154).

With iodine in aqueous sodium hydroxide solution, ethanal, propanone and phenylethanone, all of which contain a $\overset{CH_3}{{>}}C{=}O$ grouping, undergo the iodoform reaction (page 378) yielding the characteristic yellow crystals, CHI_3.

(e) Polymerisation reactions

Aldehydes have a greater tendency to polymerise than ketones; the reactions differ from aldehyde to aldehyde and with the conditions.

An aqueous solution of methanal (formaldehyde) evaporated slowly leaves a white solid polymer, polymethanal, which can be used to disinfect rooms as it yields methanal vapour on heating. Left in limewater, methanal polymerises to a mixture of sugars known as formose. Stronger alkalis, such as a concentrated solution of sodium or potassium hydroxide, convert it to a mixture of methanol and sodium or potassium methanoate (Cannizzaro's reaction).

$$2HCHO + NaOH \rightarrow CH_3OH + H.COONa$$

Methanal also forms a condensation polymer, bakelite (page 412), with phenol which is one of the *thermo-setting type*—once moulded, it cannot be remelted and remoulded.

The reactions of ethanal with alkalis also depend on the concentration of the alkali. Dilute sodium hydroxide causes dimerisation into a colourless liquid 3-hydroxybutanal, which contains both aldehyde and alcohol groups, known as aldol (the aldol condensation):

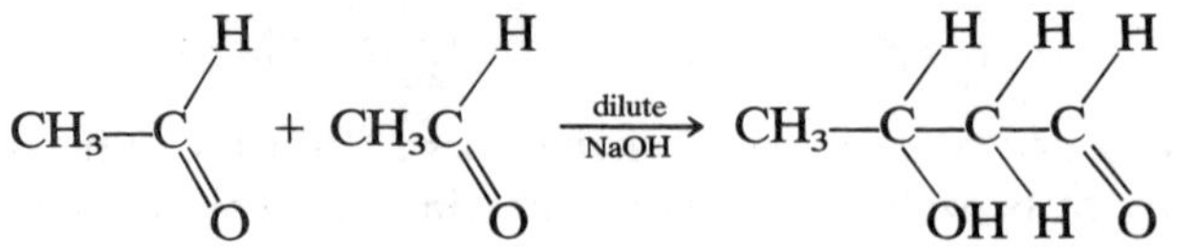

3-hydroxybutanal

More concentrated alkali, on heating the reaction mixture, leads to the formation of a brown resinous polymer of unknown constitution.

With concentrated sulphuric acid at room temperature, ethanal forms a cyclic polymer, ethanal trimer, known as paraldehyde, a colourless liquid used as a hypnotic drug in the treatment of certain mental disorders.

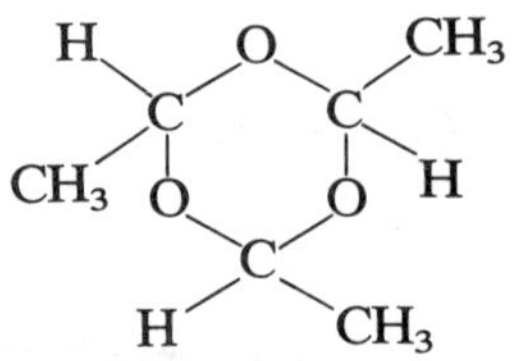

Ethanal trimer
(paraldehyde)

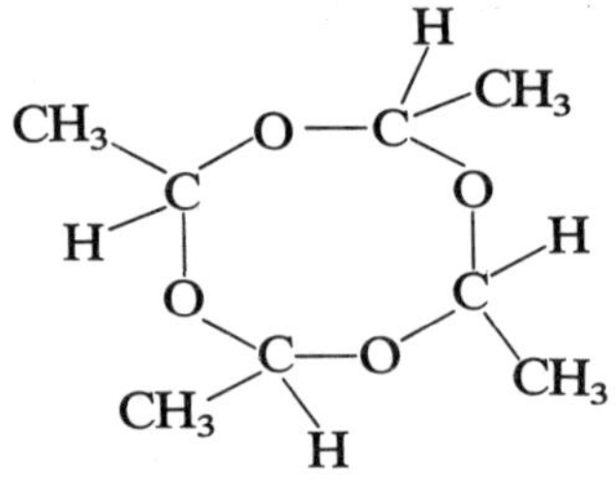

Ethanal tetramer
(metaldehyde)

Changing the conditions, dilute sulphuric acid at 0°C gives rise to ethanal tetramer, used as a solid fuel (meta fuel) and, as a spray or mixed with bran, to kill slugs.

Benzaldehyde does not polymerise to the extent of the low molecular mass alkyl aldehydes, but in alcoholic solution in the presence of sodium cyanide it dimerises to 2-hydroxy-1,2-diphenyl ethanone.

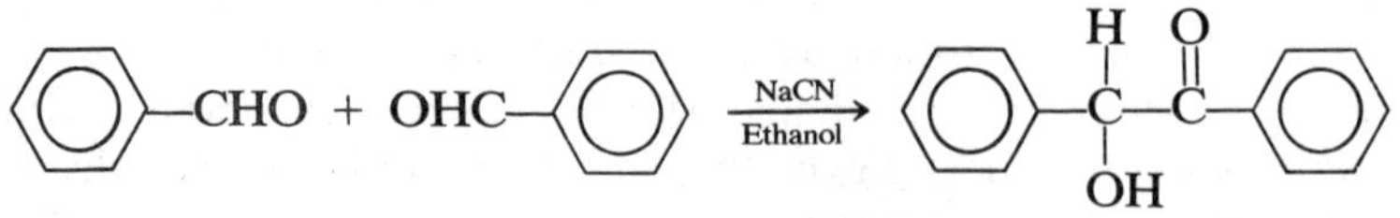

2-hydroxy-1,2-diphenylethanone
(benzoin)

With aqueous sodium hydroxide it undergoes the Cannizzaro reaction in a manner similar to methanal:

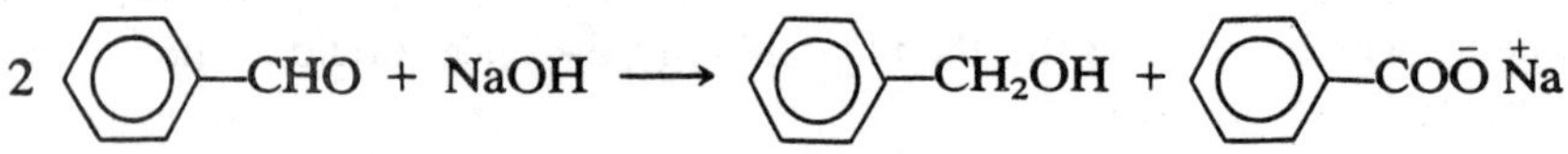

Again, in dilute aqueous solution, benzaldehyde will condense with ethanal to give 3-phenylpropenal, which is the chief constituent of cinnamon oil, and can be oxidised to 3-phenylpropenoic acid, which in turn can be used to make styrene (page 356).

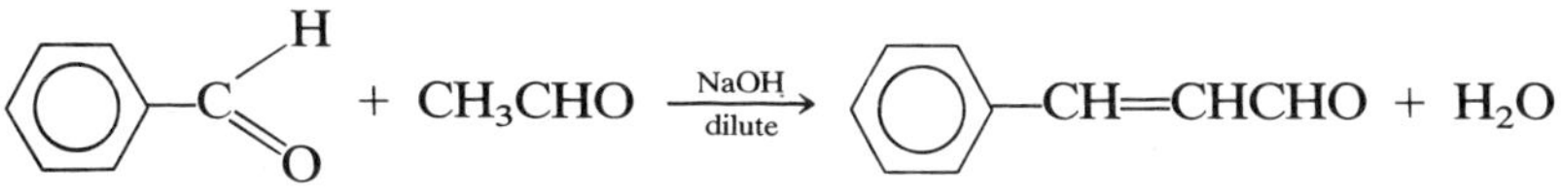

3-phenylpropenal (cinnamaldehyde)

Ketones have less tendency to polymerise than aldehydes, but propanone forms a dimer and a trimer. The dimer is formed when propanone is mixed with a dilute aqueous solution of barium hydroxide, the reaction resembling the formation of 3-hydroxybutanal.

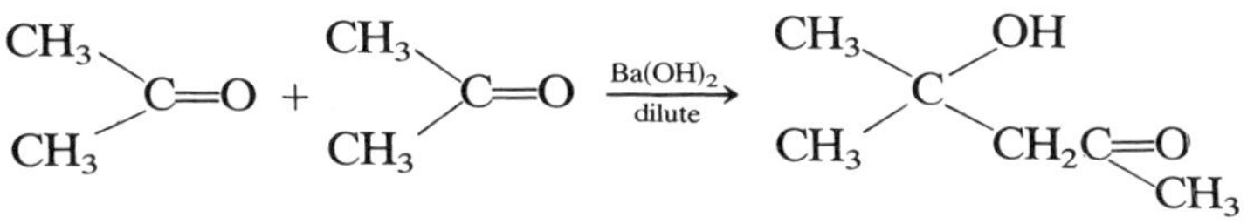

Distilled with concentrated sulphuric acid, propanone is converted to the aromatic molecule 1,3,5-trimethylbenzene by a process of autocondensation.

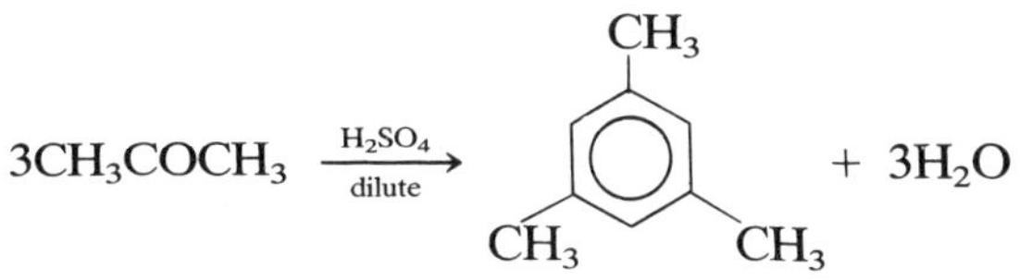

1,3,5-trimethylbenzene
(mesitylene)

26.2 Carbohydrates: classification

Carbohydrates are naturally occurring compounds of carbon, hydrogen and oxygen of general formula $C_xH_{2y}O_y$ in which the ratio of hydrogen to oxygen atoms is two to one, as in a molecule of water. All carbohydrates have this general formula, but some other compounds such as ethanoic acid ($C_2H_4O_2$) which contain hydrogen and oxygen in this ratio are not classed as carbohydrates.

The carbohydrates fall into two classes, sugars and polysaccharides. All the sugars are white crystalline substances which dissolve in water (page 230) to give sweet solutions. The simplest are the monosaccharides, which may be pentoses ($C_5H_{10}O_5$) or hexoses ($C_6H_{12}O_6$); of these the hexoses such as glucose, fructose, mannose and galactose are the more important. The disaccharides ($C_{12}H_{22}O_{11}$) such as sucrose and maltose are more complex sugars which can be hydrolysed to two hexose molecules of the same or different structural formulae.

The monosaccharides are subdivided into aldoses and ketoses according to their structures; they all have a reducing action. Disaccharides are subdivided into reducing and non-reducing sugars according to their ability to reduce Tollen's reagent and Fehling's solution.

The polysaccharides $(C_6H_{10}O_5)_n$ are amorphous compounds of very high relative molecular mass, tasteless and mostly insoluble in water. The most important are the polyhexoses (starch, cellulose and glycogen) all of which can be hydrolysed to glucose molecules and can therefore be regarded as natural polymers of glucose.

All the carbohydrates found naturally are optically active, many of them having more than one asymmetric carbon atom. A brief account of the better known carbohydrates follows.

26.3   Monosaccharides

(a) Glucose (dextrose)

Glucose occurs in the sap of plants, the blood of mammals, in honey and in fruits, particularly grapes—hence its common name, grape-sugar. It is a white crystalline solid which can be shown by X-ray analysis to have a six-membered ring structure which has two isomeric forms, α-glucose and β-glucose.

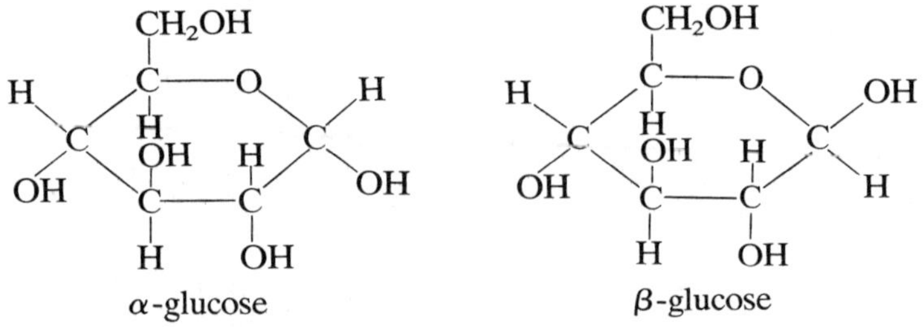

Fig. 26.1   Glucose molecules

Both these forms contain four asymmetric carbon atoms giving rise to a number of optical isomers. Naturally-occurring glucose is dextro-rotatory, having a specific rotation of +52.5°; it is thought to be a mixture of α- and β-glucose, both of which are dextro-rotatory but to different degrees. The ring structure is readily opened to give a straight chain molecule:

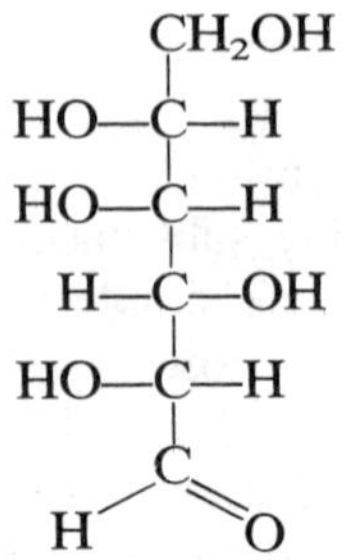

This can be seen to contain an aldehyde group, and glucose is classed as an aldose or aldohexose. The structure is confirmed by the reactions of glucose, which will reduce Fehling's or Benedict's solution, and forms a silver mirror with Tollen's reagent. It forms an addition compound with hydrogen cyanide, condensation compounds with hydroxylamine and phenyl hydrazine, and a resinous material when heated with sodium hydroxide. The presence of five hydroxyl groups in its molecule can be shown by reaction with ethanoyl chloride to form an ester containing five ethanoate groups. That the six carbon atoms are linked in a straight chain is shown by the production of n-hexane when the glucose is completely reduced. The most important reaction of glucose is its fermentation to ethanol and carbon dioxide by the action of yeast:

$$C_6H_{12}O_6 \xrightarrow{\text{yeast}} 2C_2H_5OH + 2CO_2$$

(b) Fructose (laevulose)

Fructose, which is found, mixed with glucose, in fruits or honey, is commonly called fruit-sugar. It is a ketohexose or ketose; its ring structure opens out to a straight chain structure containing a keto- group.

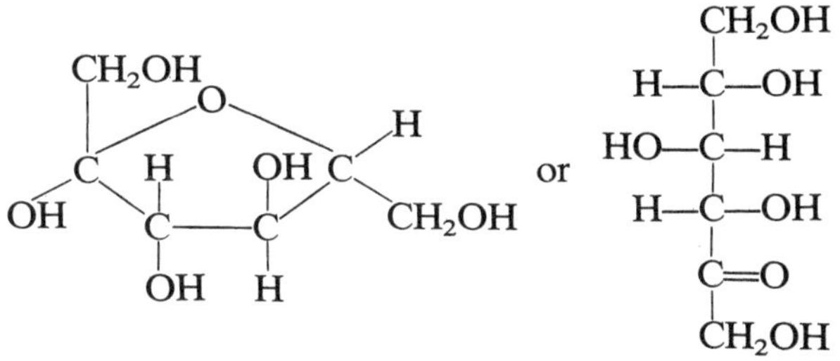

Fig. 26.2 Fructose

Unlike simple ketones, fructose reduces both Fehling's solution and Tollen's reagent, but like ketones it forms condensation products with hydroxylamine and phenyl hydrazine, and an addition compound with hydrogen cyanide. Like glucose, fructose is fermented by yeast, although less readily, to ethanol and carbon dioxide. The naturally occurring substance has three asymmetric carbon atoms and is laevo-rotatory.

(c) Mannose and galactose

Mannose and galactose are further isomers of glucose, with a chemical behaviour similar to that of glucose. Galactose is prepared by acid hydrolysis of lactose.

26.4 Disaccharides

(a) Sucrose

This is the best known of the disaccharides since it is the sugar extracted from both sugar-cane and sugar-beet. Boiled with hydrochloric acid it hydrolyses to yield an equimolecular mixture of glucose and fructose; its molecule can be shown to be made up of one of each of these monosaccharides, with the elimination of one molecule of water:

$$C_{12}H_{22}O_{11} + H_2O \rightarrow 2C_6H_{12}O_6$$

Sucrose   Fructose +
                                    Glucose

Sucrose is not a reducing agent, which suggests that the two monosaccharide molecules are linked through their aldehyde and ketone groups, thus rendering them inactive.

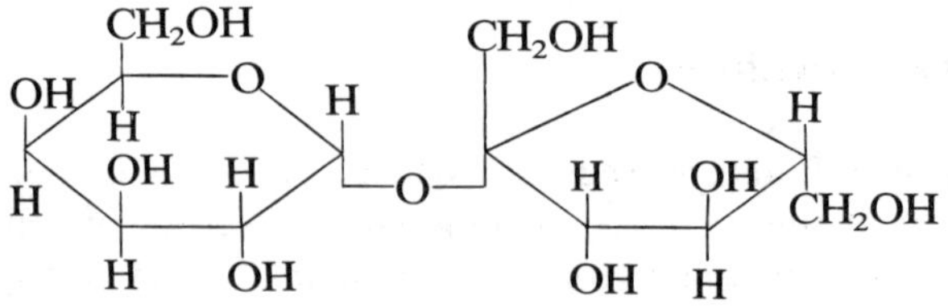

Fig. 26.3   Sucrose

Natural sucrose is dextro-rotatory; on hydrolysis it becomes laevo-rotatory since the fructose in the equimolecular mixture is more laevo-rotatory than the glucose is dextro-rotatory. This effect is known as the *inversion of cane-sugar*. The rate of inversion can be followed by placing an acidified sucrose solution in a polarimeter and taking readings at regular intervals. If a series of solutions of different acidities are used the rate of change from dextro-rotatory to laevo-rotatory will be found to depend on the concentration of hydrogen ions in the solution. The rate of inversion can be used to find the degree of ionisation of a weak acid (page 174).

(b) Maltose

Maltose is produced when starch is hydrolysed by the enzymes in saliva. It can be further hydrolysed by dilute hydrochloric acid to yield two molecules of glucose for each molecule of maltose, indicating that it is derived from two glucose molecules. Maltose is a reducing sugar, which suggests that one of the aldehyde groups of the two glucose molecules is free; consequently the linkage is thought to be between the aldehyde group of one glucose molecule and an —OH group in the other.

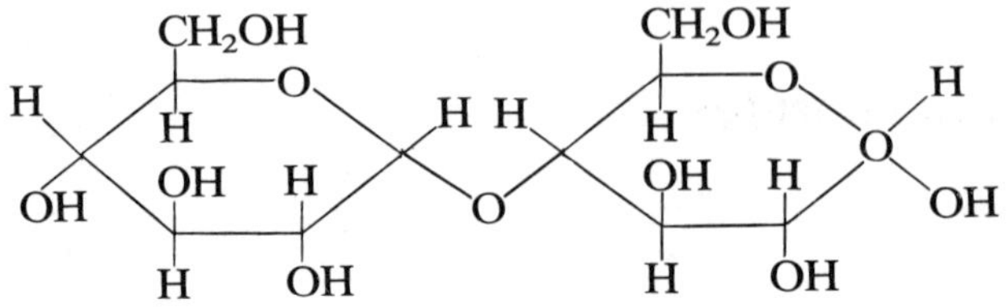

Fig. 26.4   Maltose

(c) Lactose

Lactose or milk-sugar is not found in plants but occurs in the milk of mammals. On hydrolysis it yields a mixture of glucose and galactose.

26.5 Polysaccharides

(a) Starch

Starch is the most important carbohydrate food available to animals. it occurs in cereals, potatoes, root vegetables and green plants. During digestion starch is broken down progressively by enzymes, first to maltose and then to glucose. It can also be hydrolysed to glucose by boiling with dilute hydrochloric acid:

$$(C_6H_{10}O_5)_{2n} \xrightarrow{nH_2O} n(C_{12}H_{22}O_{11}) \xrightarrow{nH_2O} 2n(C_6H_{12}O_6)$$

$$\text{Starch} \qquad\qquad \text{Maltose} \qquad\qquad \text{Glucose}$$

Thus starch is a macromolecule composed of glucose units. These units link together in chains giving rise to two different types of starch molecule, α- and β-amylose. α-Amylose, of relative molecular mass 10 000–60 000, is water soluble and made up of linear chains of glucose units with little branching or cross-linking of the chains; it occludes iodine molecules to give a blue colouration. β-Amylose is composed of molecules of relative molecular mass 50 000–100 000; the chains of glucose molecules are considerably branched and joined in a three-dimensional network; this type of starch is insoluble in water and gives a violet colouration with iodine. Both types of starch molecules are composed of α-glucose molecules linked in a giant structure:

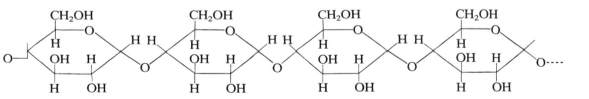

Fig. 26.5 Starch molecule composed of α-glucose units

(b) Cellulose

Cellulose, like starch, is a macromolecule made up of glucose units, but, unlike starch, it is not readily hydrolysed and therefore not so easily digested. Only mammals of the ruminant class such as cows and sheep, and other grazing animals which have the necessary digestive systems, can utilise cellulose as food. The difference between starch and cellulose is that cellulose is composed of β-glucose molecules, which gives it a structure different from that of starch.

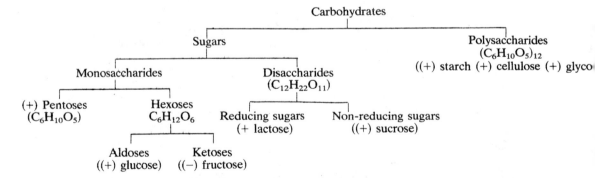

Fig. 26.6 Cellulose

Cellulose is the principal constituent of cell walls in plants; it is probably the most abundant natural carbohydrate. It does not give a blue or violet colouration with iodine.

(c) Glycogen

Excess glucose from the digestion of starch or sugar is converted in the body to the polysaccharide glycogen and stored in the liver and muscles. The liver secretes an enzyme that hydrolyses the glycogen slowly to glucose that can enter the bloodstream when it is needed. Energy for muscular contraction comes from the conversion of glycogen to 2-hydroxypropanoic acid (lactic acid).

$$(C_6H_{10}O_5)_n + nH_2O \longrightarrow 2n\left(CH_3-\overset{\displaystyle OH}{\underset{\displaystyle |}{C}}HCOOH\right) \quad \Delta H \text{ negative}$$

The lactic acid is subsequently rebuilt to glycogen.

Glycogen is a soluble white substance of relative molecular mass 100 000–2 000 000; it gives a purplish-red colour with iodine and it is sometimes known as animal starch.

Summary

Carbohydrates

Sugars — Polysaccharides $(C_6H_{10}O_5)_{12}$ ((+) starch (+) cellulose (+) glyco

Monosaccharides — Disaccharides $(C_{12}H_{22}O_{11})$

(+) Pentoses $(C_6H_{10}O_5)$ Hexoses $C_6H_{12}O_6$ Reducing sugars (+ lactose)   Non-reducing sugars ((+) sucrose)

Aldoses ((+) glucose)   Ketoses ((−) fructose)

Questions

1. Draw up a table of reactions which are shared by aldehydes and ketones. Give three reactions in which they differ.

2. Describe the reactions of ethanal and benzeldehyde, with chlorine, pointing out any resemblances or differences.

3. Give a series of reactions by which 1-hydroxybutanol might be obtained from propan-1-ol.

4. Given 2 samples known to be of propanal and propanone, describe how you could identify which was which.

5. Describe how you could obtain a sample of pure benzaldehyde from an impure sample.

6. Aldehydes, ketones and carboxylic acids contain the carbonyl group ($>C=O$). This group is often said to be responsible for the addition and condensation reactions undergone by aldehydes and ketones.

   (a) Review briefly the major addition and condensation reactions shown by aldehydes and ketones; in each case discussed, examine the role of the carbonyl group in bringing about these reactions.

   (b) Carboxylic acids do not undergo any of the addition and condensation reactions shown by aldehydes and ketones, despite the fact that they contain the carbonyl group. Give explanations for this.   (L)

7. (a) Polymerisation reactions may be classified as addition or condensation reactions. Explain the meaning of these terms, illustrating your answer with two named examples in each case. How would you demonstrate in the laboratory one such polymerisation process?

   (b) Compare the structures of the simple monosaccharides, glucose and fructose. How do solutions of these compounds affect (i) plane polarised light, and (ii) ammoniacal solutions of silver nitrate?

   Indicate by means of a block diagram the structure of a polysaccharide.

   (L)

27 Carboxylic acids and derivatives

27.1 Alkyl and aryl carboxylic acids

(a) Structure and properties

The alkyl carboxylic acids of general formula $C_nH_{2n+1}COOH$ constitute an homologous series, originally called the fatty or aliphatic acids. They occur widely in nature and several have been known since early times. Methanoic acid is the formic acid obtained from red ants (Latin: formica) and is present in the stings of other insects and nettles. Ethanoic acid (acetic acid) was obtained by the souring of wines to form vinegar (acetum) while butanoic acid (butyric acid) occurs as the glyceride in butter (butyrum). Higher molecular mass acids such as hexadecanoic (palmitic) acid and octadecanoic (stearic) acid occur as glycerides in vegetable and animal fats. The esters of the lower molecular mass acids are responsible for the taste and smell of many fruits and flowers, as are the esters of benzoic acid, the simplest of the aryl carboxylic or aromatic acids. Some examples are given in Table 27.1.

Table 27.1 Carboxylic acids

Acid	Formula	m.p. °C	b.p. °C
Methanoic (formic)	HCOOH	8	101
Ethanoic (acetic)	CH_3COOH	17	118
Propanoic	C_2H_5COOH	−21	141
Butanoic (butyric)	C_3H_7COOH	−4	164
Hexadecanoic (palmitic)	$C_{15}H_{31}COOH$	63	—
Octadecanoic (stearic)	$C_{17}H_{35}COOH$	72	—
Benzoic (systematically benzenecarboxylic acid)	⟨◯⟩—COOH	122	—

The acids of low molecular mass are colourless liquids with pungent odours (butyric acid is responsible for the smell of rancid butter and stale sweat). The boiling points of the acids are higher than might be expected from their molecular mass because of dimerisation resulting from hydrogen bonding (page 78). The lower members are completely miscible with water, but this solubility decreases as the length of the carbon chain increases and the acids of higher molecular mass are waxy solids, completely immiscible with water. Some of these are used in the manufacture of soap. Benzoic acid is also a waxy solid, soluble in hot water but only slightly so in cold. It is used as a preservative in soft drinks and in the manufacture of dyes, perfumes and rust inhibitors.

The chemical properties of the acids are those of the functional group $-C{\displaystyle {\nwarrow_{OH}^{O}}}$. The behaviour of the hydroxyl group is typical in that, for example, clouds of hydrogen chloride are released by reaction with phosphorus pentachloride, but the carbonyl group does not show the typical reactions of aldehydes and ketones. Also, unlike these compounds, carboxylic acids do not form addition compounds with sodium hydrogen sulphite or hydrogen cyanide, nor do they form condensation compounds with hydroxylamine or hydrazine. The properties of the carbonyl group are modified by the presence of the hydroxyl group; the structure of the acid is represented with the carboxylate ion as a resonance hybrid:

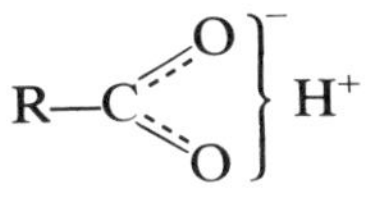

the possible canonical structures of the ion being:

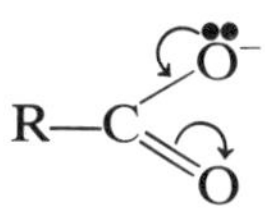

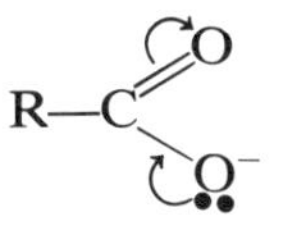

Thus the tendency to ionise as an acid is explained by the electron-attracting effect of the C=O bond in the carbonyl group making the release of the proton easier.

Ethanoic acid is not such a strong acid as methanoic, the effect of the carbonyl group being reduced by the electron-donating effect of the methyl group (page 344). On the other hand, benzoic acid is stronger than ethanoic because the phenyl group tends to withdraw electrons (page 384) thus reinforcing the effect of the carbonyl group. A comparison of the dissociation constants (page 174) for these acids illustrates the point.

Acid		$Ka\ mol\ dm^{-3}$
Methanoic	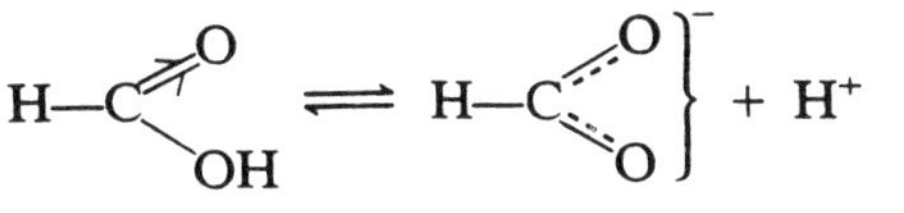	1.6×10^{-4}
Ethanoic	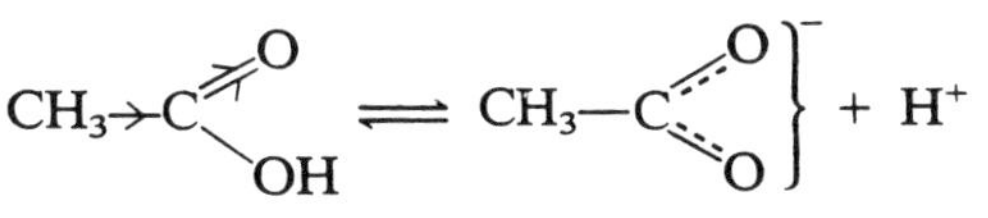	1.7×10^{-5}
Benzoic	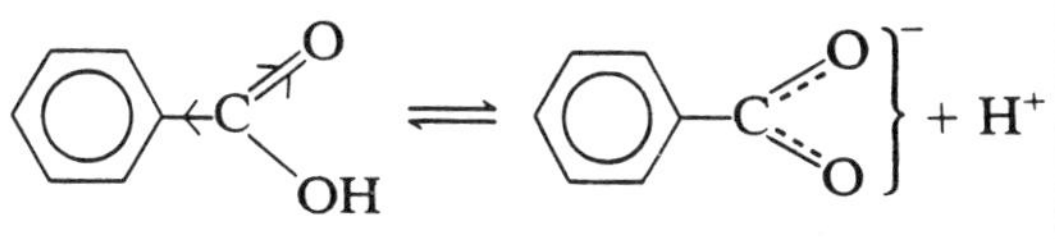	6.3×10^{-5}

(b) Reactions of the carboxylic acids

(i) As acids to form salts

The low molecular mass acids are strong enough to release carbon dioxide from carbonates, and they react with alkalis and ammonia to form salts in which the acidic hydrogen is replaced by a metal ion or the ammonium group:

$$2CH_3COOH + Na_2CO_3 \rightarrow 2CH_3COONa + CO_2 + H_2O$$

$$CH_3COOH + NH_3 \rightarrow CH_3COONH_4$$

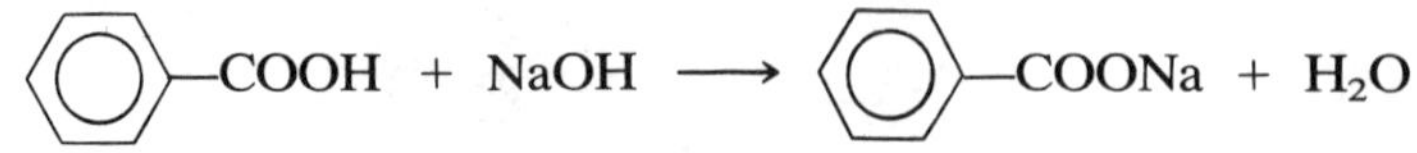

(ii) Reaction with alcohols to form esters

The carboxylic acids react with alcohols to form esters with the elimination of a molecule of water. The reactions are usually carried out in the presence of concentrated sulphuric acid which takes up the water.

$$CH_3COOH + C_2H_5OH \rightarrow CH_3COOC_2H_5 + H_2O$$
$$\text{Ethyl ethanoate}$$

Methyl benzoate

When esters are formed the hydroxyl group of the acid is replaced by the RO— group of the alcohol, in contrast to the formation of salts where the acidic hydrogen is replaced by a metal. That the acyl-oxygen bond is broken has been demonstrated by using isotopes of oxygen. An alcohol is synthesised so that some of its molecules contain the isotope of oxygen of mass eighteen (^{18}O). This is reacted with an unmarked acid in which all the oxygen atoms are of the isotope of mass sixteen (^{16}O). In the resulting reaction the water molecules produced contain only ^{16}O atoms; these must therefore have come from the acid.

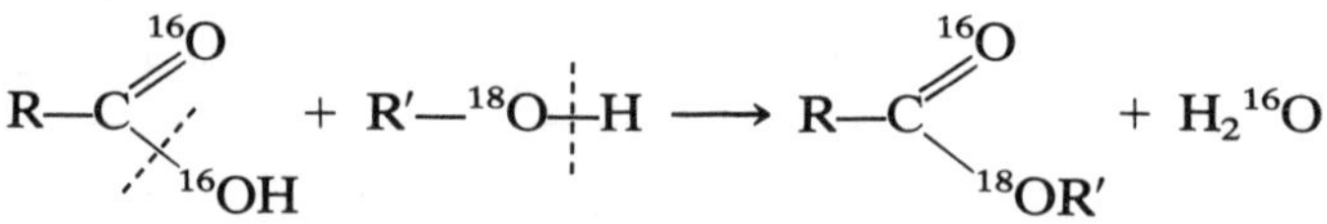

(iii) Dehydration to anhydrides

Carboxylic acids can be converted to their anhydrides by passing the vapour over heated zinc oxide, removing a molecule of water between two acid molecules:

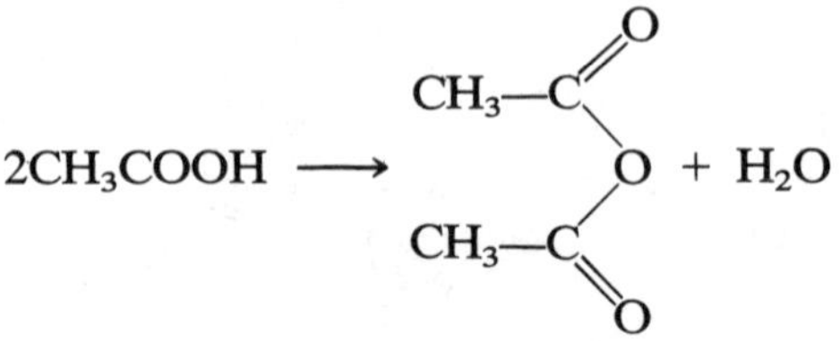

Ethanoic anhydride

Benzoic acid is dehydrated similarly by boiling the acid in ethanoic acid under reflux; the water being distilled away.

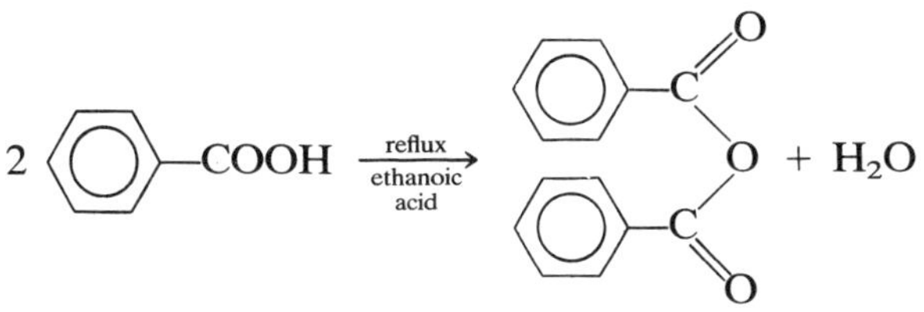

$$2 \langle \bigcirc \rangle - COOH \xrightarrow[\substack{ethanoic\\acid}]{reflux} \quad + H_2O$$

Methanoic acid, having no alkyl group in the molecule, dehydrates to give carbon monoxide and water; concentrated sulphuric acid being used to take up the water.

$$HCOOH \xrightarrow{H_2SO_4} CO + H_2O$$

(iv) Reaction with phosphorus halides

Phosphorus pentachloride and phosphorus trichloride react to replace the hydroxyl group of the acids with a chlorine atom to give acyl or benzoyl chlorides (page 443).

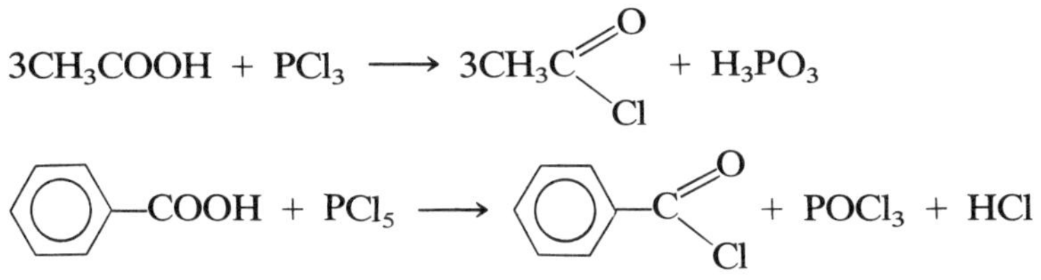

$$3CH_3COOH + PCl_3 \longrightarrow 3CH_3C{\overset{O}{\underset{Cl}{\big<}}} + H_3PO_3$$

$$\langle \bigcirc \rangle - COOH + PCl_5 \longrightarrow \langle \bigcirc \rangle - C{\overset{O}{\underset{Cl}{\big<}}} + POCl_3 + HCl$$

The formation of acyl chlorides can also be effected by using sulphur dichloride oxide.

$$CH_3COOH + SOCl_2 \rightarrow CH_3COCl + SO_2 + HCl$$

(v) Halogenation of the alkyl or aryl group

As might be expected from reactions described previously, halogens will attack the alkyl or aryl group of carboxylic acids in the presence of bright sunlight or if a halogen carrier is used as a catalyst. The reactions proceed readily with the alkyl carboxylic acids, the carbon atom next to the carboxyl group being attacked first:

$$CH_3COOH + Cl_2 \rightarrow CH_2ClCOOH + HCl$$
Chloroethanoic acid

$$CH_3CH_2COOH + Cl_2 \rightarrow CH_3CHCHClCOOH + HCl$$
2-Chloropropanoic acid

Benzoic acid undergoes nuclear substitution less readily as the carboxyl group is deactivating and, like the nitro-group in nitrobenzene, *meta*-directing (page 417).

(vi) *Oxidation*

With the exception of methanoic acid, neither the alkyl nor the aryl carboxylic acids are readily oxidised. Aliphatic acids with long carbon chains such as hexadecanoic and octadecanoic acids burn well with a yellow flame and are used in candle making.

Methanoic acid will reduce Fehling's solution and ammoniacal silver nitrate solution with difficulty, and can be oxidised to carbon dioxide and water by acidified potassium manganate(VII). These oxidation reactions of methanoic acid are consequences of the absence of alkyl groups in the molecule.

(vii) *Reduction*

Carboxylic acids resist reduction and if this is required the acid must either first be converted to the ester, anhydride or chloride, or a strong reducing agent such as lithium tetrahydridoaluminate (III) must be used; the products of reduction are alcohols.

27.2   Salts of carboxylic acids

(a) General properties

The sodium, potassium, and ammonium salts of the carboxylic acids are white crystalline solids soluble in water but not in ethoxyethane or other organic solvents. The solutions of these salts of the weakly acidic carboxylic acids are alkaline as a result of hydrolysis (page 181).

Insoluble iron salts of characteristic colours can be precipitated from neutral solutions of iron(III) chloride. The silver salts are also insoluble.

(b) Reactions

(i) *Electrolysis of sodium salts*

The electrolysis of the sodium salt of a carboxylic acid releases an alkane, together with carbon dioxide, at the anode. This reaction, Kolbe's reaction, is useful for the synthesis of pure samples of specific alkanes.

$$2RCOONa + 2H_2O \rightarrow R-R + 2CO_2 + 2NaOH + H_2$$

(ii) *Action of heat on sodium and calcium salts*

The sodium salts of carboxylic acids are stable to heat with the exception of sodium methanoate, which decomposes at 400°C with the evolution of hydrogen leaving sodium ethanedioate as a residue:

$$2HCOONa \rightarrow \begin{matrix} COONa \\ | \\ COONa \end{matrix} + H_2$$

Calcium salts, however, decompose to yield ketones, leaving a residue of calcium carbonate:

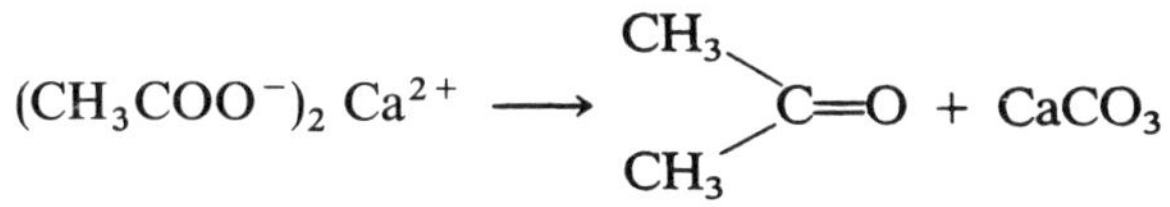

$$(CH_3COO^-)_2 \, Ca^{2+} \longrightarrow \begin{array}{c} CH_3 \\ CH_3 \end{array}\!\!\!C\!\!=\!\!O + CaCO_3$$

Calcium ethanoate   Propanone

The behaviour of the methanoate is again exceptional; methanal is evolved leaving calcium carbonate as a residue:

$$(HCOO^-)_2 \, Ca^{2+} \longrightarrow H.CHO + CaCO_3$$

Calcium methanoate

(iii) Heating sodium and calcium salts with an alkali

When heated with soda-lime both alkyl and aryl carboxylic acids yield hydrocarbons:

$$CH_3COONa + NaOH \longrightarrow CH_4 + Na_2CO_3$$
Methane

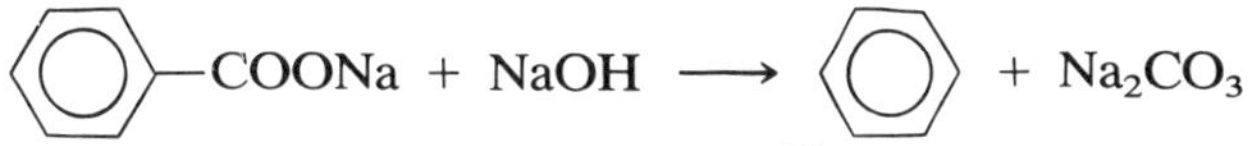

A similar decarboxylation of benzoic acid can be brought about by heating benzoic acid together with calcium oxide:

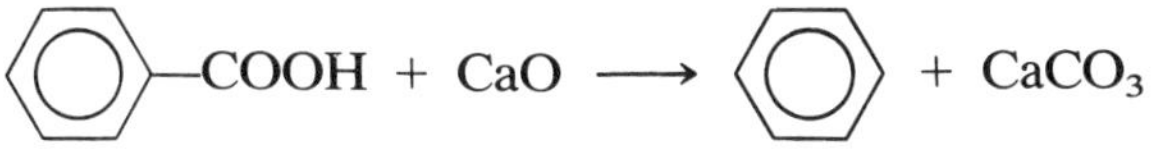

This reaction can be used to remove side-chains from the benzene nucleus.

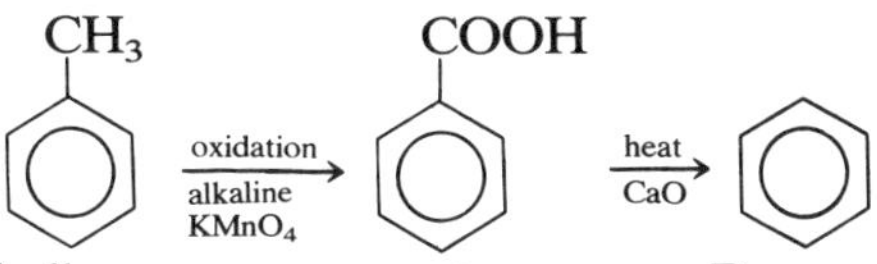

Methylbenzene   Benzene

(iv) Heating sodium salts with acyl chlorides

Anhydrides of the carboxylic acids are formed when the sodium salts are heated with acyl chlorides:

$$CH_3COONa + CH_3COCl \rightarrow (CH_3CO)_2O + NaCl$$
Ethanoic
anhydride

Benzoic anhydride

(v) *Dehydration of the ammonium salts*

The ammonium salts of the carboxylic acids are dehydrated in two stages to give first amides and then nitriles (cyanides):

$$CH_3COONH_4 \rightarrow CH_3CONH_2 \rightarrow CH_3CN$$
$$\text{Ethanamide} \quad \text{Ethanenitrile}$$

Benzamide

Benzonitrile
(phenyl cyanide)

(vi) *Reactions of silver salts*

Esters of alkyl carboxylic acids can be made by the reaction of silver salts with alkyl halides in an inert solvent such as ethoxyethane:

$$CH_3COOAg + C_2H_5Cl \rightarrow CH_3COOC_2H_5 + AgCl$$
$$\text{Ethyl ethanoate}$$

27.3   Derivatives of carboxylic acids

(a) General

The hydroxyl group of a carboxylic acid can be replaced by other functional groups giving a series of compounds of general formula R.COX, where X can be R'O—, NH_2, Cl, Br, I or R'.COO—, giving rise to esters, amides, halides or acid anyhdrides. If R is an alkyl group such as methyl or ethyl the compounds are called *acyl* derivatives; if R is C_6H_5— then the compounds are *benzoyl* derivatives. Thus the acyl derivative CH_3COCl is ethanoyl chloride and C_6H_5COCl is benzoyl chloride.

(b) Esters

The esters, in which the —OH group is replaced by the RO-group from an alcohol, are neutral, fruity-smelling liquids of relatively low boiling points. The first part of the name of a particular ester is derived from the alcohol, while the second part of the name comes from the acid, (see also Section 1.b(ii).

Esters of low molecular mass occur naturally as the flavours and scents of fruit and flowers. They are manufactured for artificial flavouring essences and cheap scents; ethyl methanoate is used in rum essence and 3-methyl-butyl ethanoate (isoamyl acetate) in pear drops. The esters are chemically unreactive, and this, together with their low boiling points, makes them useful solvents for organic compounds such as drugs and antibiotics. Ethyl ethanoate is used as a solvent for nail-varnish.

Esters of higher molecular mass occur as oils, fats and waxes; beeswax is $C_{15}H_{31}COOC_{31}H_{63}$, while palm oil, coconut oil and animal fats are the esters of propane-1,2,3-triol (glycerol). Dodecanoic acid (lauric acid) ($C_{11}H_{23}COOH$), hexadecanoic acid ($C_{15}H_{31}COOH$) and octadecanoic acid

Table 27.2   Esters

Name	Formula	b.p. °C
Methyl methanoate	$HCOOCH_3$	32
Ethyl methanoate	$HCOOC_2H_5$	53
Methyl ethanoate	CH_3COOCH_3	56
Ethyl ethanoate (ethyl acetate)	$CH_3COOC_2H_5$	77
Pentyl ethanoate (amyl acetate)	$CH_3COOC_5H_{11}$	149
Ethyl benzoate	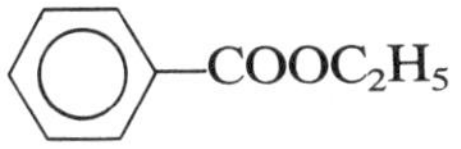 ⬡—$COOC_2H_5$	213
Phenyl benzoate	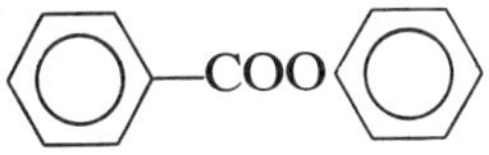 ⬡—COO—⬡	69 (m.p.)

$(C_{17}H_{35}COOH)$ occur in glycerides such as:

$$C_{17}H_{35}COO—CH_2$$
$$C_{17}H_{35}COO—CH$$
$$C_{17}H_{35}COO—CH_2$$

These are hydrolysed or *saponified* with alkali to free the acids as sodium salts for the manufacture of soaps:

$$\begin{array}{l} CH_2COOC_{17}H_{35} \\ | \\ CHCOOC_{17}H_{35} \ +3NaOH \rightarrow CH_2OHCHOHCH_2OH + 3C_{17}H_{35}COONa \\ | \\ CH_2COOC_{17}H_{35} \end{array}$$

$$\text{Propane-1,2,3-triol} \qquad \text{Sodium}$$
$$\text{(glycerol)} \qquad \text{octadecanoate}$$
$$\text{(sodium}$$
$$\text{stearate)}$$

Other types of detergent are made by freeing the acid and reducing it to the alcohol which is then sulphated; sulphated dodecyl alcohol obtained in this way is used in many shampoos:

$$C_{11}H_{23}COOH \ \rightarrow \ C_{12}H_{25}OH \ \rightarrow C_{12}H_{25}OSO_3H$$
$$\text{Dodecanoic acid} \qquad \text{Dodecan-1-ol}$$
$$\text{(lauric acid)} \qquad \text{(lauryl alcohol)}$$

The product is treated with sodium hydroxide to make sodium dodecyl sulphate; for a milder detergent an organic base is used in place of sodium hydroxide.

Some further reactions of esters are as follows:

(i) Mineral acids are used to promote the hydrolysis of esters when the product required is the free carboxylic acid rather than its sodium salt.

$$CH_3COOC_2H_5 + H_2O \xrightarrow{HCl} CH_3COOH + C_2H_5OH$$

Ethyl ethanoate   Ethanoic acid   Ethanol

(ii) Esters react with ammonia to form acyl amides. The reaction is slow and has to be carried out in either concentrated aqueous or alcoholic ammonia solution:

$$CH_3COOC_2H_5 + NH_3 \rightarrow CH_3CONH_2 + C_2H_5OH$$

Ethanamide

(iii) The effect of phosphorus pentachloride on esters is to halogenate both the acid and the alcohol, giving an acyl chloride and an alkyl chloride:

$$CH_3COOC_2H_5 + PCl_5 \rightarrow CH_3COCl + C_2H_5Cl + POCl_3$$

(iv) Esters are easily reduced, unlike the acids from which they are derived. Treatment with sodium in ethanol can be used or, more efficiently, lithium tetrahydridoaluminate (III); two alcohols result from the reduction:

$$CH_3CH_2COOC_2H_5 \xrightarrow{Na/C_2H_5OH} C_3H_7OH + C_2H_5OH$$

Ethyl propanoate   Propanol   Ethanol

(c) Acid anhydrides

The term anhydride indicates that water has been removed from an acid; anhydrides of carboxylic acids are formed when two molecules of acid lose one molecule of water.

The anhydrides of the low molecular mass alkyl carboxylic acids are colourless pungent-smelling liquids that hydrolyse readily in water. They are good acylating agents, making esters with alcohols and phenols, amides with ammonia, and acyl derivatives with amines. Benzoic anhydride, a white solid, m.p. 42°C, is less reactive than ethanoic anhydride but is still a useful benzoylating agent.

$$(CH_3CO)_2O + CH_3OH \longrightarrow CH_3COOCH_3 + CH_3COOH$$

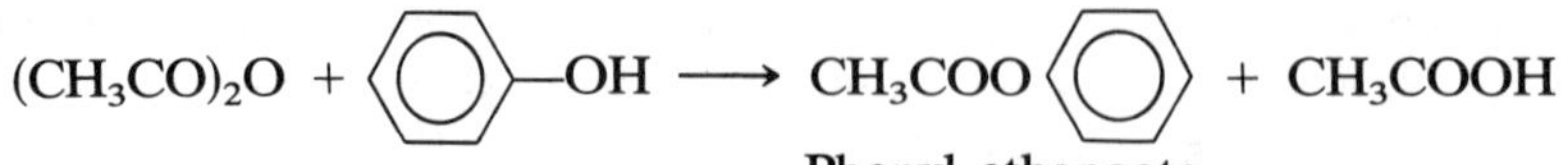

Methyl ethanoate

$$(CH_3CO)_2O + \langle\bigcirc\rangle\text{—}OH \longrightarrow CH_3COO\langle\bigcirc\rangle + CH_3COOH$$

Phenyl ethanoate

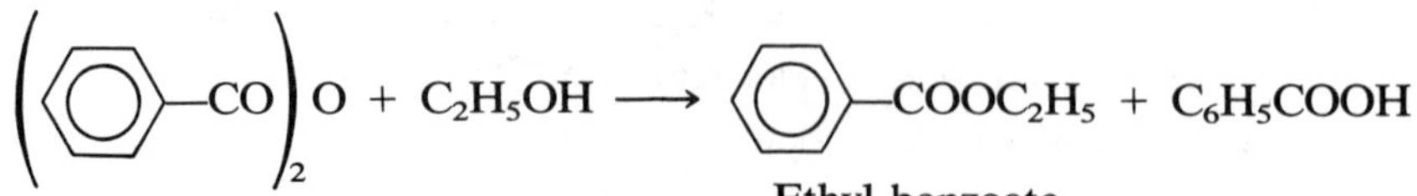

$$\left(\langle\bigcirc\rangle\text{—}CO\right)_2 O + C_2H_5OH \longrightarrow \langle\bigcirc\rangle\text{—}COOC_2H_5 + C_6H_5COOH$$

Ethyl benzoate

$$\left(\langle\bigcirc\rangle CO\right)_2 O + NH_3 \longrightarrow \langle\bigcirc\rangle CONH_2 + C_6H_5COOH$$

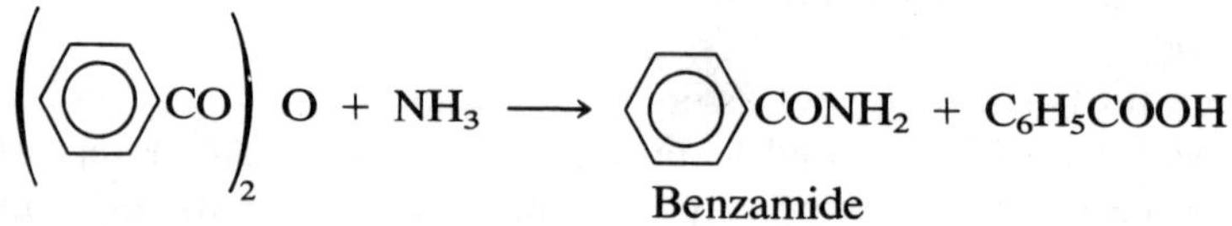

Benzamide

(d) Acyl chlorides

These compounds, in which the hydroxyl group of a carboxylic acid is replaced by a chlorine atom, are much more reactive than the corresponding acid, and make even better acylating and benzoylating agents than the anhydrides; they are of general formula $RC{\overset{O}{\underset{Cl}{\diagdown}}}$. The chemistry of acyl chlorides is described in Chapter 29.

(e) Acyl amides

The hydroxyl group of carboxylic acids can also be replaced by the $-NH_2$ group giving rise to acyl amides of general formula $RC{\overset{O}{\underset{NH_2}{\diagdown}}}$. They are white solids, soluble in water giving neutral solutions. They are described more fully in Chapter 28.

27.4 Dicarboxylic acids

The simplest of these is ethanedioic acid, which occurs in the leaves of rhubarb and sorrel (members of the *Oxalis* group of plants) and is better known by its original name of oxalic acid. The dibasic acid gives the normal reactions of the functional group, forming salts, esters, amides and acyl chlorides, but no anhydride has been isolated.

$$(COOH)_2 + 2NaOH \rightarrow (COONa)_2 + 2H_2O$$

$$(COOH)_2 + 2C_2H_5OH \rightarrow {\overset{COOC_2H_5}{\underset{COOC_2H_5}{|}}} + 2H_2O$$

In addition to these reactions ethanedioic acid is decomposed by heat to give a mixture of carbon monoxide, carbon dioxide and water:

$$(COOH)_2 \rightarrow CO + CO_2 + H_2O$$

The action of concentrated sulphuric acid on the acid or its sodium salt also yields a mixture of carbon monoxide and carbon dioxide:

$$(COONa)_2 + H_2SO_4 \rightarrow CO + CO_2 + H_2O + Na_2SO_4$$

Ethanedioic acid is oxidised quantitatively to carbon dioxide and water by acidified potassium manganate(VII) and is therefore used in volumetric analysis to standardise manganate(VII) solutions

$$2MnO_4^-(aq) + 5(COOH)_2(aq) + 6H^+(aq) \rightarrow 2Mn^{2+}(aq) + 10CO_2(g) + 8H_2O(l)$$

The dicarboxylic acids form an homologous series, the next member being propane-1,3-dioic acid (malonic acid), $COOH \cdot CH_2COOH$, whose calcium salt occurs in sugar beet. Butane-1,4-dioic acid derives its traditional name, succinic acid, from amber (latin: *succinum*) in which it occurs. Neither of these is of commercial interest, but hexane-1,6-dioic acid or adipic acid, $COOH(CH_2)_4COOH$, is made by oxidising cyclohexanol for the manufacture of nylon.

Of the unsaturated dicarboxylic acids, butenedioic acid is of interest since it provides an example of stereoisomerism arising from the presence of the double bond (page 367):

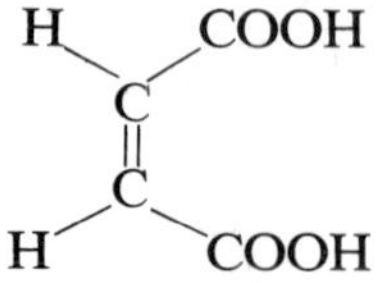

cis-butenedioic acid
(maleic acid)

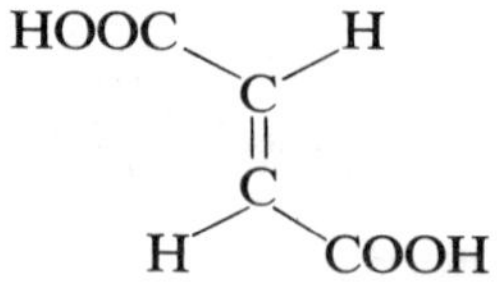

trans-butenedioic acid
(fumaric acid)

There are three isomers of benzene dicarboxylic acid, known as phthalic acids:

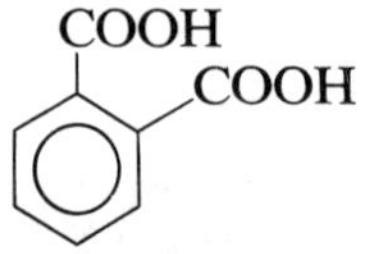

Benzene-
1,2-dicarboxylic acid
(phthalic acid)

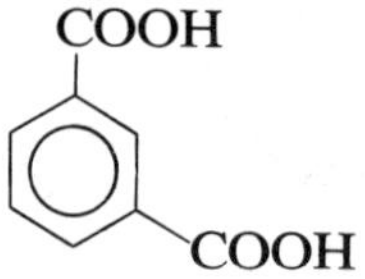

Benzene-
1,3-dicarboxylic acid
(isophthalic acid)

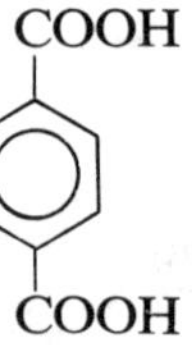

Benzene-
1,4-dicarboxylic acid
(terephthalic acid)

Of these only benzene-1,2-dicarboxylic acid can be dehydrated to form an anhydride, since the carboxyl groups are too widely separated in the other isomers:

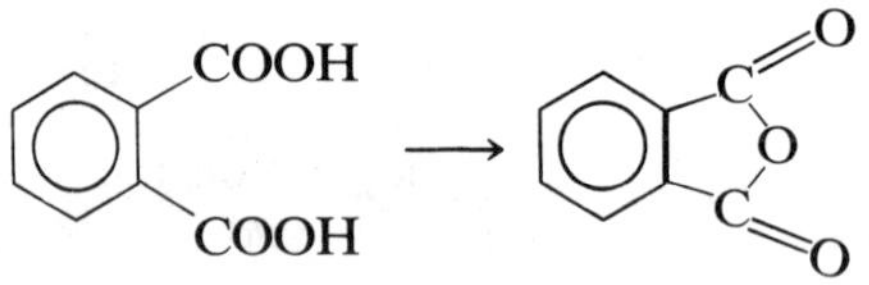

Benzene-1,2-dicarboxylic acid anhydride

The anhydride reacts with phenol to form phenolphthalein and with benzene-1,3-diol (resorcinol) to form fluorescein. Benzene-1,2-dicarboxylic acid is used together with ethane-1,2-diol to form resins which are used in the manufacture of paints. The dimethyl ester of benzene-1,2-dicarboxylic acid is an insect repellent (dimethyl phthalate). The dimethyl ester of benzene-1,4-dicarboxylic acid forms a *condensation polymer* with ethane-1,2-diol.

Condensation reactions are those in which a small molecule such as water is eliminated between two reacting molecules; when this is accompanied by polymerisation the resulting product is known as a condensation polymer. Here the eliminated molecule is methanol, and the resulting polymer is a *polyester* as can be seen from its structure. Thus:

$$\cdots\text{CH}_3\text{OOC}-\bigcirc-\text{COOCH}_3 + \text{HOCH}_2\text{CH}_2\text{OH} + \text{CH}_3\text{OOC}-\bigcirc-\text{COOCH}_3$$

$$\cdots\text{CH}_3\text{OOC}-\bigcirc-\text{COOCH}_2\text{CH}_2\text{OOC}-\bigcirc-\text{COOCH}_3\cdots + 2\text{CH}_3\text{OH}$$

Plainly further reaction with ethane-1,2-diol can occur, and under suitable conditions of temperature, pressure and concentration of reactants a long chain polymer can be made. The product, terylene, is of exceptional strength, and is widely used to make cloth, ropes, safety belts, tents and sails.

27.5 Substituted carboxylic acids

One or more of the hydrogen atoms of the alkyl or aryl group of a carboxylic acid may be substituted with a functional group, giving rise to a variety of compounds in which the properties of the carboxylic acid are modified. Thus the chloro-derivatives of ethanoic acid are stronger acids than the parent acid. The chlorine atom is electron-attracting, and facilitates the removal of the electron from the acid hydrogen atom so that ionization is promoted. The effect is shown by the acid dissociation constants:

		K_a mol dm^{-3}
Ethanoic acid	$\text{CH}_3\text{COOH} \rightleftharpoons \text{CH}_3\text{COO}^- + \text{H}^+$	1.7×10^{-5}
Monochloro-ethanoic acid	$\text{Cl}\!\leftarrow\!\text{CH}_2\text{COOH} \rightleftharpoons \text{ClCH}_2\text{COO}^- + \text{H}^+$	1.5×10^{-3}
Dichloro-ethanoic acid	$\begin{array}{c}\text{Cl}\\ \,\,\,\,\diagdown\\ \,\,\,\,\,\,\text{CHCOOH} \rightleftharpoons \text{Cl}_2\text{CHCOO}^- + \text{H}^+\\ \text{Cl}\diagup\end{array}$	5.0×10^{-2}
Trichloro-ethanoic acid	$\begin{array}{c}\text{Cl}\diagdown\\ \text{Cl}\!\leftarrow\!\text{C}-\text{COOH} \rightleftharpoons \text{Cl}_3\text{COO}^- + \text{H}^+\\ \text{Cl}\diagup\end{array}$	2.0×10^{-1}

The figures show that whereas ethanoic acid is a relatively weak acid, trichloroethanoic acid is a fairly strong acid.

The effect on the acidity of ethanoic acid resulting from the substitution of a hydroxyl group for a hydrogen atom is also to increase the strength of the acid, but not so markedly. Thus K_a for hydroxyethanoic acid (glycollic acid) is 1.5×10^{-4} mol dm^{-3}.

Similar substitution of hydrogen atoms of propanoic acid gives rise to isomerism resulting from the *position* of the substituted group.

$$CH_3C^*H(OH)COOH \qquad\qquad CH_2(OH)CH_2COOH$$

2-hydroxypropanoic acid   3-hydroxypropanoic acid
(α-hydroxypropanoic acid)   (β-hydroxypropanoic acid)

The names given in brackets refer to an older convention where the carbon atom next to the carboxylic acid group was called the α-carbon; this convention is still widely used. 2-hydroxypropanoic acid is the lactic acid of sour milk, and as it contains an asymmetric carbon atom (C^*) it forms optically active isomers (page 368). The hydroxy-propanoic acids have the properties of both alcohols (—OH) and acids (—COOH), and are similar to one another in chemical behaviour, although they react differently when heated. 2-hydroxypropanoic acid, when heated, forms a cyclic ester, a lactide, by mutual esterification between two molecules:

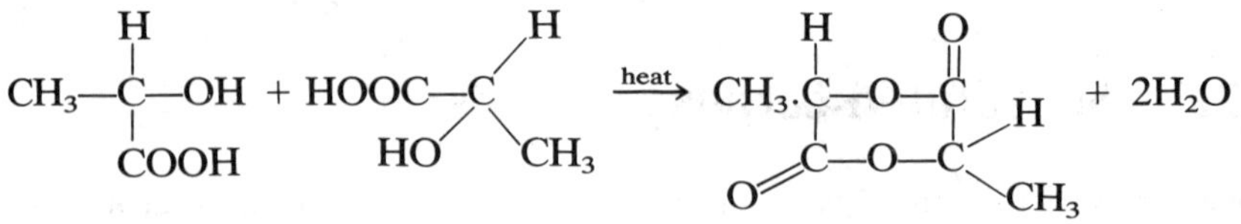

2-hydroxypropanoic acid   lactide

3-hydroxypropanoic acid (hydracrylic acid) does not form a lactide, but each molecule loses a molecule of water to form the unsaturated propenoic acid whose methyl derivative forms the polymer poly(methyl 2-methylpropenoate) (page 356).

$$HOCH_2CH_2COOH \rightarrow CH_2{=}CHCOOH + H_2O$$

3-hydroxypropanoic   Propenoic acid
acid   (acrylic acid)

Of the substituted aryl carboxylic acids, 2-hydroxybenzoic acid (salicylic acid) is of interest and importance. It occurs naturally as the methyl ester in oil of wintergreen, whose characteristic odour is well known to most of us through its use in liniments and ointments for rheumatism and sprains. A second important derivative of 2-hydroxybenzoic acid is 2-ethanoyloxy-benzenecarboxylic acid or aspirin, in which the hydroxyl group is acylated with ethanoic acid to form an ester.

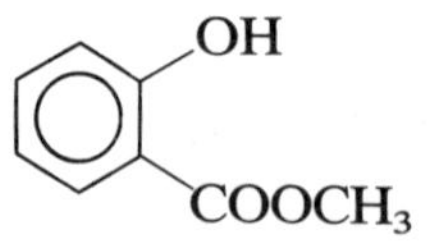

Methyl 2-hydroxybenzoate   2-ethanoyloxybenzoic acid
(methyl salicylate)   (acetyl salicylic acid)
(oil of wintergreen)   (aspirin)

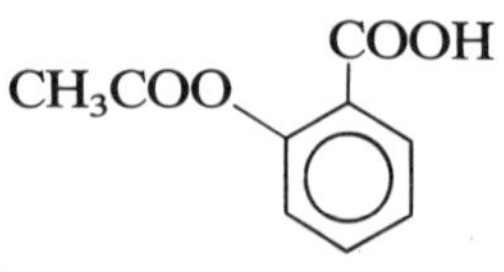

An important class of substituted carboxylic acids are the amino-acids in which a hydrogen atom is substituted with an amino group. These compounds are the molecules from which proteins are built up. They are considered in more detail in Chapter 28. The simplest is glycine or aminoethanoic acid, NH_2CH_2COOH. The amino group is basic, and its presence in the acid molecule gives rise to a compound whose aqueous solution is neutral and contains dipolar ions or 'zwitterions'.

$$NH_2CH_2COOH \rightleftharpoons {}^+NH_3CH_2COO^-$$

The polar nature of these molecules results in aminoethanoic acid being very soluble in water but insoluble in organic solvents.

Questions

1. Describe the reactions of ethanoic acid with sodium hydroxide solution, phosphorus pentachloride and chlorine, naming the products in each case.

2. Give a brief outline of how benzoyl chloride, benzamide and benzoic anyhdride may be obtained from benzoic acid

3. Suggest methods of obtaining (a) methane (b) ethane (c) propanone (d) ethenenitrile from ethanoic acid.

4. How may ethyl ethanoate be prepared from ethanol and ethanoic acid?

5. What do you understand by 'saponification'? What products would you expect to obtain from the saponification of glycerol tristearate? (propane-1,2,3-trioctadecanoate).

6. Write an equation for the reaction of sodium phenate with benzoyl chloride. What type of reaction is this? Suggest two methods by which ethyl ethanoate could be formed.

7. Identify by name and structural formula the compounds A, B, C, D and E. Give your reasons for your identifications and write equations (where possible) for the reactions which have been described.

   A is a colourless gas which burns in air with a luminous flame but does not react with an ammoniacal solution of copper(I) chloride. It combines with chlorine to give a colourless oil, B, and is absorbed by concentrated sulphuric acid to give a compound C. This product is hydrolysed by refluxing with aqueous alkali and yields an organic product D with a relative molecular mass of 60. When D is refluxed with an aqueous solution of chromic(VI) acid it is oxidised to a volatile liquid E, which is resistant to further oxidation, and mixes with water in all proportions to give a neutral solution.   (O)

8. (a) Draw structural formulae which show the spatial arrangement of the atoms in the molecules of (i) ethyne (acetylene), (ii) propanone (acetone), (iii) dichlorobenzenes, (iv) butenedioic acids, $C_2H_2(COOH)_2$, and (v) 2-hydroxypropanoic (lactic) acids. In the case of (iv) and (v) above, explain why there is more than one structure.
   (b) Explain the following:
   (i) the carbonyl group on a carboxylic acid is less reactive than that in an aldehyde or ketone;
   (ii) a dilute aqueous solution of aminoethanoic acid (glycine) is a much weaker electrolyte than an aqueous solution of ethanoic (acetic) acid of equivalent concentration.   (O)

9. An ester X $(C_{10}H_{12}O_2)$ was heated for several hours with a slight excess

of an aqueous solution of sodium hydroxide and the solution partly distilled.

(*a*) What is the name of this type of chemical reaction? The distillate contained a neutral aromatic compound Y ($C_8H_{10}O$) which, upon mild oxidation, gave compound Z (C_8H_8O). Z gave a positive triiodomethane (iodoform) reaction, but did not give a precipitate with aqueous diamminesilver ions (Tollen's reagent, 'ammoniacal silver nitrate').

(*b*) Write the full structural formula of the functional group present in Z which is responsible for the positive triiodomethane reaction.

(*c*) Write the full structural formula of a functional group which *does* give a positive reaction with Tollen's reagent.

(*d*) Hence, write a full structural formula for

  (i) the compound Z,

  (ii) the compound Y,

  (iii) an ester of molecular formula RCOOR',

  (iv) the ester X

(*e*) Name one class of organic compound to which the compound Y might belong.   (C)

10. (*a*) Write the structural formula of each of the following derivatives of ethanoic acid (acetic acid):

  (i) ethanoic anhydride (acetic anydride);

  (ii) ethyl ethanoate (ethyl acetate);

  (iii) ethanoyl chloride (acetyl chloride);

  (iv) ethanamide (acetamide).

(*b*) Each of the derivatives in (*a*) can be *hydrolysed.*

  (i) Explain what is meant by *hydrolysis.*

  (ii) Give the conditions for the hydrolysis of each derivative.

(iii) Write the derivatives in order of ease of hydrolysis, putting the easiest first.

(*c*) Stating your reasons, arrange the following in order of increasing acid strength (i.e. putting the least acidic first): ethanoic acid; monobromoethanoic acid; monochloroethanoic acid; monoiodoethanoic acid.
                                                                    (AEB 1977)

11. The structures of ethers, esters and anhydrides are similar in that each consists of two alkyl and/or acyl groups attached to an oxygen atom. Compare and contrast in tabular form the methods of preparation, physical properties and chemical reactions of an ether, an ester and an acid anhydride.   (L)

Alkyl and aryl nitrogen compounds

28.1 Nitro-compounds

Alkyl and aryl compounds containing the nitro group $-N{\Large\langle}^O_O$ do not occur naturally, but can be made in the laboratory.

The nitroalkanes containing one nitro group form a homologous series $C_nH_{2n+1}NO_2$ in which the lower molecular-mass members are liquids with a pleasant smell. They are prepared by reacting the alkanes with nitric acid vapour in sealed tubes at 350°C; the products are separated by fractional distillation.

Nitrobenzenes and methylnitrobenzenes are made by refluxing benzene or methylbenzene with a mixture of concentrated nitric and sulphuric acids which produces the nitronium ion (page 361). The compounds formed are poisonous, although having a pleasant almond-like smell.

Nitroalkanes are useful solvents for tars, paints, plastics and dyes since they have suitable boiling points and do not have obnoxious smells. The chief use of nitrobenzene is in the preparation of phenylamine (aniline) while methyl-2,4,6-trinitrobenzene is used in the manufacture of explosives.

Some typical nitro-compounds are given in Table 28.1.

The most important general reaction of the nitroalkanes and nitroarenes is their reduction to amines. This can be carried out by passing vaporised nitroalkanes mixed with hydrogen over a nickel catalyst at 150°C

$$CH_3NO_2 + 2H_2 \xrightarrow[150\,°C]{Ni} CH_3NH_2 + 2H_2O$$

Alternatively the reduction can be effected by the use of tin and concentrated hydrochloric acid. This latter method is of great importance in the preparation of phenylamine from nitrobenzene (page 455).

Attached to a benzene nucleus the nitro-group is deactivating and *meta*-directing. This results from the negative inductive effect (−I) (page 344) of the nitro-group, which tends to withdraw electrons from the benzene ring, leaving the carbon atoms at the *ortho*- and *para*- positions slightly positively charged and less susceptible to attack by electrophilic agents:

Resonance hybrid
structure for
nitrobenzene

Summary of canonical
structures for
nitrobenzene

Table 28.1 Nitro compounds

Name	Formula	m.p./b.p. °C
Nitromethane	CH_3NO_2	101
Nitroethane	$C_2H_5NO_2$	115
Nitropropane	$C_3H_7NO_2$	132
2-Nitropropane	$(CH_3)_2CHNO_2$	120
Nitrobenzene	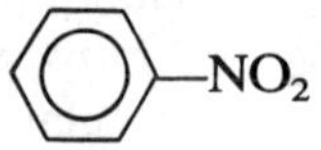 ⬡—NO_2	211
1,3-Dinitrobenzene	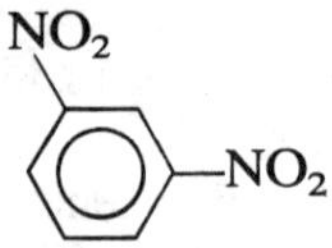 NO_2—⬡—NO_2	90 (m.p.)
1,3,5-Trinitrobenzene	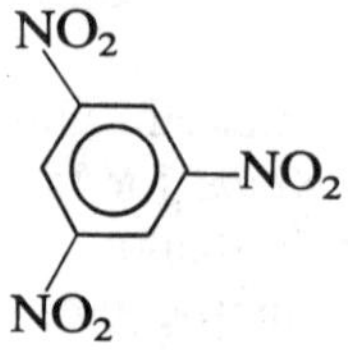 NO_2—⬡—NO_2 with NO_2	122 (m.p.)
Methyl-2-nitrobenzene	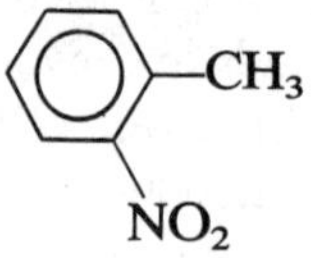 ⬡—CH_3 with NO_2	222
Methyl-2,4,6-nitrobenzene (Trinitrotoluene)	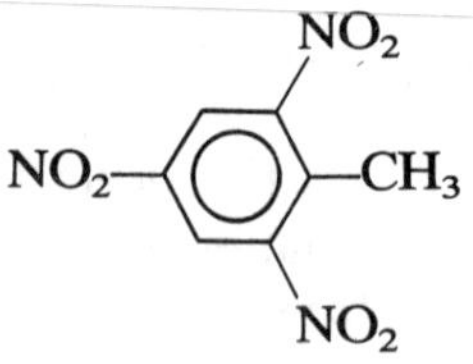 NO_2—⬡—CH_3 with NO_2 and NO_2	81 (m.p.)

The overall effect is that further substitution takes place less readily than the original nitration and at the *meta*- rather than the *ortho*- or *para*- positions. Thus to make dinitrobenzene a mixture of fuming nitric acid with concentrate sulphuric acid must be used:

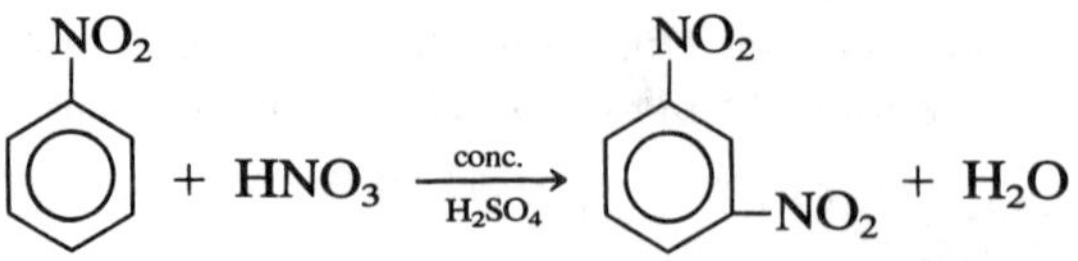

$$\text{⬡—}NO_2 + HNO_3 \xrightarrow[H_2SO_4]{conc.} NO_2\text{—⬡—}NO_2 + H_2O$$

The addition of a third nitro-group to make 1,3,5-trinitrobenzene is more difficult and takes a week for the reaction to be completed.

The (+I) inductive effect of the methyl group makes nitration of methylbenzene (toluene) easier; 1-methyl-2-nitrobenzene is formed by the action of a cold mixture of concentrated nitric and sulphuric acids (page 361). To add a second and third nitro group to form 1-methyl 2,4,6-trinitrobenzene (trinitrotoluene or TNT) more stringent conditions are needed; this preparation should not be attempted in the laboratory because of the explosive nature of TNT.

28.2   Amino-compounds

Amines have lower boiling points than the corresponding nitro-compounds; methylamine and ethylamine are both gases smelling of bad fish. Those of low molecular mass are soluble in water as well as in organic solvents; unlike ammonia, they burn well with a blue flame.

Amines can be regarded as ammonia molecules in which one or more hydrogen atoms have been replaced by alkyl or aryl groups giving primary, secondary and tertiary amino compounds. A mixture of these, together with the quaternary ammonium salt, may be made by reacting an alkyl halide with ammonia in a sealed tube:

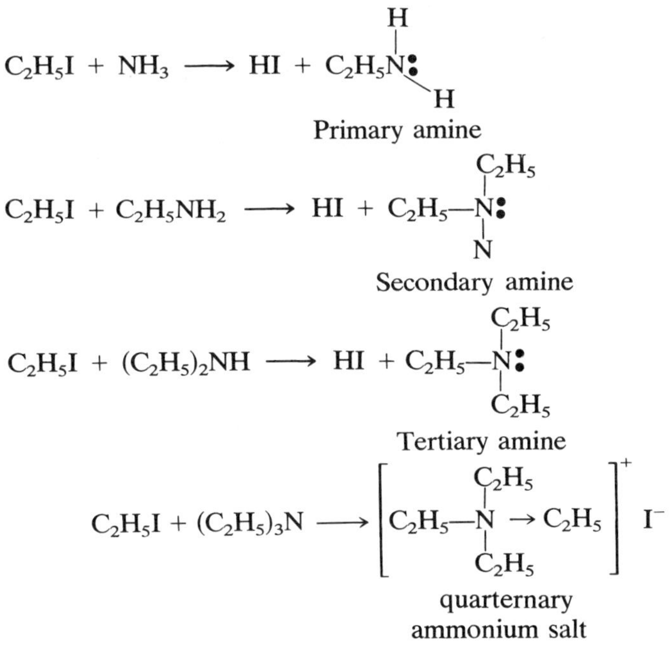

$$C_2H_5I + NH_3 \longrightarrow HI + C_2H_5\overset{\displaystyle H}{\underset{\displaystyle H}{N:}}$$

Primary amine

$$C_2H_5I + C_2H_5NH_2 \longrightarrow HI + C_2H_5-\overset{\displaystyle C_2H_5}{\underset{\displaystyle N}{N:}}$$

Secondary amine

$$C_2H_5I + (C_2H_5)_2NH \longrightarrow HI + C_2H_5-\overset{\displaystyle C_2H_5}{\underset{\displaystyle C_2H_5}{N:}}$$

Tertiary amine

$$C_2H_5I + (C_2H_5)_3N \longrightarrow \left[C_2H_5-\overset{\displaystyle C_2H_5}{\underset{\displaystyle C_2H_5}{N}} \rightarrow C_2H_5 \right]^+ I^-$$

quarternary
ammonium salt

Primary, secondary and tertiary amines with aryl groups can also be made, but aryl quaternary ammonium compounds have not been isolated. However, phenylamine will react with an alkyl halide such as methyl iodide to produce a series of compounds ending with phenyl trimethyl ammonium iodide:

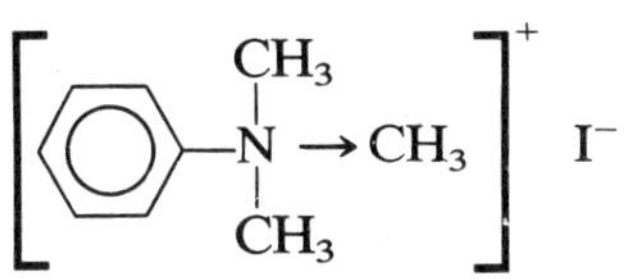

$$\left[\langle \bigcirc \rangle - \overset{\displaystyle CH_3}{\underset{\displaystyle CH_3}{N}} \rightarrow CH_3 \right]^+ I^-$$

Some reactions of amines are as follows:

(a) Like ammonia, the amines are bases:

$$CH_3NH_2 + H_2O \rightleftharpoons CH_3NH_3^+ + OH^-$$

and form salts with acids:

$$CH_3NH_2 + HCl \rightarrow (CH_3NH_3)^+Cl^-$$

The presence of the alkyl group makes them stronger bases than ammonia itself, since the $(+I)$ effect of the alkyl group (page 344) increases the electronegativity of the nitrogen atom, thus increasing its attraction for a proton; secondary amines are even more strongly basic, but tertiary amines are less basic:

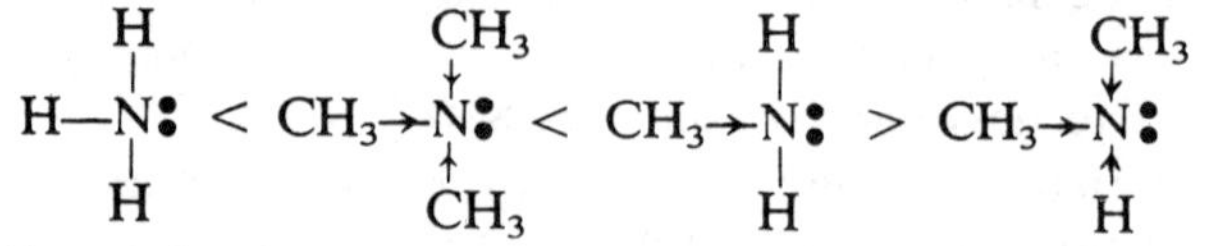

Increasing base strength of the amines $\longrightarrow$

In contrast phenylamine is a weaker base than ammonia since the phenyl group tends to withdraw electrons $(-I$ effect$)$, thus reducing the tendency of the nitrogen atom to form a coordinate bond with a proton (H^+):

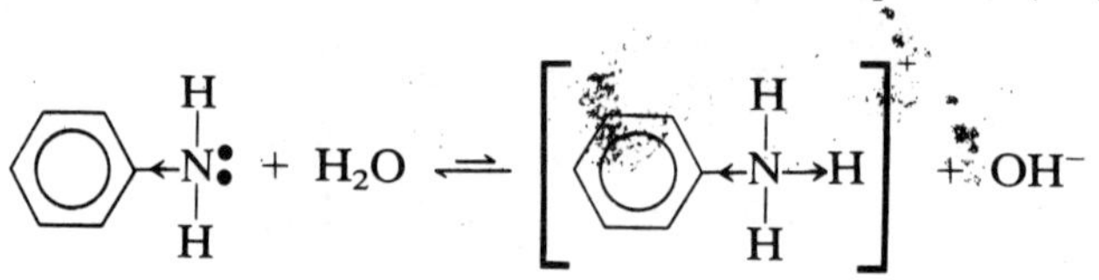

(b) Primary, secondary and tertiary amines and arylamines differ in their reactions with nitrous acid (nitric (III) acid).

Primary amines evolve nitrogen with the formation of the corresponding alcohol, although the yield of alcohol is low:

$$CH_3NH_2 + HNO_2 \rightarrow CH_3OH + H_2O + N_2$$

Secondary amines give yellow oils, nitrosoamines:

$$(CH_3)_2NH + HNO_2 \rightarrow (CH_3)_2N—N{=}O + H_2O$$

Tertiary amines give a variety of colourless products in solution, including a salt-like compound:

$$(CH_3)_3N + HNO_2 \rightarrow (CH_3)_3NH^+NO_2^-$$

Primary arylamines at room temperature react with nitrous acid to give nitrogen as do primary alkylamines; however if the temperature is kept low (5°C) *diazonium salts* are formed.

This *diazotisation reaction* is discussed in more detail later in this chapter.

(c) Both alkyl and aryl primary amines react with trichloromethane to form isocyano-compounds. Thus if phenylamine is warmed with an alcoholic solution of potassium hydroxide, isocyanobenzene is formed:

$$CHCl_3 + 3KOH + C_6H_5NH_2 \rightarrow C_6H_5NC + 3KCl + 3H_2O$$
Isocyanobenzene

This reaction is known as the *carbylamine reaction* and serves to distinguish primary from secondary and tertiary amines, the isocyanocompounds having a characteristic and extremely unpleasant smell.

(d) Primary amines in solution form complexes with metal ions similar to those formed by ammonia molecules (page 55):

$$Cu^{2+}(aq) + 4CH_3NH_2(aq) \rightarrow [Cu(CH_3NH_2)_4]^{2+}(aq)$$

The amino compounds are liganded to the metal ions by coordinate bonds utilising the lone pair of electrons on the nitrogen atom. The complex copper ions give deep blue solutions in water.

(e) Amines are readily acylated or benzoylated by reaction with acid chlorides. For example, ethylamine reacts with ethanoyl chloride to form a substituted amide:

$$C_2H_5-NH_2 + ClOCCH_3 \rightarrow C_2H_5NHOCCH_3 + HCl$$

Phenylamine reacts similarly to give *N*-phenylethanamide (acetanilide). The prefix *N*- indicates that the phenyl group is attached to the nitrogen atom and not directly to the alkyl group or the benzene nucleus.

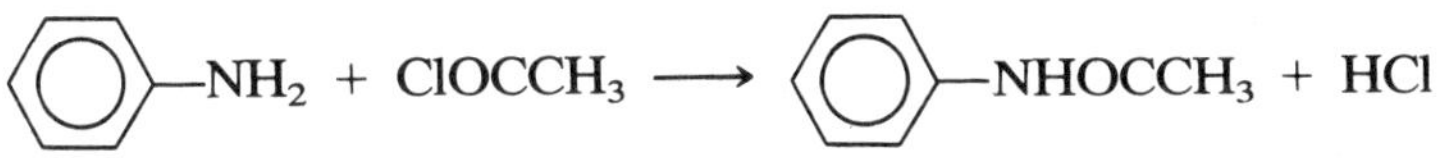

$$\bigcirc\text{—NH}_2 + \text{ClOCCH}_3 \longrightarrow \bigcirc\text{—NHOCCH}_3 + \text{HCl}$$

Phenylamine also reacts with benzoyl chloride in alkaline solution:

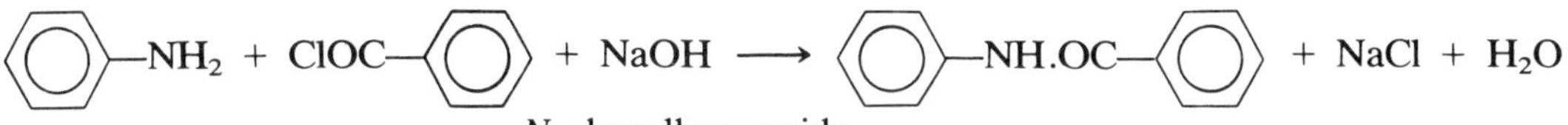

$$\bigcirc\text{—NH}_2 + \text{ClOC—}\bigcirc + \text{NaOH} \longrightarrow \bigcirc\text{—NH.OC—}\bigcirc + \text{NaCl} + \text{H}_2\text{O}$$

N-phenylbenzamide

The reactions of arylamines such as (phenyl methyl)amine are similar to those of alkylamines. The brackets indicate that the phenyl group substitutes a hydrogen atom in the methyl group not the amino-group.

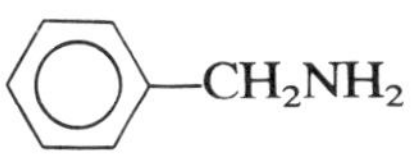

$$\bigcirc\text{—CH}_2\text{NH}_2$$

Arylamines in which the amino group is directly attached to the benzene nucleus have some individual reactions; for example reactions in which hydrogen atoms in the benzene ring are substituted.

(f) As mentioned above, (see (a)), phenylamine is a weaker base than ammonia since the phenyl group tends to draw electrons into the resonance structure of the benzene ring. This drawing of the electrons into the resonance structure of the benzene ring also makes electrophilic substitution reactions easier to that the amino group is activating and *ortho*-, *para*-directing in a similar way to the –OH group in phenols (page 384). Possible canonical structures for phenylamine are:

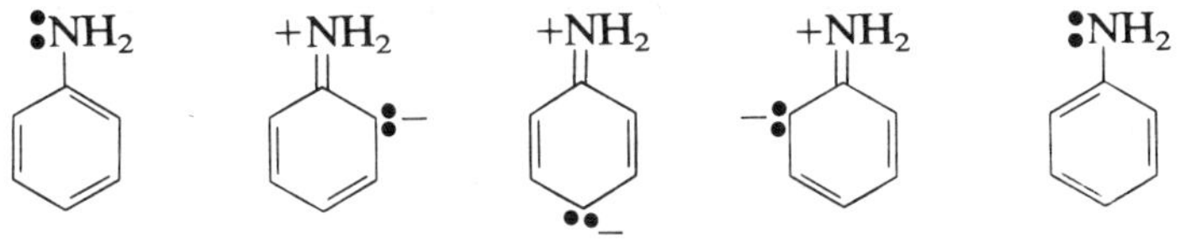

Thus phenylamine is more readily brominated or nitrated than benzene. If shaken with bromine water, 1-amino-2,4,6-tribromobenzene is rapidly formed:

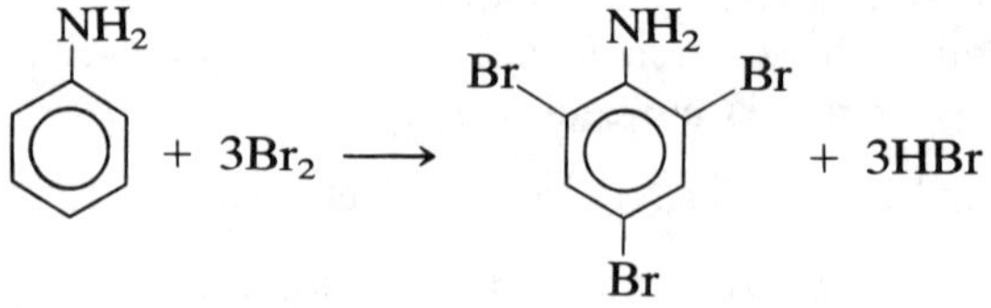

2,4,6-trinitrophenylamine can be formed by the action of dilute nitric acid. Sulphuric acid converts phenylamine to 4-aminobenzene sulphonic acid (sulphanilic acid). The reaction proceeds in two stages:

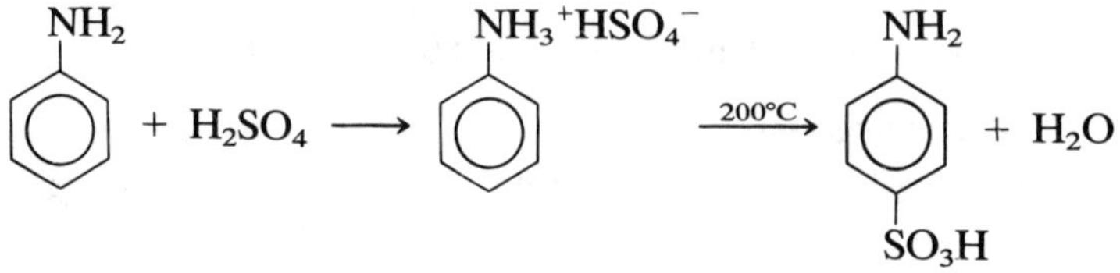

This is an important reaction since sulphanilic acid is an intermediate compound in the formation of some azo dyes and in the preparation of sulphonamide drugs.

28.3   Diazonium salts

The reaction of primary arylamines with nitrous acid to form diazonium salts takes place in acid solution at low temperatures. Thus if a solution of sodium nitrite is added to a solution of phenylamine in hydrochloric acid cooled to below 5°C, benzene diazonium chloride is formed:

$$HCl + NaNO_2 \rightarrow HNO_2 + NaCl$$

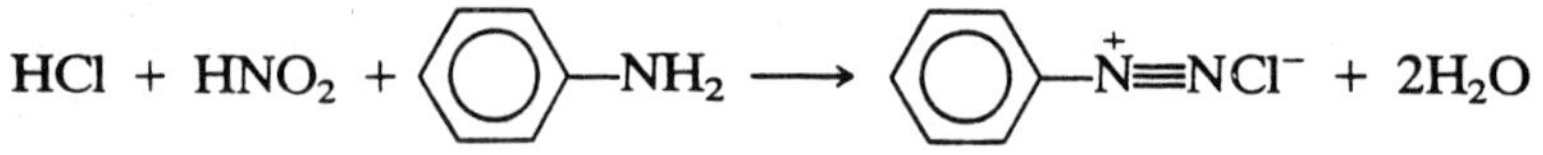

The diazonium salts are rarely crystallised out since they are explosive when dry. Further reactions are carried out in solution. These reactions fall into two classes: those leading to a great variety of new compounds, in which the diazo group is replaced by new functional groups; and those in which the diazo group is retained and coupled with phenols and aryl amines to make azo dyes.

(a)   Replacement reactions of benzene diazonium salts

(i) To make phenol
If the solution is boiled phenol is obtained with the evolution of nitrogen:

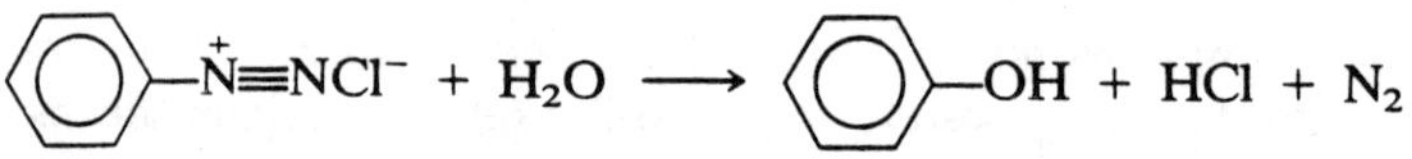

(ii) *To make aryl halogen derivatives*
Various techniques can be applied to produce different halogen compounds. Simply adding copper powder to the warm mixture will displace the nitrogen, leaving chlorobenzene:

$$C_6H_5\overset{+}{N}\equiv N\bar{C}l \xrightarrow{Cu(s)} C_6H_5Cl + N_2$$

Bromobenzene can be obtained by adding copper(I) bromide and concentrated hydrobromic acid:

$$C_6H_5\overset{+}{N}\equiv N\bar{C}l + HBr \xrightarrow{CuBr} C_6H_5Br + HCl + N_2$$

Iodobenzene can be made by warming the diazonium chloride solution with potassium iodide solution:

$$C_6H_5\overset{+}{N}\equiv N\bar{C}l + KI \rightarrow C_6H_5I + KCl + N_2$$

(iii) *To make nitriles*
The introduction of a nitrile group can be done by using copper powder in a solution of potassium cyanide:

$$C_6H_5\overset{+}{N}\equiv N\bar{C}l + KCN \xrightarrow{Cu(s)} C_6H_5CN + KCl + N_2$$

(b) Coupling reactions to make azo-dyes

(i) *With phenol*

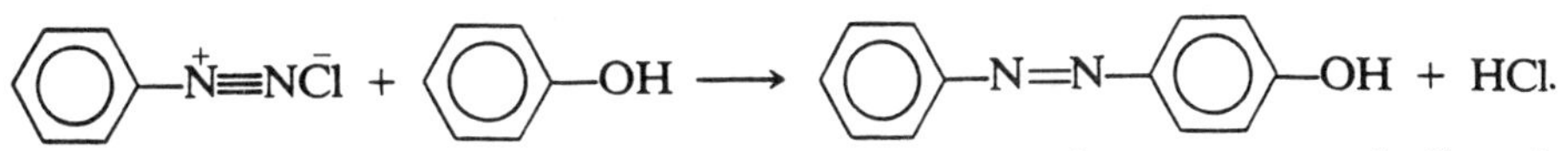

(4-hydroxyphenyl)azobenzene   (yellow dye)

(ii) *With naphthalen-2-ol*

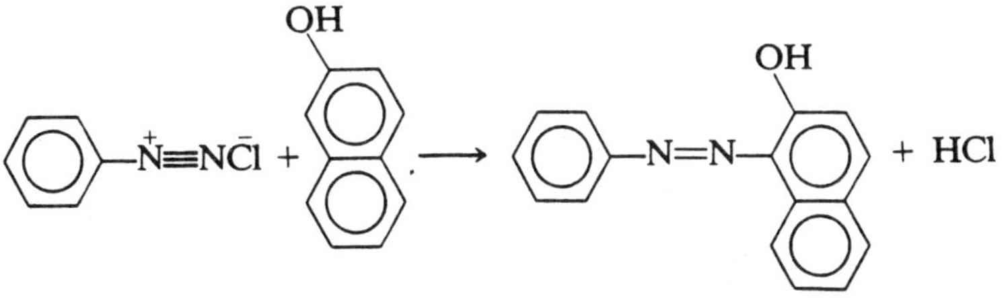

(Red dye)

(iii) *With N,N-dimethylphenylamine*

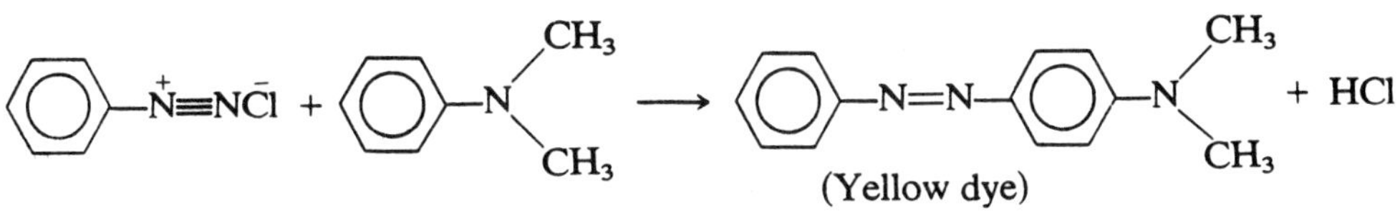

(Yellow dye)

(iv) Preparation of methyl orange

This involves the diazotisation of 4-aminobenzenesulphonic acid followed by coupling the diazo-salt with N,N-dimethylphenylamine:

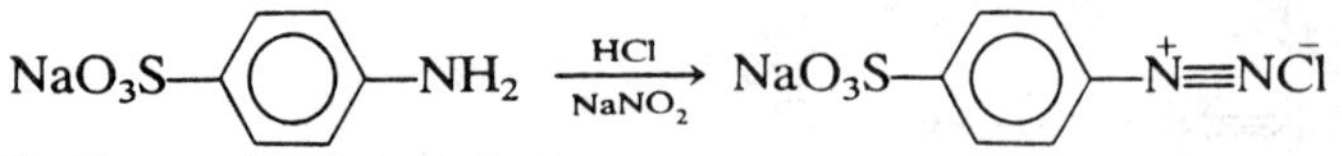

$$NaO_3S-\langle \bigcirc \rangle-NH_2 \xrightarrow[NaNO_2]{HCl} NaO_3S-\langle \bigcirc \rangle-\overset{+}{N}\equiv N\bar{C}l$$

Sodium salt of 4-aminobenzene
sulphonic acid

$$NaO_3S-\langle \bigcirc \rangle-\overset{+}{N}\equiv N\bar{C}l + \langle \bigcirc \rangle-N\overset{CH_3}{\underset{CH_3}{\big<}} \longrightarrow NaO_3S-\langle \bigcirc \rangle-N=N-\langle \bigcirc \rangle-N\overset{CH}{\underset{CH}{\big<}}$$

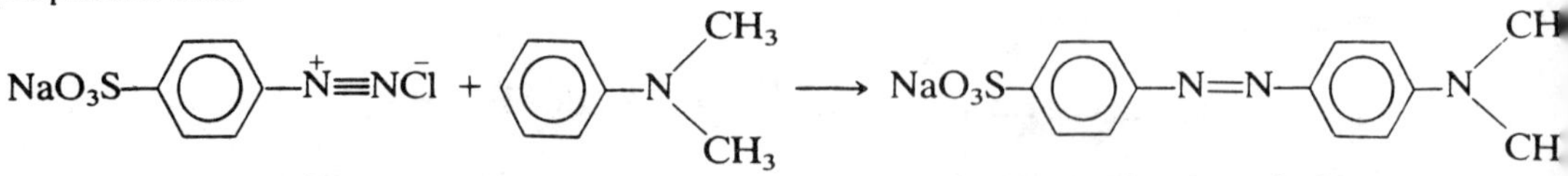

Sodium salt of methyl orange

The grouping that gives rise to colour is the azo group (—N=N—) which is responsible for the absorption of light. The actual colour depends on the other groups attached to this *chromophore*. Care should be taken in the preparation of any azo-dye since the dyes themselves tend to be carcinogens and many of the reagents such as phenol and naphthalen-2-ol are corrosive and toxic.

(c)   Reduction to hydrazines

In addition to these coupling reactions diazonium salts can be reduced to hydrazine derivatives. For example, phenylhydrazine can be made by the reduction of benzenediazonium chloride with a mixture of tin and concentrated hydrochloric acid:

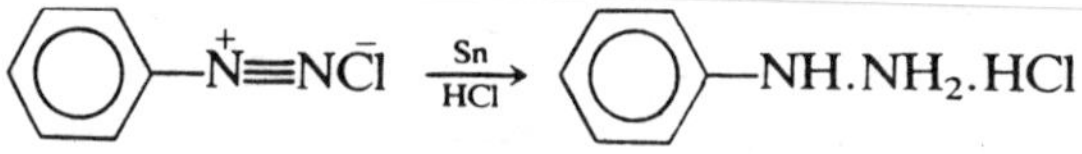

$$\langle \bigcirc \rangle-\overset{+}{N}\equiv N\bar{C}l \xrightarrow[HCl]{Sn} \langle \bigcirc \rangle-NH.NH_2.HCl$$

Phenylhydrazine hydrochloride

28.4   Amides

Amides are derivatives of carboxylic acids in which the hydroxyl group has been replaced by an amino group.

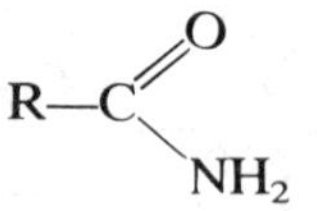

$$R-C\overset{\displaystyle O}{\underset{\displaystyle NH_2}{\big<}}$$

With the exception of methanamide, the amides are white crystalline solids, soluble in both organic solvents and water to give neutral solutions. The solubility in water and the relatively high melting points arise from the ability of amides to form hydrogen bonds.

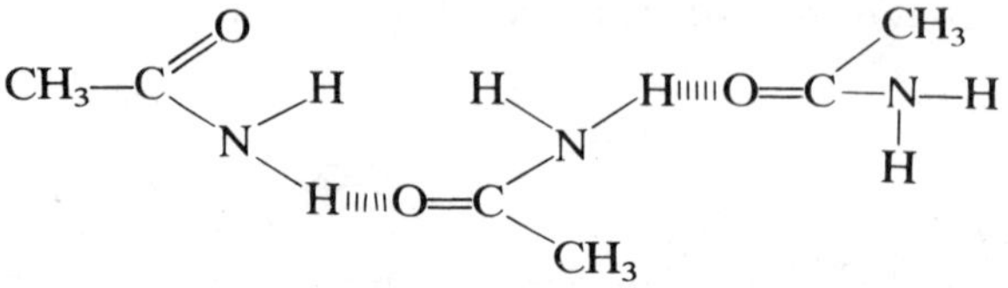

Fig. 28.1 Hydrogen bonding in ethanamide

Some reactions of amides are:
(a) The solutions in water are neutral, since amides are amphoteric, forming salts with both acids and bases:

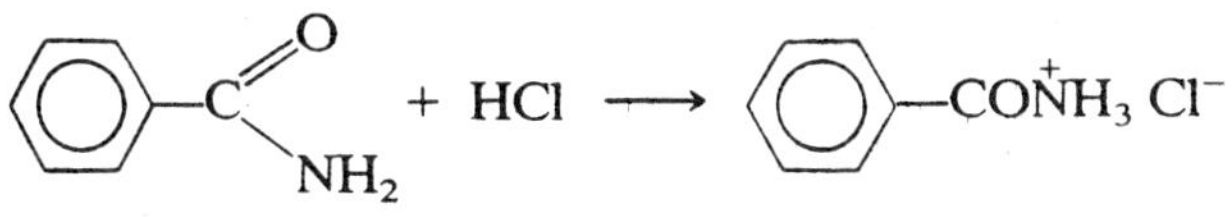

$$2C_2H_5CONH_2 + HgO \rightarrow (C_2H_5CONH)_2Hg + H_2O$$

Amides in solution are readily hydrolysed to give first an ammonium salt and then, in acid solution, a carboxylic acid or, in alkaline solution, the sodium salt of a carboxylic acid.

$$CH_3CONH_2 \xrightarrow{H_2O} CH_3CO\bar{O}\overset{+}{N}H_4 \xrightarrow[HCl]{H_2O} CH_3COOH + NH_4Cl$$

$$CH_3CO\bar{O}\overset{+}{N}H_4 \xrightarrow[NaOH]{H_2O} CH_3CO\bar{O}\overset{+}{N}a + NH_3 + H_2O$$

The release of ammonia on boiling an amide with sodium hydroxide distinguishes amides from amines, which do not undergo this reaction.
(b) Amides can also be dehydrated by heating with phosphorus(V) oxide to remove the elements of water, releasing nitriles; the nitriles will be distilled off as vapour since they are of relatively low boiling point:

$$\underset{\substack{\text{Ethanamide} \\ \text{m.p. } 82^\circ C}}{CH_3CONH_2} \xrightarrow[P_2O_5]{-H_2O} \underset{\substack{\text{Ethanenitrile} \\ \text{b.p. } 81^\circ C}}{CH_3CN}$$

This is a good method of preparing nitriles.
(c) Like amines, amides react with nitrous acid with the evolution of nitrogen, but yield carboxylic acids in place of alcohols:

$$CH_3CONH_2 + HNO_2 \rightarrow CH_3COOH + H_2O + N_2$$

(d) Alkyl amides can be degraded to amines which have one less carbon atom in the molecule by reaction with a mixture of bromine and potassium hydroxide. Thus ethanamide can be converted to methylamine:

$$CH_3CONH_2 + Br_2 + 4KOH \rightarrow CH_3NH_2 + 2KBr + K_2CO_3 + 2H_2O$$

This reaction, the Hoffmann degradation, is used both to prepare amines and to reduce the number of carbon atoms in the carbon chain. For example, butanol can be reduced to propanol:

$$C_4H_9OH \xrightarrow[H^+]{KMnO_4} C_3H_7COOH \xrightarrow{NH_3} C_3H_7CO\bar{O}\overset{+}{N}H_4 \xrightarrow[P_2O_5]{-H_2O} C_3H_7CONH_2 \xrightarrow[Br_2]{KOH} C_3H_7NH_2$$

$$C_3H_7NH_2 \xrightarrow[HCl]{NaNO_2} C_3H_7OH + N_2 + H_2O$$

An important amide is carbamide, or urea, $CO(NH_2)_2$, which is a colourless crystalline substance occurring naturally in the urine of mammals and synthesised for use as a fertiliser from ammonia and carbon dioxide:

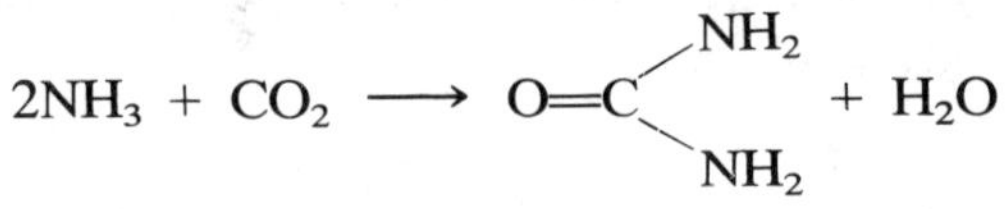

$$2NH_3 + CO_2 \longrightarrow O{=}C\begin{smallmatrix} \diagup NH_2 \\ \diagdown NH_2 \end{smallmatrix} + H_2O$$

If heated gently the molecule loses ammonia to form *biuret*.

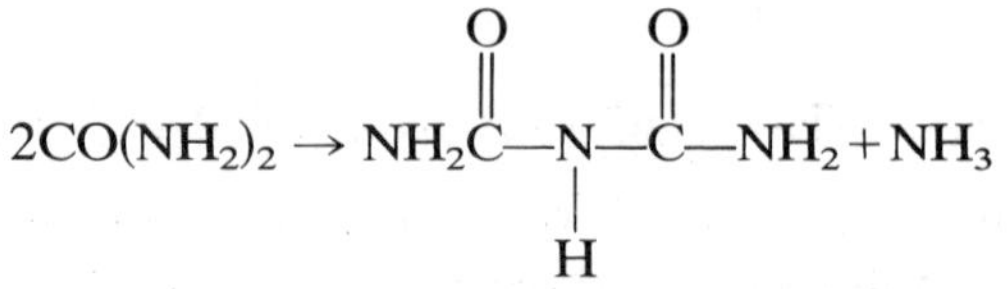

$$2CO(NH_2)_2 \rightarrow NH_2\overset{\overset{\displaystyle O}{\|}}{C}{-}\underset{\underset{\displaystyle H}{|}}{N}{-}\overset{\overset{\displaystyle O}{\|}}{C}{-}NH_2 + NH_3$$

Biuret contains the *peptide* group —NH—$\overset{\overset{\displaystyle O}{\|}}{C}$— which is a linkage group found in all proteins. An alkaline solution of biuret treated with very dilute copper(II) sulphate solution gives a violet colour. This reaction, known as the *biuret reaction*, can be used to test for proteins since all compounds containing a peptide group respond to it; the colour results from the liganding of the group to the Cu^{2+} ions.

28.5   Nitriles

Compounds containing a cyanide group —C≡N are termed nitriles. The lowest member of the homologous series of alkyl nitriles is hydrogen cyanide, HCN, the extremely poisonous gas better known as prussic acid. The succeeding members of the series are colourless pleasant-smelling liquids readily soluble in organic solvents.

		°C
Methanenitrile	HCN	26
Ethanenitrile	CH_3CN	81
Propanenitrile	C_2H_5CN	97

They are also moderately soluble in water, methanenitrile being completely miscible. They are hydrolysed when refluxed with dilute acid or alkali to yield successively amides, ammonium salts and finally carboxylic acids (see above):

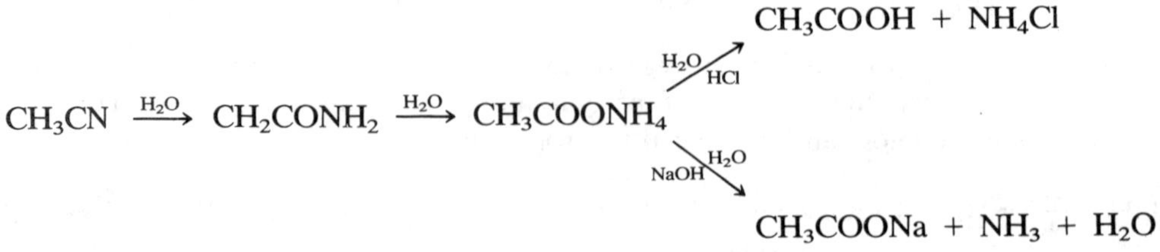

$$CH_3CN \xrightarrow{H_2O} CH_2CONH_2 \xrightarrow{H_2O} CH_3COONH_4 \begin{smallmatrix} \xrightarrow{\ \ H_2O\ \ }_{HCl}\ CH_3COOH + NH_4Cl \\ \\ \xrightarrow[NaOH]{\ \ H_2O\ \ }\ CH_3COONa + NH_3 + H_2O \end{smallmatrix}$$

Nitriles can be used to *add* a carbon atom to the carbon chain in a molecule. For example, naturally occurring fats and oils are a source of carboxylic acids, but contain only compounds with an even number of carbon atoms. So to obtain a carboxylic acid with an odd number of carbon atoms it is necessary to introduce this via a $-C\equiv N$ group:

$$C_{11}H_{23}COOH \xrightarrow{\text{LiAlH}_4} C_{12}H_{25}OH \xrightarrow{\text{PCl}_5} C_{12}H_{25}Cl \xrightarrow{\text{KCN}}$$

$$C_{12}H_{25}CN \xrightarrow[\text{HCl}]{\text{H}_2\text{O}} C_{12}H_{25}COOH$$

Alkyl nitriles also provide a convenient way of preparing amines, to which they can readily be reduced:

$$CH_3C\equiv N \xrightarrow[\text{ethanol}]{\text{Na}} CH_3CH_2NH_2$$

28.6   Amino-acids

Amino-acids are a class of compounds derived from alkyl carboxylic acids in which one or more of the hydrogen atoms of the alkyl radical has been substituted by an amino-group. The simplest of these compounds is aminoethanoic acid, NH_2CH_2COOH, commonly known as glycine. It is a white solid with a sweetish taste that dissolves in water to give a neutral solution. It is thought to exist in water as a dipolar ion or 'zwitterion' (page 415):

$$NH_2CH_2COOH + aq \rightarrow {}^+H_3NCH_2COO^-(aq)$$

Amino-acids have properties of both carboxylic acids and amines, thus glycine reacts with alkalis to give metal salts and with mineral acids to give salts in which it forms the positive ion:

$$NH_2CH_2COOH + NaOH \rightarrow NH_2CH_2COONa + H_2O$$

$$H_2NCH_2COOH + HCl \rightarrow (H_3NCH_2COOH)^+Cl^-$$

Amino-acids also react with nitrous acid with the evolution of nitrogen, and form esters and other normal derivatives of carboxylic acids.

Amino-acids are the structural units of proteins, the hydrolysis of proteins giving rise to a mixture of amino-acids. These are all of the class that have an amino-group on the carbon atom adjacent to a carboxylic group. About twenty-five of these have been isolated, ten are known to be essential to life. Some examples are given in Table 28.2 together with the approximate pH of their solutions, which varies according to the relative number of amino or carboxylic acid groups present in the molecule.

The linkage of amino-acids to build up polypeptides and finally proteins occurs by reaction between the amino-group of one molecule and the activated carboxylic acid group of another to form a peptide link. In nature the carboxylic acid group is activated by enzymes, while in the laboratory the acyl chlorides are used. For example, the amino acids:

$$H_2NRCHCOOH + H_2NR'CHCOOH + H_2NR''CHOOH$$

condense to a polypeptide with the elimination of water molecules giving:

$$\begin{array}{ccc}
 & O & O \\
 & \parallel & \parallel \\
H_2NRCH\overset{}{C}-N-R'CH\overset{}{C}-N-R''CHCOOH \\
 & | & | \\
 & H & H \\
 & \text{peptide link} & \text{peptide link}
\end{array}$$

Table 28.2 Amino acids

Amino-acid	Formula	Solution
Aminoethanoic acid (Glycine)	NH_2CH_2COOH	neutral
2-aminopropanoic acid (Alanine)	$\underset{\displaystyle CH_3\overset{\displaystyle \mid}{C}HCOOH}{\overset{\displaystyle NH_2}{}}$	neutral
2-amino-3-phenylpropanoic acid (Phenylalanine: essential)	$\underset{\displaystyle C_6H_5CH_2\overset{\displaystyle \mid}{C}HCOOH}{\overset{\displaystyle NH_2}{}}$	neutral
Aminobutanedioic acid (Aspartic acid)	$\overset{\displaystyle NH_2}{\underset{\displaystyle \underset{\displaystyle CH_2COOH}{\mid}}{\overset{\displaystyle \mid}{C}HCOOH}}$	acidic
2,6-diaminohexanoic acid (Lysine: essential)	$\overset{\displaystyle NH_2}{\underset{\displaystyle \underset{\displaystyle CH_2CH_2CH_2NH_2}{\mid}}{\overset{\displaystyle \mid}{C}H_2CHCOOH}}$	basic
2-amino-3-methylbutanoic acid (Valine: essential)	$\underset{\displaystyle CH_3}{\overset{\displaystyle CH_3}{}}{\diagdown}\underset{\displaystyle}{\overset{\displaystyle NH_2}{CHCHCOOH}}$	neutral

Clearly the polypeptide molecule still has active groups at either end so that further reaction to make a longer chain is possible. The groups R, R′ and R″ may be the same throughout the chain but usually differ as proteins are made up of a variety of amino-acids. The polypeptide chains can form globular proteins, which are water soluble or form colloidal solutions such as those in egg albumen and the glutenin of wheat, and fibrous proteins such as those in wool, silk or hair.

An immense amount of research is involved in isolating and identifying the amino-acids resulting from the hydrolysis of proteins, using methods such as paper chromatography, electrophoresis and ion exchange columns. For work in this field, and the even more difficult task of establishing the sequence of amino-acids in proteins, several Nobel Prizes have been awarded.

28.7 Nylon

This most useful and versatile synthetic fibre, the first to be prepared by polymerisation, is formed from monomers joined together by a peptide linkage. Nylon 66, one of the commonest forms, is made by condensing the acyl chloride of hexane-1,6-dioic acid with 1,6-diaminohexane:

$$Cl.OC.CH_2CH_2CH_2CH_2CO.Cl + H_2NCH_2CH_2CH_2CH_2CH_2CH_2NH_2 \rightarrow$$

$$\rightarrow ClOCCH_2CH_2CH_2CH_2\overset{\overset{\displaystyle O}{\|}}{C}-\underset{\underset{\displaystyle H}{|}}{N}-CH_2CH_2CH_2CH_2CH_2CH_2NH_2 + HCl$$

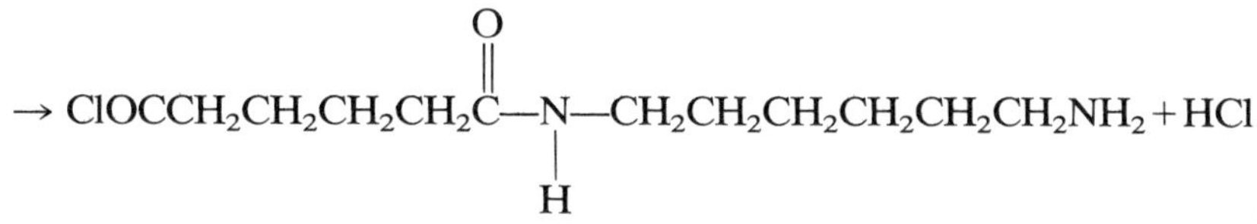

Further condensation can take place since the dimers each have acyl chloride and amino-groups at the ends of the molecule so a long chain polymer is readily built up:

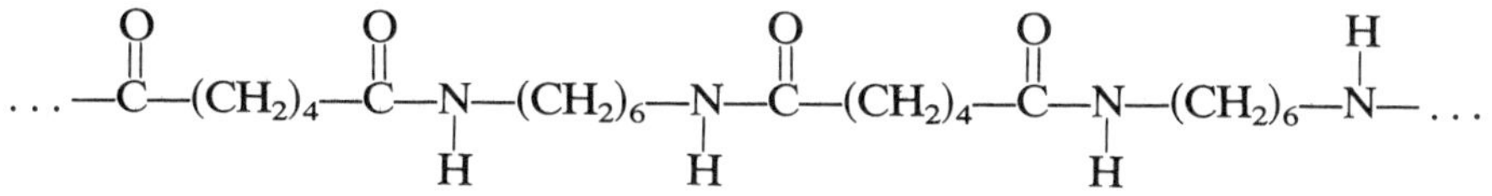

$$\ldots-\overset{\overset{\displaystyle O}{\|}}{C}-(CH_2)_4-\overset{\overset{\displaystyle O}{\|}}{C}-\underset{\underset{\displaystyle H}{|}}{N}-(CH_2)_6-\underset{\underset{\displaystyle H}{|}}{N}-\overset{\overset{\displaystyle O}{\|}}{C}-(CH_2)_4-\overset{\overset{\displaystyle O}{\|}}{C}-\underset{\underset{\displaystyle H}{|}}{N}-(CH_2)_6-\underset{\underset{\displaystyle H}{|}}{N}-\ldots$$

Nylon 6 is based on the monomer 'caprolactan', derived from hexoic acid:

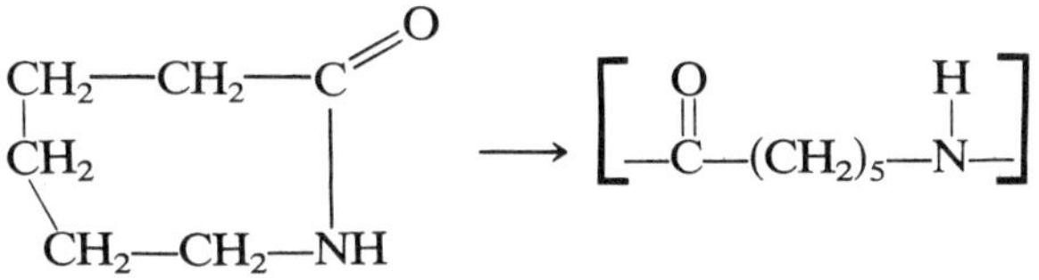

In this case the ring structure opens to give molecules capable of polymerisation to chain structures. Nylon 610 is the condensation product of an acid obtained from castor oil ($HOOC(CH_2)_8COOH$) with 1,2-diaminohexane. The slightly longer chain in the acid molecule makes a more flexible fibre. The chief source of both 1,6-diaminohexane and hexane-1,6-dioic acid is cyclohexane.

Questions

1. The reactivity of the amino ($-NH_2$) group is largely dependent on the nature of the group to which it is attached.

   Give examples which illustrate the difference in reactivity of the amino group when attached to (*a*) an alkyl group; (*b*) an aromatic nucleus; (*c*) a carbonyl group.

   How may this difference in reactivity be explained? (L)

2. The compounds phenylamine, ethylamine (aminoethane) and ethanamide each contain the $-NH_2$ group. Compare the basicities of these compounds and account for the differences in terms of their structures.

   Describe the behaviour of each of these compounds with a solution of sodium nitrite in dilute hydrochloric acid. (L)

3. (*a*) Write an equation for the reaction that occurs when an ice-cold aqueous solution of sodium nitrite is added to an ice-cold solution of phenylamine (*aniline*) in an excess of dilute hydrochloric acid.

   (*b*) Give the reagents, essential conditions, and the equations, for the conversion of the product of the reaction in (*a*) into each of the

following:
(i) Phenol. (ii) Iodobenzene. (iii) An azo dye.   (JMB)

4. (*a*) Describe briefly the preparation of aminoethanoic acid, starting from ethanoic acid.

(*b*) Write equations for *two* typical reactions of aminoethanoic acid.

(*c*) Name and give *one* example of the natural polymers which are made from substances such as aminoethanoic acid. What is the *name* and *structure* of the characteristic functional group in such polymers?

(*d*) Which of the two formulae below best represents aminoethanoic acid? Give the reason for your choice.

$$H_2NCH_2COOH \qquad H_3\overset{+}{N}CH_2COO^-$$

(*e*) Give *one* reaction which would enable you to distinguish between ethanoic acid and ethanamide.   (O)

5. Explain with the aid of *one* example in each case taken from commercial plastics, the meaning of the following terms: (*a*) *addition polymerisation*, (*b*) *condensation polymerisation*.

How do the physical properties of a polymer depend on (i) chain length, (ii) crross-linking?

Give the name and indicate the structure of *one* naturally occurring polymer.   (C)

Alkyl and aryl halogen derivatives

29.1 Introduction

Halogen derivatives of organic compounds rarely occur naturally, but are widely prepared for use in the synthesis of other compounds because of the ease of replacement of the halogen atom by other functional groups. This is particularly true of the halogenoalkanes and acyl chlorides. The halogenoarenes are less reactive but are of great value in the manufacture of insecticides.

The halogen atom is covalently bonded to the molecule in all cases, the main types of organic halogen compounds being:

(a) Halogenoalkanes (alkyl halides) of general formula $C_nH_{2n+1}X$, where X may be chlorine, bromine or iodine (fluoro-compounds have rather different properties and are dealt with separately in Section 6).

(b) Arenes in which a hydrogen atom in the benzene ring is substituted by a halogen atom, such as:

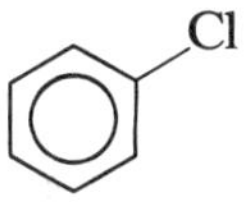

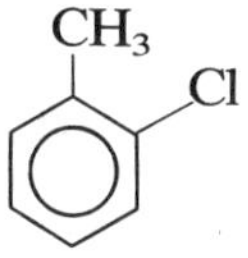

Chlorobenzene   Chloro-2-methylbenzene

These are usually chloro-derivatives; the introduction of bromine or iodine atoms is more difficult.

(c) Halogenated aryl compounds in which the halogen atom replaces a hydrogen atom in a side-chain, such as (chloromethyl) benzene:

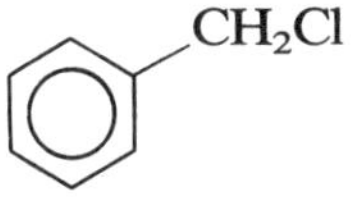

(d) Compounds in which more than one hydrogen atom has been replaced by a halogen atom, such as dichloroethane, $C_2H_4Cl_2$, tri-iodomethane CHI_3, poly(chloroethene) $(-CH_2CHCl-)_n$.

(e) Alkyl and aryl carboxylic acids in which the hydroxyl group has been replaced by a halogen atom, such as ethanoyl chloride CH_3COCl and benzoyl chloride C_6H_5COCl.

(f) Carboxylic acids in which hydrogen atoms in the alkyl radical are substituted by halogen atoms as, for example, chloroethanoic acid $CH_2ClCOOH$.

29.2 Halogenoalkanes

These are readily made from the corresponding alcohol (page 378) or alkene (page 352), those of lower molecular mass being heavy gases or dense oily liquids.

Table 29.1 Boiling points of some halogenoalkanes

Compound	Formula	Boiling Point °C
Chloromethane	CH_3Cl	-24
Bromomethane	CH_3Br	4
Iodomethane	CH_3I	43
Chloroethane	C_2H_5Cl	12
Chloropropane	C_3H_7Cl	47
1-chlorobutane	C_4H_9Cl	79
1-bromobutane	C_4H_9Br	102
1-iodobutane	C_4H_9I	131
2-bromobutane	$CH_3CH_2CHBrCH_3$	91
2-methyl-2-bromopropane	$(CH_3)_3CBr$	73

The highly electronegative halogen atoms are reactive, and, while they are covalently bonded, they induce a dipole moment in the molecule (inductive effect, page 344):

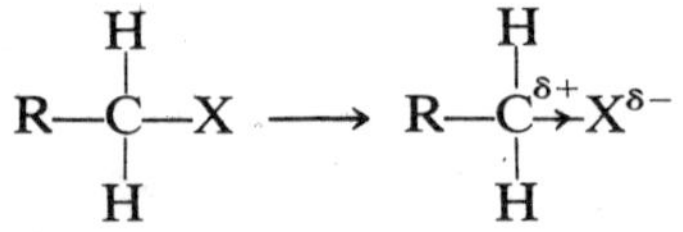

$$R-\underset{\underset{H}{|}}{\overset{\overset{H}{|}}{C}}-X \longrightarrow R-\underset{\underset{H}{|}}{\overset{\overset{H}{|}}{C}}\overset{\delta+}{\rightarrow}X^{\delta-}$$

The carbon atom with the induced charge is electron-deficient, and hence readily attacked by nucleophilic reagents such as hydroxide ions, OH^-, as in the hydrolysis of bromobutanes.

The mechanism of these reactions, resulting in the substitution of a hydroxyl group for a bromine atom, has been extensively studied. There are two sequences by which the reaction

$$C_4H_9Br(l) + OH^-(aq) \rightarrow C_4H_9OH(aq) + Br^-(aq)$$

might take place: a single-step mechanism through a transition state, or a two-step mechanism in which one step is the rate-determining step.

The single-step mechanism begins with the attack of the OH^- group on the electron-deficient carbon atom; the formation of the C—OH bond starts while the C—Br bond is being broken, giving rise to a transition state in which the negative charge is transferred from the oxygen to the bromine atom. Finally the C—Br bond is wholly broken with the transference of the two bonding electrons to the bromine, which leaves as a bromide ion while the C—OH bond is formed by the two electrons from the oxygen atom:

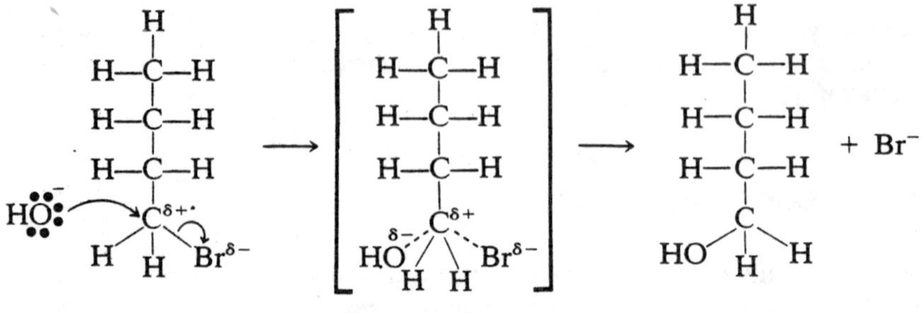

Transition state

This one-step reaction is confirmed for 1-bromobutane by experimental determination of the rate of the reaction, which is found to depend on the concentrations of both the hydroxide ions and the 1-bromobutane molecules.

$$\text{Rate} \propto k[OH^-][C_4H_9Br]$$

Since two species are involved in the rate-determining step the reaction is bimolecular; such bimolecular substitution reactions by nucleophilic reagents are designated S_N2 reactions.

The second mechanism operates in the hydrolysis of 2-bromo-2-methylpropane (isomeric with 1-bromobutane), which is found to be a unimolecular reaction whose rate depends on the concentration of 2-bromo-2-methylpropane only. The study of a model of the molecule indicates that attack on the 2-carbon atom by an OH^- ion is sterically unlikely, since it is surrounded by large groups:

$$
\begin{array}{c}
\text{H} \\
| \\
\text{H}-\text{C}-\text{H} \\
\text{H} \quad | \\
| \quad | \\
\text{H}-\text{C}-\text{C}-\text{Br} \\
| \quad | \\
\text{H} \quad | \\
\text{H}-\text{C}-\text{H} \\
| \\
\text{H}
\end{array}
$$

The first step in the mechanism is thus the formation of a bromide ion and a carbonium ion. This involves the breaking of a bond, and is the slow rate-determining step; it is followed by rapid reaction of the carbonium ion with hydroxide ions to form 2-methylpropan-2-ol:

$$
\begin{array}{c}
\text{CH}_3 \\
| \\
\text{CH}_3-\text{C}^{\delta+}-\text{Br}^{\delta-} \\
| \\
\text{CH}_3
\end{array}
\longrightarrow
\begin{array}{c}
\text{CH}_3 \\
| \\
\text{CH}_3-\text{C}^+ + \text{Br}^- \\
| \\
\text{CH}_3
\end{array}
\quad \text{(slow)}
$$

$$
\begin{array}{c}
\text{CH}_3 \\
| \\
\text{CH}_3-\text{C}^+ + \text{:}\ddot{\text{O}}\text{H}^- \\
| \\
\text{CH}_3
\end{array}
\longrightarrow
\begin{array}{c}
\text{CH}_3 \\
| \\
\text{CH}_3-\text{C}-\text{OH} \\
| \\
\text{CH}_3
\end{array}
\quad \text{(fast)}
$$

This reaction, being a substitution by nucleophilic attack but unimolecular, is designated an S_N1 reaction.

The ease of replacement of the halogen atom follows the order:

$$I > Br > Cl$$

This can readily be demonstrated by studying the rates of hydrolysis of chloro-, bromo- and iodobutane. Three drops of each compound are dissolved in one cubic centimetre of ethanol in three separate test-tubes heated to 60°C in a water bath; one cubic centimetre of silver nitrate solution is added to all three simultaneously and the mixtures shaken. Hydrolysis occurs in the water of the solution, releasing a halide ion which gives a precipitate with the silver ion:

$$C_4H_9X(l) + OH^-(aq) \rightarrow C_4H_9OH(aq) + X^-(aq)$$

$$Ag^+(aq) + X^-(aq) \rightarrow AgX(s) \quad (fast)$$

Thus the rate of hydrolysis is indicated by the appearance of a precipitate of silver halide; the order will be found to be silver iodide, silver bromide and finally silver chloride.

The relative ease of replacement of iodine results in the iodoalkanes being most useful in syntheses, and in the summary below equations for the iodocompounds are given, but it must be remembered that the bromo- and chloro-compounds will give the same reactions. Besides these substitution reactions halogenoalkanes can also be converted to alkanes and alkenes by reduction and elimination reactions.

(a) To form alkanes

To obtain a pure sample of an alkane, a halogenoalkane can be reduced by the use of a zinc/copper couple, made by placing zinc filings in a solution of copper sulphate for a short time.

$$C_2H_5I + H_2 \xrightarrow[\text{ethanol}]{\text{Zn/Cu}} C_2H_6 + HI$$

Alternatively the halogen atom can be removed by the action of sodium on the halogenoalkane dissolved in dry ether (Wurtz reaction). In this case an alkane containing twice the number of carbon atoms as the parent compound will be obtained:

$$2C_2H_5I + 2Na \xrightarrow{\text{ethoxyethane}} C_4H_{10} + 2NaI$$

(b) To form alkenes

This reaction is only applicable to halogenoalkanes with more than two carbon atoms in the molecule. The halogen compound is refluxed in an alcoholic solution of sodium hydroxide to yield an alkene (since no water is present no hydrolysis takes place):

$$C_3H_7I + NaOH \xrightarrow{\text{ethanol}} CH_3CH{=}CH_2 + NaI + H_2O$$

(c) To form ethers

Simple ethers (those with identical groups attached to the oxygen atom) are prepared by reacting the halogenoalkane with a sodium alkoxide RO^-Na^+ containing the same number of carbon atoms:

$$CH_3I + CH_3ONa \rightarrow CH_3{-}O{-}CH_3 + NaI$$

Simple ethers can also be made by reacting the halogenoalkane with *dry* silver oxide:

$$2C_2H_5I + Ag_2O \rightarrow C_2H_5-O-C_2H_5 + 2AgI$$

To obtain a mixed ether an alkoxide containing the required number of carbon atoms in the molecule must be chosen. For example, to make methyoxyethane, iodomethane and sodium ethoxide can be used:

$$CH_3I + C_2H_5ONa \rightarrow CH_3-O-C_2H_5 + NaI$$

(d) To form alcohols

To replace the halogen atom by a hydroxyl group OH^-, water must be present; either *moist* silver oxide or an aqueous solution of sodium hydroxide is used:

$$C_2H_5I + NaOH \xrightarrow{\text{aqueous}} C_2H_5OH + Na^+I^-$$

(e) To form esters

Esters are readily formed by warming a mixture of a halogenoalkane with the silver salt of the required carboxylic acid, $RCOO^-Ag^+$, in *alcoholic* solution:

$$CH_3COOAg + C_2H_5I \xrightarrow{\text{ethanol}} CH_3COOC_2H_5 + AgI$$

(f) To form nitriles

Nitriles are formed by warming a halogenoalkane with an *alcoholic* solution of potassium cyanide, K^+CN^-, to replace the halogen atom with a nitrile group:

$$CH_3I + KCN \xrightarrow{\text{ethanol}} CH_3CN + KI$$
$$\text{Methyliodide} \qquad\qquad \text{Ethanenitrile}$$

This adds a carbon atom to the molecule of the parent compound (page 426).

(g) To form amines

A mixture of amines, including quaternary ammonium salts (page 419) can be obtained by reacting a halogenoalkane with concentrated ammonia solution in a sealed tube:

$$C_2H_5I + NH_3 \rightarrow C_2H_5NH_2 + HI$$
$$C_2H_5I + C_2H_5NH_2 \rightarrow (C_2H_5)_2NH + HI$$
$$C_2H_5I + (C_2H_5)_2NH \rightarrow (C_2H_5)_3N + HI$$
$$C_2H_5I + (C_2H_5)_3N \rightarrow (C_2H_5)_4N^+I^-$$

If a pure specimen of a single primary amine is required, this is better obtained by using a Grignard reagent (see Section 3).

(h) To form nitroalkanes

A halogenoalkane refluxed with solid silver nitrite gives a good yield of nitroalkane mixed with other products from which it can be separated by fractional distillation:

$$C_2H_5I + AgNO_2 \rightarrow C_2H_5NO_2 + AgI$$
$$\text{Nitroethane}$$

(i) To form alkyl derivatives of arenes

This important reaction, known as the Friedel–Crafts reaction, is used to make homologous arenes, such as methylbenzene, by refluxing the arene with a halogenoalkane in the presence of aluminium chloride as a catalyst:

$$CH_3I + C_6H_6 \xrightarrow[\text{AlCl}_3]{} \quad \bigcirc\!\!\!-CH_3 \quad + HI$$

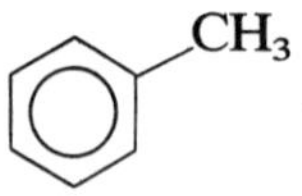

29.3   Grignard reagents

A number of the reactions of halogenoalkanes listed above are carried out more easily if the halogeno-compound is first converted to a *Grignard reagent.*

These versatile compounds are made by dissolving the halogenoalkane in dry ethoxyethane, adding magnesium turnings and refluxing the mixture over a water-bath. Iodo-compounds are most frequently used, a small crystal of iodine being added as a catalyst.

$$RX + Mg \xrightarrow{\text{ethoxyethane}} R\!-\!Mg\!-\!X$$

(where R is an alkyl radical and X is a halogen atom). The resulting compound is not isolated, but kept and used in the ethereal solution in which it was made. These reagents, for the introduction of which Grignard was awarded a Nobel prize in 1912, can be used to synthesise a great variety of compounds.

(a) To form alkanes

Grignard reagents are decomposed by water to yield alkanes having the same number of carbon atoms as the original reagent.

$$R\!-\!Mg\!-\!I + H_2O \rightarrow RH + Mg(OH)I$$

Alkanes of higher molecular mass can be made by reacting the Grignard reagent with another halogenoalkane.

$$R\!-\!Mg\!-\!I + R'I \xrightarrow{\text{ethoxyethane}} R\!-\!R' + MgI_2$$

$$C_2H_5\!-\!Mg\!-\!I + CH_3I \xrightarrow{\text{ethyoxyethane}} C_3H_8 + MgI_2$$

(b) To form primary alcohols (ascending a homologous series)

Primary alcohols containing one more carbon atom than the original compound can be made by passing methanal vapour into a solution of a Grignard reagent. An addition compound is formed which breaks up on acidification to yield an alcohol.

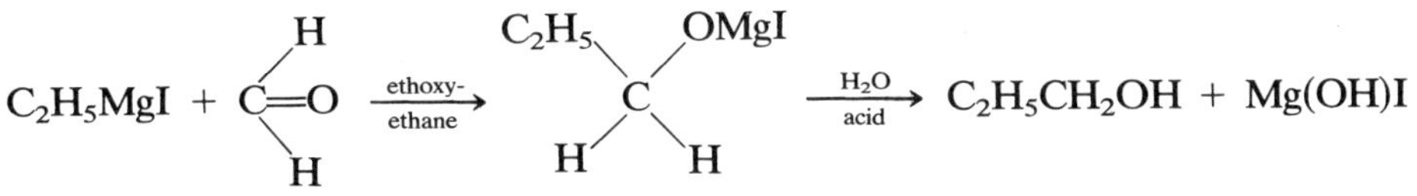

$$C_2H_5MgI + \underset{H}{\overset{H}{C}}{=}O \xrightarrow[\text{ethane}]{\text{ethoxy-}} \underset{H}{\overset{C_2H_5}{\underset{H}{C}}} \overset{OMgI}{\underset{}{}} \xrightarrow[\text{acid}]{H_2O} C_2H_5CH_2OH + Mg(OH)I$$

(c) To form secondary alcohols

Secondary alcohols are formed if aldehydes other than methanal are used; the aldehyde is dissolved in ethoxyethane and the solution added to that of the Grignard reagent:

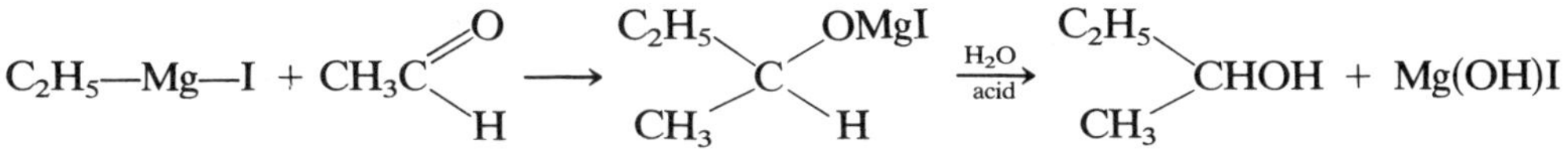

$$C_2H_5{-}Mg{-}I + CH_3C\overset{O}{\underset{H}{}} \longrightarrow \underset{CH_3}{\overset{C_2H_5}{\underset{H}{C}}}{\overset{OMgI}{}} \xrightarrow[\text{acid}]{H_2O} \underset{CH_3}{\overset{C_2H_5}{}}CHOH + Mg(OH)I$$

(d) To form tertiary alcohols

The use of ketones, in place of aldehydes, gives rise to tertiary alcohols:

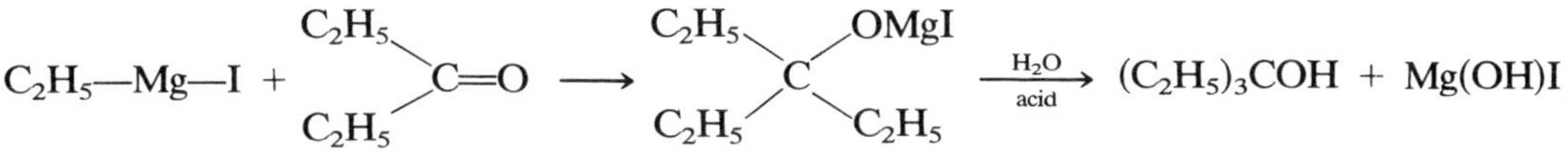

$$C_2H_5{-}Mg{-}I + \underset{C_2H_5}{\overset{C_2H_5}{}}C{=}O \longrightarrow \underset{C_2H_5}{\overset{C_2H_5}{\underset{C_2H_5}{C}}}{\overset{OMgI}{}} \xrightarrow[\text{acid}]{H_2O} (C_2H_5)_3COH + Mg(OH)I$$

(e) To form aldehydes

Aldehydes are formed on the acidification of addition compounds formed between Grignard reagents and ethyl methanoate:

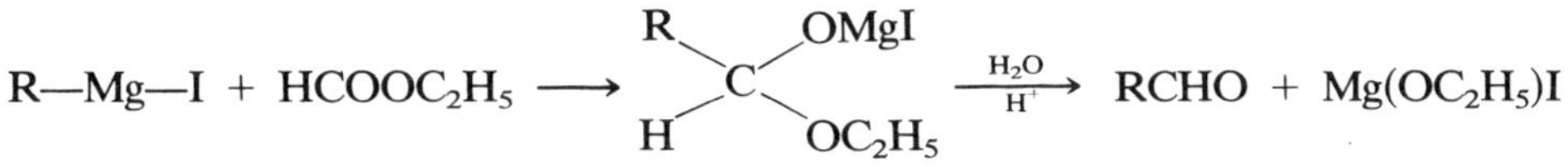

$$R{-}Mg{-}I + HCOOC_2H_5 \longrightarrow \underset{H}{\overset{R}{\underset{OC_2H_5}{C}}}{\overset{OMgI}{}} \xrightarrow[H^+]{H_2O} RCHO + Mg(OC_2H_5)I$$

(f) To form ketones

Ethereal solutions of alkyl nitriles form addition compounds with Grignard reagents which hydrolyse to yield ketones as the final products:

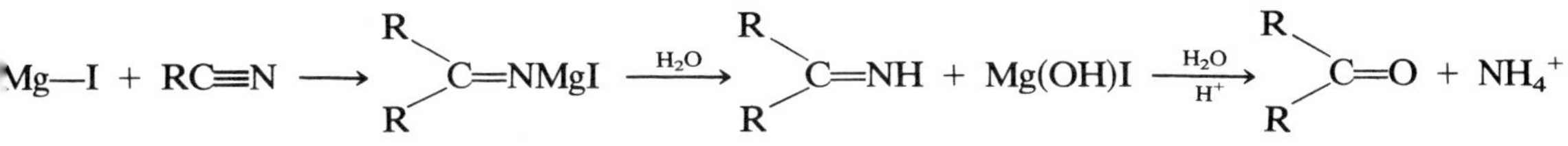

$$Mg{-}I + RC{\equiv}N \longrightarrow \underset{R}{\overset{R}{}}C{=}NMgI \xrightarrow{H_2O} \underset{R}{\overset{R}{}}C{=}NH + Mg(OH)I \xrightarrow[H^+]{H_2O} \underset{R}{\overset{R}{}}C{=}O + NH_4^+$$

(g) To form carboxylic acids (ascending a homologous series)

Solid carbon dioxide added to a cold ethereal solution of a Grignard reagent reacts to form an alkyl carboxylic acid having one more carbon atom in the carbon chain than the original molecule:

$$R-Mg-I \xrightarrow[\text{solid}]{CO_2} RC(=O)-OMgI \xrightarrow[H^+]{H_2O} R-C{\overset{\displaystyle O}{\underset{\displaystyle OH}{\big<}}} + Mg(OH)I$$

(h) To form primary amines

To obtain a sample of a single primary amine a Grignard reagent containing the required number of carbon atoms is reacted with chloramine, $ClNH_2$:

$$R-Mg-I + ClNH_2 \rightarrow RNH_2 + MgClI$$

29.4   Halogen derivatives of arenes

Halogen derivatives of benzene and other arenes are much less reactive than the halogenoalkanes; chlorobenzene, for example, is not hydrolysed even when boiled with aqueous sodium hydroxide. All three halogen derivatives are made by direct reaction between benzene and the halogen in the presence of a halogen carrier (page 359). The reaction does not take place readily with bromine or iodine, but chlorobenzene and other chloroarenes are manufactured for use in the synthesis of a number of products.

(a) To make phenol

Chlorobenzene heated under pressure to 200°C with solid sodium hydroxide in a nickel crucible gives a good yield of sodium phenate from which phenol can be obtained on acidification:

$$C_6H_5Cl + 2NaOH \rightarrow C_6H_5ONa + NaCl + H_2O$$

(b) To make phenylamine (aniline)

Phenylamine is formed when chlorobenzene is reacted under pressure at 200°C with concentrated ammonia solution, using copper(I) oxide as a catalyst:

$$C_6H_5Cl + 2NH_3 \xrightarrow{Cu_2O} C_6H_5NH_2 + NH_4Cl$$

(c) To make alkyl derivatives of benzene

An alkyl group can be added to a benzene ring by reacting an alkyl halide with chlorobenzene and sodium in ethereal solution, for example the synthesis of methylbenzene:

$$CH_3I + C_6H_5Cl + 2Na \xrightarrow{\text{ethyoxyethane}} \text{(benzene ring)}-CH_3 + NaI + NaCl$$

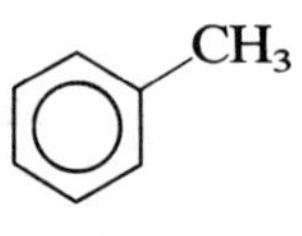

if the alkyl halide is omitted the resulting compound will be diphenyl:

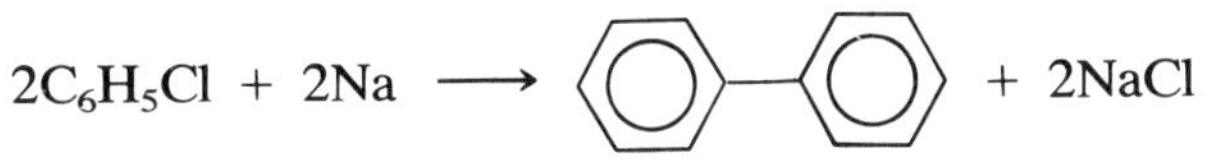

$$2C_6H_5Cl + 2Na \longrightarrow \langle\bigcirc\rangle-\langle\bigcirc\rangle + 2NaCl$$

(d) To make DDT

Chlorobenzene reacts with trichloroethanal(chloral) to make one of the most valuable insecticides, DDT. Two molecules of chlorobenzene replace the oxygen atom of the aldehyde:

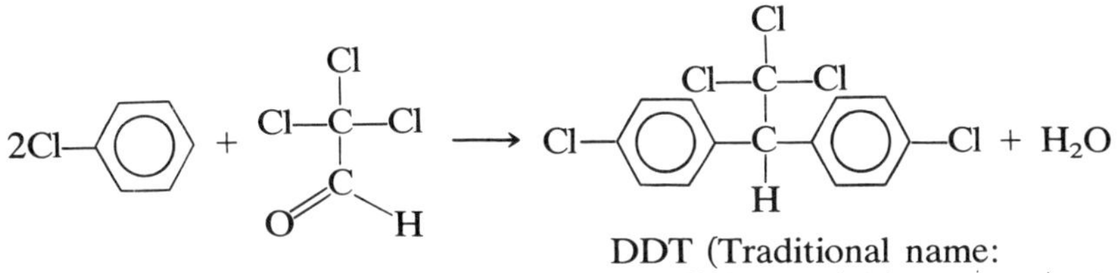

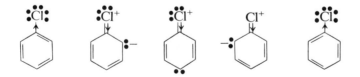

DDT (Traditional name:
*d*ichloro*d*iphenyl*t*richloroethane)

(e) Substitution products of chlorobenzene

Substitution of hydrogen atoms in the benzene nucleus of chlorobenzene by other functional groups takes place in the *ortho-* and *para-* positions. The reactions are slower than those for pure benzene since the chlorine atom is deactivating, which is unusual for an *ortho/para* directing group (page 360). Two opposing effects are involved; the chlorine atom, being electronegative, tends to withdraw electrons from the benzene nucleus, thus deactivating the carbon atoms. But, under attack by an electrophilic reagent (page 343) the inductive deactivating effect is overcome and one of the lone pairs of electrons on the chlorine atom takes part in an *ortho/para* directing reso-nance effect:

Fig. 29.1 Possible resonance structures for chlorobenzene

Thus if chlorobenzene is reacted with chlorine in the presence of aluminium chloride or iron filings 1,2-dichlorobenzene and 1,4-dichlorobenzene are slowly formed:

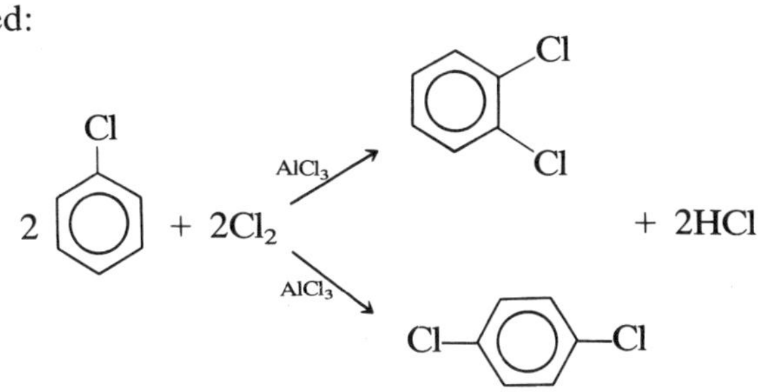

p-dichlorobenzene is manufactured for use as a moth repellent; it is more effective than the original 'moth-balls' which were made of naphthalene.

With concentrated sulphuric acid chlorobenzene is converted to chlorobenzene sulphonic acid, the product being entirely the -1,4- derivative:

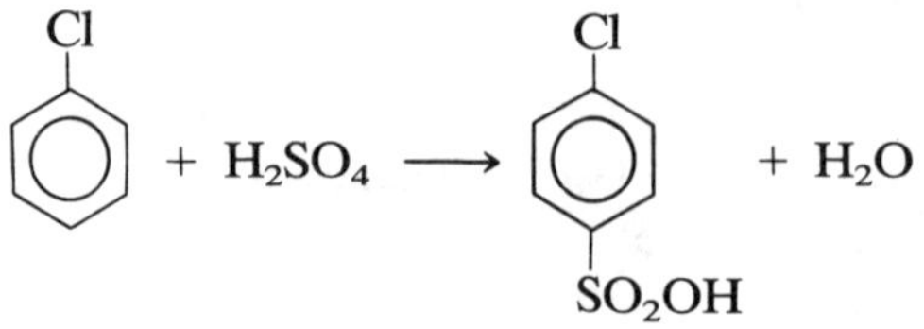

(f)   Arenes with halogenated side chains

Chloro-derivatives of methyl benzene include those in which the chlorine is attached to the benzene ring and those in which the chlorine atom is situated in the side chain. Compounds in which the halogen atom is situated in a side chain behave as halogenoalkanes:

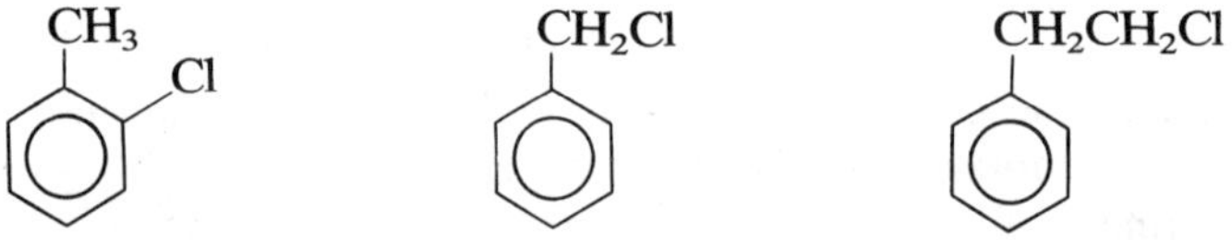

Chloro-2-methylbenzene   (Chloromethyl)benzene   (Chloroethyl)benzene

Fig. 29.2   Chlorinated aromatic compounds

Thus chloro-2-methylbenzene and (chloromethyl)benzene, each of molecular formula C_7H_7Cl, can be distinguished by boiling separately with sodium hydroxide solution; chloro-2-methylbenzene will not be hydrolysed, while (chloromethyl)benzene will be hydrolysed releasing a chloride ion. So, if the two solutions are now acidified with nitric acid and silver nitrate solution is added, a white precipitate of silver chloride will be obtained only from the (chloromethyl)benzene mixture.

29.5   Polyhalogen derivatives of alkanes

A large number of alkyl compounds with more than one halogen atom in the molecule can be made. The reactivity of the halogen atom decreases as the number in the molecule increases; for the chlorinated methanes, the reactivity of the chlorine atom diminishes from chloromethane to tetrachloromethane:

$$CH_3Cl > CH_2Cl_2 > CHCl_3 > CCl_4$$

while poly(chloroethene) (or polyvinyl chloride) is inert.

Dichloromethane is a liquid of boiling point 40°C, used as an industrial solvent. It has no isomeric forms since the tetrahedral molecule is symmetrical:

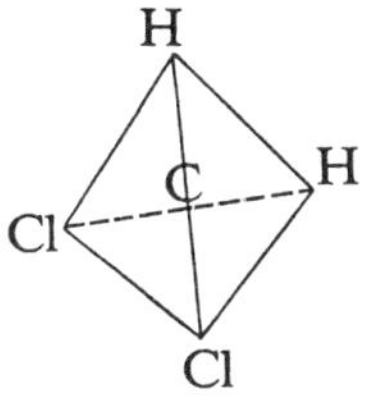

The di-halogen derivatives of ethane, however, can be of two kinds—
gem-*dihalides* in which the halogen atoms are both situated on the same
carbon atom and *vic*-*dihalides* in which the halogen atoms are each on
separate carbon atoms:

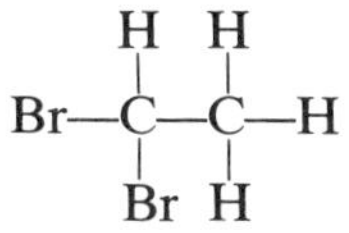

gem-dibromide 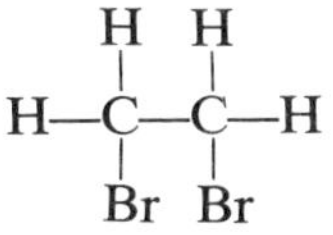   *vic*-dibromide

The *gem*-dihalides are less reactive than the *vic*-dihalides and are of
mainly theoretical interest; 1,1-dichloroethane is formed when ethanal is
reacted with phosphorus pentachloride, providing strong evidence for the
presence of the $>C{=}O$ group in aldehydes:

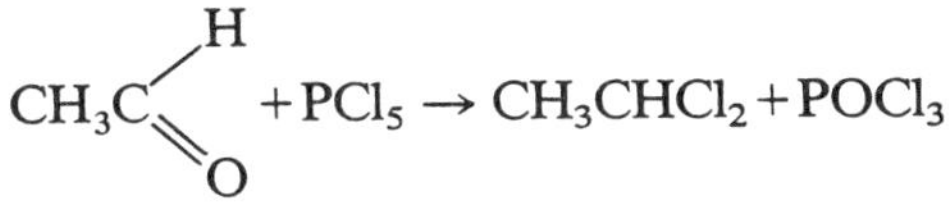

$$CH_3C\overset{H}{\underset{O}{\diagdown}} + PCl_5 \rightarrow CH_3CHCl_2 + POCl_3$$

Vic-dihalides undergo similar reactions to the monohalogenoalkanes. 1,2-
dichloroethane is readily hydrolysed by refluxing with dilute aqueous sodium
hydroxide solution to make ethane-1,2-diol (ethylene glycol, page 380).

$$CH_2ClCH_2Cl + 2NaOH \rightarrow CH_2OHCH_2OH + 2NaCl$$

Similarly a dinitrile can be formed by refluxing 1,2-dichloroethane with
potassium cyanide in alcoholic solution:

$$CH_2ClCH_2Cl + 2KCN \xrightarrow[\text{ethanol}]{} CH_2CNCH_2CN + 2KCl$$

One of the main uses of 1,2-dichloroethane is in the manufacture of PVC,
polyvinyl chloride, since it decomposes on heating to 600°C to form
chloroethene (vinyl chloride):

$$CH_2ClCH_2Cl \rightarrow CH_2{=}CHCl + HCl$$

The chloroethene is then polymerised to PVC (page 356).
    1,2-dibromoethane is added to petrol as a lead scavenger (page 263).

Important trihalogen derivatives are trichloromethane and triiodomethane. The latter is formed in the iodoform reaction which is used
to distinguish compounds containing the $\overset{CH_3}{\underset{OH}{\diagup}}C\overset{H}{\diagdown}$ or $\overset{CH_3}{\diagup}C\overset{}{\underset{O}{\diagdown}}$ groups

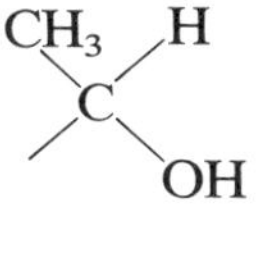

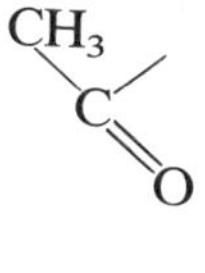

(page 378), and was widely used as antiseptic before more effective compounds were developed.

Trichloromethane, chloroform, is made by the action of bleaching powder on either ethanol or propanone in a series of reactions; if ethanol is used the chlorine in the bleaching powder first oxidises it to ethanal:

$$CH_3CH_2OH + Cl_2 \rightarrow CH_3CHO + 2HCl$$

The ethanal is then chlorinated to trichloroethanal:

$$CH_3CHO + 3Cl_2 \rightarrow CCl_3CHO + 3HCl$$

The trichloroethanal is then finally decomposed by the action of the calcium hydroxide in the bleaching powder:

$$2CCl_3CHO + Ca(OH)_2 \rightarrow 2CHCl_3 + (HCOO)_2Ca$$

Calcium
methanoate

The reaction with propanone is similar, proceeding in two stages, chlorination followed by decomposition:

$$CH_3COCH_3 + 3Cl_2 \rightarrow CCl_3COCH_3 + 3HCl$$

$$2CCl_3COCH_3 + Ca(OH)_2 \rightarrow 2CHCl_3 + (CH_3COO)_2Ca$$

Calcium
ethanoate

Trichloromethane is used to detect primary amines by the carbylamine reaction (page 420); its use as an anaesthetic, once widespread, is much less nowadays since the development of newer and less toxic compounds. It is still a valuable solvent, although for many purposes the less poisonous 1,1,1-trichloroethane is now preferred. Another trichloroderivative widely used as a solvent is 1,2,2-trichloroethene $CHCl\text{=}CCl_2$.

The tetrahalogen derivatives, tetrachloromethane and tetrachloroethane, are also industrial solvents. Tetrachloromethane is used in certain types of fire extinguisher, since its vapour is both dense and not flammable and it is safe to use on electrical fires. Its usefulness is limited because of the poisonous vapours that are formed, and it has largely been displaced by polyhalogen compounds such as bromotrifluoromethane.

29.6   Fluorocarbon compounds

Fluoroalkanes differ from the other halogenoalkanes in that they are very stable; the fluorine atom is not easily attacked or removed from the molecule. Compounds containing chains of —CF_2— units similar to the —CH_2— units of hydrocarbons have been developed. The study of these compounds is of considerable interest; they are used as oils, sealing liquids and coolants, their great stability making them particularly useful in connection with such operations as the enrichment of U_{235} by the diffusion of UF_6 (page 74) since they are not attacked by hydrogen fluoride or uranium hexafluoride and can be used in pumps etc.

Tetrafluoroethene, $CF_2\!=\!CF_2$, made by fluorinating trichloromethane, polymerises to the extremely resistant plastic polytetrafluoroethene or Teflon. Trifluorochloroethene $\begin{smallmatrix}F\\F\end{smallmatrix}\!\!>\!\!C\!=\!C\!\!<\!\!\begin{smallmatrix}F\\Cl\end{smallmatrix}$ polymerises to a plastic known as Kel F which is resistant to many substances including, particularly, hydrogen fluoride. Various fluoro-chloro-derivatives of methane and ethane are manufactured under the name of Freons; these are liquids which are excellent working fluids for use in refrigerators. They have widely replaced ammonia and sulphur dioxide, especially in refrigerator ships and domestic appliances, since their toxicity is low.

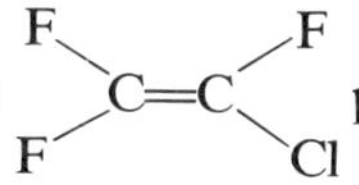

29.7 Acyl chlorides

The chloroderivatives of alkyl and aryl carboxylic acids are mentioned in Chapter 27, but it is worth reviewing their reactions here to compare them with those of the halogenoalkanes.

Acyl chlorides are rapidly hydrolysed in moist air, the reactions being violent in water.

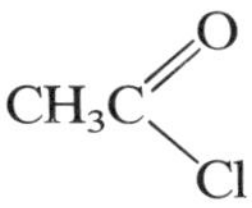

$$CH_3C\!\!\begin{smallmatrix}O\\Cl\end{smallmatrix} + H_2O \rightarrow CH_3COOH + HCl$$

Ethanoyl chloride   Ethanoic acid

Benzoyl chloride is less readily hydrolysed although it fumes in moist air. Its vapour is harmful to the eyes.

Acyl chlorides are used to introduce acyl groups into other molecules. For example, ethanol can be converted to the ester, ethyl ethanoate, by the action of ethanoyl chloride:

$$C_2H_5OH + CH_3COCl \rightarrow CH_3COOC_2H_5 + HCl$$
Ethyl ethanoate

To benzoylate compounds in a similar manner the presence of sodium hydroxide is necessary:

$$C_2H_5OH + C_6H_5COCl + NaOH \rightarrow C_6H_5COOC_2H_5 + NaCl + H_2O$$
Ethyl benzoate

Like the halogenoalkanes the acyl chlorides react with benzene in the presence of aluminium chloride (Friedel–Crafts reaction); the products are aromatic ketones:

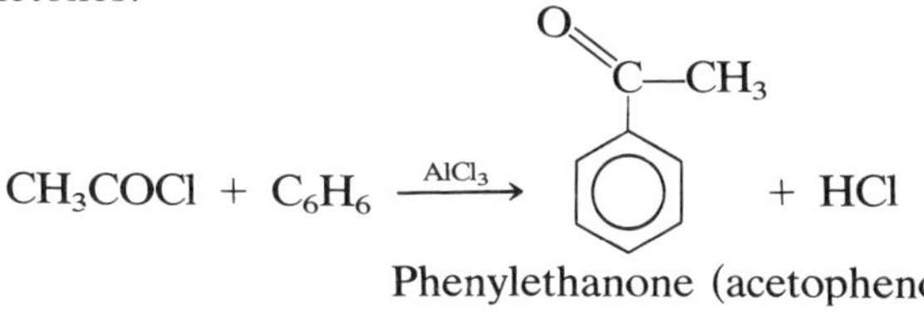

$$CH_3COCl + C_6H_6 \xrightarrow{AlCl_3} \text{(phenyl)}C(=O)CH_3 + HCl$$

Phenylethanone (acetophenone)

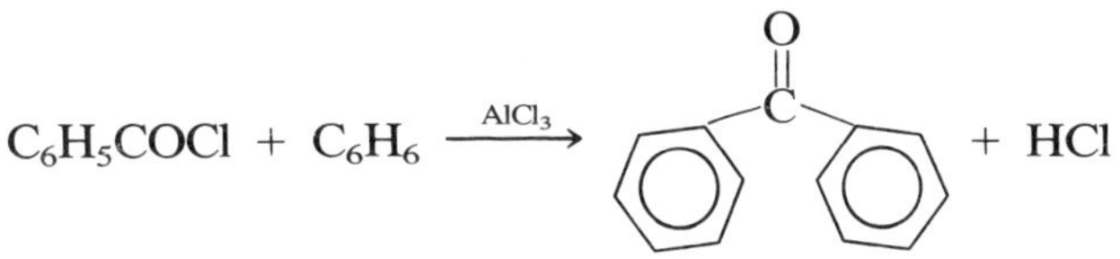

$$C_6H_5COCl + C_6H_6 \xrightarrow{AlCl_3} \text{(diphenyl ketone)} + HCl$$

Diphenylmethanone

With concentrated ammonia, acyl and benzoyl chlorides react to form amides (page 424):

$$CH_3COCl + 2NH_3 \rightarrow CH_3CONH_2 + NH_4Cl$$
Ethanoamide

$$C_6H_5COCl + 2NH_3 \rightarrow C_6H_5CONH_2 + NH_4Cl$$
Benzamide

With amines, RNH_2, substituted amides are formed. One of the most important of these is N-phenylethanamide (acetanilide):

$$CH_3COCl + C_6H_5NH_2 \rightarrow CH_3CONHC_6H_5 + HCl$$

N-phenylethanamide

Benzoyl chloride reacts similarly to make N-phenylbenzamide in the presence of sodium hydroxide:

$$C_6H_5NH_2 + NaOH + ClCOC_6H_5 \rightarrow C_6H_5NHCOC_6H_5 + NaCl + H_2O$$

N-phenyl
benzamide

With the sodium salts of carboxylic acids, acyl chlorides react to make acid anyhdrides:

Ethanoic anhydride (page 410)

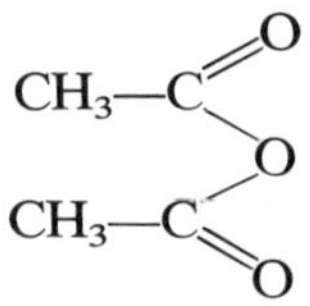

29.8   Halogen-substituted carboxylic acids

The effect of introducing halogen atoms into the alkyl group of a carboxylic acid to make compounds such as chloroethanoic acid, $CH_2ClCOOH$, has already been discussed in Chapter 27.

Questions

1. Make a list of methods by which one carbon atom can be added to the carbon chain of a member of an homologous series (methods of ascending the homologous series).

2. Suggest reactions or a series of reactions by which
(a) ethanol can be converted to nitroethane,
(b) benzene can be converted to methylbenzene,
(c) iodoethane can be converted to propane,
(d) iodoethane (as the only starting material) can be converted to pentan-3-one.

3. Describe and explain how you would distinguish by chemical experiment(s) between the members of the pairs of isomers:

(*a*)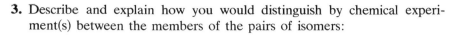

(*b*) CH_2ClCO_2H and $CH_2(OH)COCl$. (L)

4. Describe the preparation of bromoethane starting from ethanol.

Give reagents, conditions, and equations to show how you would obtain the following, starting from bromoethane: (*a*) ethylamine, (*b*) methoxyethane (ethyl methyl ether), (*c*) propanoic acid (propionic acid). (C)

5. An aromatic compound *A* with a relative molecular mass of 250 contained 33.6% carbon, 2.40% hydrogen and 64.0% bromine by mass. On refluxing *A* with aqueous sodium hydroxide, half of the bromine was removed and the resulting organic compound, *B*, was converted into *C* by the action of a mild oxidising agent. *C* gave a crystalline derivative with 2,4-dinitrophenylhydrazine and was readily oxidised, even by exposure to the air, to a monobasic acid, *D*. The latter contained the same number of carbon atoms as *C*, and 2.01 g of it neutralised 100 cm^3 of an aqueous solution of sodium hydroxide containing 0.100 moles per dm^3.

On further investigation it was found that *A* gave two isomers *E* and *F* when a radical *X* was substituted into the aromatic ring in place of one hydrogen atom.

Deduce the structural formulae of the compounds *A*, *B*, *C*, *D*, *E* and *F*, explaining fully your reasoning. (O)

Part VI
Applied Chemistry

30 Laboratory preparations of organic compounds

30.1 Introduction

In the preceding chapters many descriptions of organic reactions and the products arising from them have been given, but few formal descriptions of laboratory preparations have been included. In this chapter some classic laboratory preparations are described in order to give some idea of the methods of preparation and purification normally employed.

Details of quantities used are not given, since these vary with the size and type of apparatus available and frequently do not correspond exactly to the stoichiometric equation. Preparations should be carried out only under supervision. Careful attention should be paid to instructions on the handling of dangerous or poisonous chemicals such as concentrated acids and alkalis, bromine, phosphorus, benzene and ethyoxyethane. In some cases the products obtained vary with the conditions under which the reactions are carried out; in order to obtain good yields of pure products it is necessary to follow instructions exactly, particularly those concerning the order in which the reagents are used, the thorough mixing of these reagents and the temperatures at which the operations are conducted.

30.2 Ethene

$$C_2H_5OH \rightarrow C_2H_4 + H_2O$$

This preparation is based on the formation of ethyl hydrogensulphate, $C_2H_5.HSO_4$, when ethanol is mixed with excess concentrated sulphuric acid in the cold. The reaction mixture is then heated to about 170°C, when the ethyl hydrogensulphate decomposes to yield ethene and to regenerate sulphuric acid.

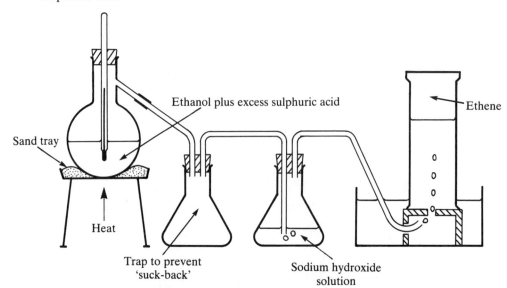

Fig. 30.1 Preparation of ethene

30.3   Ethyoxyethane

$$2C_2H_5OH \rightarrow C_2H_5{-}O{-}C_2H_5 + H_2O$$

Under different conditions from the preceding preparation, the reaction between ethanol and concentrated sulphuric acid results in the formation of ethyoxyethane rather than ethene. The two reagents are again carefully mixed in the cold in order to form ethyl hydrogensulphate, but in this case the ethanol is kept in excess. After the initial reaction the mixture is heated to 140°C; the ethyl hydrogensulphate reacts with the excess ethanol, and ethyoxyethane and water distil off. Additional alcohol is added from a tap-funnel in order to keep it in excess; the rate of addition is adjusted to be the same as that at which the distillate is collected. Great care must be taken to avoid the escape of ethoxyethane vapour into the laboratory.

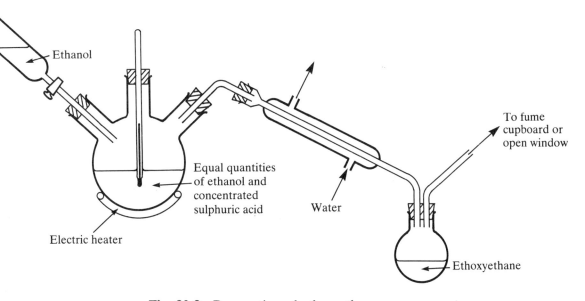

Fig. 30.2   Preparation of ethoxyethane

The crude product contains dissolved ethanol, sulphur dioxide and impurities from the methylated spirit if this has been used as a source of ethanol. To remove these the liquid is shaken with two or three separate portions of sodium hydroxide solution followed by one wash with water. The ethyoxyethane is then dried by standing over calcium chloride, filtered and redistilled; the portion boiling at 34–36°C is collected.

30.4   Bromoethane

$$C_2H_5OH + H_2SO_4 + KBr \rightarrow C_2H_5Br + KHSO_4 + H_2O$$

The bromoethane is essentially produced by the reaction of ethanol with hydrogen bromide produced *in situ* from the reaction of potassium bromide with concentrated sulphuric acid. As water is produced at the same time the sulphuric acid is kept in excess to take up as much of it as possible. The

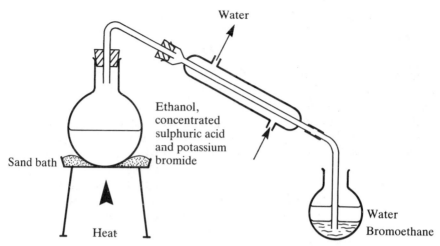

Fig. 30.3 Preparation of bromoethane

alcohol and acid are carefully mixed, keeping the temperature low, and a calculated quantity of powdered potassium bromide added, together with some anti-bumping stones. The mixture is then gently heated, and the bromoethane distils over and is collected under water, which serves the dual purpose of absorbing any hydrogen bromide that comes over and preventing the heavier but volatile bromoethane from evaporating.

The crude product is contaminated with dissolved ethanol and hydrogen bromide, which are removed by washing in a separating funnel as described for ethyoxyethane, except that sodium carbonate solution is used to avoid hydrolysing the bromoethane. The oil, which is colourless, forms the lower layer in the separating funnel. After drying over anhydrous sodium sulphate and filtering it is redistilled, the portion boiling between 36° and 39°C being collected.

30.5   1,2-dibromoethane

$$Br_2 + C_2H_4 \rightarrow CH_2Br.CH_2Br$$

The ethene for this reaction can be prepared from ethanol (Fig. 30.1) or

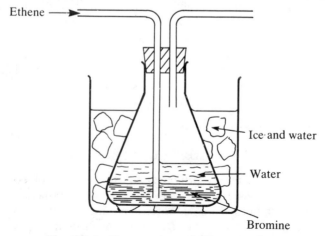

Fig. 30.4 Preparation of dibromoethane

obtained from a cylinder. Since the main reaction occurs so readily, all that is necessary is to bubble the ethene through liquid bromine, held under a layer of water to prevent the escape of bromine fumes. The reaction is continued until the colour of the bromine is discharged, leaving di-bromoethane as a heavy yellow oil which is purified in the same way as bromoethane. The distillate boiling between 130°C and 133°C is collected. Bromine should be handled with great care, preferably in a fume cupboard.

30.6   Ethanoic acid

$$C_2H_5OH \rightarrow CH_3CHO \rightarrow CH_3COOH$$

Prolonged oxidation with a strong oxidising agent is required to oxidise ethanol through ethanal, the intermediate aldehyde, to ethanoic acid. This is achieved by refluxing the ethanol with sodium dichromate(VI) dissolved in moderately concentrated sulphuric acid for about thirty minutes, during which any unreacted ethanol or ethanal are returned to the flask as condensate. The sodium dichromate(VI) changes from orange to green as the chromium is reduced from chromate(VI) to chromium(III) ions. After the reflux period the apparatus is rearranged for distillation, and a dilute solution of ethanoic acid is obtained.

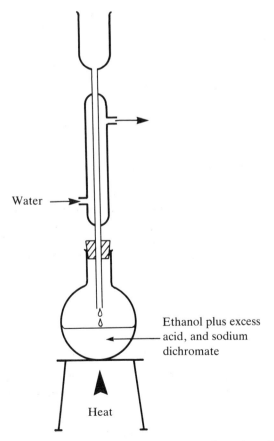

Water

Ethanol plus excess acid, and sodium dichromate

Heat

Fig. 30.5   Oxidation of ethanol to ethanoic acid

30.7   Ethyl ethanoate

$$C_2H_5OH + CH_3COOH \rightleftharpoons CH_3COOC_2H_5 + H_2O$$

The reaction between ethanol and ethanoic acid is reversible. The preparation of the ester is carried out in the presence of concentrated sulphuric acid to remove the water produced, thus driving the reaction to the right. The latter effect is enhanced by starting the reaction with an excess of alcohol, and removing the ester by distillation. The sulphuric acid also acts as a catalyst. The addition of the acid to the ethanol must be done cautiously, keeping the mixture cool and swirling the flask to ensure complete mixing.

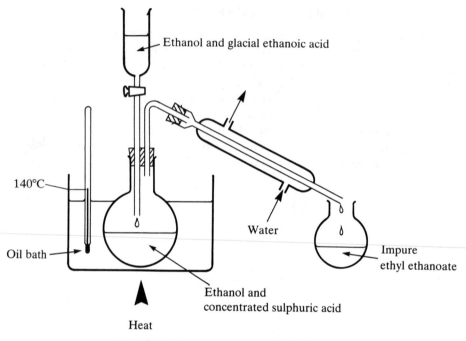

Fig. 30.6   Preparation of ethyl ethanoate

The ester is an oily liquid; since it is somewhat soluble in water it is freed from ethanoic acid and sulphur dioxide by shaking in a separating funnel with one wash of sodium carbonate solution, followed by washing with saturated brine to remove the alcohol. After drying over calcium chloride the ester is filtered and distilled to give a reasonably pure product boiling between 76°C and 78°C.

30.8   Ethanamide

$$CH_3COO^-NH_4^+ \rightarrow CH_3CONH_2 + H_2O$$

To obtain an effective yield from this reaction it is necessary to have excess glacial ethanoic acid present since the ammonium salt tends to dissociate;

$$CH_3COONH_4 \rightleftharpoons CH_3COOH + NH_3$$

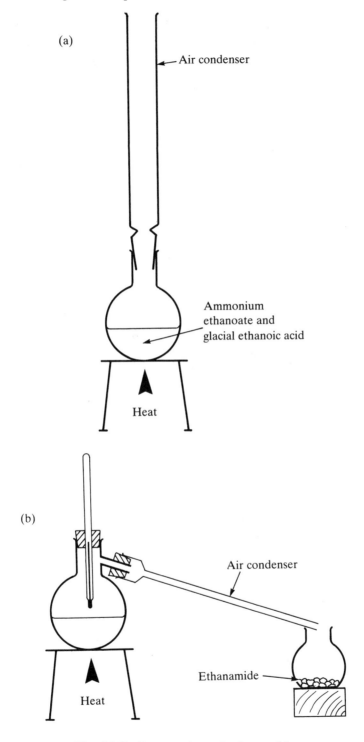

Fig. 30.7 Preparation of ethanamide

In practice excess ethanoic acid is mixed with ammonium carbonate to produce the ammonium ethanoate *in situ*. The mixture has then to be heated under reflux for about an hour to complete the reaction, using a wide-tube air condenser since ethanamide (m.p. 82°C) would condense as a solid in a water-cooled Liebig condenser. The flask is allowed to cool and the condenser rearranged for distillation. At first water and ethanoic acid distil; at a temperature of about 160°C the receiver is changed and impure ethanamide is collected up to a temperature of about 215°C. The solid product is purified by recrystallisation from alcohol (page 110).

30.9 Methylamine

$$CH_3CONH_2 + Br_2 + 4NaOH \rightarrow CH_3NH_2 + 2NaBr + Na_2CO_3 + 2H_2O$$

Ethanamide is converted to methylamine by means of Hofmann's degradation reaction (page 425). The ethanamide and bromine are carefully mixed in a flask in calculated quantities until all the ethanamide is dissolved. Again, bromine must be handled with caution. Cold dilute sodium hydroxide solution is added until the reddish-brown mixture becomes golden yellow. Finally the mixture is heated over a water-bath while concentrated sodium hydroxide solution is added dropwise from a tap funnel. Methylamine distils over to be collected in dilute hydrochloric acid as methylamine hydrochloride ($CH_3NH_3^+Cl^-$); the collection is arranged as shown in Fig. 30.8 to prevent water being sucked back into the flask as the very soluble gas dissolves.

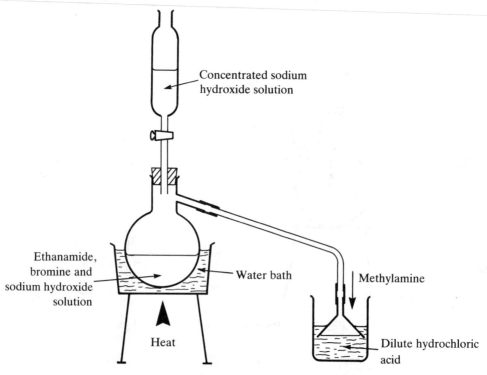

Fig. 30.8 Preparation of methylamine

30.10   Nitrobenzene

$$C_6H_6 + HNO_3 \rightarrow C_6H_5NO_2 + H_2O$$

Benzene is nitrated by nitric acid in the presence of concentrated sulphuric acid which produces the nitronium ion (page 361). The acids are first mixed cautiously, followed by the addition of benzene a little at a time, with thorough shaking to mix the reagents. The temperature is kept below 50°C during the addition to avoid the formation of any dinitrobenzene. The mixture is then heated under reflux on a water-bath at about 50–60°C for twenty to thirty minutes, during which time the flask must be shaken at intervals to keep the reagents emulsified. The contents of the flask are then cooled and poured into excess water, when the nitrobenzene separates as a yellow oil below the water. The aqueous layer is decanted off and the nitrobenzene is purified, washed and dried following the procedure for bromoethane. When it is distilled the portion boiling between 207°C and 211°C is collected. The distillation should not be taken to dryness, since traces of dinitrobenzene might decompose explosively. This reaction must be carried out in a fume cupboard because of the poisonous natures of benzene and nitrobenzene.

30.11   Phenylamine

$$C_6H_5NO_2 + 6H^+ + 6e^- \rightarrow C_6H_5NH_2 + 2H_2O$$

Phenylamine is obtained by the reduction of nitrobenzene using tin and concentrated hydrochloric acid as the reducing agent.

Nitrobenzene and granulated tin are placed in a flask fitted with a reflux condenser; concentrated hydrochloric acid is added little by little while the flask is shaken and kept cool. When the calculated quantity of acid has been added the mixture is heated on a water bath until reduction is complete. The phenylamine is present in the mixture as a salt of hexachlorostannate(IV) acid; sodium hydroxide solution is added to release the phenylamine which is then steam distilled from the mixture and purified as described on page 100.

30.12   N-phenylbenzamide (benzanilide)

$$C_6H_5CO.Cl + C_6H_5NH_2 \rightarrow C_6H_5NH.COC_6H_5 + HCl$$

Benzoyl derivatives are used to identify aromatic amines; they are prepared by the action of benzoyl chloride on a solution of the amine in sodium hydroxide. For example, phenylamine dissolved in 10% sodium hydroxide solution is shaken with benzoyl chloride in a wide-mouthed corked bottle for twenty to thirty minutes. The solid product is filtered under suction using a Buchner filter funnel and flask, washed with water and purified by recrystallisation from ethanol. The purified N-phenylbenzamide has a sharp characteristic melting point, 161°C; other aromatic amines give benzoyl derivatives with different melting points, by which they can be distinguished.

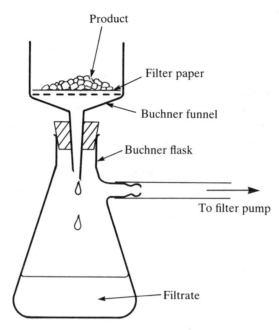

Fig. 30.9 Suction filtration

Questions

1. Outline the preparation of nitrobenzene, starting from benzene. Give the reagents, conditions and equations to show how you would carry out the following successive conversions.

$$C_6H_5NO_2 \rightarrow C_6H_5NH_2 \rightarrow C_6H_5OH \rightarrow C_6H_6 \rightarrow C_6H_5CH_3 \rightarrow C_6H_5CH_2Cl$$

(C)

2. Outline the laboratory preparation of a pure sample of bromoethane(ethyl bromide) starting from ethanol.

Give the reagents, conditions and equations to show how you would carry out the following successive conversions:

$$CH_3CH_2Br \rightarrow CH_3CH_2OH \rightarrow CH_3CHO$$

$$CH_3CH(OH)CN \xleftarrow{} CH_3CH(OH)COOH$$

(C)

3. 1,2-Dibromoethane may conveniently be prepared on a laboratory scale by passing ethene (ethylene) gas into liquid bromine covered by a layer of water.

(a) Give the equation describing the formation of 1,2-dibromoethane by the above reaction.

(b) The common laboratory source for the preparation of ethene is ethanol. Briefly explain how ethene may be prepared from ethanol, giving the appropriate equation(s).

(c) Calculate (i) the mass of bromine, (ii) the volume of ethene at s.t.p. that would theoretically be required for the preparation of 10 g of 1,2-dibromoethane as in (a) above.   (L)

Important organic reactions and synthetic pathways

31.1 Summary of reactions

Much of the business of the organic chemist is concerned with transforming one molecule into another. This may be to make a more useful product from a less useful one, or to identify an unknown compound by converting it into a readily recognisable one, or to determine how to synthesise a new product from readily available materials. These conversions usually involve the application of one or more reactions of general usefulness; some of these reactions are set out below, for convenience in alphabetical order.

(a) Acylation

The addition of an acyl group such as ethanol, $CH_3C{\displaystyle \mathop{<}^{\textstyle =O}}$, or propanoyl, $C_2H_5C{\displaystyle \mathop{<}^{\textstyle =O}}$, to an organic molecule is called *acylation*. Reactive compounds such as ethanoyl chloride (page 443) or ethanoic anhydride (page 410) are good acylating agents.

$$CH_3COCl + C_2H_5OH \longrightarrow CH_3COOC_2H_5 + HCl$$

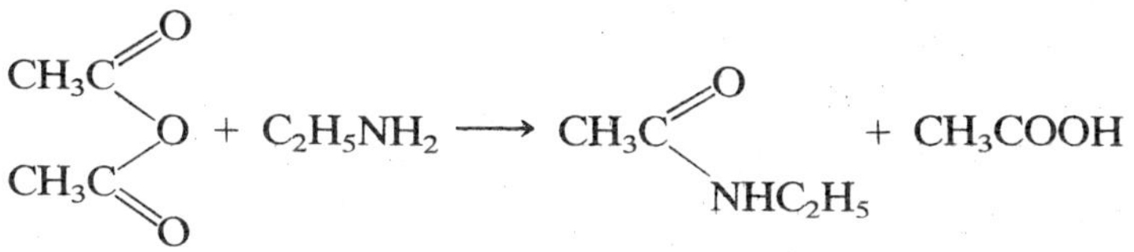

$$\begin{array}{c} CH_3C{\displaystyle \mathop{<}^{\textstyle =O}} \\ \quad\quad\quad O \\ CH_3C{\displaystyle \mathop{<}_{\textstyle =O}} \end{array} + C_2H_5NH_2 \longrightarrow CH_3C{\displaystyle \mathop{<}^{\textstyle =O}_{\textstyle NHC_2H_5}} + CH_3COOH$$

(b) Ascending a homologous series (adding a carbon atom to the carbon chain of a molecule)

This is best done by adding a nitrile group to the molecule; the $-C{\equiv}N$ group is easily hydrolysed with the formation of a carboxylic acid with one more carbon atom in the molecule than was originally present.

$$C_2H_5Br \xrightarrow{\text{KCN}} C_2H_5CN \xrightarrow{\text{H}_2\text{O/H}^+} C_2H_5COOH$$

$$CH_3CHO \xrightarrow{\text{HCN}} CH_3CH(OH)CN \xrightarrow[\text{H}^+]{\text{H}_2\text{O}} CH_3CH(OH)COOH$$

(c) Benzoylation

The introduction of a benzoyl group into a molecule is similar to the introduction of an acyl group. The reaction is usually carried out in the presence of sodium hydroxide, as in the Schotten–Baumann reaction to benzoylate phenol or phenylamine.

$$C_6H_5OH + C_6H_5COCl + NaOH \rightarrow C_6H_5\!-\!O\!-\!\underset{\underset{O}{\|}}{C}\!-\!C_6H_5 + NaCl + H_2O$$

$$C_6H_5NH_2 + C_6H_5COCl + NaOH \rightarrow C_6H_5NH\!-\!\underset{\underset{O}{\|}}{C}\!-\!C_6H_5 + NaCl + H_2O$$

(d) Condensation reactions

These involve combining two organic molecules with the elimination of a small molecule such as water (page 342). Terylene and other polyesters are made by condensation reactions, one stage of the polymerisation being:

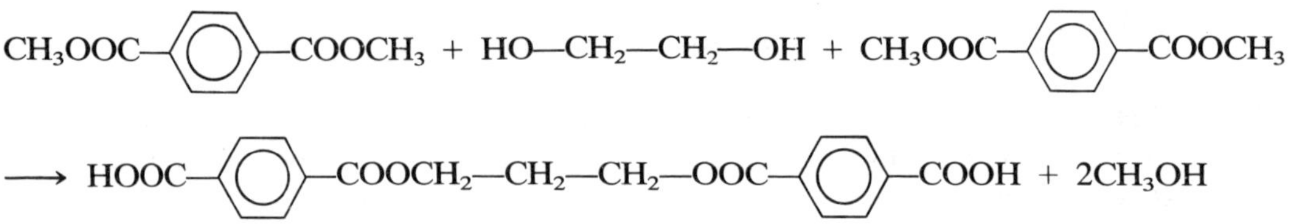

Aldehydes and ketones also undergo condensation reactions (page 391)

$$CH_3CHO + NH_2NH_2 \rightarrow CH_3CH\!=\!N.NH_2 + H_2O$$

(e) Decarboxylation

The carboxyl group can be removed from both aliphatic and aromatic acids by heating the sodium salt with soda-lime.

$$CH_3COOH + \xrightarrow{\text{NaOH}} CH_3COONa \xrightarrow[\text{heat}]{\text{NaOH}} CH_4 + Na_2CO_3$$

This is particularly useful for removing side chains from organic molecules:

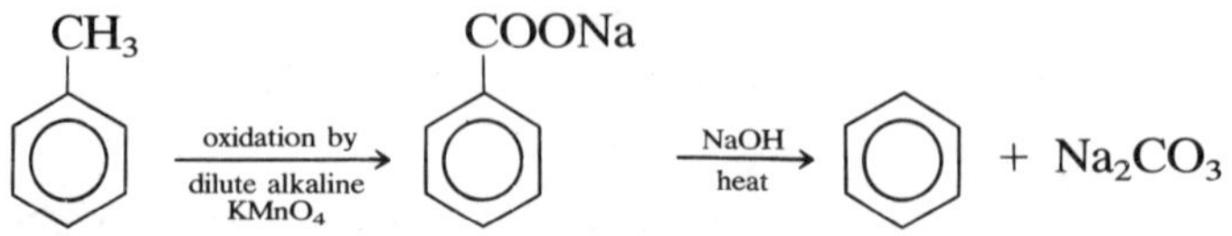

(f) Dehydration

This includes the removal of the elements of water from a *single* substance, and is effected by the use of concentrated sulphuric acid, phosphorus(V) oxide, zinc oxide or other agents:

$$C_2H_5OH \xrightarrow[H_2SO_4]{conc.} C_2H_4 + H_2O$$

$$C_6H_5CONH_2 \xrightarrow[P_2O_5]{} C_6H_5CN + H_2O$$

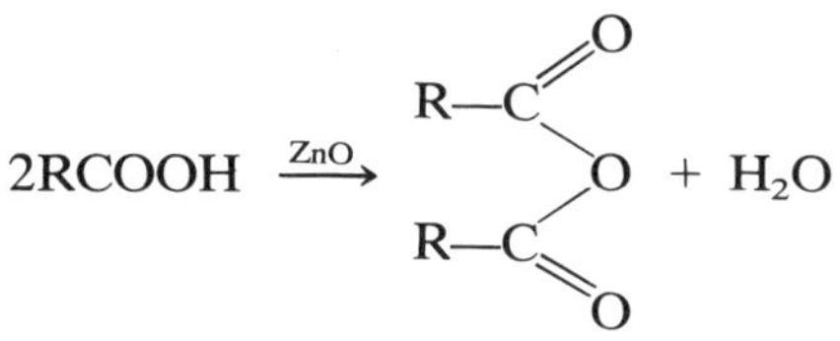

$$2RCOOH \xrightarrow{ZnO} \begin{matrix} R-C{\overset{O}{\underset{\diagdown}{\diagup}}} \\ {}O \\ R-C{\overset{\diagup}{\underset{O}{\diagdown}}} \end{matrix} + H_2O$$

(g) Dehydrogenation

Dehydrogenation is the removal of one or more hydrogen atoms from a molecule. Since this effectively increases the proportion of the more electronegative elements in the molecule it is an oxidation process.

$$C_2H_5OH \xrightarrow[catalyst]{Cu} CH_3CHO + H_2$$

$$C_6H_5C_2H_5 \xrightarrow{ZnO} C_6H_5-CH{=}CH_2 + H_2$$

(h) Descending a homologous series (removing a carbon atom from the carbon chain of a molecule)

The effect of a mixture of bromine and sodium hydroxide on an acid amide is to remove the carboxyl carbon from the molecule (page 425).

$$C_2H_5CONH_2 \xrightarrow[Br_2]{NaOH} C_2H_5NH_2$$

(i) Diazotisation

The reaction of aryl amines with nitrous acid gives rise to diazo-compounds which can be converted into a great variety of new compounds (page 422).

$$C_6H_5-NH_2 + HNO_2 \xrightarrow[HCl]{5°C} C_6H_5N{=}NCl + H_2O$$

(j) Esterification

An ester is the product of a reaction between an acid and an alcohol. Esterification of organic acids takes place more readily if an acyl chloride or anhydride is used in place of the acid.

$$C_2H_5OH + CH_3COOH \rightleftharpoons CH_3COOC_2H_5 + H_2O$$

$$C_2H_5OH + CH_3COCl \rightarrow CH_3COOC_2H_5 + HCl$$

$$C_2H_5OH + (CH_3CO)_2O \rightarrow CH_3COOC_2H_5 + CH_3COOH$$

(k) Halogenation

Halogen atoms are introduced into organic molecules in a variety of ways
(i) Addition of halogen or halogen halides to unsaturated molecules:

$$C_2H_4 + HCl \rightarrow CH_3CH_2Cl$$

$$C_2H_4 + Cl_2 \rightarrow CH_2ClCH_2Cl$$

(ii) Substitution of halogens into saturated compounds:

$$C_2H_6 + Cl_2 \xrightarrow{\text{u.v. light}} C_2H_5Cl + HCl$$

$$CH_3COOH + Cl_2 \xrightarrow{\text{u.v.}} CH_2ClCOOH + HCl$$

$+ \; Cl_2 \xrightarrow{\text{AlCl}_3}$ $+ \; HCl$

(iii) Reaction of oxygen-containing groups with phosphorus(III) chloride,
phosphorus(V) chloride or sulphur dichloride oxide or other corresponding
halide compounds.

$$3CH_3COOH + PCl_3 \rightarrow 3CH_3COCl + H_3PO_3$$
$$CH_3CH_2OH + PCl_5 \rightarrow C_2H_5Cl + POCl_3 + HCl$$
$$(CH_3)_2C{=}O + PCl_5 \rightarrow CH_3C(Cl_2)CH_3 + POCl_3$$
$$C_2H_5OH + SOCl_2 \rightarrow C_2H_5Cl + HCl + SO_2$$

(l) Hydration

This is the addition of a water molecule or molecules to an unsaturated
compound. Thus alcohols are formed from alkenes:

$$C_2H_4 + H_2O \xrightarrow[\text{high pressure}]{300^\circ C} C_2H_5OH$$

Aldehydes also are produced from alkynes, the conditions in this case being
milder; the alkyne is bubbled into a dilute solution of sulphuric acid
containing mercury(II) sulphate:

$$C_2H_2 + H_2O \xrightarrow[\text{HgSO}_4]{\text{H}_2\text{SO}_4} CH_3CHO$$

(m) Hydrogenation

Hydrogenation is the addition of hydrogen to a suitable molecule; thus
alkenes, alkynes and arenes are readily hydrogenated using a nickel catalyst:

$$C_2H_4 + H_2 \xrightarrow[150^\circ C]{Ni} C_2H_6$$

$$C_2H_2 + H_2 \xrightarrow[150^\circ C]{Ni} C_2H_4 \xrightarrow[Ni]{H_2} C_2H_6$$

$$C_6H_6 + 3H_2 \xrightarrow[150^\circ C]{Ni} C_6H_{12}$$

Nitriles can be saturated by the addition of hydrogen generated by sodium in an alcoholic solution of the nitrile:

$$CH_3C\equiv N + 2H_2 \xrightarrow[C_2H_5OH]{Na} CH_3CH_2NH_2$$

Similarly aldehydes and ketones yield primary and secondary alcohols respectively:

$$CH_3C{\overset{H}{\underset{O}{\diagup}}} + H_2 \xrightarrow[C_2H_5OH]{Na} CH_3CH_2OH$$

$$(CH_3)_2C{=}O + H_2 \xrightarrow[C_2H_5OH]{Na} (CH_3)_2CHOH$$

All these additions of hydrogen are, in effect, reductions.

(n) Hydrolysis

This is the chemical decomposition of a compound by reaction with water:

(i) Halogenoalkanes

$$C_4H_9Br + H_2O \xrightarrow[NaOH]{dilute} C_4H_9OH + HBr$$

(ii) Acyl chlorides

$$RCOCl + H_2O \rightarrow RCOOH + HCl$$

(iii) Esters

$$CH_3COOC_2H_5 + H_2O \xrightarrow[\substack{or\ NaOH \\ solution}]{HCl} CH_3COOH + C_2H_5OH$$

(iv) Alkyl Nitriles

$$RCN \xrightarrow[HCl]{H_2O} RCONH_2 \xrightarrow[HCl]{H_2O} RCOONH_4 \xrightarrow[HCl]{H_2O} RCOOH + NH_4Cl$$

(o) Nitration

The nitro group can be introduced into alkane or arene molecules by the action of nitric acid, the conditions varying with the compound to be nitrated:

$$C_2H_6 + HNO_3 \xrightarrow[\substack{sealed \\ vessels}]{350°C} C_2H_5NO_2 + H_2O$$

$$C_6H_6 + HNO_3 \xrightarrow[reflux]{H_2SO_4} C_6H_5NO_2 + H_2O$$

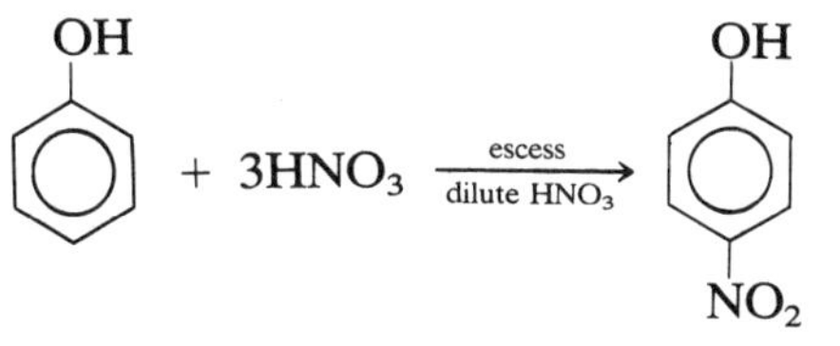

Alternatively metal nitrate(III) (nitrite) compounds can be used:

$$C_2H_5Br + AgNO_2 \longrightarrow C_2H_5NO_2 + AgBr$$

$$ClCH_2COOH \xrightarrow{KNO_2} NO_2CH_2COOH \xrightarrow{heat} CH_3NO_2 + CO_2$$

(p) Oxidation

A reaction in which the proportion of the more electronegative elements in the molecule is increased is classed as an oxidation:

$$CH_4 + Cl_2 \longrightarrow CH_3Cl + HCl$$

$$C_2H_5OH \xrightarrow[H_2SO_4]{Cr_2O_7^{2-}} CH_3CHO + H_2O$$

(q) Reduction

A reaction in which the proportion of the more electronegative elements in the molecule is decreased is classed as a reduction. This includes the direct addition of hydrogen (see above, hydrogenation). Further examples of reduction are:

$$CH_3COOH \xrightarrow[ether]{LiAlH_4} CH_3CHO \quad (poor\ yield)$$

$$CH_3COOC_2H_5 \xrightarrow[C_2H_5OH]{Na} CH_3CHO + C_2H_5OH \quad (good\ yield)$$

$$C_6H_5NO_2 \xrightarrow[HCl]{Sn} C_6H_5NH_2$$

(r) Sulphonation

Sulphonation, the introduction of the sulphonic acid group —SO_3H into the molecule, is readily effected with the arenes by the action of concentrated sulphuric acid or of sulphur dichloride dioxide:

$$C_6H_6 \xrightarrow[reflux]{H_2SO_4} C_6H_5SO_3H$$

$$C_6H_6 \xrightarrow{SO_2Cl_2} C_6H_5SO_2Cl \xrightarrow{H_2O} C_6H_5SO_3H + HCl$$

(s) Sulphation

This is the formation of an ester by the reaction of sulphuric acid with an alkene or alcohol. In contrast to sulphonated compounds the sulphur atom is not directly bonded to the carbon atom.

$$CH_2{=}CH_2 + H_2SO_4 \rightarrow CH_3CH_2OSO_2OH$$

$$CH_3CH_2OH + H_2SO_4 \rightarrow CH_3CH_2OSO_2OH$$
$$(CH_3CH_2HSO_4)$$

31.2 Some synthetic routes

Working out how to convert one molecule into another using some of the reactions listed above provides useful exercise in the application of organic chemistry. The conversions are effected in steps which can be represented in a 'flow diagram'. Reactants and products are shown in unbalanced equations with reagents and/or conditions over or under the arrow. Some typical examples are given below.

(a) Alcohol to higher amine

To obtain propylamine starting from ethanol, and thus adding a carbon atom to the carbon chain, the steps required are:

$$C_2H_5OH \xrightarrow{PBr_3} C_2H_5Br \xrightarrow[ethanol]{KCN} C_2H_5CN \xrightarrow[Na/ethanol]{} C_2H_5CH_2NH_2$$

(b) Alcohol to higher amide

Starting from alcohol the first two steps of flow diagram (a) are required, followed by treatment of the nitrile by partial hydrolysis:

$$- - - - C_2H_5CN \xrightarrow[H_2O]{H^+} C_2H_5CONH_2$$

(c) Alcohol to higher alcohol

(i) Again the steps of (a) can be used to obtain a nitrile with one more carbon atom than the original alcohol, and in full the flow diagram will be:

$$ROH \xrightarrow{PBr_3} RBr \xrightarrow[ethanol]{KCN} RCN \xrightarrow[H_2O]{H^+} RCOOH \xrightarrow[ether]{LiAlH_4} RCH_2OH$$

(ii) Alternatively a Grignard reagent may be used to add the extra carbon atom:

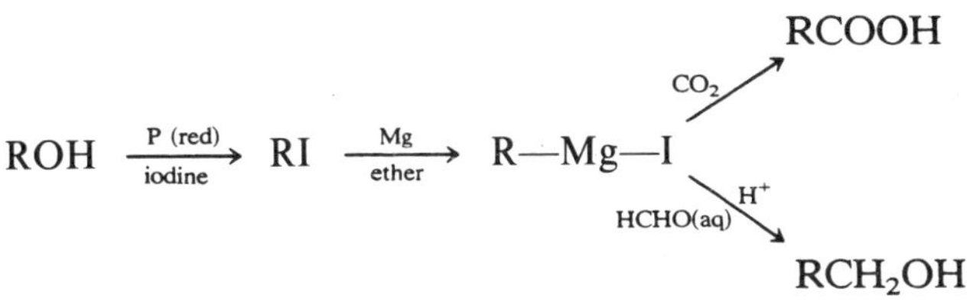

$$ROH \xrightarrow[iodine]{P\ (red)} RI \xrightarrow[ether]{Mg} R\text{—}Mg\text{—}I \begin{array}{c} \xrightarrow{CO_2} RCOOH \\ \xrightarrow[HCHO(aq)]{H^+} RCH_2OH \end{array}$$

(d) Halogenoalkane to hydroxy acid

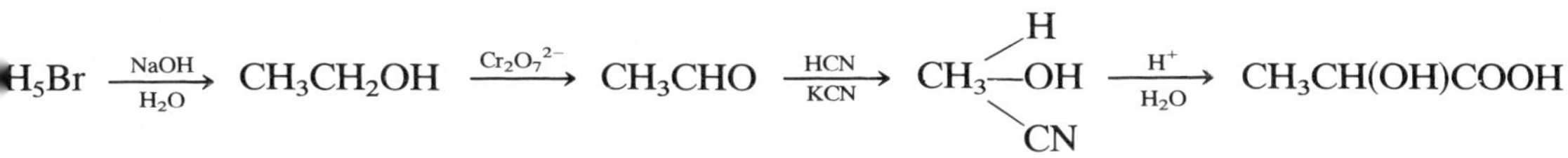

$$H_5Br \xrightarrow[H_2O]{NaOH} CH_3CH_2OH \xrightarrow{Cr_2O_7{}^{2-}} CH_3CHO \xrightarrow[KCN]{HCN} CH_3\text{—}\overset{H}{\underset{CN}{C}}\text{—}OH \xrightarrow[H_2O]{H^+} CH_3CH(OH)COOH$$

(e) Primary alcohol to secondary alcohol

$$RCH_2OH \xrightarrow[conc.]{,H_2SO_4} R'CH{=}CH_2 \xrightarrow{HBr} R'CHBrCH_3 \xrightarrow[H_2O]{NaOH} R'CH(OH)CH_3$$

(f) Carboxylic acid to amino acid

$$RCH_2COOH \xrightarrow[Cl_2]{u.v.} RCHClCOOH \xrightarrow[pressure]{NH_3} RCH(NH_2)COOH$$

(g) Benzoic acid to phenylamine

$$C_6H_5COOH \xrightarrow[lime]{soda} C_6H_6 \xrightarrow[.H_2SO_4]{HNO_3} C_6H_6NO_2 \xrightarrow[HCl]{Sn} C_6H_6NH_2$$

(h) Phenylamine to methylbenzene

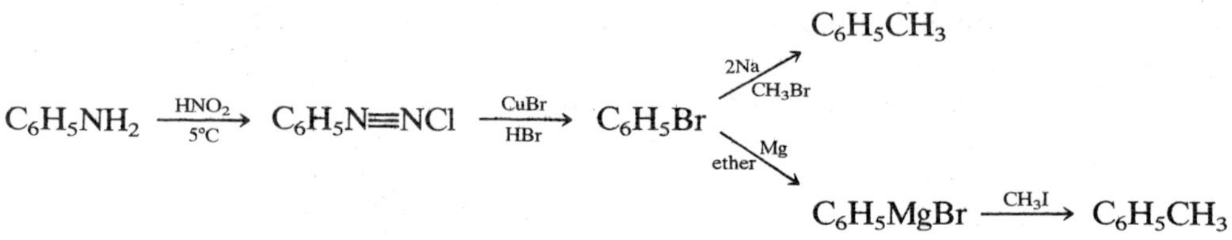

Reagents used to change an aromatic molecule containing a benzene ring with more than one functional group have to be chosen with care to make sure that only the required reaction is effected. The following examples illustrate this point:

(i)

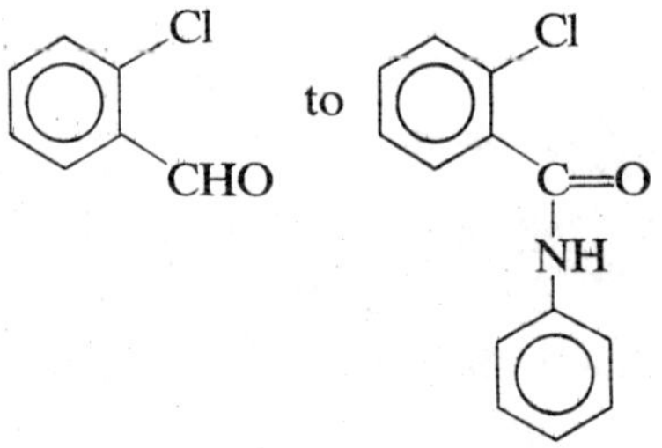

In this case reagents must be chosen to react with the aldehyde group leaving the chloro- group untouched:

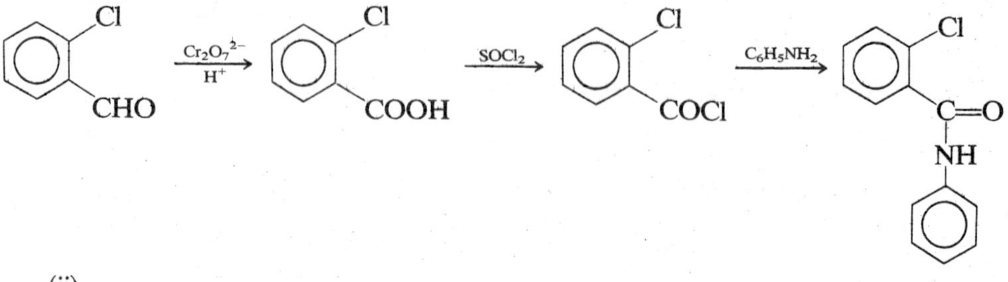

(ii)

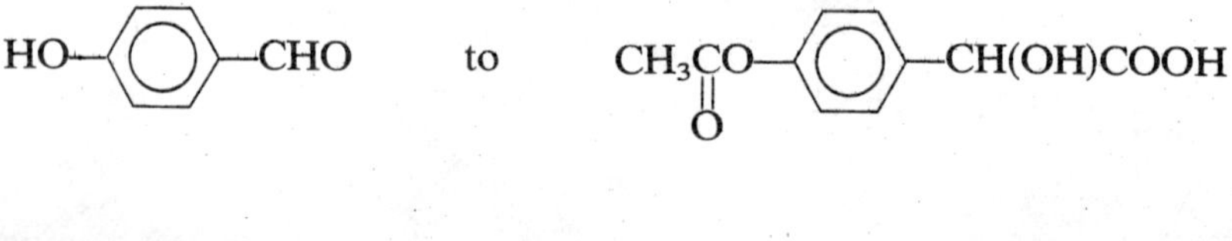

Here both functional groups must be altered using a succession of suitable reagents.

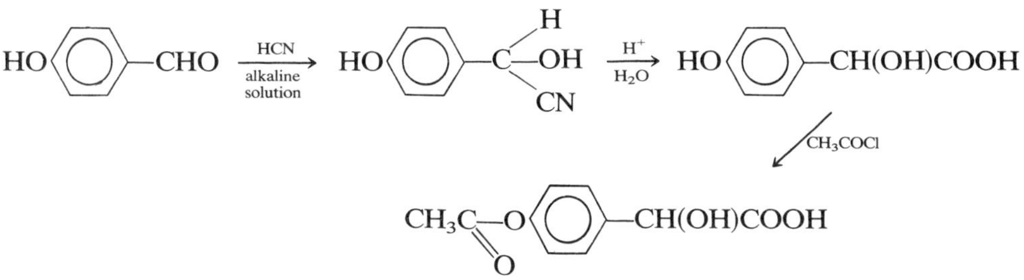

HO—⬡—CHO $\xrightarrow[\text{solution}]{\text{HCN alkaline}}$ HO—⬡—C(H)(OH)(CN) $\xrightarrow[\text{H}_2\text{O}]{\text{H}^+}$ HO—⬡—CH(OH)COOH

$\downarrow$ CH$_3$COCl

CH$_3$C(=O)—O—⬡—CH(OH)COOH

The aldehyde group is tackled before the phenolic group to avoid hydrolysis of the resulting ester by alkali or acid treatment during the conversion of —CHO to —CH(OH)COOH.

(iii)   NH$_2$-phenyl   to   NH$_2$-phenyl-NO$_2$

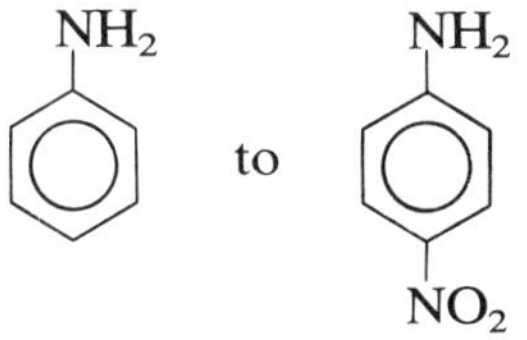

The strongly oxidising condition necessary for nitration will affect the —NH$_2$ group of the phenylamine; this must first be protected by acylation. Nitration can then be carried out followed by recovery of the required compound by hydrolysis of the acylated molecule:

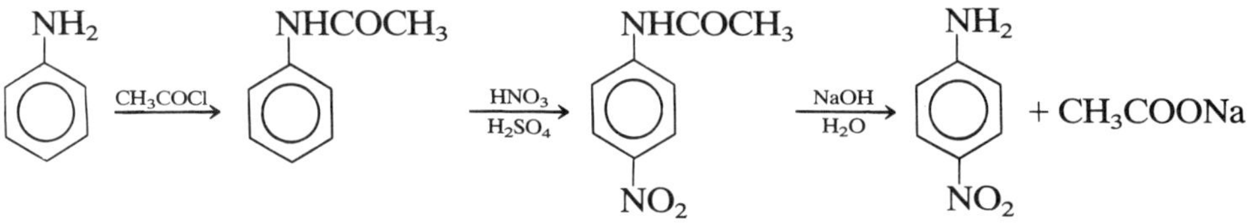

NH$_2$ $\xrightarrow{\text{CH}_3\text{COCl}}$ NHCOCH$_3$ $\xrightarrow[\text{H}_2\text{SO}_4]{\text{HNO}_3}$ NHCOCH$_3$(NO$_2$) $\xrightarrow[\text{H}_2\text{O}]{\text{NaOH}}$ NH$_2$(NO$_2$) + CH$_3$COONa

The reaction will be complicated by the formation of some 2-nitrophenylamine which must be separated from the required 4-nitro derivative by fractional crystallisation.

Questions

1. The following represent important examples of different types of organic reactions:

A. C_6H_5—COONa + NaOH → C_6H_6 + Na$_2$CO$_3$

B. CH_3—CHO + NH$_3$ → CH_3—CH(OH)—NH$_2$

C. CH_3—CO—CH$_3$ + NH$_2$OH → $(CH_3)_2$—C=NOH + H$_2$O

D. CH_3—CH$_2$—I + H$_2$O → CH_3—CH$_2$—OH + HI

E. Styrene $\xrightarrow[\text{peroxide}]{\text{dodecanoyl}}$ Polystyrene
(phenylethene)   (polyphenylethene)

F. C_6H_6 + CH$_3$—Cl → C_6H_5—CH$_3$ + HCl

(a) Explaining your reasons, place each of these reactions in the appropriate category from the following list:
(i) polymerization, (ii) decarboxylation, (iii) substitution (displacement), (iv) condensation, (v) addition, (vi) oxidation, (vii) hydrolysis.
(b) Give one example, other than any of those above, of each of the following:
 (i) a condensation reaction;
 (ii) a substitution reaction;
(iii) a polymerisation reaction.
(c) Give the conditions of reaction F.
(d) What is the function of dodecanoyl peroxide in reaction E?
(e) State, giving a reason, whether a similar reaction to B takes place with propanone (acetone).   (AEB 1977)

2. (a) Outline with essential experimental conditions how the following conversions can be carried out:

   (i) $CH_3CH_2OH \rightarrow CH_3CH_2CH_2OH$

   (ii) $CH_3CH_2OH \rightarrow CH_3CH(OH)CH_3$

   (iii) $CH_3CH_2OH \rightarrow CH_3CH_2OCH_2CH_3$

   (iv) $CH_3CHO \rightarrow CH_3CH(OH)COOH$

   (v) $CH_3CH_2COOH \rightarrow CH_3COOH$

   (b) What chemical tests would you perform to show that conversions (ii) and (iv) had taken place?
   (c) Which of the organic substances above exhibit *optical isomerism*?
                                                               (AEB 1979)

3. The scheme below gives the interrelation of five organic compounds A, B, C, D and E whose structural or molecular formulae are as shown

(a) Give the structural formulae of A, C, D and E.
(b) Give the reagents and reacting conditions which could be used for converting B to A; C to D; D to E; E to D.
(c) Explain why the isomer you have chosen for D predominates as the product of the conversion of C to D.   (L)

4. Describe briefly how each of the following can be prepared in the laboratory, starting from methylbenzene (*toluene*) and any other organic

or inorganic reagents. Give equations and outline the mechanisms of the reaction where possible.

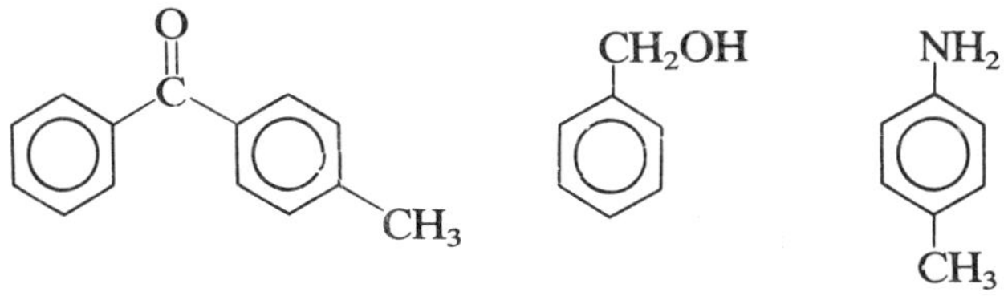

(A)   (B)   (C)

State briefly how you would carry out experiments to demonstrate each of the following:

(i) that the hydroxyl group of substance (B) is in the side chain and not directly attached to the ring;

(ii) that substance (C) is a primary aromatic amine.   (JMB)

5. (*a*) Describe how you would carry out the following conversions, giving the essential conditions for the reactions but no details of the apparatus or of the processes whereby the products are separated and purified:

(i) $C_2H_5OH \rightarrow CH_3COCl$;

(ii) $CH_3COOH \rightarrow CH_2(NH_2)COOH$;

(iii) $C_6H_6 \rightarrow C_6H_5COCH_3$

  (*b*) In the case of the product in (*a*)(ii), explain why the concentration of an aqueous solution of the acid cannot be determined by direct titration with an alkali.

(*c*) What is the mechanism of the reaction in (*a*)(iii)?   (O)

32   Important industrial processes

32.1   Extraction of metals

(a) General

The methods used to extract metals depend substantially on the position of the metal in the electrochemical series. The very reactive metals, such as potassium, sodium or aluminium, with high negative electrode potentials, have to be prepared by electrolysis of fused compounds, since their oxides are not easily reduced.

Metals moderately reactive, with less negative potentials, can be obtained by smelting their oxides with carbon, since these can be reduced by carbon or carbon monoxide; zinc, lead and iron are all prepared by this method.

Copper, with a positive electrode potential, is readily extracted from its ores; the carbonate ores such as malachite are converted to copper oxide before reduction with carbon, while the sulphide ores are reduced by roasting in a reverberatory furnace. Mercury is also obtained by roasting the sulphide, cinnabar.

Both copper and mercury are sometimes found native, that is, as the free metal; silver, gold and other noble metals are also found uncombined. The extraction of such metals is mainly a concentrating and purifying process.

(b) Extraction of aluminium

Aluminium is prepared from the oxide, Al_2O_3, which occurs, with iron(III) oxide and silica impurities, as bauxite. Since the purification of aluminium metal is technically difficult the ore is carefully purified before extraction of the metal, by digestion with caustic soda under pressure (to maintain a suitable temperature). This dissolves the aluminium oxide and the silica, and enables the iron oxide to be filtered off:

$$2NaOH(aq) + Al_2O_3(s) \rightarrow 2NaAlO_2(aq) + H_2O(l)$$

$$2NaOH(aq) + SiO_2(s) \rightarrow Na_2SiO_3(aq) + H_2O(l)$$

The mixture is diluted and 'seeded' with some aluminium hydroxide which brings about the precipitation of most of the aluminium as the hydroxide:

$$NaAlO_2(aq) + 2H_2O(l) \rightarrow Al(OH)_3(s) + NaOH(aq)$$

The purified hydroxide is roasted, dried and fed into a cell containing molten cryolite, Na_3AlF_6, in which it dissolves, and which carries the current for electrolysis. A required temperature of 900–1000°C is maintained by the heating effect of the heavy current. The graphite lining of the cell provides the carbon cathode at which aluminium ions are discharged.

$$2Al^{3+} + 6e^- \rightarrow 2Al$$

The anodes are also made of carbon, in the form of a hard-baked mixture of pitch and coke; they have to be continually replaced as they are eroded by the oxygen released in electrolysis.

Fluoride ions are discharged initially at the anode, but the fluorine immediately attacks the aluminium oxide, releasing oxygen gas:

$$6F^- \rightarrow 3F_2 + 6e^-$$

$$3F_2 + Al_2O_3 \rightarrow 2Al^{3+} + 6F^- + 1\tfrac{1}{2}O_2$$

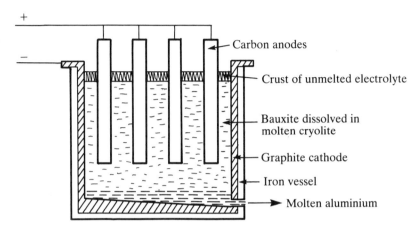

Fig. 32.1 Extraction of aluminium

(c) Extraction of iron

Iron is extracted from its oxides in a blast furnace. The oxide ores haematite, Fe_2O_3, and magnetite, Fe_3O_4, are used directly, but the carbonate ore siderite, $FeCO_3$, is first roasted to convert it to the oxide.

The extraction process is continuous, with charges of ore, limestone and coke being fed in at the top of the furnace as required, while molten iron and slag are tapped off at the base.

At the bottom of the furnace a blast of hot air is introduced through nozzles or 'tuyères'. The coke burns exothermically to form carbon dioxide, raising the temperature to about 1400–1500°C:

$$C + O_2 \rightarrow CO_2$$

At this temperature there is also some direct reduction of iron oxides by coke.

$$FeO + C \rightarrow Fe + CO$$

The carbon dioxide is reduced to carbon monoxide as it passes up through more of the hot coke; this reaction is endothermic, reducing the temperature to about 800–900°C, at which iron(II) oxide is readily reduced by carbon monoxide.

$$FeO + CO \rightarrow Fe + CO_2$$

At about the same temperature the infusible silicon dioxide present in the ore is converted to fusible slag by the action of the limestone.

$$CaCO_3 \rightarrow CaO + CO_2$$

$$CaO + SiO_2 \rightarrow CaSiO_3$$

Iron and slag trickle to the base of the furnace where the fused slag floats on the iron; they are both tapped off from time to time.

Further up the furnace, at temperatures of about 500–600°C, the reduction of the oxides to iron(II) oxide takes place.

$$Fe_2O_3 + CO \rightleftharpoons 2FeO + CO_2$$

$$Fe_3O_4 + CO \rightleftharpoons 3FeO + CO_2$$

At this temperature some iron(III) oxide is also reduced completely to iron in a reaction:

$$Fe_2O_3 + 3CO \rightleftharpoons 2Fe + 3CO_2$$

which is exothermic and helps to maintain the temperature.

At the top of the furnace the temperature is in the region of 250–300°C, drying the incoming ore and making it more porous. The waste gases, largely carbon monoxide, carbon dioxide and nitrogen, are circulated through stoves used to preheat the air for the blast, additional air being injected to burn the carbon monoxide. Some of the waste gas may also be burned to generate power for the blowers.

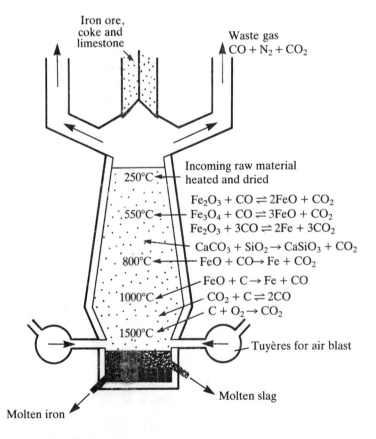

Fig. 32.2 Blast furnace for manufacture of iron

The iron obtained directly from the blast furnace, pig iron, contains a high percentage of impurities such as carbon (as iron carbide, Fe_3C, or as graphite), phosphorus, silicon and manganese. Remelted, it is used to produce cast iron guttering, railings, fire grates and other objects that can be cast in sand moulds, but most is converted into steel.

Steel is made by removing the impurities from molten pig iron by oxidation and then adding calculated amounts of carbon together with small amounts of the transition metals such as manganese, cobalt, nickel, chromium or tungsten to produce the grade of steel required.

Modern converters, such as the Kaldo or the L.D. converters, use a blast of pure oxygen to burn away the carbon. In the L.D. converter, Fig. 32.3, the molten iron is placed in a steel container lined with a refractory material and supported on trunnions which allow the converter to be tipped.

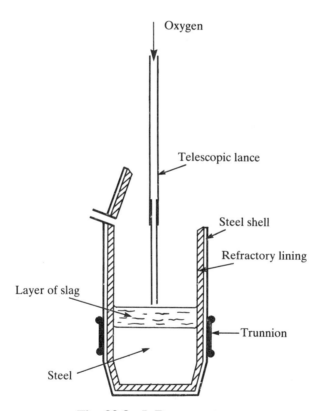

Fig. 32.3 L.D. converter

Oxygen is blown through the lance at supersonic speeds, which allows it to penetrate the slag and enter the molten metal. The impurities are oxidised, and flames erupt from the mouth of the converter as the carbon monoxide burns. The oxides of sulphur, phosphorus and silicon combine with the converter lining to form a slag which floats on the metal. For iron from ores with a high phosphorus content a basic lining made from calcined dolomite (calcium and magnesium oxides) is necessary, but for iron containing less phosphorus a lining of silica (silicon dioxide) is used. The use of pure oxygen avoids the absorption of nitrogen by the steel, which was apt to happen with the air-blown Bessemer converters, from which the modern converters have been developed.

Conversion may also be carried out in the Open Hearth or Siemens–Martin furnaces, where the pig iron is melted with scrap steel and iron ore in a shallow hearth furnace. The impurities are oxidised by air and by the iron ore; additional oxygen may also be used. The furnace is lined with calcined dolomite or with silica, as in the converters described above.

(d) Extraction of copper

Copper is mainly extracted from its sulphide ores, although some is obtained from cuprite, Cu_2O, and some occurs native.

The extraction from copper pyrites, $CuFeS_2$, starts with the pulverisation of the ore, which is then cleaned from impurities such as sand and mud by a flotation process. The pulverised ore is mixed with water containing chemical foaming agents, and air is blown into the mixture with the result that the metal sulphides rise up to the surface and can be skimmed off in the froth. The cleaned sulphides are then roasted to convert the iron into iron(II) oxide:

$$2CuFeS_2(s) + 4O_2(g) \rightarrow Cu_2S(s) + 2FeO(s) + 3SO_2(g)$$

Silica is added to the heated mixture, out of contact with air, to convert the iron(II) oxide to molten slag, which is poured away.

$$FeO(s) + SiO_2(s) \rightarrow FeSiO_3(s)$$

The copper sulphide remaining is roasted with a limited amount of oxygen to convert it to copper:

$$2Cu_2S(s) + 3O_2(g) \rightarrow 2Cu_2O(s) + 2SO_2(g)$$

$$Cu_2S(s) + 2Cu_2O(s) \rightarrow 6Cu(s) + SO_2(g)$$

The crude copper thus made is known as 'blister copper' because the escaping sulphur dioxide leaves blisters on the surface of the cooling metal. Copper for electric cables and telephone wire must be very pure, so the blister copper is refined electrolytically. Anodes of impure copper are placed in a bath of copper sulphate solution and the pure copper is deposited at the cathode. Some of the copper that occurs native as boulder copper is refined electrolytically directly from the material of the deposit in a similar manner.

32.2   The Solvay process

This method of manufacturing sodium carbonate is one of the most elegant and well thought out processes in industrial chemistry. The raw materials brine and limestone are among the cheapest available, and by careful management virtually the only waste material produced is calcium chloride.

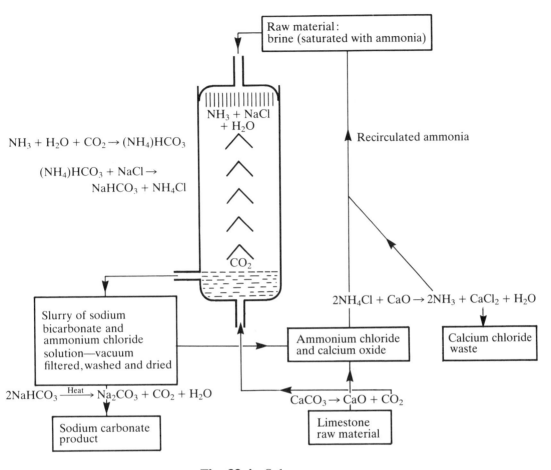

Fig. 32.4 Solvay process

The reaction in the tower results from the formation of the hydrogen carbonate ion from the carbon dioxide which is introduced at the bottom:

$$CO_2(g) + H_2O(l) \rightleftharpoons HCO_3^-(aq) + H^+(aq)$$

followed by the formation of the ammonium ion from the ammonia which has entered the top of the tower dissolved in brine:

$$NH_3(aq) + H^+(aq) \rightleftharpoons NH_4^+(aq)$$

From the ions now present, Na^+, NH_4^+, Cl^- and HCO_3^-, the least soluble product that can form is sodium hydrogen carbonate, this is consequently deposited as a slurry in a solution of ammonium chloride, from which it is filtered, washed and dried. The hydrogen carbonate is roasted to give the desired product, sodium carbonate:

$$2NaHCO_3(s) \rightarrow Na_2CO_3(s) + H_2O(l) + CO_2(g)$$

The carbon dioxide from this reaction is passed into the tower together with fresh supplies obtained by heating limestone:

$$CaCO_3(s) \rightarrow CaO(s) + CO_2(g)$$

The quicklime from this reaction is used to recover ammonia from the ammonium chloride:

$$CaO(s) + 2NH_4Cl(s) \rightarrow CaCl_2(s) + H_2O(l) + 2NH_3(g)$$

The recovered ammonia is recirculated, leaving calcium chloride as the final and only waste product.

32.3 Electrolysis of brine

Brine is again the main raw material in this process, which yields sodium hydroxide, chlorine and hydrogen. The process depends on the high over-voltage of hydrogen at a mercury cathode (page 205) which results in the release of sodium metal at the cathode.

The electrolysis is carried out as a continuous process in the Castner–Kellner cell, Fig. 32.5.

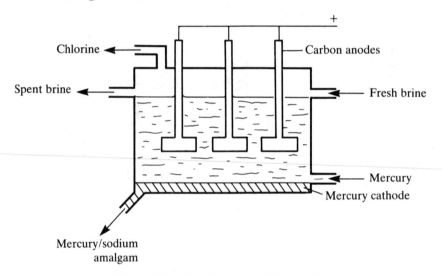

Fig. 32.5 The Castner–Kellner cell

Mercury is made to flow along the bottom of the cell, acting as cathode. The discharged sodium dissolved in the mercury is removed as an amalgam:

$$Na^+ + e^- \rightarrow Na$$

The amalgam is treated with water in cells containing a lattice of iron, at which hydrogen does not have a high overvoltage, and sodium hydroxide is formed with the release of hydrogen:

$$2Na(s) + 2H_2O(aq) \rightarrow 2NaOH(aq) + H_2(g)$$

The sodium hydroxide solution is concentrated and sold as solution or evaporated to give a solid product. The hydrogen is collected as a valuable by-product and the mercury recirculated to the cell. The anodes are of graphite to resist corrosion by the chlorine, which is discharged and collected:

$$2Cl^- \rightarrow Cl_2 + 2e^-$$

The Castner–Kellner cell is also used to produce potassium hydroxide by replacing the brine with a solution of potassium chloride.

32.4   Oxygen and nitrogen from liquid air

Vast quantities of oxygen for the steel industry and of nitrogen for the production of ammonia and nitric acid are obtained from the air.

Air is freed from dust and passed over soda-lime to remove both carbon dioxide and moisture; the purified air is compressed to 150–200 atmospheres and cooled in heat exchangers using the cold gases from the final distillation stage. The cooled air is finally passed through an expansion valve and a proportion is liquefied by the cooling effect of the expansion. The liquid is drawn off to be separated by fractional distillation. Nitrogen, boiling at 77 K, is collected first, contaminated with some argon, followed by oxygen, boiling at 90 K.

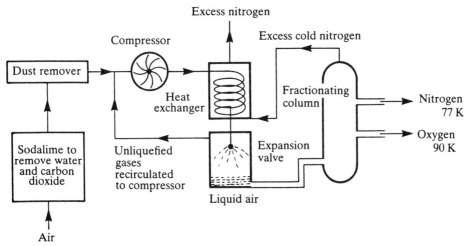

Fig. 32.6   Manufacture of oxygen and nitrogen

32.5   The Haber process

The synthesis of ammonia from nitrogen and hydrogen is summarised by the equation:

$$N_2(g) + 3H_2(g) \rightleftharpoons 2NH_3(g) \quad \Delta H = -92 \text{ kJ mol}^{-1}$$

Applying Le Chatelier's principle (page 164) to the equilibrium it can be seen that the production of ammonia is promoted by increasing the pressure on the gaseous mixture, since the forward reaction leads to a reduction in volume of the gases present; high pressure also favours the rapid establishment of equilibrium, since the rate of gaseous reactions is increased with increased pressure (page 146). In practice the high cost of pressure equipment has to be set against the proportion of ammonia formed, and moderate pressures of 250–350 atmospheres are found to give an economical yield.

Moderate temperatures of about 550°C are also used since, although a high conversion could be expected at a much lower temperature, the rate of reaction would be so slow that equilibrium conditions would not be reached at an economical rate, and the yield of ammonia would be negligible.

The nitrogen for the process is obtained from the fractionation of liquid air, while the hydrogen is mainly obtained from the petroleum industry or from methane by reaction with steam, using a nickel catalyst:

$$CH_4(g) + H_2O(g) \rightarrow CO(g) + 3H_2(g)$$

The gases must be dried and purified, especially from oxygen, sulphur or arsenic, which poison the iron catalyst; a mixture of one volume of nitrogen to three of hydrogen is preheated and passed into the reactor where the exothermic reaction helps to maintain the catalyst temperature. Equilibrium conditions are rapidly established giving a mixture containing about 15% ammonia which is extracted either by liquefaction or by solution in water. The unreacted gases are recirculated for further conversion.

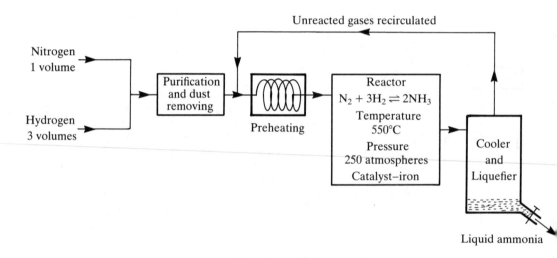

Fig. 32.7 Haber process

32.6   The Ostwald process for nitric acid

Much of the ammonia from the Haber process is used to produce nitric acid, since it is quite readily oxidised to nitrogen monoxide in the presence of a catalyst.

A mixture of one volume of ammonia and ten volumes of air is heated to about 700°C and passed over a platinum-rhodium gauze, which forms the catalyst:

$$4NH_3(g) + 5O_2(g) \rightarrow 4NO(g) + 6H_2O(aq) \ \Delta H^{\ominus}_{298} = -900 \text{ kJ mol}^{-1}$$

The highly exothermic reaction causes the temperature to rise to about 1000°C; air is admitted to the reaction chamber to cool the gases and to oxidise the nitrogen monoxide to nitrogen dioxide:

$$2NO(g) + O_2(g) \rightleftharpoons 2NO_2(g)$$

The mixture of gases, now comprising nitrogen dioxide, nitrogen monoxide, nitrogen and excess oxygen, is passed into steel absorption towers where the nitrogen dioxide is absorbed in sprays of water:

$$3NO_2(g) + H_2O(l) \rightarrow 2HNO_3(aq) + NO(g)$$

The nitrogen monoxide is recirculated with further oxygen until the formation of nitric acid is complete.

The solution contains 50% nitric acid, and is distilled to yield the constant boiling mixture containing 68% acid, which is sold as commercial concentrated nitric acid.

32.7   The contact process for sulphuric acid

In this process the key reaction is the oxidation of sulphur dioxide to sulphur(VI) oxide by the oxygen of the air. The sulphur dioxide is prepared by burning sulphur or roasting sulphide ores in excess of air. The mixture of gases, carefully dried and purified, especially from arsenic compounds, is passed over a vanadium(V) oxide catalyst at a temperature of about 450°C:

$$2SO_2(g) + O_2(g) \rightleftharpoons 2SO_3(g) \quad \Delta H_{298}^{\ominus} = -192 \text{ kJ mol}^{-1}.$$

The exothermic reaction maintains the catalyst temperature so that further heating is not necessary. At this temperature and normal pressure the equilibrium mixture yields 95–98% sulphur(VI) oxide; on Le Chatelier's principle a lower temperature would favour a higher yield from the exothermic reaction, but the decrease in reaction rate would make the conversion uneconomically slow. Nor is it worthwhile to use pressure apparatus since the already high conversion of sulphur dioxide to sulphur(VI) oxide is not greatly increased by applied pressure; the contraction of volume during the reaction is only equivalent to that of the oxygen consumed—less than one-tenth of the reaction mixture, which contains all the nitrogen from the air used. In practice, excess air is also used to keep the equilibrium well to the right.

The absorption of the sulphur(VI) oxide by water is impracticable since it results in a fine mist which is very difficult to condense. The sulphur(VI) oxide is therefore absorbed by concentrated sulphuric acid to form oleum or fuming sulphuric acid:

$$SO_3(g) + H_2SO_4(l) \rightarrow H_2S_2O_7(l)$$

The oleum is sold as such or is suitably diluted as required.

32.8   Extraction of phosphorus

Phosphorus is manufactured as white phosphorus by reduction of calcium phosphate ores in an electric arc furnace. The process is not electrolytic, the arc is used to attain the very high temperature required. A mixture of crushed ore, coke and sand is placed in a furnace lined with carbon and containing carbon electrodes. The arc is struck between the lining of the base and the electrodes, producing a temperature in the region of 1450°C, which is sufficient to promote a reaction between calcium phosphate and silicon dioxide, releasing phosphorus(V) oxide:

$$2Ca_3(PO_4)_2(s) + 6SiO_2(s) \rightarrow 6CaSiO_3(s) + P_4O_{10}(s)$$

The phosphorus(V) oxide reacts with the coke, yielding a vapour of phosphorus mixed with carbon monoxide which is passed into water where the phosphorus condenses:

$$P_4O_{10}(s) + 10C(s) \rightarrow P_4(s) + 10CO(g)$$

The calcium silicate slag is tapped off from the bottom of the furnace as a liquid.

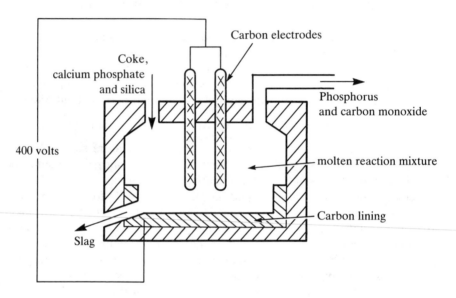

Fig. 32.8   Manufacture of phosphorus

If red phosphorus is required, it is obtained by heating white phosphorus, out of contact with air, at 270°C for several days, when conversion to the more stable allotrope occurs. The conversion is exothermic and may become explosive if the temperature is allowed to rise.

32.9   The petroleum industry

Crude oil, or petroleum, not only supplies the petrol, burning oil, diesel oil, lubricating oil, wax and fuel oils that modern society demands, but also provides the raw material for many carbon-based goods manufactured in the petrochemical industry, such as plastics, synthetic fibres, synthetic rubber, paints, varnishes, solvents, detergents and insecticides.

The fractional distillation of crude oil separates the various hydrocarbons into 'fractions' of specified boiling point ranges, as summarised in Fig. 32.9. The various fractions usually have to be purified from sulphur and other undesirable components before distribution for use; they are frequently redistilled.

The petroleum gases are sources of methane (natural gas) and propane-butane (Calor gas); petroleum ether is a solvent, gasoline yields petrol for motor-car engines; kerosine provides the heavier fuel for jet engines as well as the paraffin for oil stoves; diesel oil is a grade of gas oil, which is also used for domestic and industrial heating. The heavier oils may be burned as such as fuel oil, or may be separated by vacuum distillation into lubricating oils and waxes, while the final residue of bitumen is used for roofing felts and sealing materials, and as asphalt for road surfaces.

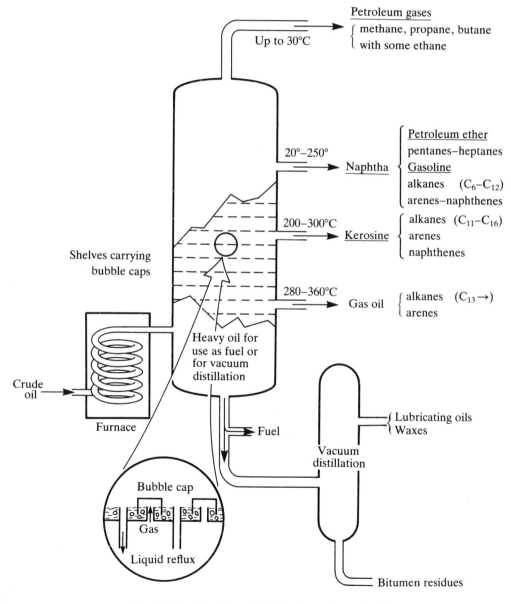

Fig. 32.9 Fractional distillation of crude oil

The various fractions may be subjected to further processes to modify or improve their characteristics, or to provide more of a particular type of product, since the quantities of the various fractions initially produced do not always meet the demand. One of these processes, involving molecular rearrangements, is 'cracking' (page 349); this may be used to produce more petrol suitable for internal combustion engines from higher boiling fractions, or to 'reform' the natural alkanes in the petrol fraction to compounds that give a better performance in such engines. Cracking also provides the vast quantities of ethene, butene and butadiene needed as starting materials for products from the petrochemical industry. Most of these have already been mentioned in the relevant chapters.

Methane from fractional distillation and from natural gas is used in the manufacture of methanol by oxidation using a limited amount of oxygen:

$$2CH_4(g) + O_2(g) \xrightarrow[\substack{\text{high} \\ \text{pressure}}]{\text{Cu}} 2CH_3OH(l)$$
(kept in excess)

Reacted with steam a mixture of carbon and hydrogen is obtained:

$$CH_4(g) + H_2O(g) \xrightarrow[750°C]{\text{Ni}} CO(g) + 3H_2(g)$$

This mixture can be reacted further with steam to produce very pure hydrogen:

$$CO(g) + 3H_2(g) + H_2O(g) \rightarrow CO_2(g) + 4H_2(g)$$

The very soluble carbon dioxide is washed out.

Questions

1. (a) Describe the manufacture of sulphuric acid, explaining carefully the reasons for the chemical and physical conditions used in each stage of the process. Details of the industrial plant are *not* required.
   (b) Give two industrial uses of sulphuric acid. (O)
2. Outline the commercial preparation of chlorine and give two large scale uses for the element. (C)
3. (a) The Haber process for the manufacture of ammonia can be represented by the following equation:

$$N_2(g) + 3H_2(g) \rightleftharpoons 2NH_3(g) \Delta H^\ominus = -92 \text{ kJ}$$

   (i) From chemical theory, show how the most suitable conditions for the production of ammonia can be predicted.
   (ii) State the conditions used in practice and give reasons why these are chosen.
   (b) Describe how ammonia is converted into nitric acid on a large scale.
   (AEB 1979)
4. *Outline* the chemistry involved in the extraction of phosphorus from calcium orthophosphate and state which allotrope is produced.
   (AEB 1979)

5. For *either* aluminium *or* iron:

Give a brief account of the occurrence of the metal or its compounds in nature, and of the method by which the metal is extracted from its ores.

Summarise the advantages and disadvantages of the metal in everyday use.

Describe with essential practical details how you would convert the metal into the chloride in which it has oxidation number +3. Certain physical properties of the metal chloride are set out at the end of the question. What conclusions can you draw from them about the chloride of the metal you have chosen?

	Aluminium Chloride	*Iron(III) Chloride*
Melting Point	190°C	290°C
Boiling Point	180°C	Not given in tables
	Soluble in benzene	Soluble in benzene
Molecular Weight	267 in benzene, and also in vapour just above b.p.	324 in benzene 167 in ethanol
pH of aqueous solution	less than 7	less than 7

$$Al = 27, \quad Fe = 56, \quad Cl = 35.5 \qquad \text{(L)}$$

6. Petroleum is the major industrial source of alkanes, alkenes and aromatic hydrocarbons which, in turn, are the starting materials for the production of many important chemical compounds and materials of use and benefit to man.

Outline the methods by which hydrocarbons are obtained from crude petroleum on an industrial scale and indicate what particular hydrocarbons are isolated. (L)

Glossary of traditional names in common use

TRADITIONAL NAME	RECOMMENDED NAME
acetaldehyde	ethanal
acetanilide	N-phenylethanamide
acetic acid	ethanoic acid
acetone	propanone
acetylene	ethyne
acetylides	dicarbides
acrylic acid	propenoic acid
adipic acid	hexane-1,6-dioic acid
alanine	2-aminopropanoic acid
aniline	phenylamine
benzyl alcohol	phenylmethanol
bicarbonate, bisulphate etc.	hydrogencarbonate etc.
carbon tetrachloride	tetrachloromethane
chloral	trichloroethanal
chloroform	trichloromethane
citric acid	2-hydroxypropane-1,2,3-tricarboxylic acid
ether	ethoxyethane
ethylene glycol	ethane-1,2-diol
ethylene oxide	epoxyethane
ferricyanide ion	hexacyanoferrate(III) ion
ferrocyanide ion	hexacyanoferrate(II) ion
formaldehyde	methanal
formic acid	methanoic acid
glycerine (glycerol)	propane-1,2,3-triol
glycollic acid	hydroxyethanoic acid
hexamethylenediamine	hexane-1,6-diamine
hypochlorite ion	chlorate(I) ion
iodoform	triiodomethane
isoamyl acetate	3-methylbutyl ethanoate
lactic acid	2-hydroxypropanoic acid
lead dioxide	lead(IV) oxide
lead monoxide	lead(II) oxide
malic acid	2-hydroxybutanedioic acid
malonic acid	propanedioic acid
mercurous ion	dimercury(I) ion
mercaptans	thiols
metaldehyde	ethanal tetramer
metavanadate ion	polytrioxovanadate(V) ion
neoprene	poly(2-chlorobuta-1,3-diene)
nitric oxide	nitrogen monoxide
nitrous oxide	dinitrogen oxide
olefins	alkenes

TRADITIONAL NAME	RECOMMENDED NAME
oxalic acid	ethanedioic acid
paraldehyde	ethanal trimer
per-acids	peroxoacids
(containing (—O—O—) group)	
permangate ion	manganate(VII) ion
phosgene	carbonyl chloride
phosphorous acid	phosphonic acid
pyrophosphate ion	heptaoxodiphosphate(V) ion
phthalic acid	benzene-1,2-dicarboxylic acid
pyrogallol	benzene-1,2,3-triol
red lead	dilead(II)lead(IV) oxide
salicylic acid	2-hydroxybenzoic acid
stannic compounds	tin(IV) compounds
stannous compounds	tin(II) compounds
stearic acid	octadecanoic acid
styrene	phenylethene
succinic acid	butanedioic acid
sulphuryl chloride	sulphur dichloride dioxide
tartaric acid	2,3-dihydroxybutanedioic acid
terephthalic acid	benzene-1,4-dicarboxylic acid
thionyl chloride	sulphur dichloride oxide
toluene	methyl benzene
vinyl acetate	ethenyl ethanoate
vinyl chloride	chloroethene
o-xylene	1,2–dimethylbenzene

Answers to numerical questions

Chapter 1 (page 7)

1. 64
2. 142
3. 7.1 g
4. 0.1
5. 0.1 M
6. (a) 0.1 M (b) 0.2 M
7. (a) 0.004 (b) 0.0025 (c) 0.005
8. 2.24 l.

Chapter 3 (page 40)

5. (d) 7/8 m, 2.6×10^{22}
6. (c)(ii) 14, 28
7. (a)(i) 18, 19, 19, 18
     (ii) W and Z, X and Y
   (b)(ii) 40, 22
   (c)(i) 2

Chapter 5 (page 87)

1. 126 (PF_5)
6. (e) 40 kP_a (hydrogen), 60 kP_a (helium)

Chapter 7 (page 123)

3. (d) 364
4. Relative molecular mass in water 68.7, in vapour 142
5. (i) 0.032 (ii) 25.7 kP_a (iii) 77.17°C
6. −0.64°C
7. 1700

Chapter 9 (page 159)

2. (b) 3
3. 1
4. (a)(ii)2 (iii) 1, (b)(ii) 3, (c) A
5. (a) 0 and 2, (c) 0.01 l mol^{-1} s^{-1}

Chapter 10 (page 170)

1. $K_p = 0.40$ atm, 0.52 atm
2. (a) $K_p = p_{NO_2}^2/p_{N_2O_4}$
     (c)(i) 25% (ii) 40% (iii) $K_p = 2.69 \times 10^4$ Nm^{-2}
3. (a) 0.018 g l^{-1}
4. (b)(ii) 63.5 g
5. (a)(iii) 1×10^{-10} mol^2 l^{-2}
     (b) 7.8 g l^{-1}

Chapter 11 (page 184)

4. (*a*) 2   (*b*) 2.02   (*c*) 12.6
5. (*a*) 3.7   (*b*) 10.3
8. (*b*)(i) 7.08×10^{-3} M
10. (*a*) 1   (*b*)(i) 5×10^{-3} (ii) 10^{-1} (iii) 5.6×10^{-4} (*c*) 1.3
11. (*c*)(i) 4.75   (ii) 0.42%

Chapter 12 (page 202)

1. (i) +0.46 V
4. (*c*) +0.637 V

Chapter 13 (page 212)

3. (*a*) 0.0790 g   (*b*) 13.9 cm^3
5. (i) 0.013 (1.3%)   (ii) 2.89
6. (ii) 2.012 g   (iii) Silver 0.007 M, nitric acid 0.093 M
7. (*b*)(i) 151 ohm^{-1} cm^2 mol^{-1}
     (ii) 1.59×10^{-2} mol dm^{-3}

Chapter 20 (page 304)

4. (*a*) $x = 1$ (*b*) +4

Chapter 22 (page 337)

7. (*a*) $x = 5$, $y = 3$

Chapter 25 (page 386)

6. (*a*) C_2H_6O (*b*) 46   (*c*) C_2H_6O

Chapter 30 (page 456)

3. (*c*)(i) 8.51 g   (ii) 1.19 l.

Index

Absorption coefficient 111
Acetanilide *see* N-phenylethanamide
Acetic acid *see* ethanoic acid
Acetic anhydride *see* ethanoic anhydride
Acetophenone *see* phenylethanone
Acetylene *see* ethyne
Acetylsalicylic acid *see* 2-ethanoyloxybenzoic acid
Acid/alkali titrations 179
Acid/base indicators 177
Acid dissociation constants 175
Acidic oxides 291
Acids 173
  strong/weak 174
Acrilan 356
Acrylic acid *see* propenoic acid
Acrylonitrile *see* propenenitrile
Actinides 2
Activation energy 156
Acyl amides 411
Acylation 410, 411, 443, 457
Acyl chlorides 443
Acyl derivatives 408
Addition polymers 354
Addition reactions 341, 389
Adipic acid *see* hexanedioic acid
Adsorption 104, 253
Alanine *see* 2-aminopropanoic acid
Alcohols 374*ff*, 435
Aldehydes 388
Aldol *see* 3-hydroxybutanal
Aldohexose 396
Aldoses 396
Aliphatic compounds 340
Alkali metals 2 *see also* s-block elements
Alkaline earth metals 2, 236
Alkaline potassium manganate(VII) 377
Alkanes 340, 344*ff*, 434
Alkenes 348*ff*, 434
Alkyl group 361
Alkyl halides 431
Alkynes 348*ff*
Allotropes 83, 91, 257
Alumina 104
Aluminate(III) ion 252
Aluminium 247, 255
Aluminium chloride 253

hydroxide 252
oxide 206, 252
potassium sulphate-12-water 254
sulphate 254
Aluminium, manufacture 468
Aluminium triethyl 354
Alums 254
Amalgams 336
Americum 38
Amides 424
Amines 419, 435
Amino-acids 427
4-aminobenzenesulphonic acid 422
Aminobutanedioic acid 428
Amino-compounds 419
Amino-ethanoic acid 415, 427
2-amino-3-methylbutanoic acid 428
2-aminopropanoic acid 370, 428
Ammonia 48, 175, 276, 475
Ammonia-boron trichloride 273
  trifluoride 55, 247, 273
Ammoniacal silver nitrate 406
Ammonium alum 254
  carbonate 277
  chloride 277
  compounds 277
  ion 226, 277
  metavanadate *see* ammonium vanadate(V)
  nitrate 277
  phosphate 277
  sulphate 277
  vanadate(V) 325
Amorphous carbon 260
Amorphous sulphur 293
Amphoteric hydroxides 252
  oxides 252, 291
Amyl acetate *see* pentyl ethanoate
α- and β-amylose 399
Anhydrides 404
Anhydrite 236
Aniline *see* phenylamine
Animal starch *see* glycogen
Anions 43, 204
Anodising 250
Anti-bonding orbitals 50
Antimony 272, 286
α-particles 31

rays 31
rgon 2, 28
renes 340, 356*ff*
rsenic 272, 286
ryl halides *see* halogenoarenes
scending a homologous series 426, 438
spartic acid *see* aminobutanedioic acid
spirin 414
ssociation of molecules 72, 101, 122
statine 2, 305
ston's mass spectrograph 12
tomic bomb 39
tomic crystal 45, 82
tomic hydrogen torch 224
tomisation, enthalpy of 129
tomic number 2, 12, 33
tomic orbitals 26
tomic radii 217
tomic volume 218
utocatalysis 327
vogadro constant 4, 31, 86, 208
vogadro's law 6
zeotropic mixtures 98
zo-dyes 423

king powder 246, 286
lmer 18
arium 234
compounds 239*ff*
chromate 326
peroxide 292
ase, dissociation constant 175
ases 173
asic oxides 291
auxite 252
ecquerel 31
eckmann thermometer 114
enedict's solution 389
enzaldehyde 362
enzamide 408, 410
enzene 52, 356*ff*, 383
enzene carbaldehyde *see* benzaldhyde
enzene carboxamide *see* benzamide
enzene carboxylic acid *see* benzoic acid
enzenediazonium chloride 422
enzenedicarboxylic acids 412
enzenesulphonic acid 361
enzene-1,2,3-triol, 382
enzoic acid 175
enzoin *see* 2-hydroxy-1,2-diphenylethanone

Benzoly chloride 393, 443
Benzoyl derivatives 408
Benzoylation 443, 457
Benzyl alcohol *see* phenylmethanol
Benzylamine *see* (phenylmethyl)amine
Berkeley and Hartley 120
Beryllium 2, 234, 255
Beryllium chloride 4, 8, 48
Bidentate ligands, 56
Bis[di(carboxymethyl)amino]ethane *see* edta
Bismuth 2, 272, 286
    (III)nitrate 286
Biuret 426
Blast furnace 469
Bleaching powder 315, 442
Blister copper 472
Body-centred cubic packing 79
Bohr 18
Boiler scale 230
Boiling point/composition diagrams 96
Boiling point constant 114
Boiling point elevation 113
Boltzmann 75
Bomb calorimeter 138
Bond angle 48
Bond energy 131
Bonding
    coordinate 55
    covalent 45
    dative covalent 55
    electrovalent 43
    hybridisation 48, 51
    hydrogen 61
    ionic 43
    metallic 61
    orbitals 49
Boracic acid *see* trioxoboric(III)acid
Borate ion *see* tetrahydroxoborate(III)
Borax 249
Born–Haber cycle 134
Boron, 2, 247
    nitride 249
    oxide 249
    trifluoride 48, 249
Borosilicate glass 267
Boyle's law 66
Brackett 18
Bragg equation 85
Brand, Hennig 281
β-rays 31

Brine, electrolysis of 206, 474
Bromates 315
Bromides *see* Halides
Bromine 305
Bromoalkanes *see* halogenoalkanes
Bromobutane 432
Bromoethane 352, 449
Bromothymol blue 178
Brønsted and Lowry 173
Bronze 262
Buffer solutions 182
Butane 345
Butanedioic acid 411
Butanedione dioxime 333
Butan-1-ol 432
Butan-2-ol 380
Butanoic acid 402
Butenes 350
Butene dioic acid (*cis* and *trans*) 412
Butyric acid *see* butanoic acid

Cadmium 335
Caesium 2, 234
Caesium chloride 81
Calcite 83, 368
Calcium 234
    bicarbonate *see* hydrogencarbonate
    carbonate 83, 231
    chloride 240
    compounds 239
    ethanoate 407
    fluoride 241
    hydrogencarbonate 231
    hydroxide 240
    oxide 240
Calgon 286
Calomel electrode 193
Calorimetry 138*ff*
Cannizzaro's reaction 393
Canonical forms 60, 357
Carbamide 425
Carbanion 343
Carbides *see* dicarbide ion
Carbohydrates 395
Carbolic acid *see* phenol
Carbon 2, 260
    dating 40
    dioxide 90, 266
    disulphide 263
    monoxide 265, 333

    tetrachloride *see* tetrachloromethane
Carbonates 266
Carbonium ion 343
Carbonyl chloride 289
Carbonyl compounds 59, 265
Carbonyl group 388
Carboxylate ion 403
Carboxylic acids 402*ff*
    salts 406
Carbylamine reaction *see* isocyano reaction
Cassiterite 262
Castner–Kellner cell 474
Catalyst 146, 166, 334
Catenation 263, 295
Cations 43, 204
Cellulose 399
Cement 243
Chadwick 38
Chain reaction 348
Chalk 236
Charcoal 260
Charles' law 66
Chelate compounds 56
Chile saltpetre 236, 308
Chloramine 438
Chloral *see* trichloroethanal
Chlorate ions 316
Chloric(I) acid 175 315
Chloric(V)acid 315
Chlorides 221
Chlorine 13, 305
    oxides 314
Chlorobenzene 359, 431, 438
Chlorobutane 432
Chloroethanoic acid 405, 413
Chloroethene 356, 441
Chloroform *see* trichloromethane
(Chloromethyl)benzene 360, 440
Chloro-2-methylbenzene 360, 431
Chlorophyll 60
Chlorosilanes 267
Chlorosulphonic acid 302
Chromate(VI) ion 326
Chromatography 102*ff*, 371
Chrome alum (Chromium(III)potassium-
    sulphate-12-water) 254
Chromic(VI)acid 326
Chromium 321
Chromium(III) compounds 326
Chromium(III)chloride-6-water 331

romophore 424
nabar 336
namaldehyde *see* 3-phenylpropenal
ck reactions 158
al 261
balt 322
balt-60 40
balt(II) & (III) ions 329
lligative properties 108
lorimeter 158
mbustion
  nthalpy of 129
mmon ion effect 169
mplex ions 56, 330
ncentration cells 195
ncrete 243
ndensation polymers 412, 429
ndensation reactions 342, 391, 458
nductivity 159, 209
ell 209
vater 209
njugate pairs 174
nstant boiling mixtures 99
ntact process 477
nvergence limits 24
oling curve 146
ordinate bonds 54
ordination compounds 330
ordination number 56, 81
pper 279, 322, 472
pper(I) compounds 329
pper(II) compounds 329
pper pyrites 472
pper(II) sulphide 297
rundum 252
urtelle 356
valent bonding 45
valent compounds 45
valent crystals 82, 84
valent hydrides 227
valent radius 86
acking' reactions 349
tical constants 76
tical mass 38
olite 206
oscopic constant 116
stal gardens 266
stal lattice 79
stallisation
  ractional 110

water of 230
Crystals, metal 79
Cubic, close packing 79
Curie, Marie 31
Cyanides *see* nitriles
Cyanohydrins 390
Cyclohexane 346
Cyclohexanol 385
Cyclopropane 346

Dalton's law of partial pressures 68
Daniell cell 189
Dative bonds *see* coordinate bonds
d-block elements 216, 220, 319
Decarboxylation 407, 458
Decay, radioactive 35
DDT 439
Degradation reaction 425
Degree of dissociation 174, 212
Degree of ionisation, 175, 212
Dehydrogenation, 459
Deionisation 232
Delocalisation of electrons 60
Density of gases 74
Depression of freezing point 115
Descending chromatography 104
Descent of homologous series 425, 459
Detergents 409
Dettol 382
Deuterium 12, 39
Deuterons 37
Dextrose 396
Diagonal relationship 244, 255
1-6-diaminohexane 429
Diamminesilver ion 56
Diamond 45, 83, 260
Diamond lattice 83
Diazonium salts 422*ff*
Dibromoethane 351, 450
Dicarbide ion 264
Dicarboxylic acids 411
Dichloroethane 392
Dichloroethanoic acid 413
Dichloromethane 348, 440
Dichlorotetraquochromium(III) ions 331
Dichlorotetraminechromium(III) ions 332
Diethylether *see* ethoxyethane
Diffraction grating 84
Diffusion of gases 73
Dihydric alcohols 380

2,3-dihydroxybutanedioic acid 370
Dilead(II) lead(IV) oxide 268, 292
Dimercury(I) chloride 193
  compounds 336
Dimerisation 78
1,2-dimethylbenzene 359
1,3-dimethylbenzene 359
1,4-dimethylbenzene 359
2,2-dimethylbutane 346
Dimethyl glyoxime 333
Dinitrogen oxide 279
Dinitrogen tetraoxide 279
2,4-dinitrophenylhydrizine 392
Diphenyl 439
Dipole moment 47, 77
Dipole–dipole attraction 77
Disaccharides 397
Discharge of ions 205
Disproportionation 201, 316, 327, 329
Dissociation 122
  constants, acid 174
  constants, base 175
  degree of 212
  in solvents 101, 122
  pressure 167
  thermal 72
Distillation
  fractional 96
  steam 100
Distribution constant *see* partition coefficient
Distribution law *see* partition law
Disulphur dichloride 303
Dodecanoic acid 409
d orbitals, 21, 27
Double bonds, 45, 349
Dumas 70
Dyeing 253

Ebullioscopic constant, *see* boiling point constant
edta 56
Effusion, 74
Einstein 33
Electrochemical cell 187
Electrochemical equivalent 208
Electrochemical series 191
Electrode potential 187
Electrodes 187
Electrolysis 204, 236
  of brine 206, 474

Electrolytes 204
  strong 211
  weak 211
Electrolytic conductivity 209
Electromagnetic spectrum 17
Electronegativity 46
Electron 8
  acceptor 184
  affinity 134
  charge/mass ratio 10
  delocalisation 60
  density 26
  diffraction patterns 10
  wave nature 10
Electronegativity 46
Electron impact method 25
Electron repulsion theory 47
Electron pairs 47
  inert pair effect 248
Electrophilic reagents 343
Electrophoresis 428
Electrovalent compounds 43
Element 2
Elevation of boiling point 113
Elimination reactions 341
Elution chromatography 104
Emission spectra 17, 237
En *see* ethane-1,2 diamine
Enantiomorphs 370
Enantiotropy 91
Endothermic reaction 128, 137, 141
Energy, activation 156
Energy levels 21
Enthalpy
  of atomisation, 129
  of combustion, 129
  of formation 129
  of ionisation 134
  of hydration 137
  of hydrogenation 129
  of neutralisation 137
  of reaction 128
  of solution 137
Entropy 141
Epoxyethane 380
Equilibrium constants 163
Esterification 459
Esters 404, 408, 435
Ethanamide 424, 452
Ethanal 391

ane 345*ff*
ane-1,2-diamene 56, 332
anedioic acid 411
ane-1,2-diol 353, 380
anenitrile 426
anoic acid 403, 451
merisation 78
ssociation constant 175
anoic anhydride 404
anol 375
  actions 377*ff*
anoyl chloride 443
hanoyloxybenzoic acid 414
ne 51, 348, 379, 448
rs 385, 434
oxyethane 378, 385, 449
ylamine *see* amino compounds
ylbenzene 358, 391
yl benzoate 410
yl ethanoate 408, 452
yl group 345
yl hydrogen sulphate 378
ylenediaminetetraacetic acid 56
ylene glycol *see* ethane-1,2-diol
yne 348
ectics 93*ff*
action of metals 468

e-centred cubic close packing 79
an's rule 46
aday's law 207
adays's constant 208
-breeder reactor 39
ock elements 216
ding's solution 329, 389
omagnetism 319
ilisers 277, 286
  extinguishers 266, 308
t order reactions 148
t 267
vers of sulphur, 294
oride ion, 310
orides, solubility of 309
orine, 2, 317
orocarbons 442*ff*
orspar 308
bitals 21
maldehyde *see* methanal
mamide *see* methanamide
mation enthalpy of 129

Formic acid *see* methanoic acid
Fossil fuels 261
Fractional crystallisation 31, 110
Fractional distillation 96
  of crude oil 479
  of liquid air 475
Frasch process 293
Free energy 143
Freezing point constant 116
Freezing point depression 115
Free radical 343
Freons 443
Friedel–Crafts reaction 361, 436, 443
Fructose 397
Fumaric acid *see* butenedioic acid
Functional group 374

Galactose 397
Galena 262
Gallium 248
Galvanised iron 324, 335
Gamma rays 31
Gammexane *see* hexachlorocyclohexane
Gas chromatography 104
Gas constant 68
Gas syringe 71
Geiger and Marsden 11
Geiger counter 36
gem-dihalides 441
Germanium 226, 259, 262
  compounds 264*ff*
Glass 267
Glucose 396
Glycerides 409
Glycerol *see* propane-1,2,3-triol
Glycine *see* aminoethanoic acid
Glycogen 400
Glycollic acid *see* hydroxyethanoic acid
Graham's law 73
Gram-molecular volume *see* molar volume
Graphite 45, 83, 260
Green vitriol, *see* iron(II) sulphate(VI)-7-water
Grey tin 257
Grignard reagent 436*ff*
Group 0 2
Group I 2, 234
Group II 2, 234
Group III 2, 247
Group IV 2, 257
Group V 2, 272

Group VI 2, 288
Group VII 2, 305
Group displacement law 33
Groups, in periodic table 2, 217
Gypsum 243

Haber process 165, 475
Haematite 327
Haemoglobin 60, 265
Half cell 188
Half life 35, 153
Halides 309
Halogen carrier 359
Halogenation 460
Halogenoalkanes 431*ff*
Halogenoarenes 438
Halogenoxides 314
Halogens 2, 305
Hard water 230
Head elements 236
Heat content *see* enthalpy
Heat of *see* enthalpy of
Heat-proof glass 267
Heisenberg's uncertainty principle 10
Helium 28, 50
   nuclei 31
Helmholtz double layer 187
Henry's law 111
Hertz 17
Hess's law 129
Heterogenous catalysis 147
Heterolytic fission 342
Hexaammine chromate(III) ions 58
Hexaammine cobalt(III) ions 57
Hexachlorocyclohexane 360
Hexacyanoferrate(II) ions 57
Hexacyanoferrate(III) ions 57
Hexadecanoic acid 402
Hexadentate ligands 56
Hexafluoroaluminate ion 253
Hexafluorosilicate(IV)ion 269
Hexagonal close packing 79
Hexagonal crystals 80
Hexane-1,6-dioic acid 429
Hexoses 395
Hoffmann degradation 425
Homogenous catalyst 147
Homologons series 340, 345
Homolytic fission 342
Hund's rule 22, 49

Hydrated ions 55
Hydrates 55
Hydration 460
Hydrazine 175, 278, 391
Hydrazones 392
Hydride ion 225
Hydrides 221, 228
   covalent 227
   ionic 227
   interstitial 227
Hydriodic acid 313
Hydrocarbons 340
Hydrochloric acid 313
Hydrocyanic acid 175
Hydrogen 224
Hydrogen bomb 39
Hydrogen bonds 61, 78, 227
   bromide 299
   chloride 311
   cyanide 390, 426
   electrode 189
   fluoride 312
   halides 225
   iodide 299, 314
   ion 226
   peroxide 292
   sulphide 295
Hydrogenation 226, 460
Hydrolysis 181, 232, 434, 461
Hydroxybenzenesulphonic acid 385
2-hydroxybenzoic acid 414
3-hydroxybutanal 394
2-hydroxy-1,2-diphenylethanone 394
Hydroxyethanoic acid 413
Hydroxylamine 278, 391
2-hydroxypropanoic acid 390, 414
'Hypo' *see* sodium thiosulphate
Hypochlorous acid *see* chloric acids

Ice 229
Iceland spar 242, 368
Ideal gas equation 66
Ideal solutions 95
Immiscible liquids 100
Indicators 177
Indium 248
Inductive effect 344
Inert pair effect 248
Infra-red radiation 17
Interhalogen compounds 316

nuclear distances 86
stitial carbides 324
stitial hydrides 227, 324
rides 275, 324
sion of cane sugar 398
e(I) ions 316
e(V) ions 316
les 309
e 305
stals 84
ptafluoride 317
nochloride 317
ntoxide 315
chloride 317
butane 433
ethane 434
form reaction 378, 441
xchange 231
compounds 43
stals 79
drides 227
lius 85, 218
ation constants 174
gree of 212
ergy 23
thalpy of 134
3, 43

m 254
bonyl *see* pentacarbonyl iron(0) 59
mplex ions 331
) compounds 328
I) compounds 327
) diiron(III) oxide 292, 328
) ions, detection 330
I) ions, detection 330
nufacture 290, 468
sting 323
) sulphate(VI)-7-water 301
ano reaction 420
ectronic species 251, 265
erism 364*ff*
ers 345
orphs 255
smotic solutions 122
nic solutions 122
pes 4, 12, 31

t, Joliot-Curie 38

lé formula 357

Ketohexoses 397
Ketones 388
Ketoses 397
Kieselguhr 104
Kinetic energy 66
Kinetic theory 66*ff*
Kohlrausch's law 211
Krypton 28

Lactic acid *see* 2-hydroxypropanoic acid
Lanthanoids 2, 216
Lattice, crystal 79
    energy 134
Lauric acid *see* dodecanoic acid
Lawrencium 2, 38
L.D. converter 471
Lead 2, 94, 262
    chromate 269, 326
    compounds 264
    ethanoate 269
    extraction 262
    poisoning 262
    sulphide 297
    tetraethyl 263
Le Chatelier's principle 164
Lewis acids 184
Ligands 56
Lime 242
Limestone 236
Linear molecules 48, 53, 266
Line spectrum 17, 24
Liquid air 475
Lister 382
Lithium 234
    aluminium hydride *see* tetrahydridoaluminate
        (III) ion
    chloride 46
    compounds 239
    diagonal relationship 244
Litmus 178
Lothar Meyer 218
Lyman series 18
Lysine 428

Macromolecules 45
Magnesium 2, 234
    compounds 239*ff*
    hydroxide 240
    oxide, lattice energy 135
Magnetite *see* iron(II) diiron(III) oxide

Malachite 329
Maleic acid *see* butenedioic acid
Malonic acid 411
Maltose 398
Manganate(VI) ion 327
Manganate(VII) 327
Manganese 321
  (II)compounds 327
  (IV) compounds 327
  (II) sulphide 297
Mannose 397
Marble 261
Margarine 226
Markownikov's rule 353
Mass concentration 108
Mass defect 33
Mass number 33
Mass spectrometer 14
Maximum boiling point mixtures 99
Maxwell's distribution law 75
Mechanism of reactions 154, 432
Mendeleyev 2, 262
Mercury 335
  (I) compounds *see also* dimercury(I)
  (II) compounds 336
  (II) sulphide 297, 336
Mesitylene *see* 3,5-trimethylbenzene
Metaborate ion *see* tetrahydroxoborate(III) ion
Metaldehyde *see* ethanal tetramer
Metal crystals 79
Metallic bonding 61
Metalloids 3, 257
Metals, extraction 468
Metamerism 367, 386
Metastability 91
Meta-xylene *see* 1,3-dimethylbenzene
Methanal 388, 393
Methanamide *see* formamide
Methane 340, 345
Methanoic acid 175, 402, 405
Methanol 264, 265, 375, 480
Methylamine 425, 454
Methylated spirit 375
Methylbenzene 358, 362, 407
2-methylbutane 345
Methyl cyanide *see* ethanenitrile
Methyl group 345
Methyl 2-hydroxybenzoate 414
Methyl methacrylate (methyl 2-methylpro-
  penoate) 356

Methyl-2-nitrobenzene 361
Methyl orange 178, 422
Methyl red 178
Methyl salicylate *see* methyl-2-hydroxybenzo
Methyl-2,4,6-trinitrobenzene 361
Millikan 9
  oil drop 10
Minimum boiling points 99
Molar conductivity 210
Molar critical volume 76
Molar depression constant 116
Molar elevation constant 114
Molar solutions 5
Molar volume 6
Mole 4
Molecular crystal 84
Molecular orbitals 48
Molecular velocities 75
Molecularity, of reaction 155
Molecules 3
  shapes of, 47
Mole fraction 108
Monastral blue 333
Monoclinic crystals 80
Monoclinic sulphur 92, 293
Monodentate ligands 56
Monomers 356
Monosaccharides 396
Monotropy 91
Mordant 253
Moseley 12

Naphthalene 84
Naphthalen-2-ol (β-naphthol) 424
Natural gas 479
Negative hydrogen ion 225
Neon 28
Neptunium 39
Nernst equation 194
Neutral oxides 291
Neutron 8, 12, 38
  absorber 38
Nickel 321
  butanedione dioxime 333
  carbonyl *see* tetracarbonyl nickel(0)
  sulphide 297
Nicol prism
Nitrate(V) ion 280
Nitration 461
Nitrates 280

ic(III) acid 175, 279
ic(V) acid 280, 476
ides 273
iles 425, 435
oalkanes 417
obenzene 417
oethane 418
ogen 2, 274
ycle 274
ioxide 278
quid air 474
olecule 54
onoxide 278
ichloride 273
ophenol 384
ophos 286
otoluene, see methyl-2-nitrobenzene
ous acid see nitric(III) acid
ole gases 2, 28
mal oxides 291
clear fission 38
clear reactions 37
cleons 8
leophilic reagents 343
leus 11
lides 13, 34
on 428

adecanoic acid 409
ahedral molecules 48
of vitriol 301
of wintergreen 414
um 447
ical isomers 333, 368, 380
itals, electronic 18
er of reactions 148
ho-boric acid see trioxoboric(III) acid
horhombic crystals 80
ho-xylene see 1,2-dimethylbenzene
nosis 117ff
notic pressure 117
wald's dilution law 174
wald's process for nitric acid 476
rvoltage 205
lic acid see ethanedioic acid
dation 196, 462
umber 198
ates 320
des 221, 291
dising agent 197

Oximes 391
Oxonium ion 226
Oxygen, 2, 290
    difluoride 314
    from liquid air 475
Oxyhydrogen flame 224
Ozone 290
Ozonolysis 291, 354

Palmitic acid see hexadecanoic acid
Paper chromatography 103
Paradichlorobenzene see 1,4-dichlorobenzene
Paraffins, see alkanes
Paraldehyde see ethanal trimer
Paramagnetism 319
Para-xylene see 1,4-dimethylbenzene
Partial pressure 68
Partition coefficient see also distribution constant
    101
Partition law 101
Pasteur 371
Pauli's exclusion principle 49
Pauling 46, 85
p-block elements 247
p electrons 49
Pentacarbonyl iron(0) 59
Pentane 345
Pentyl ethanoate 409
Peptide group 426
Periodicity 216
Permutit 232
Peroxides 292
Peroxodisulphuric acid 301
Perspex 356
Petrochemicals 478
Petroleum 478
Pewter 262
Pfeffer 119
Pfund 18
Phase diagrams and equilibria 89ff
Phenol 175, 382ff, 438
Phenolphthalein 178
Phenylalanine 428
Phenylamine 175, 421, 455
N-phenylbenzamide 444, 455
Phenyl benzoate 383
N-phenylethanamide 421, 444
Phenylethanoate 383
Phenylethanone 389, 391, 443
Phenylethene 356, 358

Phenylhydrazine 424
Phenyl group 358
Phenylmethanol 381
3-phenylpropenal 395
Phenyltrimethylammonium iodide 419
pH meter 201
pH values 175
Phosgene *see* carbonyl chloride
Phosphine 282
Phosphonium ion 282
Phosphor bronze 282
Phosphoric(V) acid 181, 282
Phosphorus
  allotropes 93, 282
  hydrides 282
  manufacture 477
  oxoacids 285
  pentachloride 283, 377
  (V) oxide (pentoxide) 282
  trichloride 283
Photography 308
Phthalic acid *see* benzene-1,2-dicarboxylic acid
  anhydride *see* benzene-1,2-dicarboxylic anhyd-
  ride
π-bonds 48
Pitchblende 31
pK values 175, 178
Planar molecules 48, 52
Planck's constant 10, 19, 25
Plane polarised light 368
Plastic, molecular mass of 122
Plastics 356
Platinum black 189
Plumbate(II) & (IV) ions 268
Plutonium 38
Polar covalent molecules 46, 47
Polarimeter 159, 369
Polarising power 46
Polonium 2, 303
Polyatomic ions 60
Poly(chloroethene) 356
Polydentate ligands 56
Polyesters 413
Polyhydric alcohols 381
Polymerisation 354
Polymers, addition 354
  condensation 393, 413
Polymethylmethacrylate *see* perspex
Polymorphism 91
Polypeptides 427

Poly(phenylethene) 356
Polypropylene 356
Polysaccharides 395, 399*ff*
Polystyrene *see* poly(phenylethene)
Polytetrafluoroethene 443
Polythene 356
Polyvinyl chloride *see* poly(chloroethene)
p orbitals 21, 27, 49
Potash alum *see* aluminium potassium sulph:
  12-water
Potassium, 2, 234
  chlorate (V) 316
  chlorate(VII) 316
  compounds 239*ff*
  cyanide 435
  hexacyanoferrate(III) 58
  iodate(V) 316
  iodide 316
  manganate(VI) 327
  manganate(VII) (permanganate) 198, 327,
  peroxide 237
  peroxodisulphate 301, 334
Potentiometric titrations 201
Priestley, Joseph 290
Primary alcohols 379, 437
Primary amines 419*ff*
Propane 345
Propanone 391
Propane-1,2,3-triol 380, 409
Propene 350
Propenoic acid 414
Propenenitrile 356
Propyl group 345
Proteins, molecular mass 122
Proton 8
  acceptor 173
  donor 173
Prussic acid *see* hydrogen cyanide
PTFE *see* poly(tetrafluoroethene)
Pyridine 375
Pyrogallol *see* benzene-1,2,3-triol
PVC *see* poly(chloroethene)

Quantum numbers 21
  spin 21, 49
Quartz *see* silicon(IV) oxide
Quaternary ammonium salts 419
Quicklime *see* calcium oxide

Racemic mixtures 371

...dioactive decay 33
...onstant 36
...dioactive tracers 40
...dioactivity 31
...dio isotopes 40
...dium 2, 31
...dius, cationic/anionic ratio 81
...don 31
...oult's law 95, 112
...te constant *see* velocity constant
...te determining step 154
...te equation 147
...te expression 147
...te of reaction 146
...action, enthalpy of 128
...mechanism 154, 342
...eversible 163
...actor, nuclear 39
...arrangement reactions 340
...crystallisation 110
...d lead *see* dilead(II) lead(IV) oxide
...d phosphorus 93, 282
...dox potentials 196
...dox reactions 187
...ducing agent 197
...ducing sugar 396
...duction 197, 462
...frigerating liquids 276
...gnault 70
...(gas constant) 68
...lative atomic mass 4
...lative molecular mass 4
...sistivity 209
...sonance energy 358
...sonance hybrid 60, 357
...tention factor 103
...versible reactions 163
...ombic sulphur 92, 293
...ck phosphate 281
...bidium 234
...sting 323
...therford 11, 33
...dberg constant 18

...licyclic acid *see* 2-hydroxybenzoic acid
...ltpetre 236
...l volatile 277
...ponification 409
...turated hydrocarbons *see* alkanes

Saturated solutions 109
Saturation vapour pressure 111
s-block elements 234
Scandium 320
Second order reactions 148
s electrons 49
Semi-conductors 262, 288
Selenium 303
Sidgwick *see* electron pair repulsion theory
Siemens–Martin furnace 472
σ bonds 48
Silanes *see* silicon hydrides
Silica *see* silicon(IV) oxide
Silica gel 266
Silica glass 267
Silicates 261
Silicon 261
Silicones 267
Silicon dioxide *see* silicon(IV) oxide
Silicon hydrides 264
Silicon(IV) oxide 45, 83, 261, 266
Silicon tetrachloride (silicon(IV) chloride) 259
Silicon tetrafluoride (silicon(IV) fluoride) 259
Silver
    bromide 309
    chloride 309
    chromate 326
    diammine complex ion 56
    ethanoate 169
    fluoride 309, 317
    nitrate 309, 389
    oxide 435
Single bonds 45
Slaked lime 240
S_N1/S_N2 reactions 433
Soap 409
Soda glass 267
Soddy 33
Sodium 234
    ammonium tartrate 376
    compounds 239*ff*
    dichromate(VI) 451
    dihydrogenphosphate(V) 286
    ethanedioate 407
    iodate(V) 308
    methanoate 406
    nitrite 279
    oxalate *see* ethanedioate
    phenate 382
    salts of carboxylic acids 406

Sodium—*continued*
   tetrahydridoborate(III) 226
   thiosulphate 301
Solder 262
Solubility 108*ff*
   product 168, 296
Solution, enthalpy of 137
Solution, of gases 111
Solvay process 472
Solvent extraction 102
s orbitals 21, 49
Sørensen 176
sp, sp^2, *sp*^{3} bond hybridisation, 51*ff* 247, 259
Spectrometer, mass 14
Spin quantum number 21, 49
Stability band 35
Stability constants for complex ions 58
Standard electrode potentials 189
Stannate ions 268
Starch 399
Stationary phase 103
Stationary states 19
Steam distillation 100
Stearic acid *see* octadecanoic acid
Steel 471
Stereoisomerism 367
Strontium 234*ff*
Structural isomerism 364
Styrene 358
Sublimation 90*ff*
Substitution reactions 341
Succinic acid *see* butanedioic acid
Sucrose 395
Sulphanilic acid *see* 4-aminobenzenesulphonic
   · acid
Sulphate(IV) ions *see* sulphite ions
Sulphate(VI) ions 300
Sulphides 297
Sulphite ions 298
Sulphonamide drugs 422
Sulphonation 462
Sulphur
   allotropes 92, 293
   compounds 295
   dichloride oxide 302, 377
   hexafluoride 302
   oxides 297
   tetrachloride 303
Sulphuric(IV) acid (sulphurous acid) 298
Sulphuric(VI) 299, 477

Superphosphate 286
Supersaturated solutions 109

Tartaric acid *see* 2,3-dihydroxybutanedioic aci◖
Tautomerism 367
TCP 382
Tellurium 303
Temperature/composition diagrams 93
Terephthalic acid *see* benzenedicarboxylic acic
Terylene 413
Tetraamminecopper(II) ion 57, 330
Tetraaquacopper(II) ion 55, 330
Tetrabromoethane 352
Tetracarbonyl nickel(0) 59, 333
Tetrachlorocuprate(II) ion 57
Tetrachlorocobaltate(II) ion 57, 332
Tetrachloromethane 259, 269, 348
Tetrafluoroethene 356, 443
Tetrahedral molecules 48
Tetrahydridoaluminate(III), ion 226, 251, 390
Tetrahydridoborate(III) ion 249
Tetrahydroxoborate(III) ion (borate ion) 249
Tetrathionate ion 302
Thallium 248
Thermal dissociation 72
Thermonuclear fusion 39
Thermoplastic compounds 355
Thermosetting plastics 393
Thio compounds 289
Thiocyanate ion 331
Thionyl chloride *see* sulphur dichloride oxide
Thiosulphuric acid 301
Thompson 2, 8, 9
Thorium 31
Tin 94, 262
   allotropes 257
   amalgam 336
   chloride 336
   compounds 364*ff*
   plating 262
Titanium 321
   compounds 322
   tetrachloride 354
TNT 361
Tollen's reagent 389
Toluene *see* methylbenzene
Tracers, radioactive 40
Transistors 262
Transition elements 2, 319

ransition temperature 91
4,6-tribromophenol 383
richloroethanal 393
1,1-trichloroethane 442
richloroethanoic acid 413
richloromethane 348, 442
riethyl aluminium 354
riiodomethane 441
3,5-trimethylbenzene 395
rilead tetraoxide *see* dilead(II)lead(IV) oxide
rinitrotoluene *see* methyl-2,4,6-trinitrobenzene
rioxoboric(III) acid 250
rioxonitric(V) acid *see* nitric(V) acid
riple bond 45, 53
riple point 89
swett 104
ungsten 226
ype-metal 262

ltra-violet radiation 17
nit cell 79
ranium 31, 33
hexafluoride 39
rea *see* carbamide

alency electrons 43
aline *see* 2-amino-3-methylbutanoic acid
anadium 321
oxidation states 325
anadium(V) oxide 325, 477
an der Waals
forces 76*ff*
radius 86
apour density 69
apour pressure/composition diagram 95

Vapour pressure, lowering of 111
Velocity constant 147
Victor Meyer 70
Vinyl chloride *see* chloroethene

Water
of crystallisation 230
as a solvent 229
hard 230
hydrogen bonding in 228
Water gas 260, 264
Water glass 266
Water molecule 229
Wave nature of electrons 10
Wöhler 340
Wurzite structure 84, 229

Xenon 28
X-rays 10, 84, 237
Xylene *see* dimethylbenzene

Yeast 397

Zeolites 232
Ziegler catalysts 354
Zinc
amalgam 336
blende 84
compounds 335*ff*
electrode 187
galvanising 324, 335
sulphide 8, 84, 297
Zinc/copper couple 434
Zwitterion 415, 427

Relative atomic masses
of selected elements (approximate values)

Element	Symbol	Atomic number Z	Relative atomic mass A_r	Element	Symbol	Atomic number Z	Relative atomic mass A_r
Aluminium	Al	13	26.9	Manganese	Mn	25	54.9
Antimony	Sb	51	121.8	Mercury	Hg	80	200.6
Argon	Ar	18	39.9	Molybdenum	Mo	42	95.9
Astatine	At	85	210.0	Neon	Ne	10	20.2
Arsenic	As	33	74.9	Nickel	Ni	28	58.7
Barium	Ba	56	137.3	Nitrogen	N	7	14.0
Beryllium	Be	4	9.0	Oxygen	O	8	15.9
Bismuth	Bi	83	208.9	Phosphorus	P	15	30.9
Boron	B	5	10.8	Platinum	Pt	78	195.1
Bromine	Br	35	79.9	Potassium	K	19	39.1
Cadmium	Cd	48	112.4	Radium	Ra	88	226.1
Calcium	Ca	20	40.1	Radon	Rn	86	222.0
Carbon	C	6	12.0	Rubidium	Rb	37	85.5
Cerium	Ce	58	140.1	Scandium	Sc	21	44.9
Chlorine	Cl	17	35.5	Selenium	Se	34	78.9
Chromium	Cr	24	52.0	Silicon	Si	14	28.1
Cobalt	Co	27	58.9	Silver	Ag	47	107.9
Copper	Cu	29	63.5	Sodium	Na	11	22.9
Fluorine	F	9	19.0	Strontium	Sr	38	87.6
Gallium	Ga	31	69.7	Sulphur	S	16	32.1
Germanium	Ge	32	72.6	Tellurium	Te	52	127.6
Gold	Au	79	196.9	Thallium	Tl	81	204.4
Helium	He	2	4.0	Thorium	Th	90	232.0
Hydrogen	H	1	1.0	Tin	Sn	50	118.7
Iodine	I	53	126.9	Titanium	Ti	22	47.9
Iron	Fe	26	55.8	Tungsten	W	74	183.8
Krypton	Kr	36	83.8	Uranium	U	92	238.0
Lead	Pb	82	207.2	Vanadium	V	23	50.9
Lithium	Li	3	6.9	Xenon	Xe	54	131.3
Magnesium	Mg	12	24.3	Zinc	Zn	30	65.4